Regression Techniques for Data Analysis

Mike Nguyen

Regression Techniques for Data Analysis

Springer

Mike Nguyen
Data Sciences and Operations
University of Southern California
Los Angeles, CA, USA

ISBN 978-3-032-01834-2 ISBN 978-3-032-01835-9 (eBook)
https://doi.org/10.1007/978-3-032-01835-9

Mathematics Subject Classification: 62J07, 62J12, 62J86, 62P20, 62J02, 62J05, 62G08, 62M10

This Springer imprint is published by the registered company Springer Nature Switzerland AG
The registered company address is: Gewerbestrasse 11, 6330 Cham, Switzerland

Declarations

Competing Interests The author has no competing interests to declare that are relevant to the content of this manuscript.

Contents

Chapter 1
Introduction

Welcome to this volume dedicated to **regression**. Whether you are a seasoned researcher, an aspiring data scientist, or simply curious about the foundations of quantitative modeling, regression is a cornerstone technique worth mastering. Here, we will delve into:

- **Linear regression**
- **Nonlinear regression**
- **Generalized linear models**
- **Linear mixed models**
- **Nonlinear and generalized linear mixed models**
- **Nonparametric regression**

This volume offers a wide-ranging survey of methods to help you gain deeper insights from your data.

Why Regression?
Think of regression as a powerful lens that helps us interpret relationships hidden within numbers. By fitting mathematical functions to data, regression serves to:

- **Illuminate patterns**: Discover trends and relationships between variables
- **Quantify uncertainty**: Assess confidence in the relationships observed
- **Make predictions**: Project future outcomes based on historical trends

Asking the Right Questions
Regression enables you to explore sophisticated questions, such as:

- Which predictors matter most in explaining a response?
- How does a change in one variable affect another?
- How well can we predict future outcomes from the past?

M. Nguyen, *Regression Techniques for Data Analysis*,
https://doi.org/10.1007/978-3-032-01835-9_1

What Regression *Is*—and *Isn't*

While regression is a versatile and powerful tool, it is important to clarify its scope. Here is what you can expect to learn—and what lies outside the realm of regression:

1. What you will get:

- **Foundational understanding**: How regression works, its assumptions, and its applications
- **Practical skills**: Building, diagnosing, and interpreting regression models
- **Tools for inference and prediction**: Quantifying relationships and projecting outcomes
- **Handling real-world data challenges**: Addressing nonlinearity, multicollinearity, missing data, and more

2. What regression isn't:

- **Causal analysis by itself**: Regression shows associations, not causation. Establishing causality requires careful design (e.g., experiments) or specialized methods (e.g., instrumental variables).
- **A catch-all solution**: Regression is one tool in a broader data analysis toolkit. Problems like clustering, time series forecasting, or unsupervised learning often require alternative approaches.
- **Guaranteed answers**: The quality of insights depends on the quality of your data, model selection, and validation efforts.
- **A plug-and-play tool**: Effective use of regression requires understanding the data, model assumptions, and diagnostics—automating regression without context often leads to flawed conclusions.

This book will empower you with the knowledge to use regression effectively while recognizing its limitations. By the end of this book, you will be equipped with both **theoretical knowledge** and **practical skills** for building and interpreting regression models in a variety of contexts.

The Philosophy of Regression

As you explore regression, remember these key themes:

- **Model assumptions**: Ensure the assumptions of your model align with your data.
- **Data transformations**: Use transformations thoughtfully to reveal relationships.
- **Diagnostic checks**: Evaluate model fit and residual patterns.
- **Complexity versus interpretability**: Balance the need for detail with the importance of clarity.

Through it all, the goal remains the same: to tell a **story with data**—one that is rigorous, evidence-based, and useful for informed decision-making.

Code Replication

This book was built with R version 4.4.3 (2025-02-28 ucrt) and the following packages:

Package	Version	Source
abind	1.4-8	CRAN (R 4.4.1)
agridat	1.24	CRAN (R 4.4.3)
ape	5.8-1	CRAN (R 4.4.3)
assertthat	0.2.1	CRAN (R 4.2.3)
backports	1.5.0	CRAN (R 4.4.0)
bookdown	0.43	CRAN (R 4.4.3)
boot	1.3-31	CRAN (R 4.4.3)
broom	1.0.7	CRAN (R 4.4.2)
bslib	0.9.0	CRAN (R 4.4.2)
cachem	1.1.0	CRAN (R 4.4.2)
callr	3.7.6	CRAN (R 4.4.2)
car	3.1-3	CRAN (R 4.4.2)
carData	3.0-5	CRAN (R 4.4.2)
cellranger	1.1.0	CRAN (R 4.4.2)
cli	3.6.3	CRAN (R 4.4.2)
coda	0.19-4.1	CRAN (R 4.4.2)
colorspace	2.1-1	CRAN (R 4.4.2)
corpcor	1.6.10	CRAN (R 4.4.0)
crayon	1.5.3	CRAN (R 4.4.2)
cubature	2.1.1	CRAN (R 4.4.2)
curl	6.2.1	CRAN (R 4.4.3)
data.table	1.16.4	CRAN (R 4.4.2)
DBI	1.2.3	CRAN (R 4.4.2)
dbplyr	2.5.0	CRAN (R 4.4.2)
desc	1.4.3	CRAN (R 4.4.2)
devtools	2.4.5	CRAN (R 4.4.2)
digest	0.6.37	CRAN (R 4.4.2)
dplyr	1.1.4	CRAN (R 4.4.2)
ellipsis	0.3.2	CRAN (R 4.4.2)
evaluate	1.0.3	CRAN (R 4.4.2)
extrafont	NA	NA
extrafontdb	NA	NA
fansi	1.0.6	CRAN (R 4.4.2)
faraway	1.0.9	CRAN (R 4.4.3)
fastmap	1.2.0	CRAN (R 4.4.2)
forcats	1.0.0	CRAN (R 4.4.2)
foreign	0.8-88	CRAN (R 4.4.3)
fs	1.6.5	CRAN (R 4.4.2)
generics	0.1.3	CRAN (R 4.4.2)

(continued)

Package	Version	Source
ggplot2	3.5.1	CRAN (R 4.4.2)
glue	1.8.0	CRAN (R 4.4.2)
gtable	0.3.6	CRAN (R 4.4.2)
haven	2.5.4	CRAN (R 4.4.2)
Hmisc	5.2-3	CRAN (R 4.4.3)
hms	1.1.3	CRAN (R 4.4.2)
htmltools	0.5.8.1	CRAN (R 4.4.2)
htmlwidgets	1.6.4	CRAN (R 4.4.2)
httr	1.4.7	CRAN (R 4.4.2)
investr	1.4.2	CRAN (R 4.4.3)
jpeg	0.1-10	CRAN (R 4.4.0)
jquerylib	0.1.4	CRAN (R 4.4.2)
jsonlite	1.9.1	CRAN (R 4.4.3)
kableExtra	1.4.0	CRAN (R 4.4.3)
knitr	1.50	CRAN (R 4.4.3)
lattice	0.22-6	CRAN (R 4.4.3)
latticeExtra	0.6-30	CRAN (R 4.4.2)
lifecycle	1.0.4	CRAN (R 4.4.2)
lme4	1.1-36	CRAN (R 4.4.2)
lmerTest	3.1-3	CRAN (R 4.4.2)
lsr	0.5.2	CRAN (R 4.4.3)
ltm	1.2-0	CRAN (R 4.4.3)
lubridate	1.9.4	CRAN (R 4.4.2)
magrittr	2.0.3	CRAN (R 4.4.2)
MASS	7.3-65	CRAN (R 4.4.3)
matlib	1.0.0	CRAN (R 4.4.3)
Matrix	1.7-2	CRAN (R 4.4.3)
MCMCglmm	2.36	CRAN (R 4.4.2)
memoise	2.0.1	CRAN (R 4.4.2)
mgcv	1.9-1	CRAN (R 4.4.3)
minqa	1.2.8	CRAN (R 4.4.2)
modelr	0.1.11	CRAN (R 4.4.2)
munsell	0.5.1	CRAN (R 4.4.2)
nlme	3.1-167	CRAN (R 4.4.3)
nloptr	2.2.1	CRAN (R 4.4.3)
nlstools	2.1-0	CRAN (R 4.4.2)
nnet	7.3-20	CRAN (R 4.4.3)
numDeriv	2016.8-1.1	CRAN (R 4.4.0)
openxlsx	4.2.8	CRAN (R 4.4.3)
pbkrtest	0.5.3	CRAN (R 4.4.2)
pillar	1.10.2	CRAN (R 4.4.3)
pkgbuild	1.4.6	CRAN (R 4.4.2)
pkgconfig	2.0.3	CRAN (R 4.4.2)
pkgload	1.4.0	CRAN (R 4.4.2)
png	0.1-8	CRAN (R 4.4.0)
ppsr	0.0.5	CRAN (R 4.4.3)

(continued)

Package	Version	Source
prettyunits	1.2.0	CRAN (R 4.4.2)
processx	3.8.6	CRAN (R 4.4.3)
ps	1.9.0	CRAN (R 4.4.3)
pscl	1.5.9	CRAN (R 4.4.2)
purrr	1.0.2	CRAN (R 4.4.2)
R6	2.6.1	CRAN (R 4.4.3)
RColorBrewer	1.1-3	CRAN (R 4.4.0)
Rcpp	1.0.14	CRAN (R 4.4.2)
readr	2.1.5	CRAN (R 4.4.2)
readxl	1.4.5	CRAN (R 4.4.3)
remotes	2.5.0	CRAN (R 4.4.2)
reprex	2.1.1	CRAN (R 4.4.2)
rgl	1.3.17	CRAN (R 4.4.2)
rio	1.2.3	CRAN (R 4.4.2)
rlang	1.1.5	CRAN (R 4.4.2)
RLRsim	3.1-8	CRAN (R 4.4.3)
rmarkdown	2.29	CRAN (R 4.4.2)
rprojroot	2.0.4	CRAN (R 4.4.2)
rstudioapi	0.17.1	CRAN (R 4.4.2)
Rttf2pt1	NA	NA
rvest	1.0.4	CRAN (R 4.4.2)
sass	0.4.9	CRAN (R 4.4.2)
scales	1.3.0	CRAN (R 4.4.2)
sessioninfo	1.2.3	CRAN (R 4.4.3)
stringi	1.8.4	CRAN (R 4.4.0)
stringr	1.5.1	CRAN (R 4.4.2)
svglite	2.1.3	CRAN (R 4.4.2)
systemfonts	1.2.1	CRAN (R 4.4.2)
tensorA	0.36.2.1	CRAN (R 4.4.0)
testthat	3.2.3	CRAN (R 4.4.2)
tibble	3.2.1	CRAN (R 4.4.2)
tidyr	1.3.1	CRAN (R 4.4.2)
tidyselect	1.2.1	CRAN (R 4.4.2)
tidyverse	2.0.0	CRAN (R 4.4.2)
tzdb	0.5.0	CRAN (R 4.4.3)
usethis	3.1.0	CRAN (R 4.4.2)
utf8	1.2.4	CRAN (R 4.4.2)
vctrs	0.6.5	CRAN (R 4.4.2)
viridisLite	0.4.2	CRAN (R 4.4.2)
webshot	0.5.5	CRAN (R 4.4.3)
withr	3.0.2	CRAN (R 4.4.2)
xfun	0.51	CRAN (R 4.4.3)
xml2	1.3.8	CRAN (R 4.4.3)
xtable	1.8-4	CRAN (R 4.4.2)
yaml	2.3.10	CRAN (R 4.4.2)
zip	2.3.2	CRAN (R 4.4.3)

```
#> - Session info -------------------------------------------
#>  setting  value
#>  version  R version 4.4.3 (2025-02-28 ucrt)
#>  os       Windows 11 x64 (build 26100)
#>  system   x86_64, mingw32
#>  ui       RTerm
#>  language (EN)
#>  collate  English_United States.utf8
#>  ctype    English_United States.utf8
#>  tz       America/Los_Angeles
#>  date     2025-05-07
#>  pandoc   3.1.1
  @ C:/Program Files/RStudio/resources/app/bin/quarto/bin/tools/(via rmarkdown)
#>  quarto   NA
  @ C:\\PROGRA~1\\RStudio\\resources\\app\\bin\\quarto\\bin\\quarto.exe
#>
#> - Packages -----------------------------------------------
#>  package     * version date (UTC) lib source
#>  bookdown      0.43    2025-04-15 [1] CRAN (R 4.4.3)
#>  cachem        1.1.0   2024-05-16 [1] CRAN (R 4.4.2)
#>  cli           3.6.3   2024-06-21 [1] CRAN (R 4.4.2)
#>  codetools     0.2-20  2024-03-31 [1] CRAN (R 4.4.3)
#>  desc          1.4.3   2023-12-10 [1] CRAN (R 4.4.2)
#>  devtools      2.4.5   2022-10-11 [1] CRAN (R 4.4.2)
#>  digest        0.6.37  2024-08-19 [1] CRAN (R 4.4.2)
#>  ellipsis      0.3.2   2021-04-29 [1] CRAN (R 4.4.2)
#>  evaluate      1.0.3   2025-01-10 [1] CRAN (R 4.4.2)
#>  fastmap       1.2.0   2024-05-15 [1] CRAN (R 4.4.2)
#>  fs            1.6.5   2024-10-30 [1] CRAN (R 4.4.2)
#>  glue          1.8.0   2024-09-30 [1] CRAN (R 4.4.2)
#>  htmltools     0.5.8.1 2024-04-04 [1] CRAN (R 4.4.2)
#>  htmlwidgets   1.6.4   2023-12-06 [1] CRAN (R 4.4.2)
#>  httpuv        1.6.15  2024-03-26 [1] CRAN (R 4.4.2)
#>  jpeg        * 0.1-10  2022-11-29 [1] CRAN (R 4.4.0)
#>  knitr         1.50    2025-03-16 [1] CRAN (R 4.4.3)
#>  later         1.4.1   2024-11-27 [1] CRAN (R 4.4.2)
#>  lifecycle     1.0.4   2023-11-07 [1] CRAN (R 4.4.2)
#>  magrittr      2.0.3   2022-03-30 [1] CRAN (R 4.4.2)
#>  memoise       2.0.1   2021-11-26 [1] CRAN (R 4.4.2)
#>  mime          0.12    2021-09-28 [1] CRAN (R 4.4.0)
#>  miniUI        0.1.1.1 2018-05-18 [1] CRAN (R 4.4.2)
#>  pillar        1.10.2  2025-04-05 [1] CRAN (R 4.4.3)
#>  pkgbuild      1.4.6   2025-01-16 [1] CRAN (R 4.4.2)
#>  pkgconfig     2.0.3   2019-09-22 [1] CRAN (R 4.4.2)
#>  pkgload       1.4.0   2024-06-28 [1] CRAN (R 4.4.2)
#>  profvis       0.4.0   2024-09-20 [1] CRAN (R 4.4.2)
#>  promises      1.3.2   2024-11-28 [1] CRAN (R 4.4.2)
#>  purrr         1.0.2   2023-08-10 [1] CRAN (R 4.4.2)
#>  R6            2.6.1   2025-02-15 [1] CRAN (R 4.4.3)
#>  Rcpp          1.0.14  2025-01-12 [1] CRAN (R 4.4.2)
#>  remotes       2.5.0   2024-03-17 [1] CRAN (R 4.4.2)
#>  rlang         1.1.5   2025-01-17 [1] CRAN (R 4.4.2)
#>  rmarkdown     2.29    2024-11-04 [1] CRAN (R 4.4.2)
#>  rstudioapi    0.17.1  2024-10-22 [1] CRAN (R 4.4.2)
#>  sessioninfo   1.2.3   2025-02-05 [1] CRAN (R 4.4.3)
#>  shiny         1.10.0  2024-12-14 [1] CRAN (R 4.4.2)
#>  tibble        3.2.1   2023-03-20 [1] CRAN (R 4.4.2)
#>  tinytex       0.57    2025-04-15 [1] CRAN (R 4.4.3)
#>  urlchecker    1.0.1   2021-11-30 [1] CRAN (R 4.4.2)
#>  usethis       3.1.0   2024-11-26 [1] CRAN (R 4.4.2)
#>  vctrs         0.6.5   2023-12-01 [1] CRAN (R 4.4.2)
#>  xfun          0.51    2025-02-19 [1] CRAN (R 4.4.3)
```

```
#>  xtable        1.8-4   2019-04-21 [1] CRAN (R 4.4.2)
#>  yaml          2.3.10  2024-07-26 [1] CRAN (R 4.4.2)
#>
#>  [1] C:/Program Files/R/R-4.4.3/library
#>  * -- Packages attached to the search path.
#>
#> ----------------------------------------------------------
```

Chapter 2
Linear Regression

This chapter develops the classical linear model as the cornerstone of regression methodology. It presents ordinary least squares, its geometric and probabilistic interpretations, and the Gauss–Markov theorem. Diagnostics for multicollinearity, heteroskedasticity, and influential observations are paired with remedies such as generalized least squares and robust variance estimation. Extensions to maximum-likelihood estimation unify treatment of continuous and discrete outcomes, while penalized approaches—including ridge, lasso, and elastic net—address high-dimensional feature spaces. Finally, we introduce partial least squares for dimension reduction and demonstrate applications in marketing analytics, experimental psychology, and economic forecasting.

Linear regression is one of the most fundamental tools in statistics and econometrics, widely used for modeling relationships between variables. It forms the cornerstone of predictive analysis, enabling us to understand and quantify how changes in one or more explanatory variables are associated with a dependent variable. Its simplicity and versatility make it an essential tool in fields ranging from economics and marketing to healthcare and environmental studies.

At its core, linear regression addresses questions about associations rather than causation. For example:

- How are advertising expenditures associated with sales performance?
- What is the relationship between a company's revenue and its stock price?
- How does the level of education correlate with income?

These questions are about patterns in data—not necessarily causal effects. While regression can provide insights into potential causal relationships, establishing causality requires more than just regression analysis. It requires careful consideration of the study design, assumptions, and potential confounding factors.

So, why is it called "linear?" The term refers to the structure of the model, where the dependent variable (outcome) is modeled as a linear combination of one or more

M. Nguyen, *Regression Techniques for Data Analysis*,
https://doi.org/10.1007/978-3-032-01835-9_2

independent variables (predictors). For example, in simple linear regression, the relationship is represented as:

$$Y = \beta_0 + \beta_1 X + \epsilon,$$

where Y is the dependent variable, X is the independent variable, β_0 and β_1 are parameters to be estimated, and ϵ is the error term capturing randomness or unobserved factors.

Linear regression serves as a foundation for much of applied data analysis because of its wide-ranging applications:

1. **Understanding patterns in data:** Regression provides a framework to summarize and explore relationships between variables. It allows us to identify patterns such as trends or associations, which can guide further analysis or decision-making.
2. **Prediction:** Beyond exploring relationships, regression is widely used for making predictions. For instance, given historical data, we can use a regression model to predict future outcomes like sales, prices, or demand.
3. **Building blocks for advanced techniques:** Linear regression is foundational for many advanced statistical and machine learning models, such as logistic regression, ridge regression, and neural networks. Mastering linear regression equips you with the skills to tackle more complex methods.

Regression and Causality: A Crucial Distinction

It is essential to remember that regression alone does not establish causation. For instance, a regression model might show a strong association between advertising and sales, but this does not prove that advertising directly causes sales to increase. Other factors—such as seasonality, market trends, or unobserved variables—could also influence the results.

Establishing causality requires additional steps, such as controlled experiments, instrumental variable techniques, or careful observational study designs. As we work through the details of linear regression, we will revisit this distinction and highlight scenarios where causality might or might not be inferred.

What Is an Estimator?

At the heart of regression lies the process of **estimation**—the act of using data to determine the unknown characteristics of a population or model (Fig. 2.1).

An **estimator** is a mathematical rule or formula used to calculate an estimate of an unknown quantity based on observed data. For example, when we calculate the average height of a sample to estimate the average height of a population, the sample mean is the estimator.

In the context of regression, the quantities we typically estimate are:

- **Parameters**: Fixed, unknown values that describe the relationship between variables (e.g., coefficients in a regression equation)

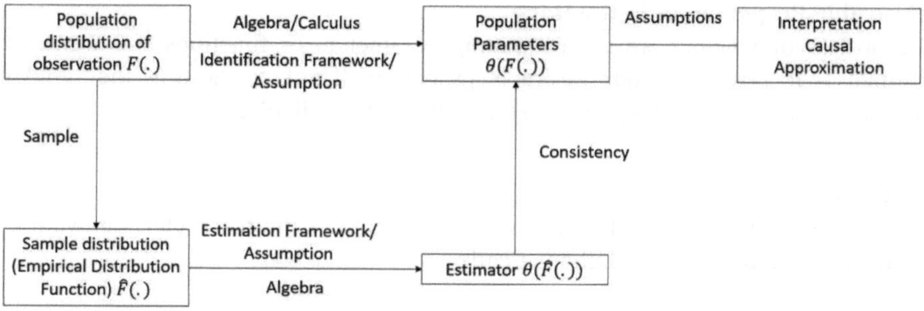

Fig. 2.1 Framework for statistical inference and causal interpretation

  - **Estimating parameters** → Parametric models (finite parameters, e.g., coefficients in regression)

- **Functions**: Unknown relationships or patterns in the data, often modeled without assuming a fixed functional form
  - **Estimating functions** → Nonparametric models (focus on shapes or trends, not a fixed number of parameters)

Types of Estimators

To better understand the estimation process, let us introduce two broad categories of estimators that we will work with:

1. **Parametric Estimators**
   Parametric estimation focuses on a finite set of parameters that define a model. For example, in a simple linear regression:
   $$Y = \beta_0 + \beta_1 X + \epsilon,$$
   the task is to estimate the parameters β_0 (intercept) and β_1 (slope). Parametric estimators rely on specific assumptions about the form of the model (e.g., linearity) and the distribution of the error term (e.g., normality).
2. **Nonparametric Estimators**
   Nonparametric estimation avoids assuming a specific functional form for the relationship between variables. Instead, it focuses on estimating patterns or trends directly from the data, for example, using a scatterplot smoothing technique to visualize how sales vary with advertising spend without imposing a linear or quadratic relationship.

These two categories reflect a fundamental trade-off in statistical analysis: **Parametric models** are often simpler and more interpretable but require strong assumptions, while **nonparametric models** are more flexible but may require more data and computational resources.

Desirable Properties of Estimators

Regardless of whether we are estimating parameters or functions, we want our estimators to possess certain desirable properties. Think of these as the "golden standards" that help us judge whether an estimator is reliable:

1. **Unbiasedness**
   An estimator is unbiased if it hits the true value of the parameter, on average, over repeated samples. Mathematically:

   $$E[\hat{\beta}] = \beta.$$

   This means that, across multiple samples, the estimator does not systematically overestimate or underestimate the true parameter.
2. **Consistency**
   Consistency ensures that as the sample size increases, the estimator converges to the true value of the parameter. Formally:

   $$plim\ \hat{\beta}_n = \beta.$$

   This property relies on the [Law of Large Numbers], which guarantees that larger samples reduce random fluctuations, leading to more precise estimates.
3. **Efficiency**
   Among all unbiased estimators, an efficient estimator has the smallest variance.
   - The ordinary least squares method is efficient because it is the **best linear unbiased estimator (BLUE)** under the Gauss–Markov theorem.
   - For estimators that meet specific distributional assumptions (e.g., normality), maximum-likelihood estimators are asymptotically efficient, meaning they achieve the lowest possible variance as the sample size grows.

Why These Properties Matter

Understanding these properties is crucial because they ensure that the methods we use for estimation are reliable, precise, and robust (Table 2.1). Whether we are estimating coefficients in a regression model or uncovering a complex pattern in the data, these properties provide the foundation for statistical inference and decision-making.

Now that we have established what estimators are, the types we will encounter, and their desirable properties, we can move on to understanding how these concepts apply specifically to the ordinary least squares method—the backbone of linear regression.

Table 2.1 Comparison of estimation methods in regression analysis

Estimator	Key assumptions	Strengths	Limitations
Ordinary least squares	Errors are independent, identically distributed (i.i.d.) with mean 0 and constant variance. Linear relationship between predictors and response	Simple, well-understood method. Minimizes residual sum of squares (easy to interpret coefficients)	Sensitive to outliers and violations of normality. Can perform poorly if predictors are highly correlated (multicollinearity)
Generalized least squares	Errors have a known correlation structure or heteroskedasticity structure that can be modeled	Handles correlated or non-constant-variance errors. More flexible than OLS when noise structure is known	Requires specifying (or estimating) the error covariance structure. Misspecification can lead to biased estimates
Maximum likelihood	Underlying probability distribution (e.g., normal) must be specified correctly	Provides a general framework for estimating parameters under well-defined probability models. Can extend to complex likelihoods	Highly sensitive to model misspecification. May require more computation than OLS or GLS
Penalized (regularized) estimators	Coefficients assumed to be shrinkable; model typically allows coefficient penalization	Controls overfitting via regularization. Handles high-dimensional data or many predictors. Can perform feature selection (e.g., Lasso)	Requires choosing tuning parameter(s) (e.g., regularization strength). Interpretation of coefficients becomes less straightforward
Robust estimators	Less sensitive to heavy-tailed or outlier-prone distributions (weaker assumptions on the error structure)	Resistant to large deviations or outliers in data. Often maintains good performance under mild model misspecifications	Less efficient if errors are truly normal. Choice of robust method and tuning can be subjective
Partial least squares	Predictors may be highly correlated; dimension reduction is desired	Simultaneously reduces dimensionality and fits regression. Works well with collinear, high-dimensional data	Can be harder to interpret than OLS (latent components instead of original predictors). Requires choosing the number of components

2.1 Ordinary Least Squares

Ordinary least squares (OLS) is the backbone of statistical modeling, a method so foundational that it often serves as the starting point for understanding data relationships. Whether predicting sales, estimating economic trends, or uncovering patterns in scientific research, OLS remains a critical tool. Its appeal lies in simplicity: OLS models the relationship between a dependent variable and one or more predictors by minimizing the squared differences between observed and predicted values.

Why OLS Works: Linear and Nonlinear Relationships
OLS rests on the conditional expectation function (CEF), $E[Y|X]$, which describes the expected value of Y given X. Regression shines in two key scenarios:

1. **Perfect fit (linear CEF)**
   If $E[Y_i|X_{1i}, \dots, X_{Ki}] = a + \sum_{k=1}^{K} b_k X_{ki}$, the regression of Y_i on $X_{1i}, \dots, X_{Ki}$ exactly equals the CEF. In other words, the regression gives the true average relationship between Y and X.
   If the true relationship is linear, regression delivers the exact CEF. For instance, imagine you are estimating the relationship between advertising spend and sales revenue. If the true impact is linear, OLS will perfectly capture it.
2. **Approximation (nonlinear CEF)**
   If $E[Y_i|X_{1i}, \dots, X_{Ki}]$ is nonlinear, OLS provides the best linear approximation to this relationship. Specifically, it minimizes the expected squared deviation between the linear regression line and the nonlinear CEF.
   For example, the effect of advertising diminishes at higher spending levels? OLS still works, providing the best linear approximation to this nonlinear relationship by minimizing the squared deviations between predictions and the true (but unknown) CEF.

In other words, regression is not just a tool for "linear" relationships—it is a workhorse that adapts remarkably well to messy, real-world data.

2.1.1 *Simple Regression (Basic) Model*

The simplest form of regression is a straight line:

$$Y_i = \beta_0 + \beta_1 X_i + \epsilon_i$$

where

- Y_i: the dependent variable or outcome we are trying to predict (e.g., sales, temperature)
- X_i: the independent variable or predictor (e.g., advertising spend, time)

- β_0: the intercept—where the line crosses the Y-axis when $X = 0$
- β_1: the slope, representing the change in Y for a one-unit increase in X
- ϵ_i: the error term, accounting for random factors that X cannot explain

Assumptions about the error term (ϵ_i):

$$E(\epsilon_i) = 0$$

$$\text{Var}(\epsilon_i) = \sigma^2$$

$$\text{Cov}(\epsilon_i, \epsilon_j) = 0 \quad \text{for all } i \neq j$$

Since ϵ_i is random, Y_i is also random:

$$\begin{aligned} E(Y_i) &= E(\beta_0 + \beta_1 X_i + \epsilon_i) \\ &= \beta_0 + \beta_1 X_i \end{aligned}$$

$$\begin{aligned} \text{Var}(Y_i) &= \text{Var}(\beta_0 + \beta_1 X_i + \epsilon_i) \\ &= \text{Var}(\epsilon_i) \\ &= \sigma^2. \end{aligned}$$

Since $\text{Cov}(\epsilon_i, \epsilon_j) = 0$, the outcomes across observations are independent. Hence, Y_i and Y_j are uncorrelated as well, conditioned on the Xs.

2.1.1.1 Estimation in Ordinary Least Squares

The goal of OLS is to estimate the regression parameters (β_0, β_1) that best describe the relationship between the dependent variable Y and the independent variable X. To achieve this, we minimize the sum of squared deviations between observed values of Y_i and their expected values predicted by the model.

The deviation of an observed value Y_i from its expected value, based on the regression model, is:

$$Y_i - E(Y_i) = Y_i - (\beta_0 + \beta_1 X_i).$$

This deviation represents the error in prediction for the ith observation.

To ensure that the errors do not cancel each other out and to prioritize larger deviations, we consider the squared deviations. The sum of squared deviations, denoted by Q, is defined as:

$$Q = \sum_{i=1}^{n} (Y_i - \beta_0 - \beta_1 X_i)^2.$$

The goal of OLS is to find the values of β_0 and β_1 that minimize Q. These values are called the **OLS estimators**.

To minimize Q, we take partial derivatives with respect to β_0 and β_1, set them to zero, and solve the resulting system of equations. After simplifying, the estimators for the slope (b_1) and intercept (b_0) are obtained as follows:

Slope (b_1):

$$b_1 = \frac{\sum_{i=1}^{n}(X_i - \bar{X})(Y_i - \bar{Y})}{\sum_{i=1}^{n}(X_i - \bar{X})^2}.$$

Here, $\bar{X}$ and $\bar{Y}$ represent the means of X and Y, respectively. This formula reveals that the slope is proportional to the covariance between X and Y, scaled by the variance of X.

Intercept (b_0):

$$b_0 = \frac{1}{n}\left(\sum_{i=1}^{n} Y_i - b_1 \sum_{i=1}^{n} X_i\right) = \bar{Y} - b_1\bar{X}.$$

The intercept is determined by aligning the regression line with the center of the data.

Intuition Behind the Estimators

- b_1 (slope): This measures the average change in Y for a one-unit increase in X. The formula uses deviations from the mean to ensure that the relationship captures the joint variability of X and Y.
- b_0 (intercept): This ensures that the regression line passes through the mean of the data points $(\bar{X}, \bar{Y})$, anchoring the model in the center of the observed data. Equivalently, we can also write these parameters in terms of covariances.

The covariance between two variables is defined as:

$$\text{Cov}(X_i, Y_i) = E[(X_i - E[X_i])(Y_i - E[Y_i])]$$

Properties of covariance:

1. $\text{Cov}(X_i, X_i) = \sigma_X^2$.
2. If $E(X_i) = 0$ or $E(Y_i) = 0$, then $\text{Cov}(X_i, Y_i) = E[X_i Y_i]$.
3. For $W_i = a + bX_i$ and $Z_i = c + dY_i$,
   $\text{Cov}(W_i, Z_i) = bd \cdot \text{Cov}(X_i, Y_i)$.

For a bivariate regression, the slope β in a bivariate regression is given by:

$$\beta = \frac{\text{Cov}(Y_i, X_i)}{\text{Var}(X_i)}.$$

For a multivariate case, the slope for X_k is:

$$\beta_k = \frac{\text{Cov}(Y_i, \tilde{X}_{ki})}{\text{Var}(\tilde{X}_{ki})}$$

where $\tilde{X}_{ki}$ represents the residual from a regression of X_{ki} on the $K - 1$ other covariates in the model.

The intercept is:

$$\beta_0 = E[Y_i] - \beta_1 E(X_i).$$

Note:

- OLS does not require the assumption of a specific distribution for the variables. Its robustness is based on the minimization of squared errors (i.e., no distributional assumptions).

2.1.1.2 Properties of Least Squares Estimators

The properties of the ordinary least squares estimators (b_0 and b_1) are derived based on their statistical behavior. These properties provide insights into the accuracy, variability, and reliability of the estimates.

2.1.1.2.1 Expectation of the OLS Estimators

The OLS estimators b_0 (intercept) and b_1 (slope) are unbiased. This means their expected values equal the true population parameters:

$$E(b_1) = \beta_1,$$

$$E(b_0) = E(\bar{Y}) - \bar{X}\beta_1.$$

Since the expected value of the sample mean of Y, $E(\bar{Y})$, is:

$$E(\bar{Y}) = \beta_0 + \beta_1 \bar{X},$$

the expected value of b_0 simplifies to:

$$E(b_0) = \beta_0.$$

Thus, b_0 and b_1 are unbiased estimators of their respective population parameters β_0 and β_1.

2.1.1.2.2 Variance of the OLS Estimators

The variability of the OLS estimators depends on the spread of the predictor variable X and the error variance σ^2. The variances are given by:
Variance of b_1 (slope):

$$\text{Var}(b_1) = \frac{\sigma^2}{\sum_{i=1}^{n}(X_i - \bar{X})^2}.$$

Variance of b_0 (intercept):

$$\text{Var}(b_0) = \sigma^2 \left(\frac{1}{n} + \frac{\bar{X}^2}{\sum_{i=1}^{n}(X_i - \bar{X})^2} \right).$$

These formulas highlight that:

- $\text{Var}(b_1) \to 0$ as the number of observations increases, provided X_i values are distributed around their mean $\bar{X}$.
- $\text{Var}(b_0) \to 0$ as n increases, assuming X_i values are appropriately selected (i.e., not all clustered near the mean).

2.1.1.3 Mean Square Error (MSE)

The mean square error (MSE) quantifies the average squared residual (error) in the model:

$$MSE = \frac{SSE}{n-2} = \frac{\sum_{i=1}^{n} e_i^2}{n-2} = \frac{\sum_{i=1}^{n}(Y_i - \hat{Y}_i)^2}{n-2},$$

where SSE is the sum of squared errors and $n-2$ represents the degrees of freedom for a simple linear regression model (two parameters estimated: β_0 and β_1).

The expected value of the MSE equals the error variance (i.e., unbiased estimator of MSE):

$$E(MSE) = \sigma^2.$$

2.1.1.4 Estimating Variance of the OLS Coefficients

The sample-based estimates of the variances of b_0 and b_1 are expressed as follows:
Estimated variance of b_1 (slope):

$$s^2(b_1) = \widehat{\text{Var}}(b_1) = \frac{MSE}{\sum_{i=1}^{n}(X_i - \bar{X})^2}.$$

Estimated Variance of b_0 (Intercept):

$$s^2(b_0) = \widehat{\text{Var}}(b_0) = MSE \left(\frac{1}{n} + \frac{\bar{X}^2}{\sum_{i=1}^{n}(X_i - \bar{X})^2} \right).$$

These estimates rely on the MSE to approximate σ^2.

The variance estimates are unbiased:

$$E(s^2(b_1)) = \text{Var}(b_1),$$
$$E(s^2(b_0)) = \text{Var}(b_0).$$

Implications of These Properties

1. **Unbiasedness:** The unbiased nature of b_0 and b_1 ensures that, on average, the regression model accurately reflects the true relationship in the population.
2. **Decreasing variance:** As the sample size n increases or as the spread of X_i values grows, the variances of b_0 and b_1 decrease, leading to more precise estimates.
3. **Error estimation with MSE:** MSE provides a reliable estimate of the error variance σ^2, which feeds directly into assessing the reliability of b_0 and b_1.

2.1.1.5 Residuals in Ordinary Least Squares

Residuals are the differences between observed values (Y_i) and their predicted counterparts ($\hat{Y}_i$). They play a central role in assessing model fit and ensuring the assumptions of OLS are met.

The residual for the ith observation is defined as:

$$e_i = Y_i - \hat{Y}_i = Y_i - (b_0 + b_1 X_i),$$

where:

- e_i: residual for the ith observation
- $\hat{Y}_i$: predicted value based on the regression model
- Y_i: actual observed value

Residuals estimate the unobservable error terms ϵ_i:

- e_i is an estimate of $\epsilon_i = Y_i - E(Y_i)$.
- ϵ_i is always unknown because we do not know the true values of β_0 and β_1.

2.1.1.5.1 Key Properties of Residuals

Residuals exhibit several mathematical properties that align with the OLS estimation process:

1. **Sum of Residuals**
   The residuals sum to zero:

$$\sum_{i=1}^{n} e_i = 0.$$

   This ensures that the regression line passes through the centroid of the data, $(\bar{X}, \bar{Y})$.
2. **Orthogonality of Residuals to Predictors** The residuals are orthogonal (uncorrelated) to the predictor variable X:

$$\sum_{i=1}^{n} X_i e_i = 0.$$

   This reflects the fact that the OLS minimizes the squared deviations of residuals along the Y-axis, not the X-axis.

2.1.1.5.2 Expected Values of Residuals

The expected values of residuals reinforce the unbiased nature of OLS:

1. **Mean of Residuals**
   The residuals have an expected value of zero:

$$E[e_i] = 0.$$

2. **Orthogonality to Predictors and Fitted Values**
   Residuals are uncorrelated with both the predictor variables and the fitted values:

$$E[X_i e_i] = 0,$$
$$E[\hat{Y}_i e_i] = 0.$$

These properties highlight that residuals do not contain systematic information about the predictors or the fitted values, reinforcing the idea that the model has captured the underlying relationship effectively.

2.1.1.5.3 Practical Importance of Residuals

1. **Model Diagnostics**
   Residuals are analyzed to check the assumptions of OLS, including linearity, homoskedasticity (constant variance), and independence of errors. Patterns in residual plots can signal issues such as nonlinearity or heteroskedasticity.

2. **Goodness-of-Fit**
   The sum of squared residuals, $\sum e_i^2$, measures the total unexplained variation in Y. A smaller sum indicates a better fit.
3. **Influence Analysis**
   Large residuals may indicate outliers or influential points that disproportionately affect the regression line.

2.1.1.6 Inference in Ordinary Least Squares

Inference allows us to make probabilistic statements about the regression parameters (β_0, β_1) and predictions (Y_h). To perform valid inference, certain assumptions about the distribution of errors are necessary.
Normality Assumption

- OLS estimation itself does **not** require the assumption of normality.
- However, to conduct hypothesis tests or construct confidence intervals for β_0, β_1 and predictions, distributional assumptions are necessary.
- Inference on β_0 and β_1 is **robust** to moderate departures from normality, especially in large samples due to the [central limit theorem].
- Inference on predicted values, Y_{pred}, is more sensitive to normality violations.

When we assume a **normal error model**, the response variable Y_i is modeled as:

$$Y_i \sim N(\beta_0 + \beta_1 X_i, \sigma^2),$$

where:

- $\beta_0 + \beta_1 X_i$: mean response
- σ^2: variance of the errors

Under this model, the sampling distributions of the OLS estimators, b_0 and b_1, can be derived.

2.1.1.6.1 Inference for β_1 (Slope)

Under the normal error model:

1. **Sampling Distribution of** b_1

$$b_1 \sim N\left(\beta_1, \frac{\sigma^2}{\sum_{i=1}^n (X_i - \bar{X})^2}\right).$$

   This indicates that b_1 is an unbiased estimator of β_1 with variance proportional to σ^2.

2. **Test Statistic**

$$t = \frac{b_1 - \beta_1}{s(b_1)} \sim t_{n-2},$$

where $s(b_1)$ is the standard error of b_1:

$$s(b_1) = \sqrt{\frac{MSE}{\sum_{i=1}^{n}(X_i - \bar{X})^2}}.$$

3. **Confidence Interval**
A $(1 - \alpha)100\%$ confidence interval for β_1 is:

$$b_1 \pm t_{1-\alpha/2;n-2} \cdot s(b_1).$$

2.1.1.6.2 Inference for β_0 (Intercept)

1. **Sampling Distribution of b_0**
Under the normal error model, the sampling distribution of b_0 is:

$$b_0 \sim N\left(\beta_0, \sigma^2\left(\frac{1}{n} + \frac{\bar{X}^2}{\sum_{i=1}^{n}(X_i - \bar{X})^2}\right)\right).$$

2. **Test Statistic**

$$t = \frac{b_0 - \beta_0}{s(b_0)} \sim t_{n-2},$$

where $s(b_0)$ is the standard error of b_0:

$$s(b_0) = \sqrt{MSE\left(\frac{1}{n} + \frac{\bar{X}^2}{\sum_{i=1}^{n}(X_i - \bar{X})^2}\right)}.$$

3. **Confidence Interval**
A $(1 - \alpha)100\%$ confidence interval for β_0 is:

$$b_0 \pm t_{1-\alpha/2;n-2} \cdot s(b_0).$$

2.1.1.6.3 Mean Response

In regression, we often estimate the mean response of the dependent variable Y for a given level of the predictor variable X, denoted as X_h. This estimation provides a predicted average outcome for a specific value of X based on the fitted regression model.

- Let X_h represent the level of X for which we want to estimate the mean response.
- The mean response when $X = X_h$ is denoted as $E(Y_h)$.
- A point estimator for $E(Y_h)$ is $\hat{Y}_h$, which is the predicted value from the regression model:

$$\hat{Y}_h = b_0 + b_1 X_h.$$

The estimator $\hat{Y}_h$ is unbiased because its expected value equals the true mean response $E(Y_h)$:

$$\begin{aligned} E(\hat{Y}_h) &= E(b_0 + b_1 X_h) \\ &= \beta_0 + \beta_1 X_h \\ &= E(Y_h). \end{aligned}$$

Thus, $\hat{Y}_h$ provides a reliable estimate of the mean response at X_h.

The variance of $\hat{Y}_h$ reflects the uncertainty in the estimate of the mean response:

$$\begin{aligned} \text{Var}(\hat{Y}_h) &= \text{Var}(b_0 + b_1 X_h) \quad (\text{definition of } \hat{Y}_h) \\ &= \text{Var}\left((\bar{Y} - b_1\bar{X}) + b_1 X_h\right) \quad (\text{since } b_0 = \bar{Y} - b_1\bar{X}) \\ &= \text{Var}\left(\bar{Y} + b_1(X_h - \bar{X})\right) \quad (\text{factor out } b_1) \\ &= \text{Var}\left(\bar{Y} + b_1(X_h - \bar{X})\right) \\ &= \text{Var}(\bar{Y}) + (X_h - \bar{X})^2\text{Var}(b_1) + 2(X_h - \bar{X})\text{Cov}(\bar{Y}, b_1). \end{aligned}$$

Since $\text{Cov}(\bar{Y}, b_1) = 0$ (due to the independence of the errors, ϵ_i), the variance simplifies to:

$$\text{Var}(\hat{Y}_h) = \frac{\sigma^2}{n} + (X_h - \bar{X})^2 \frac{\sigma^2}{\sum_{i=1}^{n}(X_i - \bar{X})^2}.$$

This can also be expressed as:

$$\text{Var}(\hat{Y}_h) = \sigma^2 \left(\frac{1}{n} + \frac{(X_h - \bar{X})^2}{\sum_{i=1}^{n}(X_i - \bar{X})^2}\right).$$

To estimate the variance of $\hat{Y}_h$, we replace σ^2 with MSE, the mean squared error from the regression:

$$s^2(\hat{Y}_h) = MSE\left(\frac{1}{n} + \frac{(X_h - \bar{X})^2}{\sum_{i=1}^{n}(X_i - \bar{X})^2}\right).$$

Under the normal error model, the sampling distribution of $\hat{Y}_h$ is:

$$\hat{Y}_h \sim N\left(E(Y_h), \text{Var}(\hat{Y}_h)\right),$$

$$\frac{\hat{Y}_h - E(Y_h)}{s(\hat{Y}_h)} \sim t_{n-2}.$$

This result follows because $\hat{Y}_h$ is a linear combination of normally distributed random variables, and its variance is estimated using $s^2(\hat{Y}_h)$.

A $100(1-\alpha)\%$ confidence interval for the mean response $E(Y_h)$ is given by:

$$\hat{Y}_h \pm t_{1-\alpha/2;n-2} \cdot s(\hat{Y}_h),$$

where:

- $\hat{Y}_h$: point estimate of the mean response
- $s(\hat{Y}_h)$: estimated standard error of the mean response
- $t_{1-\alpha/2;n-2}$: critical value from the t-distribution with $n-2$ degrees of freedom

2.1.1.6.4 Prediction of a New Observation

When analyzing regression results, it is important to distinguish between:

1. **Estimating the mean response** at a particular value of X
2. **Predicting an individual outcome** for a particular value of X

Mean Response Versus Individual Outcome

- **Same point estimate**
  The formula for both the estimated mean response and the predicted individual outcome at $X = X_h$ is identical:

$$\hat{Y}_{pred} = \hat{Y}_h = b_0 + b_1 X_h.$$

- **Different variance**
  Although the point estimates are the same, the level of uncertainty differs. When predicting an individual outcome, we must consider not only the uncertainty in estimating the mean response ($\hat{Y}_h$) but also the additional random variation within the distribution of Y.

Therefore, **prediction intervals** (for individual outcomes) account for more uncertainty and are consequently wider than **confidence intervals** (for the mean response).

To predict an individual outcome for a given X_h, we combine the mean response with the random error:

$$Y_{pred} = \beta_0 + \beta_1 X_h + \epsilon.$$

Using the least squares predictor:

$$\hat{Y}_{pred} = b_0 + b_1 X_h,$$

since $E(\epsilon) = 0$.
The variance of the predicted value for a new observation, Y_{pred}, includes both:

1. Variance of the estimated mean response:

$$\sigma^2 \left(\frac{1}{n} + \frac{(X_h - \bar{X})^2}{\sum_{i=1}^{n} (X_i - \bar{X})^2} \right)$$

2. Variance of the error term, ϵ, which is σ^2

Thus, the total variance is:

$$\begin{aligned}
\text{Var}(Y_{pred}) &= \text{Var}(b_0 + b_1 X_h + \epsilon) \\
&= \text{Var}(b_0 + b_1 X_h) + \text{Var}(\epsilon) \\
&= \sigma^2 \left(\frac{1}{n} + \frac{(X_h - \bar{X})^2}{\sum_{i=1}^{n} (X_i - \bar{X})^2} \right) + \sigma^2 \\
&= \sigma^2 \left(1 + \frac{1}{n} + \frac{(X_h - \bar{X})^2}{\sum_{i=1}^{n} (X_i - \bar{X})^2} \right).
\end{aligned}$$

We estimate the variance of the prediction using MSE, the mean squared error:

$$s^2(pred) = MSE \left(1 + \frac{1}{n} + \frac{(X_h - \bar{X})^2}{\sum_{i=1}^{n} (X_i - \bar{X})^2} \right).$$

Under the normal error model, the standardized predicted value follows a t-distribution with $n - 2$ degrees of freedom:

$$\frac{Y_{pred} - \hat{Y}_h}{s(pred)} \sim t_{n-2}.$$

A $100(1-\alpha)\%$ prediction interval for Y_{pred} is:

$$\hat{Y}_{pred} \pm t_{1-\alpha/2;n-2} \cdot s(pred).$$

2.1.1.6.5 Confidence Band

In regression analysis, we often want to evaluate the uncertainty around the entire regression line, not just at a single value of the predictor variable X. This is achieved using a **confidence band**, which provides a confidence interval for the mean response, $E(Y) = \beta_0 + \beta_1 X$, over the entire range of X values.

The Working–Hotelling confidence band is a method to construct simultaneous confidence intervals for the regression line. For a given X_h, the confidence band is expressed as:

$$\hat{Y}_h \pm W s(\hat{Y}_h),$$

where:

- $W^2 = 2F_{1-\alpha;2,n-2}$
  - $F_{1-\alpha;2,n-2}$ is the critical value from the F-distribution with 2 and $n-2$ degrees of freedom.
- $s(\hat{Y}_h)$ is the standard error of the estimated mean response at X_h:

$$s^2(\hat{Y}_h) = MSE\left(\frac{1}{n} + \frac{(X_h - \bar{X})^2}{\sum_{i=1}^{n}(X_i - \bar{X})^2}\right)$$

Key Properties of the Confidence Band

1. **Width of the interval**
   - The width of the confidence band changes with X_h because $s(\hat{Y}_h)$ depends on how far X_h is from the mean of X ($\bar{X}$).
   - The interval is narrowest at $X = \bar{X}$, where the variance of the estimated mean response is minimized.
2. **Shape of the band**
   - The boundaries of the confidence band form a hyperbolic shape around the regression line.
   - This reflects the increasing uncertainty in the mean response as X_h moves farther from $\bar{X}$.

3. **Simultaneous coverage**

   - The Working–Hotelling band ensures that the true regression line $E(Y) = \beta_0+\beta_1 X$ lies within the band across all values of X with a specified confidence level (e.g., 95%).

2.1.1.7 Analysis of Variance (ANOVA) in Regression

ANOVA in regression decomposes the total variability in the response variable (Y) into components attributed to the regression model and residual error. In the context of regression, ANOVA provides a mechanism to assess the fit of the model and test hypotheses about the relationship between X and Y.

The **corrected total sum of squares (SSTO)** quantifies the total variation in Y:

$$SSTO = \sum_{i=1}^{n}(Y_i - \bar{Y})^2,$$

where $\bar{Y}$ is the mean of the response variable. The term "corrected" refers to the fact that the sum of squares is calculated relative to the mean (i.e., the uncorrected total sum of squares is given by $\sum Y_i^2$).

Using the fitted regression model $\hat{Y}_i = b_0 + b_1 X_i$, we estimate the conditional mean of Y at X_i. The total sum of squares can be decomposed as:

$$\begin{aligned}\sum_{i=1}^{n}(Y_i - \bar{Y})^2 &= \sum_{i=1}^{n}(Y_i - \hat{Y}_i + \hat{Y}_i - \bar{Y})^2 \\ &= \sum_{i=1}^{n}(Y_i - \hat{Y}_i)^2 + \sum_{i=1}^{n}(\hat{Y}_i - \bar{Y})^2 + 2\sum_{i=1}^{n}(Y_i - \hat{Y}_i)(\hat{Y}_i - \bar{Y}) \\ &= \sum_{i=1}^{n}(Y_i - \hat{Y}_i)^2 + \sum_{i=1}^{n}(\hat{Y}_i - \bar{Y})^2\end{aligned}$$

- The cross-product term is zero, as shown below.
- This decomposition simplifies to:

$$SSTO = SSE + SSR,$$

where:

- $SSE = \sum_{i=1}^{n}(Y_i - \hat{Y}_i)^2$: error sum of squares (variation unexplained by the model)

- $SSR = \sum_{i=1}^{n}(\hat{Y}_i - \bar{Y})^2$: regression sum of squares (variation explained by the model), which measure how the conditional mean varies about a central value

Degrees of freedom are partitioned as:

$$SSTO = SSR + SSE$$
$$(n-1) = (1) + (n-2).$$

To confirm that the cross-product term is zero:

$$\begin{aligned}
\sum_{i=1}^{n}(Y_i-\hat{Y}_i)(\hat{Y}_i-\bar{Y}) &= \sum_{i=1}^{n}(Y_i - \bar{Y} - b_1(X_i - \bar{X}))(\bar{Y} + b_1(X_i - \bar{X}) - \bar{Y}) \\
&\quad \text{(Expand } Y_i - \hat{Y}_i \text{ and } \hat{Y}_i - \bar{Y}\text{)} \\
&= \sum_{i=1}^{n}(Y_i - \bar{Y} - b_1(X_i - \bar{X}))(b_1(X_i - \bar{X})) \\
&= b_1\sum_{i=1}^{n}(Y_i - \bar{Y})(X_i - \bar{X}) - b_1^2\sum_{i=1}^{n}(X_i - \bar{X})^2 \\
&\quad \text{(Distribute terms in the product)} \\
&= b_1\frac{\sum_{i=1}^{n}(Y_i-\bar{Y})(X_i-\bar{X})}{\sum_{i=1}^{n}(X_i-\bar{X})^2}\sum_{i=1}^{n}(X_i-\bar{X})^2 - b_1^2\sum_{i=1}^{n}(X_i-\bar{X})^2 \\
&\quad \text{(Substitute } b_1 \text{ definition)} \\
&= b_1^2\sum_{i=1}^{n}(X_i - \bar{X})^2 - b_1^2\sum_{i=1}^{n}(X_i - \bar{X})^2 \\
&= 0.
\end{aligned}$$

Table 2.2 presents the analysis of variance decomposition for the simple linear regression model, including sums of squares, mean squares, and the F-test for overall model significance.

Table 2.2 ANOVA Decomposition for Simple Linear Regression

Source of variation	Sum of squares	df	Mean square	F-statistic
Regression (model)	SSR	1	$MSR = \frac{SSR}{1}$	$F = \frac{MSR}{MSE}$
Error	SSE	$n-2$	$MSE = \frac{SSE}{n-2}$	
Total (corrected)	$SSTO$	$n-1$		

The expected values of the mean squares are:

$$E(MSE) = \sigma^2,$$

$$E(MSR) = \sigma^2 + \beta_1^2 \sum_{i=1}^{n} (X_i - \bar{X})^2.$$

- **If** $\beta_1 = 0$:
  - The regression model does not explain any variation in Y beyond the mean, and $E(MSR) = E(MSE) = \sigma^2$.
  - This condition corresponds to the null hypothesis, $H_0 : \beta_1 = 0$.
- **If** $\beta_1 \neq 0$:
  - The regression model explains some variation in Y, and $E(MSR) > E(MSE)$.
  - The additional term $\beta_1^2 \sum_{i=1}^{n} (X_i - \bar{X})^2$ represents the variance explained by the predictor X.

The difference between $E(MSR)$ and $E(MSE)$ allows us to infer whether $\beta_1 \neq 0$ by comparing their ratio.
Assuming the errors ϵ_i are independent and identically distributed as $N(0, \sigma^2)$ and under the null hypothesis $H_0 : \beta_1 = 0$, we have:

1. The scaled MSE follows a chi-square distribution with $n-2$ degrees of freedom:

$$\frac{MSE}{\sigma^2} \sim \chi^2_{n-2}.$$

2. The scaled MSR follows a chi-square distribution with 1 degree of freedom:

$$\frac{MSR}{\sigma^2} \sim \chi^2_1.$$

3. These two chi-square random variables are independent.

The ratio of two independent chi-square random variables, scaled by their respective degrees of freedom, follows an F-distribution. Therefore, under H_0:

$$F = \frac{MSR}{MSE} \sim F_{1,n-2}.$$

The F-statistic tests whether the regression model provides a significant improvement over the null model (constant $E(Y)$).

The hypotheses for the F-test are:

- **Null hypothesis** (H_0): $\beta_1 = 0$ (no relationship between X and Y).

- **Alternative hypothesis** (H_a): $\beta_1 \neq 0$ (a significant relationship exists between X and Y).

The rejection rule for H_0 at significance level α is:

$$F > F_{1-\alpha;1,n-2},$$

where $F_{1-\alpha;1,n-2}$ is the critical value from the F-distribution with 1 and $n-2$ degrees of freedom.

1. **If** $F \leq F_{1-\alpha;1,n-2}$:
   - Fail to reject H_0. There is insufficient evidence to conclude that X significantly explains variation in Y.
2. **If** $F > F_{1-\alpha;1,n-2}$:
   - Reject H_0. There is significant evidence that X explains some of the variation in Y.

2.1.1.8 Coefficient of Determination (R^2)

The **coefficient of determination** (R^2) measures how well the linear regression model accounts for the variability in the response variable Y. It is defined as:

$$R^2 = \frac{SSR}{SSTO} = 1 - \frac{SSE}{SSTO},$$

where:

- SSR: regression sum of squares (variation explained by the model)
- $SSTO$: total sum of squares (total variation in Y about its mean)
- SSE: error sum of squares (variation unexplained by the model)

Properties of R^2
1. **Range**:

$$0 \leq R^2 \leq 1.$$

   - $R^2 = 0$: The model explains none of the variability in Y (e.g., $\beta_1 = 0$).
   - $R^2 = 1$: The model explains all the variability in Y (perfect fit).

2. **Proportionate reduction in variance**: R^2 represents the proportionate reduction in the total variation of Y after fitting the model. It quantifies how much better the model predicts Y compared to simply using $\bar{Y}$.

3. **Potential misinterpretation**: It is not really correct to say R^2 is the "variation in Y explained by X." The term "variation explained" assumes a causative or deterministic explanation, which is not always correct. For example:
   - R^2 shows how much variance in Y is accounted for by the regression model, but it does not imply causation.
   - In cases with confounding variables or spurious correlations, R^2 can still be high, even if there's no direct causal link between X and Y.

For simple linear regression, R^2 is the square of the Pearson correlation coefficient, r:

$$R^2 = (r)^2,$$

where:

- $r = \text{corr}(X, Y)$ is the sample correlation coefficient

The relationship between b_1 (the slope of the regression line) and r is given by:

$$b_1 = \left(\frac{\sum_{i=1}^n (Y_i - \bar{Y})^2}{\sum_{i=1}^n (X_i - \bar{X})^2} \right)^{1/2}.$$

Additionally, r can be expressed as:

$$r = \frac{s_y}{s_x} \cdot r,$$

where s_y and s_x are the sample standard deviations of Y and X, respectively.

2.1.1.9 Lack of Fit in Regression

The **lack of fit** test evaluates whether the chosen regression model adequately captures the relationship between the predictor variable X and the response variable Y. When there are repeated observations at specific values of X, we can partition the error sum of squares (SSE) into two components:

1. **Pure error**
2. **Lack of fit**

Given the observations:

- Y_{ij}: the jth replicate for the ith distinct value of X
  - $Y_{11}, Y_{21}, \ldots, Y_{n_1,1}$: n_1 repeated observations of X_1
  - $Y_{1c}, Y_{2c}, \ldots, Y_{n_c,c}$: n_c repeated observations of X_c

- $\bar{Y}_j$: the mean response for replicates at X_j
- $\hat{Y}_{ij}$: the predicted value from the regression model for X_j

the error sum of squares (SSE) can be decomposed as:

$$\begin{aligned}\sum_i\sum_j(Y_{ij}-\hat{Y}_{ij})^2 &= \sum_i\sum_j(Y_{ij}-\bar{Y}_j+\bar{Y}_j-\hat{Y}_{ij})^2\\ &= \sum_i\sum_j(Y_{ij}-\bar{Y}_j)^2+\sum_j n_j(\bar{Y}_j-\hat{Y}_{ij})^2+\text{cross product term}\\ &= \sum_i\sum_j(Y_{ij}-\bar{Y}_j)^2+\sum_j n_j(\bar{Y}_j-\hat{Y}_{ij})^2.\end{aligned}$$

- The **cross-product term** is zero because the deviations within replicates and the deviations between replicates are orthogonal.
- This simplifies to:

$$SSE = SSPE + SSLF,$$

where:

- $SSPE$ (pure error sum of squares): Variation within replicates for the same X_j, reflecting natural variability in the response
  - Degrees of freedom: $df_{pe} = n-c$, where n is the total number of observations, and c is the number of distinct X values.
- $SSLF$ (lack of fit sum of squares): Variation between the replicate means $\bar{Y}_j$ and the model-predicted values $\hat{Y}_{ij}$. If SSLF is large, it suggests the model may not adequately describe the relationship between X and Y.
  - Degrees of freedom: $df_{lf} = c-2$, where 2 accounts for the parameters in the linear regression model (β_0 and β_1).

- **Mean square for pure error (MSPE):**

$$MSPE = \frac{SSPE}{df_{pe}} = \frac{SSPE}{n-c}.$$

- **Mean square for lack of fit (MSLF):**

$$MSLF = \frac{SSLF}{df_{lf}} = \frac{SSLF}{c-2}.$$

2.1.1.9.1 The F-Test for Lack of Fit

The **F-test for lack of fit** evaluates whether the chosen regression model adequately captures the relationship between the predictor variable X and the response variable Y. Specifically, it tests whether any systematic deviations from the model exist that are not accounted for by random error.

- **Null hypothesis** (H_0):
  The regression model is adequate:

$$H_0 : Y_{ij} = \beta_0 + \beta_1 X_i + \epsilon_{ij}, \quad \epsilon_{ij} \sim \text{i.i.d. } N(0, \sigma^2).$$

- **Alternative hypothesis** (H_a):
  The regression model is not adequate and includes an additional function $f(X_i, Z_1, \ldots)$ to account for the lack of fit:

$$H_a : Y_{ij} = \alpha_0 + \alpha_1 X_i + f(X_i, Z_1, \ldots) + \epsilon^*_{ij}, \quad \epsilon^*_{ij} \sim \text{i.i.d. } N(0, \sigma^2).$$

Expected Mean Squares

- The expected mean square for pure error (MSPE) is the same under both H_0 and H_a:

$$E(MSPE) = \sigma^2.$$

- The expected mean square for lack of fit (MSLF) depends on whether H_0 is true:
  - Under H_0 (model is adequate):

$$E(MSLF) = \sigma^2.$$

  - Under H_a (model is not adequate):

$$E(MSLF) = \sigma^2 + \frac{\sum n_j f(X_i, Z_1, \ldots)^2}{n-2}.$$

The test statistic for the lack of fit test is:

$$F = \frac{MSLF}{MSPE},$$

where:

- $MSLF = \frac{SSLF}{c-2}$,
  and $SSLF$ is the lack of fit sum of squares.
- $MSPE = \frac{SSPE}{n-c}$,
  and $SSPE$ is the pure error sum of squares.

Table 2.3 ANOVA table for linear regression with lack of fit test

Source of variation	Sum of squares	df	Mean square	F statistic
Regression	SSR	1	$MSR = \frac{SSR}{1}$	$F = \frac{MSR}{MSE}$
Error	SSE	$n-2$	$MSE = \frac{SSE}{n-2}$	
Lack of fit	SSLF	$c-2$	$MSLF = \frac{SSLF}{c-2}$	$F = \frac{MSLF}{MSPE}$
Pure error	SSPE	$n-c$	$MSPE = \frac{SSPE}{n-c}$	
Total (corrected)	SSTO	$n-1$		

Under H_0, the F-statistic follows an F-distribution:

$$F \sim F_{c-2,n-c}.$$

Decision Rule

- Reject H_0 at significance level α if:

$$F > F_{1-\alpha;c-2,n-c}.$$

- **Failing to reject** H_0:
  - Indicates that there is no evidence of lack of fit.
  - Does not imply the model is "true," but it suggests that the model provides a reasonable approximation to the true relationship.

To summarize, when repeat observations exist at some levels of X, the **error sum of squares (SSE)** can be further partitioned into **lack of fit (SSLF)** and **pure error (SSPE)**. This leads to an extended ANOVA table (Table 2.3):
Repeat observations have important implications for the coefficient of determination (R^2):

1. R^2 Cannot attain 1 with repeat observations:
   - With repeat observations, SSE (error sum of squares) cannot be reduced to 0 because $SSPE > 0$ (variability within replicates).
2. **Maximum** R^2:
   - The maximum attainable R^2 in the presence of repeat observations is:

$$R^2_{\max} = \frac{SSTO - SSPE}{SSTO}.$$

3. **Importance of repeat observations:**
   - Not all levels of X need repeat observations, but their presence enables the separation of pure error from lack of fit, making the F-test for lack of fit possible.

Estimation of σ^2 with Repeat Observations

1. **Use of MSE:**
   - When H_0 is appropriate (the model fits well), MSE is typically used as the estimate of σ^2 instead of $MSPE$ because it has more degrees of freedom and provides a more reliable estimate.
2. **Pooling estimates:**
   - In practice, MSE and $MSPE$ may be pooled if H_0 holds, resulting in a more precise estimate of σ^2.

2.1.1.10 Joint Inference for Regression Parameters

Joint inference considers the simultaneous coverage of confidence intervals for multiple regression parameters, such as β_0 (intercept) and β_1 (slope). Ensuring adequate confidence for both parameters together requires adjustments to maintain the desired family-wise confidence level.
Let:

- $\bar{A}_1$: The event that the confidence interval for β_0 covers its true value.
- $\bar{A}_2$: The event that the confidence interval for β_1 covers its true value.

The individual confidence levels are:

$$P(\bar{A}_1) = 1 - \alpha,$$
$$P(\bar{A}_2) = 1 - \alpha.$$

The joint confidence coefficient, $P(\bar{A}_1 \cap \bar{A}_2)$, is:

$$\begin{aligned} P(\bar{A}_1 \cap \bar{A}_2) &= 1 - P(\bar{A}_1 \cup \bar{A}_2), \\ &= 1 - P(A_1) - P(A_2) + P(A_1 \cap A_2), \\ &\geq 1 - P(A_1) - P(A_2), \\ &= 1 - 2\alpha. \end{aligned}$$

This means that if α is the significance level for each parameter, the joint confidence coefficient is at least $1-2\alpha$. This inequality is known as the **Bonferroni inequality**.

Bonferroni Confidence Intervals
To ensure a desired joint confidence level of $(1 - \alpha)$ for both β_0 and β_1, the Bonferroni method adjusts the confidence level for each parameter by dividing α by the number of parameters. For two parameters:

1. The confidence level for each parameter is $(1 - \alpha/2)$.
2. The resulting **Bonferroni-adjusted confidence intervals** are:

$$b_0 \pm B \cdot s(b_0),$$
$$b_1 \pm B \cdot s(b_1),$$

   where $B = t_{1-\alpha/4;n-2}$ is the critical value from the t-distribution with $n - 2$ degrees of freedom.

Interpretation of Bonferroni Confidence Intervals
1. **Coverage probability:**
   - If repeated samples were taken, $(1 - \alpha)100\%$ of the joint intervals would contain the true values of (β_0, β_1).
   - This implies that $\alpha \times 100\%$ of the samples would miss at least one of the true parameter values.
2. **Conservatism:**
   - The Bonferroni method ensures the family-wise confidence level but is **conservative**. The actual joint confidence level is often higher than $(1 - \alpha)100\%$.
   - This conservatism reduces statistical power.

```
# Load necessary libraries
library(ggplot2)
library(MASS)

# Set seed for reproducibility
set.seed(123)

# Generate synthetic data
n <- 100  # Number of observations
x <- rnorm(n, mean = 0, sd = 1)  # Predictor
beta_0 <- 2  # True intercept
beta_1 <- 3  # True slope
sigma <- 1  # Standard deviation of error
y <-
    beta_0 + beta_1 * x + rnorm(n, mean = 0, sd = sigma)  #
↪  Response
```

```r
# Fit linear model
model <- lm(y ~ x)
summary(model)
#>
#> Call:
#> lm(formula = y ~ x)
#>
#> Residuals:
#>     Min      1Q  Median      3Q     Max
#> -1.9073 -0.6835 -0.0875  0.5806  3.2904
#>
#> Coefficients:
#>             Estimate Std. Error t value Pr(>|t|)
#> (Intercept)  1.89720    0.09755   19.45   <2e-16 ***
#> x            2.94753    0.10688   27.58   <2e-16 ***
#> ---
#> Signif. codes:
#> 0 '***' 0.001 '**' 0.01 '*' 0.05 '.' 0.1 ' ' 1
#>
#> Residual standard error: 0.9707 on 98 degrees of freedom
#> Multiple R-squared:  0.8859, Adjusted R-squared:  0.8847
#> F-statistic: 760.6 on 1 and 98 DF,  p-value: < 2.2e-16

# Extract coefficients and standard errors
b0_hat <- coef(model)[1]
b1_hat <- coef(model)[2]
s_b0 <-
    summary(model)$coefficients[1, 2]  # Standard error of
    ↪ intercept
s_b1 <- summary(model)$coefficients[2, 2]  # Standard error of
↪ slope

# Desired confidence level
alpha <- 0.05  # Overall significance level

# Bonferroni correction
adjusted_alpha <- alpha / 2  # Adjusted alpha for each
↪ parameter

# Critical t-value for Bonferroni adjustment
t_crit <-
    qt(1 - adjusted_alpha, df = n - 2)  # n-2 degrees of
    ↪ freedom
```

```
# Bonferroni confidence intervals
ci_b0 <- c(b0_hat - t_crit * s_b0, b0_hat + t_crit * s_b0)
ci_b1 <- c(b1_hat - t_crit * s_b1, b1_hat + t_crit * s_b1)

# Print results
cat("Bonferroni Confidence Intervals:\n")
#> Bonferroni Confidence Intervals:
cat("Intercept (beta_0): [",
    round(ci_b0[1], 2),
    ",",
    round(ci_b0[2], 2),
    "]\n")
#> Intercept (beta_0): [ 1.7 , 2.09 ]
cat("Slope (beta_1): [",
    round(ci_b1[1], 2),
    ",",
    round(ci_b1[2], 2),
    "]\n")
#> Slope (beta_1): [ 2.74 , 3.16 ]

# Calculate the covariance matrix of coefficients
cov_matrix <- vcov(model)

# Generate points for confidence ellipse
ellipse_points <-
    MASS::mvrnorm(n = 1000,
                  mu = coef(model),
                  Sigma = cov_matrix)

# Convert to data frame for plotting
ellipse_df <- as.data.frame(ellipse_points)
colnames(ellipse_df) <- c("beta_0", "beta_1")
```

```
# Plot confidence intervals and ellipse
p <- ggplot() +
    # Confidence ellipse
    geom_point(
        data = ellipse_df,
        aes(x = beta_0, y = beta_1),
        alpha = 0.1,
        color = "grey"
    ) +
```

```
    # Point estimate
    geom_point(aes(x = b0_hat, y = b1_hat),
               color = "red",
               size = 3) +
    # Bonferroni confidence intervals
    geom_errorbar(aes(x = b0_hat, ymin = ci_b1[1], ymax =
    ↪  ci_b1[2]),
                  width = 0.1,
                  color = "blue") +
    geom_errorbarh(aes(y = b1_hat, xmin = ci_b0[1], xmax =
    ↪  ci_b0[2]),
                   height = 0.1,
                   color = "blue") +
    labs(title = "Bonferroni Confidence Intervals and Joint
    ↪  Confidence Region",
         x = "Intercept (beta_0)",
         y = "Slope (beta_1)") +
    theme_minimal()

print(p)
```

Figure 2.2 shows the Bonferroni confidence internals:

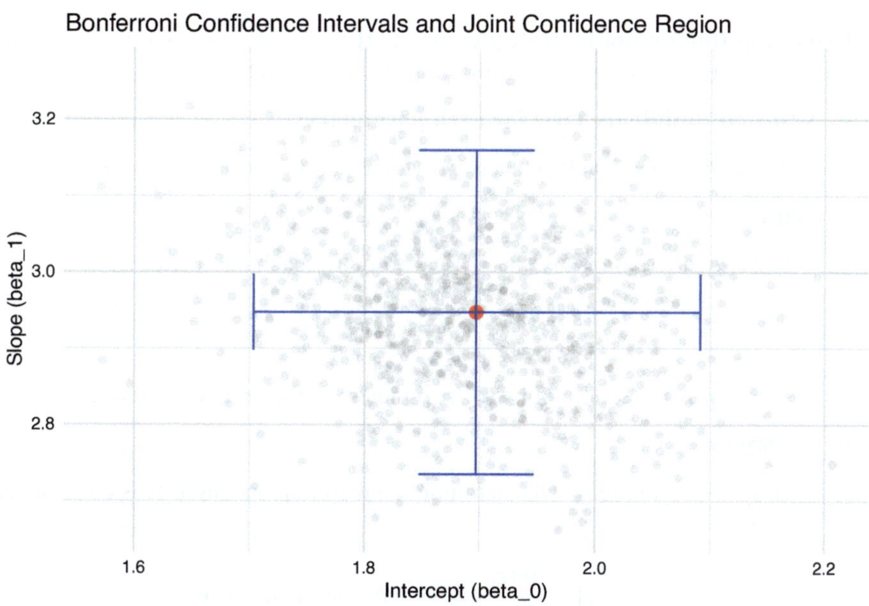

Fig. 2.2 Bonferroni confidence intervals and joint confidence region

1. The red point represents the estimated coefficients (b0_hat, b1_hat).
2. The blue lines represent the Bonferroni-adjusted confidence intervals for beta_0 and beta_1.
3. The grey points represent the joint confidence region based on the covariance matrix of coefficients.
4. The Bonferroni intervals ensure family-wise confidence level but are conservative.
5. Simulation results demonstrate how often the true values are captured in the intervals when repeated samples are drawn .

Notes

1. Conservatism of Bonferroni intervals

   - The **Bonferroni interval is conservative**:
     - The joint confidence level is a lower bound, ensuring family-wise coverage of at least $(1-\alpha)100\%$.
     - This conservatism results in wider intervals, reducing the statistical power of the test.
   - Adjustments for conservatism:
     - Practitioners often choose a larger α (e.g., $\alpha = 0.1$) to reduce the width of the intervals in Bonferroni joint tests.
     - A higher α allows for a better balance between confidence and precision, especially for exploratory analyses.

2. **Extending Bonferroni to multiple parameters**: The Bonferroni method is not limited to two parameters. For testing g parameters, such as $\beta_0, \beta_1, \ldots, \beta_{g-1}$:

   - **Adjusted confidence level for each parameter:**
     - The confidence level for each individual parameter is $(1 - \alpha/g)$.
   - **Critical** t-value:
     - For two-sided intervals, the critical value for each parameter is:

$$t_{1-\frac{\alpha}{2g};n-p},$$

       where p is the total number of parameters in the regression model.
   - **Example:**
     - If $\alpha = 0.05$ and $g = 10$, each individual confidence interval is constructed at the:

$$(1 - \frac{0.05}{10}) = 99.5\% \text{ confidence level.}$$

  - This corresponds to using $t_{1-\frac{0.005}{2};n-p}$ in the formula for the confidence intervals.

3. **Limitations** for large g
   - **Wide intervals:**
     - As g increases, the intervals become excessively wide, often leading to reduced usefulness in practical applications.
     - This issue stems from the conservatism of the Bonferroni method, which prioritizes family-wise error control.
   - **Suitability for small** g:
     - The Bonferroni procedure works well when g is relatively small (e.g., $g \leq 5$).
     - For larger g, alternative methods (discussed below) are more efficient.
4. Correlation between parameters—orrelation of b_0 and b_1:
   - The estimated regression coefficients b_0 and b_1 are often correlated:
     - **Negative correlation** if $\bar{X} > 0$
     - **Positive correlation** if $\bar{X} < 0$
   - This correlation can complicate joint inference but does not affect the validity of Bonferroni-adjusted intervals.
5. Alternatives to Bonferroni

Several alternative procedures provide more precise joint inference, especially for larger g:

1. **Scheffé's method:**
   - Constructs simultaneous confidence regions for all possible linear combinations of parameters
   - Suitable for exploratory analyses but may result in even wider intervals than Bonferroni
2. **Tukey's honest significant difference:**
   - Designed for pairwise comparisons in ANOVA but can be adapted for regression parameters
3. **Holm's step-down procedure:**
   - A sequential testing procedure that is less conservative than Bonferroni while still controlling the family-wise error rate
4. **Likelihood ratio tests:**
   - Construct joint confidence regions based on the likelihood function, offering more precision for large g

Table 2.4 Classical assumptions of the linear regression model

Assumption	Description
Linearity	Linear relationship between predictors and response
Independence of errors	Errors are independent (important in time series/clustering)
Homoskedasticity	Constant variance of residuals across predictors
Normality of errors	Residuals are normally distributed
No multicollinearity	Predictors are not highly correlated
No outliers/leverage points	No undue influence from outliers or high-leverage points
Exogeneity	Predictors are uncorrelated with the error term (no endogeneity)
Full rank	Predictors are linearly independent (no perfect multicollinearity)

2.1.1.11 Assumptions of Linear Regression

To ensure valid inference and reliable predictions in linear regression, the following assumptions must hold. We will cover them in depth in the next section (Table 2.4).

2.1.1.12 Diagnostics for Model Assumptions

Constant Variance

- To check homoskedasticity:
  - Plot residuals versus fitted values or residuals versus predictors.
  - Look for patterns or a funnel-shaped spread indicating heteroskedasticity.

Outliers

- Detect outliers using:
  - Residual versus predictor plot
  - Box plots
  - Stem-and-leaf plots
  - Scatter plots

Standardized Residuals
Residuals can be standardized to have unit variance, known as **studentized residuals**:

$$r_i = \frac{e_i}{s(e_i)}.$$

Semi-studentized Residuals
A simplified standardization using the mean squared error (MSE):

$$e_i^* = \frac{e_i}{\sqrt{MSE}}.$$

Nonindependent Error Terms

- To detect nonindependence:
  - Plot residuals versus time for time-series data.
  - Residuals e_i are not independent because they depend on $\hat{Y}_i$, which is derived from the same regression function.
- Detect dependency by plotting the residual for the ith response versus the $(i-1)$-th.

Nonnormality of Error Terms

- To assess normality:
  - Plot distribution of residuals.
  - Create box plots, stem-and-leaf plots, or normal probability plots.
- Issues such as an incorrect regression function or nonconstant error variance can distort the residual distribution.
- Normality tests require relatively large sample sizes to detect deviations.

Normality of Residuals

- Use tests based on the empirical cumulative distribution function (ECDF) (check [normality assessment])

Constancy of Error Variance

- Statistical tests for homoskedasticity:
  - **Brown–Forsythe test (modified Levene test)**:

    Robust against nonnormality, examines the variance of residuals across levels of predictors
  - **Breusch–Pagan test (Cook–Weisberg test)**:

    Tests for heteroskedasticity by regressing squared residuals on predictors

2.1.1.13 Remedial Measures for Violations of Assumptions

When the assumptions of simple linear regression are violated, appropriate remedial measures can be applied to address these issues (Table 2.5). Below is a list of measures for specific deviations from the assumptions.

Table 2.5 Common regression issues and their remedies

Issue	Remedy	Explanation
Nonlinearity	Apply transformations (e.g., log, square root)	Transformation of variables can help linearize the relationship between X and Y
	Use more complicated models (e.g., polynomial regression, splines)	Higher-order terms or nonlinear models can capture nonlinear relationships
Nonconstant error variance	Apply **weighted least squares**	WLS assigns weights to observations based on the inverse of their variance
	Use transformations (e.g., log, square root)	Transformations can stabilize error variance
Correlated errors	Use serially correlated error models (e.g., ARIMA for time-series data)	Time-series models account for serial dependence in the errors
Nonnormality of errors	Transform Y or use nonparametric methods	Transformations can normalize residuals; nonparametric methods do not assume normality
Omitted variables	Use multiple regression to include additional relevant predictors	Adding relevant variables reduces omitted variable bias and improves model accuracy
Outliers	Apply robust estimation techniques (e.g., Huber regression, M-estimation)	Robust methods reduce the influence of outliers on parameter estimates

2.1.1.13.1 General Remedies

- Use more **complicated models** (e.g., nonlinear models, generalized linear models).
- Apply **transformations** (see [variable transformation]) on X and/or Y to stabilize variance, linearize relationships, or normalize residuals. Note that transformations may not always yield "optimal" results.

2.1.1.13.2 Specific Remedies for Assumption Violations

2.1.1.13.3 Remedies in Detail

1. **Nonlinearity:**
   - Transformations: Apply transformations to the response variable Y or the predictor variable X. Common transformations include:
     - Logarithmic transformation: $Y' = \log(Y)$ or $X' = \log(X)$.
     - Polynomial terms: Include X^2, X^3, etc., to capture curvature.

- Alternative models:
  - Polynomial regression or splines for flexibility in modeling nonlinear relationships.

2. **Nonconstant error variance:**
   - Weighted least squares:
     - Assigns weights to observations inversely proportional to their variance.
   - Transformations:
     - Use a log or square root transformation to stabilize variance.

3. **Correlated errors:**
   - For time-series data:
     - Use serially correlated error models such as AR(1) or ARIMA.
     - These models explicitly account for dependency in residuals over time.

4. **Nonnormality:**
   - Transformations:
     - Apply a transformation to Y (e.g., log or square root) to make the residuals approximately normal.
   - Nonparametric regression:
     - Methods like LOESS or Theil–Sen regression do not require the normality assumption.

5. **Omitted variables:**
   - Introduce additional predictors:
     - Use multiple regression to include all relevant independent variables.
   - Check for multicollinearity when adding new variables.

6. **Outliers:**
   - Robust regression:
     - Use methods such as Huber regression or M-estimation to reduce the impact of outliers on model coefficients.
   - Diagnostics:
     - Identify outliers using Cook's distance, leverage statistics, or studentized residuals.

2.1.1.14 Transformations in Regression Analysis

Transformations involve modifying one or both variables to address issues such as nonlinearity, nonconstant variance, or nonnormality. However, it's important to note that the properties of least-squares estimates apply to the **transformed model**, not the original variables.

When transforming the dependent variable Y, we fit the model as:

$$g(Y_i) = b_0 + b_1 X_i,$$

where $g(Y_i)$ is the transformed response. To interpret the regression results in terms of the original Y, we need to **transform back**:

$$\hat{Y}_i = g^{-1}(b_0 + b_1 X_i).$$

Direct back-transformation of predictions can introduce **bias**. For example, in a log-transformed model:

$$\log(Y_i) = b_0 + b_1 X_i,$$

the unbiased back-transformed prediction of Y_i is:

$$\hat{Y}_i = \exp(b_0 + b_1 X_i + \frac{\sigma^2}{2}),$$

where $\frac{\sigma^2}{2}$ accounts for the bias correction due to the log transformation.

2.1.1.14.1 Box–Cox Family of Transformations

The **Box–Cox transformation** is a versatile family of transformations defined as:

$$Y' = \begin{cases} \frac{Y^\lambda - 1}{\lambda}, & \text{if } \lambda \neq 0, \\ \ln(Y), & \text{if } \lambda = 0. \end{cases}$$

This transformation introduces a parameter λ that is estimated from the data (Table 2.6). Common transformations include:

Choosing the Transformation Parameter λ

Table 2.6 Box–Cox transformations for different values of lambda

λ	Transformation Y'
2	Y^2
0.5	$\sqrt{Y}$
0	$\ln(Y)$
−0.5	$1/\sqrt{Y}$
−1	$1/Y$

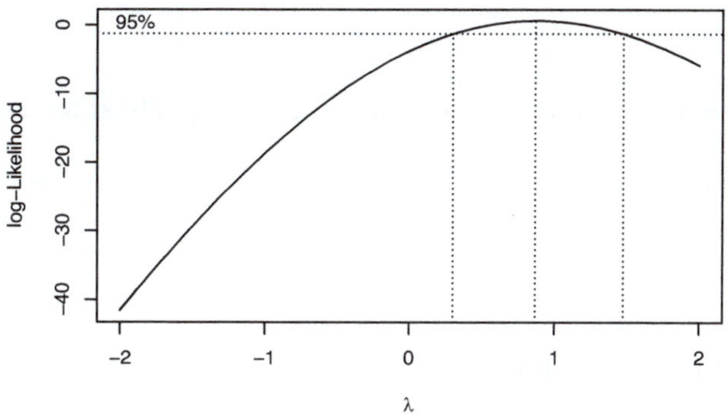

Fig. 2.3 Log-likelihood profile with 95% confidence interval for lambda

The value of λ can be selected using one of the following methods:

1. **Trial and error**:
   - Apply different transformations and compare the residual plots or model fit statistics (e.g., R^2 or AIC).
2. Maximum likelihood estimation:
   - Choose λ to maximize the likelihood function under the assumption of normally distributed errors.
3. **Numerical search**:
   - Use computational optimization techniques to minimize the residual sum of squares (RSS) or another goodness-of-fit criterion (Fig. 2.3).

```
# Install and load the necessary library
if (!require("MASS")) install.packages("MASS")
library(MASS)

# Fit a linear model
set.seed(123)
n <- 50
x <- rnorm(n, mean = 5, sd = 2)
y <- 3 + 2 * x + rnorm(n, mean = 0, sd = 2)
model <- lm(y ~ x)

# Apply Box-Cox Transformation
boxcox_result <- boxcox(model, lambda = seq(-2, 2, 0.1),
↪ plotit = TRUE)
```

```
# Find the optimal lambda
optimal_lambda <- boxcox_result$x[which.max(boxcox_result$y)]
cat("Optimal lambda for Box-Cox transformation:",
↪ optimal_lambda, "\n")
#> Optimal lambda for Box-Cox transformation: 0.8686869
```

Notes

1. **Benefits of transformations**:
   - **Stabilize variance**: helps address heteroskedasticity
   - **Linearize relationships**: useful for nonlinear data
   - **Normalize residuals**: addresses nonnormality issues
2. **Caveats**:
   - Interpretability: Transformed variables may complicate interpretation.
   - Over-transformation: Excessive transformations can distort the relationship between variables.
3. **Applicability**:
   - Transformations are most effective for issues like nonlinearity or nonconstant variance. They are less effective for correcting independence violations or omitted variables.

2.1.1.14.2 Variance-Stabilizing Transformations

Variance-stabilizing transformations are used when the standard deviation of the response variable depends on its mean. The **delta method**, which applies a Taylor series expansion, provides a systematic approach to find such transformations.

Given that the standard deviation of Y is a function of its mean:

$$\sigma = \sqrt{\text{var}(Y)} = f(\mu),$$

where $\mu = E(Y)$ and $f(\mu)$ is a smooth function of the mean, we aim to find a transformation $h(Y)$ such that the variance of the transformed variable $h(Y)$ is constant for all values of μ.

Expanding $h(Y)$ in a [Taylor expansion] series around μ:

$$h(Y) = h(\mu) + h'(\mu)(Y - \mu) + \text{higher-order terms}.$$

Ignoring higher-order terms, the variance of $h(Y)$ can be approximated as:

$$\text{var}(h(Y)) = \text{var}(h(\mu) + h'(\mu)(Y - \mu)).$$

Since $h(\mu)$ is a constant:

$$\text{var}(h(Y)) = \left(h'(\mu)\right)^2 \text{var}(Y).$$

Substituting $\text{var}(Y) = (f(\mu))^2$, we get:

$$\text{var}(h(Y)) = \left(h'(\mu)\right)^2 (f(\mu))^2 .$$

To stabilize the variance (make it constant for all μ), we require:

$$\left(h'(\mu)\right)^2 (f(\mu))^2 = \text{constant}.$$

Thus, the derivative of $h(\mu)$ must be proportional to the inverse of $f(\mu)$:

$$h'(\mu) \propto \frac{1}{f(\mu)}.$$

Integrating both sides gives:

$$h(\mu) = \int \frac{1}{f(\mu)} \, d\mu.$$

The specific form of $h(\mu)$ depends on the function $f(\mu)$, which describes the relationship between the standard deviation and the mean (Table 2.7).
Variance-stabilizing transformations are particularly useful for:

1. **Poisson-distributed data**: Use $h(Y) = 2\sqrt{Y}$ to stabilize variance.
2. **Exponential or multiplicative models**: Use $h(Y) = \ln(Y)$ for stabilization.
3. **Power law relationships**: Use transformations like $h(Y) = Y^{-1}$ or other forms derived from $f(\mu)$.

Example (Variance-Stabilizing Transformation for the Poisson Distribution)
For a **Poisson distribution**, the variance of Y is equal to its mean:

$$\sigma^2 = \text{var}(Y) = E(Y) = \mu.$$

Table 2.7 Examples of variance-stabilizing transformations

$f(\mu)$	Transformation $h(Y)$	Purpose
$\sqrt{\mu}$	$\int \frac{1}{\sqrt{\mu}} d\mu = 2\sqrt{Y}$	Stabilizes variance for Poisson data
μ	$\int \frac{1}{\mu} d\mu = \ln(Y)$	Stabilizes variance for exponential or multiplicative models
μ^2	$\int \frac{1}{\mu^2} d\mu = -\frac{1}{Y}$	Stabilizes variance for certain power law data

Thus, the standard deviation is:

$$\sigma = f(\mu) = \sqrt{\mu}.$$

Using the relationship for variance-stabilizing transformations:

$$h'(\mu) \propto \frac{1}{f(\mu)} = \mu^{-0.5}.$$

Integrating $h'(\mu)$ gives the variance-stabilizing transformation:

$$h(\mu) = \int \mu^{-0.5}\, d\mu = 2\sqrt{\mu}.$$

Hence, the variance-stabilizing transformation is:

$$h(Y) = \sqrt{Y}.$$

This transformation is widely used in Poisson regression to stabilize the variance of the response variable.

```
# Simulate Poisson data
set.seed(123)
n <- 500
x <- rnorm(n, mean = 5, sd = 2)
y <- rpois(n, lambda = exp(1 + 0.3 * x))  #
↪  Poisson-distributed Y

# Fit linear model without transformation
model_raw <- lm(y ~ x)

# Apply square root transformation
y_trans <- sqrt(y)
model_trans <- lm(y_trans ~ x)

# Compare residual plots
# par(mfrow = c(2, 1), mar = c(4, 4, 2, 1))

# Residual plot for raw data
plot(
    fitted(model_raw),
    resid(model_raw),
    main = "Residuals Raw Data",
    xlab = "Fitted Values",
    ylab = "Residuals"
```

```
)
abline(h = 0, col = "red", lty = 2)
```

```
# Residual plot for transformed data
plot(
    fitted(model_trans),
    resid(model_trans),
    main = "Residuals: Transformed Data (sqrt(Y))",
    xlab = "Fitted Values",
    ylab = "Residuals"
)
abline(h = 0, col = "blue", lty = 2)
```

2.1.1.14.3 General Strategy When $f(\mu)$ Is Unknown

If the relationship between var(Y) and μ (i.e., $f(\mu)$) is unknown, the following steps can help:

1. **Trial and error**:
   - Apply common transformations (e.g., $\log(Y)$, $\sqrt{Y}$) and examine residual plots (Fig. 2.4).
   - Select the transformation that results in stabilized variance (residuals show no pattern in plots) (Fig. 2.5).

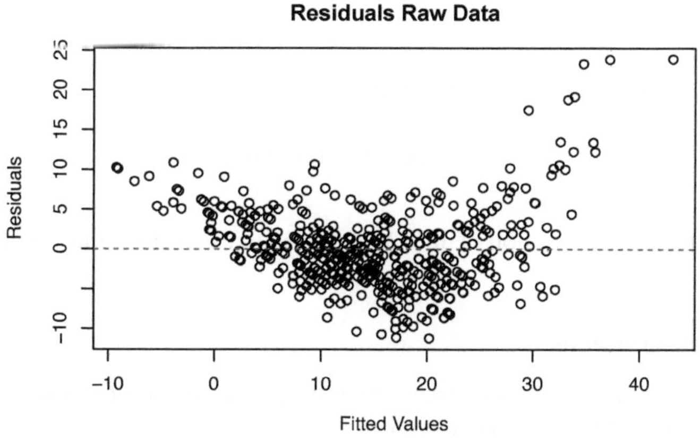

Fig. 2.4 Residual raw data

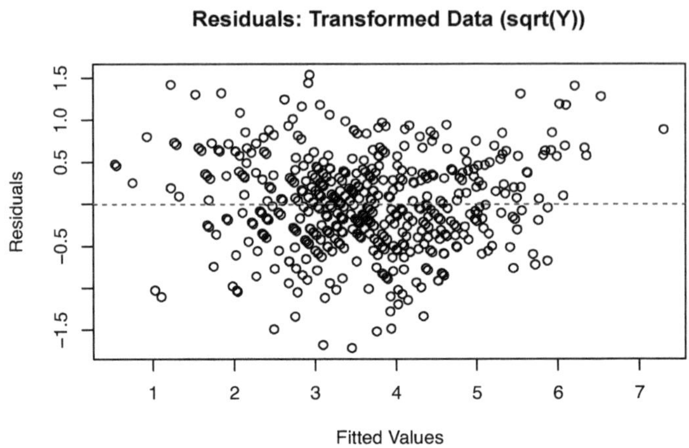

Fig. 2.5 Residuals: transformed data

2. **Leverage prior research**:
   - Consult researchers or literature on similar experiments to determine the transformations typically used.
3. **Analyze observations with the same predictor value**:
   - If multiple observations Y_{ij} are available at the same X value:
     - Compute the mean $\bar{Y}_i$ and standard deviation s_i for each group.
     - Check if $s_i \propto \bar{Y}_i^\lambda$.
       * For example, assume:

$$s_i = a\bar{Y}_i^\lambda.$$

       * Taking the natural logarithm:

$$\ln(s_i) = \ln(a) + \lambda \ln(\bar{Y}_i).$$

     - Perform a regression of $\ln(s_i)$ on $\ln(\bar{Y}_i)$ to estimate λ and suggest the form of $f(\mu)$.
4. **Group observations**:
   - If individual observations are sparse, try grouping similar observations by X values to compute $\bar{Y}_i$ and s_i for each group.

Table 2.8 Recommended transformations based on variance structure and data type

Transformation	Situation	Comments
$\sqrt{Y}$	$var(\epsilon_i) = k\,E(Y_i)$	For counts following a Poisson distribution
$\sqrt{Y} + \sqrt{Y+1}$	$var(\epsilon_i) = k\,E(Y_i)$	Useful for small counts or datasets with zeros
$\log(Y)$	$var(\epsilon_i) = k\,(E(Y_i))^2$	Appropriate for positive integers with a wide range
$\log(Y+1)$	$var(\epsilon_i) = k\,(E(Y_i))^2$	Used when the data includes zero counts
$1/Y$	$var(\epsilon_i) = k\,(E(Y_i))^4$	For responses mostly near zero, with occasional large values
$\arcsin(\sqrt{Y})$	$var(\epsilon_i) = k\,E(Y_i)(1 - E(Y_i))$	Suitable for binomial proportions or percentage data

2.1.1.14.4 Common Transformations and Their Applications

The table below summarizes common transformations used to stabilize variance under various conditions, along with their appropriate contexts and comments (Table 2.8):

1. **Choosing the transformation**:
   - Start by identifying the relationship between the variance of the residuals ($var(\epsilon_i)$) and the mean of the response variable ($E(Y_i)$).
   - Select the transformation that matches the identified variance structure.
2. **Transformations for zero values**:
   - For data with zeros, transformations like $\sqrt{Y+1}$ or $\log(Y+1)$ can be used to avoid undefined values. But this will seriously jeopardize model assumption (Chen & Roth, 2024).
3. **Use in regression models**:
   - Apply these transformations to the dependent variable Y in the regression model.
   - Always check residual plots to confirm that the transformation stabilizes variance and resolves nonlinearity.
4. **Interpretation after transformation**:
   - After transforming Y, interpret the results in terms of the transformed variable.
   - For practical interpretation, back-transform predictions and account for any associated bias.

2.1.2 Multiple Linear Regression

The geometry of least squares regression involves projecting the response vector $\mathbf{y}$ onto the space spanned by the columns of the design matrix $\mathbf{X}$. The fitted values $\hat{\mathbf{y}}$ can be expressed as:

$$\begin{aligned}\hat{\mathbf{y}} &= \mathbf{Xb} \\ &= \mathbf{X}(\mathbf{X}'\mathbf{X})^{-1}\mathbf{X}'\mathbf{y} \\ &= \mathbf{Hy},\end{aligned}$$

where:

- $\mathbf{H} = \mathbf{X}(\mathbf{X}'\mathbf{X})^{-1}\mathbf{X}'$ is the projection operator (sometimes denoted as $\mathbf{P}$)
- $\hat{\mathbf{y}}$ is the projection of $\mathbf{y}$ onto the linear space spanned by the columns of $\mathbf{X}$ (the model space)

The dimension of the model space is equal to the rank of $\mathbf{X}$ (i.e., the number of linearly independent columns in $\mathbf{X}$).
Properties of the Projection Matrix $\mathbf{H}$

1. Symmetry:

   - The projection matrix $\mathbf{H}$ is symmetric:

   $$\mathbf{H} = \mathbf{H}'.$$

2. Idempotence:

   - Applying $\mathbf{H}$ twice gives the same result:

   $$\mathbf{HH} = \mathbf{H}.$$

   Proof:

   $$\begin{aligned}\mathbf{HH} &= \mathbf{X}(\mathbf{X}'\mathbf{X})^{-1}\mathbf{X}'\mathbf{X}(\mathbf{X}'\mathbf{X})^{-1}\mathbf{X}' \\ &= \mathbf{X}(\mathbf{X}'\mathbf{X})^{-1}\mathbf{I}\mathbf{X}' \\ &= \mathbf{X}(\mathbf{X}'\mathbf{X})^{-1}\mathbf{X}' \\ &= \mathbf{H}.\end{aligned}$$

3. Dimensionality:

   - $\mathbf{H}$ is an $n \times n$ matrix (where n is the number of observations).
   - The rank of $\mathbf{H}$ is equal to the rank of $\mathbf{X}$, which is typically the number of predictors (including the intercept).

4. Orthogonal complement:
   - The matrix $(\mathbf{I}-\mathbf{H})$, where:

$$\mathbf{I}-\mathbf{H}=\mathbf{I}-\mathbf{X}(\mathbf{X}'\mathbf{X})^{-1}\mathbf{X}',$$

   is also a projection operator.
   - It projects onto the orthogonal complement of the space spanned by the columns of $\mathbf{X}$ (i.e., the space orthogonal to the model space).
5. Orthogonality of projections:
   - $\mathbf{H}$ and $(\mathbf{I}-\mathbf{H})$ are orthogonal:

$$\mathbf{H}(\mathbf{I}-\mathbf{H})=\mathbf{0}.$$

   - Similarly:

$$(\mathbf{I}-\mathbf{H})\mathbf{H}=\mathbf{0}.$$

Intuition for $\mathbf{H}$ and $(\mathbf{I}-\mathbf{H})$

- $\mathbf{H}$: Projects $\mathbf{y}$ onto the model space, giving the fitted values $\hat{\mathbf{y}}$.
- $\mathbf{I}-\mathbf{H}$: Projects $\mathbf{y}$ onto the residual space, giving the residuals $\mathbf{e}$:

$$\mathbf{e}=(\mathbf{I}-\mathbf{H})\mathbf{y}.$$

- $\mathbf{H}$ and $(\mathbf{I}-\mathbf{H})$ divide the response vector $\mathbf{y}$ into two components:

$$\mathbf{y}=\hat{\mathbf{y}}+\mathbf{e}.$$

  - $\hat{\mathbf{y}}=\mathbf{H}\mathbf{y}$ (fitted values).
  - $\mathbf{e}=(\mathbf{I}-\mathbf{H})\mathbf{y}$ (residuals).
- The properties of $\mathbf{H}$ (symmetry, idempotence, dimensionality) reflect its role as a linear transformation that projects vectors onto the model space.

This geometric perspective provides insight into the mechanics of least squares regression, particularly how the response variable $\mathbf{y}$ is decomposed into fitted values and residuals.
Similar to simple regression, the total sum of squares in multiple regression analysis can be partitioned into components corresponding to the regression (model fit) and the residuals (errors).

The uncorrected total sum of squares is:

$$\mathbf{y}'\mathbf{y}=\hat{\mathbf{y}}'\hat{\mathbf{y}}+\mathbf{e}'\mathbf{e},$$

where:

- $\hat{\mathbf{y}} = \mathbf{H}\mathbf{y}$ (fitted values, projected onto the model space)
- $\mathbf{e} = (\mathbf{I} - \mathbf{H})\mathbf{y}$ (residuals, projected onto the orthogonal complement of the model space)

Expanding this using projection matrices:

$$\begin{aligned}\mathbf{y}'\mathbf{y} &= (\mathbf{H}\mathbf{y})'(\mathbf{H}\mathbf{y}) + ((\mathbf{I} - \mathbf{H})\mathbf{y})'((\mathbf{I} - \mathbf{H})\mathbf{y}) \\ &= \mathbf{y}'\mathbf{H}'\mathbf{H}\mathbf{y} + \mathbf{y}'(\mathbf{I} - \mathbf{H})'(\mathbf{I} - \mathbf{H})\mathbf{y} \\ &= \mathbf{y}'\mathbf{H}\mathbf{y} + \mathbf{y}'(\mathbf{I} - \mathbf{H})\mathbf{y}.\end{aligned}$$

This equation shows the partition of $\mathbf{y}'\mathbf{y}$ into components explained by the model ($\hat{\mathbf{y}}$) and the unexplained variation (residuals).

For the corrected total sum of squares, we adjust for the mean (using the projection matrix $\mathbf{H}_1$):

$$\mathbf{y}'(\mathbf{I} - \mathbf{H}_1)\mathbf{y} = \mathbf{y}'(\mathbf{H} - \mathbf{H}_1)\mathbf{y} + \mathbf{y}'(\mathbf{I} - \mathbf{H})\mathbf{y}.$$

Here:

- $\mathbf{H}_1 = \frac{1}{n}\mathbf{J}$, where $\mathbf{J}$ is an $n \times n$ matrix of ones.
- $\mathbf{H} - \mathbf{H}_1$ projects onto the subspace explained by the predictors after centering.

Why the Correction Matters (Table 2.9)

Table 2.9 Comparison of uncorrected and corrected total sum of squares in linear regression

Aspect	Uncorrected total sum of squares ($\mathbf{y}'\mathbf{y}$)	Corrected total sum of squares ($\mathbf{y}'(\mathbf{I} - \mathbf{H}_1)\mathbf{y}$)
Definition	Total variation in y relative to the origin	Total variation in y relative to its mean (centered data)
Adjustment	No adjustment for the mean of y	Adjusts for the mean of y by centering it
Equation	$\mathbf{y}'\mathbf{y} = \hat{\mathbf{y}}'\hat{\mathbf{y}} + \mathbf{e}'\mathbf{e}$	$\mathbf{y}'(\mathbf{I}-\mathbf{H}_1)\mathbf{y} = \mathbf{y}'(\mathbf{H}-\mathbf{H}_1)\mathbf{y}+\mathbf{y}'(\mathbf{I}-\mathbf{H})\mathbf{y}$
Projection matrices	$\mathbf{H}$: projects onto model space $\mathbf{I} - \mathbf{H}$: Projects onto residuals	$\mathbf{H}_1 = \frac{1}{n}\mathbf{J}$: adjusts for the mean $\mathbf{H} - \mathbf{H}_1$: projects onto predictors after centering $\mathbf{I} - \mathbf{H}$: projects onto residuals
Interpretation	Includes variation due to the mean of y	Focuses on variation in y around its mean
Usage	Suitable for raw, uncentered data	Common in regression and ANOVA to isolate variability explained by predictors
Application	Measures total variability in y, including overall level (mean)	Measures variability explained by predictors relative to the mean

Table 2.10 ANOVA decomposition of the corrected total sum of squares

Source	SS	df	MS	F
Regression	$SSR = \mathbf{y}'(\mathbf{H} - \frac{1}{n}\mathbf{J})\mathbf{y}$	$p-1$	$MSR = SSR/(p-1)$	MSR/MSE
Error	$SSE = \mathbf{y}'(\mathbf{I} - \mathbf{H})\mathbf{y}$	$n-p$	$MSE = SSE/(n-p)$	
Total	$SST = \mathbf{y}'(\mathbf{I} - \mathbf{H}_1)\mathbf{y}$	$n-1$		

- In ANOVA and regression, removing the contribution of the mean helps isolate the variability explained by predictors from the overall level of the response variable.
- Corrected sums of squares are more common when comparing models or computing R^2, which requires centering to ensure consistency in proportionate variance explained.

The corrected total sum of squares can be decomposed into the sum of squares for regression (SSR) and the sum of squares for error (SSE) (Table 2.10):

where:

- p: number of parameters (including intercept)
- n: number of observations

Alternatively, the regression model can be expressed as:

$$\mathbf{Y} = \mathbf{X}\hat{\boldsymbol{\beta}} + (\mathbf{Y} - \mathbf{X}\hat{\boldsymbol{\beta}}),$$

where:

- $\hat{\mathbf{Y}} = X\hat{\boldsymbol{\beta}}$: vector of fitted values (in the subspace spanned by $\mathbf{X}$).
- $\mathbf{e} = \mathbf{Y} - \mathbf{X}\hat{\boldsymbol{\beta}}$: vector of residuals (in the orthogonal complement of the subspace spanned by $\mathbf{X}$).
- $\mathbf{Y}$ is an $n \times 1$ vector in the n-dimensional space $\mathbb{R}^n$.
- $\mathbf{X}$ is an $n \times p$ full-rank matrix, with its columns generating a p-dimensional subspace of $\mathbb{R}^n$. Hence, any estimator $\mathbf{X}\hat{\boldsymbol{\beta}}$ is also in this subspace.

In linear regression, the ordinary least squares estimator $\hat{\beta}$ minimizes the squared Euclidean distance $\|\mathbf{Y} - \mathbf{X}\beta\|^2$ between the observed response vector $\mathbf{Y}$ and the fitted values $\mathbf{X}\beta$. This minimization corresponds to the orthogonal projection of $\mathbf{Y}$ onto the column space of $\mathbf{X}$.

We solve the optimization problem:

$$\min_{\beta} \|\mathbf{Y} - \mathbf{X}\beta\|^2.$$

The objective function can be expanded as:

$$\|\mathbf{Y} - \mathbf{X}\beta\|^2 = (\mathbf{Y} - \mathbf{X}\beta)^\top(\mathbf{Y} - \mathbf{X}\beta).$$

Perform the multiplication:

$$(\mathbf{Y}-\mathbf{X}\beta)^\top(\mathbf{Y}-\mathbf{X}\beta)=\mathbf{Y}^\top\mathbf{Y}-\mathbf{Y}^\top\mathbf{X}\beta-\beta^\top\mathbf{X}^\top\mathbf{Y}+\beta^\top(\mathbf{X}^\top\mathbf{X})\beta.$$

Since $\mathbf{Y}^\top\mathbf{X}\beta$ is a scalar, it equals $\beta^\top\mathbf{X}^\top\mathbf{Y}$. Therefore, the expanded expression becomes:

$$\|\mathbf{Y}-\mathbf{X}\beta\|^2=\mathbf{Y}^\top\mathbf{Y}-2\beta^\top\mathbf{X}^\top\mathbf{Y}+\beta^\top(\mathbf{X}^\top\mathbf{X})\beta.$$

To find the β that minimizes this expression, take the derivative with respect to β and set it to 0:

$$\frac{\partial}{\partial\beta}\left[\mathbf{Y}^\top\mathbf{Y}-2\beta^\top\mathbf{X}^\top\mathbf{Y}+\beta^\top(\mathbf{X}^\top\mathbf{X})\beta\right]=0.$$

Computing the gradient:

$$\frac{\partial}{\partial\beta}=-2\mathbf{X}^\top\mathbf{Y}+2(\mathbf{X}^\top\mathbf{X})\beta.$$

Setting this to zero:

$$-2\mathbf{X}^\top\mathbf{Y}+2\mathbf{X}^\top\mathbf{X}\beta=0.$$

Simplify:

$$\mathbf{X}^\top\mathbf{X}\beta=\mathbf{X}^\top\mathbf{Y}.$$

If $\mathbf{X}^\top\mathbf{X}$ is invertible, the solution is:

$$\hat{\beta}=(\mathbf{X}^\top\mathbf{X})^{-1}\mathbf{X}^\top\mathbf{Y}.$$

Orthogonal Projection Interpretation

The fitted values are:

$$\hat{\mathbf{Y}}=\mathbf{X}\hat{\beta}.$$

From the normal equations, $\mathbf{X}^\top(\mathbf{Y}-\mathbf{X}\hat{\beta})=0$, which implies that the residual vector $\mathbf{Y}-\hat{\mathbf{Y}}$ is orthogonal to every column of $\mathbf{X}$. Therefore:

- $\hat{\mathbf{Y}}=\mathbf{X}\hat{\beta}$ is the orthogonal projection of $\mathbf{Y}$ onto $\mathrm{Col}(\mathbf{X})$.
- $\mathbf{Y}-\mathbf{X}\hat{\beta}$ lies in the orthogonal complement of $\mathrm{Col}(\mathbf{X})$.

Pythagoras Decomposition

The geometric interpretation gives us the decomposition:

$$\mathbf{Y} = \mathbf{X}\hat{\beta} + (\mathbf{Y} - \mathbf{X}\hat{\beta}),$$

where:

- $\mathbf{X}\hat{\beta}$ is the projection of $\mathbf{Y}$ onto the column space of $\mathbf{X}$
- $(\mathbf{Y} - \mathbf{X}\hat{\beta})$ is the residual vector, orthogonal to $\mathbf{X}\hat{\beta}$

Since the two components are orthogonal, their squared norms satisfy:

$$\|\mathbf{Y}\|^2 = \mathbf{Y}^\top \mathbf{Y}$$

(definition of norm squared)

$$= (\mathbf{Y} - \mathbf{X}\hat{\beta} + \mathbf{X}\hat{\beta})^\top (\mathbf{Y} - \mathbf{X}\hat{\beta} + \mathbf{X}\hat{\beta})$$

(add and subtract $\mathbf{X}\hat{\beta}$)

$$= (\mathbf{Y} - \mathbf{X}\hat{\beta})^\top (\mathbf{Y} - \mathbf{X}\hat{\beta}) + 2\,(\mathbf{X}\hat{\beta})^\top (\mathbf{Y} - \mathbf{X}\hat{\beta}) + (\mathbf{X}\hat{\beta})^\top (\mathbf{X}\hat{\beta})$$

(expand $(a+b)^\top (a+b)$)

$$= \|\mathbf{Y} - \mathbf{X}\hat{\beta}\|^2 + 2\,(\mathbf{X}\hat{\beta})^\top (\mathbf{Y} - \mathbf{X}\hat{\beta}) + \|\mathbf{X}\hat{\beta}\|^2$$

(rewrite each quadratic form as a norm)

$$= \|\mathbf{Y} - \mathbf{X}\hat{\beta}\|^2 + \|\mathbf{X}\hat{\beta}\|^2$$

(orthogonality: $(\mathbf{X}\hat{\beta})^\top (\mathbf{Y} - \mathbf{X}\hat{\beta}) = 0$)

$$= \|\mathbf{X}\hat{\beta}\|^2 + \|\mathbf{Y} - \mathbf{X}\hat{\beta}\|^2$$

where the norm of a vector $\mathbf{a}$ in $\mathbb{R}^p$ is defined as:

$$\|\mathbf{a}\| = \sqrt{\mathbf{a}^\top \mathbf{a}} = \sqrt{\sum_{i=1}^{p} a_i^2}.$$

We are saying that $\mathbf{Y}$ is decomposed into two orthogonal components:

1. $\mathbf{X}\hat{\beta}$ (the projection onto Col($\mathbf{X}$)
   - $\|\mathbf{X}\hat{\beta}\|$ measures the part of $\mathbf{Y}$ explained by the model.
2. $\mathbf{Y} - \mathbf{X}\hat{\beta}$ (the residual lying in the orthogonal complement).
   - $\|\mathbf{Y} - \mathbf{X}\hat{\beta}\|$ measures the residual error.

This geometric interpretation (projection plus orthogonal remainder) is exactly why we call $\mathbf{X}\hat{\beta}$ the *orthogonal projection* of $\mathbf{Y}$ onto the column space of $\mathbf{X}$. This decomposition also underlies the analysis of variance (ANOVA) in regression.
The coefficient of multiple determination, denoted R^2, measures the proportion of the total variation in the response variable ($\mathbf{Y}$) that is explained by the regression model. It is defined as:

$$R^2 = \frac{SSR}{SSTO} = 1 - \frac{SSE}{SSTO},$$

where:

- SSR: regression sum of squares (variation explained by the model)
- SSE: error sum of squares (unexplained variation)
- $SSTO$: total sum of squares (total variation in $\mathbf{Y}$)

The adjusted R^2 adjusts R^2 for the number of predictors in the model, penalizing for adding predictors that do not improve the model's fit substantially. It is defined as:

$$R_a^2 = 1 - \frac{SSE/(n-p)}{SSTO/(n-1)} = 1 - \frac{(n-1)SSE}{(n-p)SSTO},$$

where:

- n: number of observations
- p: number of parameters (including the intercept)

In multiple regression, R_a^2 provides a more reliable measure of model fit, especially when comparing models with different numbers of predictors (Table 2.11).
In a regression model with coefficients $\beta = (\beta_0, \beta_1, \dots, \beta_{p-1})^\top$, the sums of squares are used to evaluate the contribution of predictors to explaining the variation in the response variable.

Table 2.11 Key differences between R^2 and R_a^2

Aspect	R^2	R_a^2
Behavior with predictors	Always increases (or remains constant) when more predictors are added, even if they are not statistically significant	Includes a penalty for the number of predictors. May decrease if added predictors do not improve the model sufficiently
Interpretation	Proportion of the total variation in $\mathbf{Y}$ explained by the regression model	Adjusted measure of explained variance, accounting for model complexity
Range	Ranges between 0 and 1	Can be lower than R^2, particularly when the model includes irrelevant predictors
Usefulness	Useful for understanding the overall fit of the model	Useful for comparing models with different numbers of predictors

1. Model sums of squares:

   - SSM: total model sum of squares, capturing the variation explained by all predictors:

$$SSM = SS(\beta_0, \beta_1, \ldots, \beta_{p-1}).$$

2. Marginal contribution:

   - SSM_m: conditional model sum of squares, capturing the variation explained by predictors after accounting for others:

$$SSM_m = SS(\beta_0, \beta_1, \ldots, \beta_{p-1} | \beta_0).$$

Decompositions of SSM_m (Table 2.12)

1. Sequential sums of squares (Type I SS)

- Definition:
  - Sequential SS depends on the order in which predictors are added to the model.
  - It represents the additional contribution of each predictor given only the predictors that precede it in the sequence.
- Formula:

$$SSM_m = SS(\beta_1|\beta_0) + SS(\beta_2|\beta_0, \beta_1) + \cdots + SS(\beta_{p-1}|\beta_0, \ldots, \beta_{p-2}).$$

Table 2.12 Comparison of sequential, marginal, and partial SS

Aspect	Sequential SS (Type I)	Marginal SS (Type II)	Partial SS (Type III)
Dependency	Depends on the order in which predictors are entered	Independent of order; adjusts for noncollinear predictors	Independent of order; evaluates unique contributions
Usage	Default in software functions like `anova()` (Type I SS)	Models without interactions or hierarchical dependencies	Commonly used for hypothesis testing
Interpretation	Measures the additional contribution of predictors in sequence	Measures the contribution of a predictor, ignoring collinear terms	Measures the unique contribution of each predictor
Uniqueness	Not unique; changes with predictor order	Unique for a given model without interactions	Unique for a given model

- Key points:
  - Sequential SS is not unique; it depends on the order of the predictors.
  - Default in many statistical software functions (e.g., `anova()` in R).

2. Marginal sums of squares (Type II SS)

- Definition:
  - Marginal SS evaluates the contribution of a predictor after accounting for all other predictors except those with which it is collinear.
  - It ignores hierarchical relationships or interactions, focusing on independent contributions.
- Formula: $SSM_m = SS(\beta_j|\beta_1, \ldots, \beta_{j-1}, \beta_{j+1}, \ldots, \beta_{p-1})$ where Type II SS evaluates the contribution of β_j while excluding any terms collinear with β_j.
- Key points:
  - Type II SS is independent of predictor order.
  - Suitable for models without interaction terms or when predictors are balanced.

3. Partial sums of squares (Type III SS)

- Definition:
  - Partial SS evaluates the contribution of each predictor after accounting for all other predictors in the model.
  - It quantifies the unique contribution of a predictor, controlling for the presence of others.
- Formula:

$$SSM_m = SS(\beta_1|\beta_0, \beta_2, \ldots, \beta_{p-1}) + \cdots + SS(\beta_{p-1}|\beta_0, \beta_1, \ldots, \beta_{p-2}).$$

- Key points:
  - Partial SS is unique for a given model.
  - More commonly used in practice for assessing individual predictor importance.

Practical notes

- Use Type III SS (Partial SS) when:
  - The focus is on individual predictor contributions while accounting for all others
  - Conducting hypothesis tests on predictors in complex models with interactions or hierarchical structures

- Use Type II SS (Marginal SS) when:
  - Working with balanced datasets or models without interaction terms
  - Ignoring interactions and focusing on independent effects
- Use Type I SS (Sequential SS) when:
  - Interested in understanding the incremental contribution of predictors based on a specific order of entry (e.g., stepwise regression)

2.1.2.1 OLS Assumptions

- A1 Linearity
- A2 Full Rank
- A3 Exogeneity of Independent Variables
- A4 Homoskedasticity
- A5 Data Generation (Random Sampling)
- A6 Normal Distribution

2.1.2.1.1 A1 Linearity

The linear regression model is expressed as:

$$A1 : y = \mathbf{x}\beta + \epsilon.$$

This assumption is not restrictive since x can include nonlinear transformations (e.g., interactions, natural logarithms, quadratic terms).

However, when combined with A3 Exogeneity of Independent Variables, linearity can become restrictive.

2.1.2.1.2 Log Model Variants

Logarithmic transformations of variables allow for flexible modeling of nonlinear relationships. Common log model forms include (Table 2.13):

2.1.2.1.3 Higher-Order Models

Higher-order terms allow the effect of x_1 on y to depend on the level of x_1. For example:

$$y = \beta_0 + x_1\beta_1 + x_1^2\beta_2 + \epsilon.$$

The partial derivative of y with respect to x_1 is:

Table 2.13 Interpretation of coefficients in different functional forms of regression models

Model	Form	Interpretation of β	In words
Level-level	$y = \beta_0 + \beta_1 x + \epsilon$	$\Delta y = \beta_1 \Delta x$	A unit change in x results in a β_1 unit change in y
Log-level	$\ln(y) = \beta_0 + \beta_1 x + \epsilon$	$\%\Delta y = 100\beta_1 \Delta x$	A unit change in x results in a $100\beta_1\%$ change in y
Level-log	$y = \beta_0 + \beta_1 \ln(x) + \epsilon$	$\Delta y = (\beta_1/100)\%\Delta x$	A 1% change in x results in a $(\beta_1/100)$ unit change in y
Log-log	$\ln(y) = \beta_0 + \beta_1 \ln(x) + \epsilon$	$\%\Delta y = \beta_1 \%\Delta x$	A 1% change in x results in a $\beta_1\%$ change in y

$$\frac{\partial y}{\partial x_1} = \beta_1 + 2x_1\beta_2.$$

- The effect of x_1 on y depends on the value of x_1.
- Partial effect at the average: $\beta_1 + 2E(x_1)\beta_2$.
- Average partial effect: $E(\beta_1 + 2x_1\beta_2)$.

2.1.2.1.4 Interaction Terms

Interactions capture the joint effect of two variables. For example:

$$y = \beta_0 + x_1\beta_1 + x_2\beta_2 + x_1x_2\beta_3 + \epsilon.$$

- β_1 is the average effect of a unit change in x_1 on y when $x_2 = 0$.
- The partial effect of x_1 on y, which depends on the level of x_2, is:

$$\beta_1 + x_2\beta_3.$$

2.1.2.1.5 A2 Full Rank

The full rank assumption ensures the uniqueness and existence of the parameter estimates in the population regression equation. It is expressed as:

$$A2 : \text{rank}(E(\mathbf{x}'\mathbf{x})) = k.$$

This assumption is also known as the identification condition.

Key Points

1. No perfect multicollinearity:
   - The columns of $\mathbf{x}$ (the matrix of predictors) must be linearly independent.

- No column in $\mathbf{x}$ can be written as a linear combination of other columns.

2. Implications:

- Ensures that each parameter in the regression equation is identifiable and unique
- Prevents computational issues, such as the inability to invert $\mathbf{x}'\mathbf{x}$, which is required for estimating $\hat{\beta}$

Example of Violation

If two predictors, x_1 and x_2, are perfectly correlated (e.g., $x_2 = 2x_1$), the rank of $\mathbf{x}$ is reduced, and $\mathbf{x}'\mathbf{x}$ becomes singular. In such cases:

- The regression coefficients cannot be uniquely estimated.
- The model fails to satisfy the full rank assumption.

2.1.2.1.6 A3 Exogeneity of Independent Variables

The exogeneity assumption ensures that the independent variables ($\mathbf{x}$) are not systematically related to the error term (ϵ). It is expressed as:

$$A3 : E[\epsilon|x_1, x_2, \ldots, x_k] = E[\epsilon|\mathbf{x}] = 0.$$

This assumption is often referred to as strict exogeneity or mean independence (see [Correlation and Independence]).

Key Points

1. Strict exogeneity:

- Independent variables carry no information about the error term ϵ.
- By the [law of iterated expectations], $E(\epsilon) = 0$, which can be satisfied by always including an intercept in the regression model.

2. Implication:

- A3 implies:

$$E(y|\mathbf{x}) = \mathbf{x}\beta,$$

meaning the conditional mean function is a linear function of $\mathbf{x}$. This aligns with A1 Linearity.

3. Relationship with independence:

- Also referred to as mean independence, which is a weaker condition than full independence (see [Correlation and Independence])

Table 2.14 Comparison between A3 and A3a

Aspect	A3 (strict exogeneity)	A3a (weaker exogeneity)
Definition	$E(\epsilon\|\mathbf{x}) = 0$	$E(\mathbf{x}_i'\epsilon_i) = 0$
Strength	Stronger assumption; implies A3a	Weaker assumption; does not imply A3
Interpretation	Predictors provide no information about ϵ	Predictors are uncorrelated with ϵ
Causality	Enables causal interpretation	Does not allow causal interpretations

2.1.2.1.7 A3a: Weaker Exogeneity Assumption

A weaker version of the exogeneity assumption is:

$$A3a : E(\mathbf{x_i'}\epsilon_i) = 0$$

This implies:

- The independent variables ($\mathbf{x}_i$) are uncorrelated with the error term (ϵ_i).
- Weaker than mean independence in A3 (Table 2.14).

Notes on Practical Relevance

- Checking for exogeneity:
  - Strict exogeneity cannot be tested directly, but violations can manifest as omitted variable bias, endogeneity, or measurement error.
  - Including all relevant predictors and ensuring accurate measurement can help satisfy this assumption.
- Violations of exogeneity:
  - If A3 is violated, standard OLS estimates are biased and inconsistent.
  - In such cases, instrumental variable (IV) methods or other approaches may be required to correct for endogeneity.

2.1.2.1.8 A4 Homoskedasticity

The homoskedasticity assumption ensures that the variance of the error term (ϵ) is constant across all levels of the independent variables ($\mathbf{x}$). It is expressed as:

$$A4 : \text{Var}(\epsilon|\mathbf{x}) = \text{Var}(\epsilon) = \sigma^2.$$

Key Points

1. Definition:
   - The variance of the disturbance term ϵ is the same for all observations, regardless of the values of the predictors $\mathbf{x}$.

2. Practical implication:
   - Homoskedasticity ensures that the errors do not systematically vary with the predictors.
   - This is critical for valid inference, as the standard errors of the coefficients rely on this assumption.
3. Violation (heteroskedasticity):
   - If the variance of ϵ depends on **x**, the assumption is violated.
   - Common signs include funnel-shaped patterns in residual plots or varying error sizes.

2.1.2.1.9 A5 Data Generation (Random Sampling)

The random sampling assumption ensures that the observations $(y_i, x_{i1}, \dots, x_{ik-1})$ are drawn independently and identically distributed (iid) from the joint distribution of $(y, \mathbf{x})$. It is expressed as:

$$A5 : \{y_i, x_{i1}, \dots, x_{ik-1} : i = 1, \dots, n\}.$$

Key Points

1. Random sampling:
   - The dataset is assumed to be a random sample from the population.
   - Each observation is independent of others and follows the same probability distribution.
2. Implications:
   - With A3 (exogeneity of independent variables) and A4 (homoskedasticity), random sampling implies:
     - Strict exogeneity:

       $$E(\epsilon_i | x_1, \dots, x_n) = 0.$$

       Independent variables do not contain information for predicting ϵ.
     - Non-autocorrelation:

       $$E(\epsilon_i \epsilon_j | x_1, \dots, x_n) = 0 \quad \text{for } i \neq j.$$

       The error terms are uncorrelated across observations, conditional on the independent variables.
     - Variance of errors:

       $$\text{Var}(\epsilon | \mathbf{X}) = \text{Var}(\epsilon) = \sigma^2 \mathbf{I}_n.$$

3. When A5 may not hold:
   - In time series data, where observations are often autocorrelated
   - In spatial data, where neighboring observations may not be independent

Practical Considerations

- Time series data:
  - Use methods such as autoregressive models or generalized least squares (GLS) to address dependency in observations.
- Spatial data:
  - Spatial econometric models may be required to handle correlation across geographic locations.
- Checking random sampling:
  - While true randomness cannot always be verified, exploratory analysis of the residuals (e.g., for patterns or autocorrelation) can help detect violations.

2.1.2.1.10 A5a: Stationarity in Stochastic Processes

A stochastic process $\{x_t\}_{t=1}^T$ is stationary if, for every collection of time indices $\{t_1, t_2, \ldots, t_m\}$, the joint distribution of:

$$x_{t_1}, x_{t_2}, \ldots, x_{t_m}$$

is the same as the joint distribution of:

$$x_{t_1+h}, x_{t_2+h}, \ldots, x_{t_m+h}$$

for any $h \geq 1$.
Key Points on Stationarity

1. Definition:
   - A stationary process has statistical properties (mean, variance, and covariance) that are invariant over time.
   - For example, the joint distribution for the first ten observations is identical to the joint distribution for the next ten observations, regardless of their position in time.
2. Implication:
   - Stationarity ensures that the relationships observed in the data remain consistent over time.
   - Independent draws automatically satisfy the stationarity condition.

Table 2.15 Differences between stationarity and weak stationarity

Aspect	Stationarity	Weak stationarity
Joint distribution	Entire joint distribution remains unchanged over time	Focuses only on the first two moments: mean and covariance
Dependence over time	Observations at all lags are equally distributed	Observations far apart are "almost independent"
Application	Ensures strong consistency in time-series processes	More practical for many time-series applications

Weak Stationarity
A stochastic process $\{x_t\}_{t=1}^{T}$ is weakly stationary if:

- The covariance between x_t and x_{t+h} depends only on the lag h and not on t.
- As $h \to \infty$, the covariance diminishes, meaning x_t and x_{t+h} become "almost independent."

Weak stationarity is often sufficient for many time-series analyses, especially when focusing on correlations and trends rather than full distributions (Table 2.15).

Common Weakly Dependent Processes

1. Moving average process of order 1 (MA(1))
   An MA(1) process models the dependent variable y_t as a function of the current and one-period lagged stochastic error term:

$$y_t = u_t + \alpha_1 u_{t-1},$$

   where u_t is white noise, independently and identically distributed (iid) with variance σ^2.

Key Properties

1. Mean:

$$E(y_t) = E(u_t) + \alpha_1 E(u_{t-1}) = 0$$

2. Variance:

$$\begin{aligned}\mathrm{Var}(y_t) &= \mathrm{Var}(u_t) + \alpha_1^2 \mathrm{Var}(u_{t-1}) \\ &= \sigma^2 + \alpha_1^2 \sigma^2 \\ &= \sigma^2(1 + \alpha_1^2)\end{aligned}$$

   An increase in the absolute value of α_1 increases the variance.

3. Autocovariance:

   - For lag 1:

$$\text{Cov}(y_t, y_{t-1}) = \alpha_1 \text{Var}(u_{t-1}) = \alpha_1 \sigma^2.$$

   - For lag 2 or greater:

$$\text{Cov}(y_t, y_{t-2}) = 0.$$

The MA(1) process is invertible if $|\alpha_1| < 1$, allowing it to be rewritten as an autoregressive (AR) representation:

$$u_t = y_t - \alpha_1 u_{t-1}.$$

Invertibility implies that we can express the current observation in terms of past observations.
An MA(q) process generalizes the MA(1) process to include q lags:

$$y_t = u_t + \alpha_1 u_{t-1} + \cdots + \alpha_q u_{t-q},$$

where $u_t \sim WN(0, \sigma^2)$ (white noise with mean 0 and variance σ^2).
Key characteristics:

1. Covariance stationary:
   - An MA(q) process is covariance stationary irrespective of the parameter values.
2. Invertibility:
   - An MA(q) process is invertible if the parameters satisfy certain conditions (e.g., $|\alpha_i| < 1$).
3. Autocorrelations:
   - The autocorrelations are nonzero for lags up to q but are 0 for lags beyond q.
4. Conditional mean:
   - The conditional mean of y_t depends on the q lags, indicating "long-term memory."

Example

1. **Autocovariance of an MA(1)**

1. Lag 1:

$$\begin{aligned}\mathrm{Cov}(y_t, y_{t-1}) &= \mathrm{Cov}(u_t + \alpha_1 u_{t-1}, u_{t-1} + \alpha_1 u_{t-2})\\ &= \alpha_1 \mathrm{Var}(u_{t-1})\\ &= \alpha_1 \sigma^2.\end{aligned}$$

2. Lag 2:

$$\begin{aligned}\mathrm{Cov}(y_t, y_{t-2}) &= \mathrm{Cov}(u_t + \alpha_1 u_{t-1}, u_{t-2} + \alpha_1 u_{t-3})\\ &= 0.\end{aligned}$$

An MA process captures a linear relationship between the dependent variable y_t and the current and past values of a stochastic error term u_t. Its properties make it useful for modeling time-series data with limited memory and short-term dependencies.

2. **Autoregressive process of order 1 (AR(1))**

An autoregressive process of order 1 (AR(1)) is defined as:

$$y_t = \rho y_{t-1} + u_t, \quad |\rho| < 1$$

where u_t represents independent and identically distributed (i.i.d.) random noise over t with variance σ^2.
Covariance at lag 1:

$$\begin{aligned}Cov(y_t, y_{t-1}) &= Cov(\rho y_{t-1} + u_t, y_{t-1})\\ &= \rho Var(y_{t-1})\\ &= \rho \frac{\sigma^2}{1-\rho^2}.\end{aligned}$$

Covariance at lag h:

$$Cov(y_t, y_{t-h}) = \rho^h \frac{\sigma^2}{1-\rho^2}.$$

Stationarity implies that the distribution of y_t does not change over time, requiring constant mean and variance. For this process:

1. Mean stationarity:
   Assuming $E(y_t) = 0$, we have:

$$y_t = \rho^t y_0 + \rho^{t-1} u_1 + \rho^{t-2} u_2 + \cdots + \rho u_{t-1} + u_t.$$

   If the initial observation $y_0 = 0$, then y_t is simply a weighted sum of the random shocks u_t from all prior time periods. Thus, $E(y_t) = 0$ for all t.

2. Variance stationarity:
   The variance is computed as:

$$Var(y_t) = Var(\rho y_{t-1} + u_t).$$

   Expanding and simplifying:

$$Var(y_t) = \rho^2 Var(y_{t-1}) + Var(u_t) + 2\rho Cov(y_{t-1}, u_t).$$

   Since u_t is independent of y_{t-1}, $Cov(y_{t-1}, u_t) = 0$, giving:

$$Var(y_t) = \rho^2 Var(y_{t-1}) + \sigma^2.$$

   Solving recursively, we find:

$$Var(y_t) = \frac{\sigma^2}{1 - \rho^2}.$$

For the variance to remain constant over time, it is required that $|\rho| < 1$ and $p \notin \{1, -1\}$.
Key insights on stationarity

1. Stationarity requirement:
   The condition $|\rho| < 1$ ensures stationarity, as this guarantees that both the mean and variance of the process are constant over time.
2. Weak dependence:
   As $|\rho| < 1$, the dependency between observations diminishes with increasing lag h, as seen from the covariance:

$$Cov(y_t, y_{t-h}) = \rho^h \frac{\sigma^2}{1 - \rho^2}.$$

To estimate an AR(1) process, we utilize the Yule-Walker equations, which relate the autocovariances of the process to its parameters.

Starting with the AR(1) process:

$$y_t = \epsilon_t + \phi y_{t-1},$$

multiplying both sides by $y_{t-\tau}$ and taking expectations, we get:

$$y_t y_{t-\tau} = \epsilon_t y_{t-\tau} + \phi y_{t-1} y_{t-\tau}.$$

For $\tau \geq 1$, the autocovariance $\gamma(\tau)$ satisfies:

$$\gamma(\tau) = \phi \gamma(\tau - 1).$$

Dividing through by the variance $\gamma(0)$, we obtain the autocorrelation:

$$\rho_\tau = \phi^\tau.$$

Thus, the autocorrelations decay geometrically as ϕ^τ, where $|\phi| < 1$ ensures stationarity and decay over time.

Generalizing to AR(p)

An AR(p) process extends the AR(1) structure to include p lags:

$$y_t = \phi_1 y_{t-1} + \phi_2 y_{t-2} + \cdots + \phi_p y_{t-p} + \epsilon_t.$$

Here, ϵ_t is white noise with $E(\epsilon_t) = 0$ and $Var(\epsilon_t) = \sigma^2$.

The AR(p) process is covariance stationary if the roots of the characteristic equation lie outside the unit circle:

$$1 - \phi_1 z - \phi_2 z^2 - \cdots - \phi_p z^p = 0.$$

For the AR(p) process, the autocorrelations ρ_τ decay more complexly compared to the AR(1). However, they still diminish over time, ensuring weak dependence among distant observations.

The Yule–Walker equations for an AR(p) process provide a system of linear equations to estimate the parameters $\phi_1, \phi_2, \ldots, \phi_p$:

$$\gamma(\tau) = \phi_1\gamma(\tau - 1) + \phi_2\gamma(\tau - 2) + \cdots + \phi_p\gamma(\tau - p), \quad \tau \geq 1.$$

This system can be written in matrix form for $\tau = 1, \ldots, p$ as:

$$\begin{bmatrix} \gamma(1) \\ \gamma(2) \\ \vdots \\ \gamma(p) \end{bmatrix} = \begin{bmatrix} \gamma(0) & \gamma(1) & \ldots & \gamma(p-1) \\ \gamma(1) & \gamma(0) & \ldots & \gamma(p-2) \\ \vdots & \vdots & \ddots & \vdots \\ \gamma(p-1) & \gamma(p-2) & \ldots & \gamma(0) \end{bmatrix} \begin{bmatrix} \phi_1 \\ \phi_2 \\ \vdots \\ \phi_p \end{bmatrix}.$$

This system is solved to estimate the coefficients $\phi_1, \phi_2, \ldots, \phi_p$.

3. **ARMA(p, q) Process**

An ARMA(p, q) process combines autoregressive (AR) and moving average (MA) components to model time series data effectively. The general form is:

$$y_t = \phi_1 y_{t-1} + \phi_2 y_{t-2} + \cdots + \phi_p y_{t-p} + \epsilon_t + \alpha_1\epsilon_{t-1} + \alpha_2\epsilon_{t-2} + \cdots + \alpha_q\epsilon_{t-q}.$$

A simple case of ARMA(1, 1) is given by:

$$y_t = \phi y_{t-1} + \epsilon_t + \alpha\epsilon_{t-1},$$

where:

- ϕ captures the autoregressive behavior
- α controls the moving average component
- ϵ_t represents white noise

ARMA processes can capture seasonality and more complex dependencies than pure AR or MA models.

4. **Random Walk Process**

A **random walk** is a nonstationary process defined as:

$$y_t = y_0 + \sum_{s=1}^{t} u_s,$$

where u_s are i.i.d. random variables. Properties:

1. **Nonstationarity**:
   - If $y_0 = 0$, then $E(y_t) = 0$, but the variance increases over time:

$$Var(y_t) = t\sigma^2.$$

2. **Not weakly dependent**:
   - The covariance of the process does not diminish with increasing lag h:

$$Cov\left(\sum_{s=1}^{t} u_s, \sum_{s=1}^{t-h} u_s\right) = (t-h)\sigma^2.$$

   As h increases, the covariance remains large, violating the condition for weak dependence.

2.1.2.1.11 A5a: Stationarity and Weak Dependence in Time Series

For time series data, the set $\{y_t, x_{t1}, \ldots, x_{tk-1}\}$, where $t = 1, \ldots, T$, must satisfy the conditions of **stationarity** and **weak dependence**. These properties are essential to ensure the consistency and efficiency of estimators in time-series regression models.

1. **Stationarity**: A stationary process has statistical properties (e.g., mean, variance, autocovariance) that remain constant over time. This ensures that the relationships in the data do not change as t progresses, making it possible to draw meaningful inferences.
2. **Weak dependence**: Weak dependence implies that observations far apart in time are "almost independent." While there may be short-term correlations, these

diminish as the time lag increases. This property ensures that the sample averages are representative of the population mean.

The [weak law of large numbers] provides a foundation for the consistency of sample means. If $\{z_t\}$ is a weakly dependent, stationary process with $E(|z_t|) < \infty$ and $E(z_t) = \mu$, then:

$$\frac{1}{T}\sum_{t=1}^{T} z_t \xrightarrow{p} \mu.$$

Interpretation:

- As the sample size $T \to \infty$, the sample mean $\bar{z} = \frac{1}{T}\sum_{t=1}^{T} z_t$ converges in probability to the true mean μ.
- This ensures the **consistency** of estimators based on time-series data.

The [central limit theorem] extends the WLLN by describing the distribution of the sample mean. Under additional regularity conditions (e.g., finite variance) (Greene, 1990), the sample mean $\bar{z}$ satisfies:

$$\sqrt{T}(\bar{z} - \mu) \xrightarrow{d} N(0, B),$$

where:

$$B = \text{Var}(z_t) + 2\sum_{h=1}^{\infty} \text{Cov}(z_t, z_{t-h}).$$

Interpretation:

- The sample mean $\bar{z}$ is approximately normally distributed for large T.
- The variance of the limiting distribution, B, depends not only on the variance of z_t but also on the covariances between z_t and its past values.

2.1.2.1.12 A6 Normal Distribution

A6: $\epsilon|\mathbf{x} \sim N(0, \sigma^2 I_n)$

The assumption here implies that the error term ϵ is normally distributed with mean zero and variance $\sigma^2 I_n$. This assumption is fundamental for statistical inference in linear regression models.

Using assumptions A1 Linearity, A2 Full Rank, and A3 Exogeneity of Independent Variables, we derive the identification (or orthogonality condition) for the population parameter β:

$$\begin{aligned}
y &= x\beta + \epsilon && \text{(A1: Model Specification)}\\
x'y &= x'x\beta + x'\epsilon && \text{(Multiply both sides by } x'\text{)}\\
E(x'y) &= E(x'x)\beta + E(x'\epsilon) && \text{(Taking expectation)}\\
E(x'y) &= E(x'x)\beta && \text{(A3: Exogeneity, } E(x'\epsilon) = 0\text{)}\\
[E(x'x)]^{-1}E(x'y) &= [E(x'x)]^{-1}E(x'x)\beta && \text{(Invertibility of } E(x'x)\text{, A2)}\\
[E(x'x)]^{-1}E(x'y) &= \beta && \text{(Simplified solution for } \beta\text{)}
\end{aligned}$$

Thus, β is identified as the vector of parameters that minimizes the expected squared error.

To find β, we minimize the expected value of the squared error:

$$\underset{\gamma}{\operatorname{argmin}}\, E\big((y - x\gamma)^2\big).$$

The first-order condition is derived by taking the derivative of the objective function with respect to γ and setting it to zero:

$$\begin{aligned}
\frac{\partial E\big((y - x\gamma)^2\big)}{\partial \gamma} &= 0\\
-2E(x'(y - x\gamma)) &= 0\\
E(x'y) - E(x'x\gamma) &= 0\\
E(x'y) &= E(x'x)\gamma\\
(E(x'x))^{-1}E(x'y) &= \gamma.
\end{aligned}$$

This confirms that $\gamma = \beta$.

The second-order condition ensures that the solution minimizes the objective function. Taking the second derivative:

$$\frac{\partial^2 E\big((y - x\gamma)^2\big)}{\partial \gamma'^2} = 0 = 2E(x'x).$$

If assumption A3 Exogeneity of Independent Variables holds, $E(x'x)$ is positive semi-definite (PSD). Thus, $2E(x'x)$ is also PSD, ensuring a minimum.

2.1.2.1.13 Hierarchy of OLS Assumptions

Table 2.16 summarizes the hierarchical nature of assumptions required to derive different properties of the OLS estimator.

Table 2.16 Mapping linear model assumptions to estimator properties and inference validity

Assumption	Identification data description	Unbiasedness inference consistency	Gauss–Markov (BLUE) asymptotic inference (z and chi-squared)	Classical LM (BUE) small-sample inference (t and F)
A2 Full Rank variation in **X**	✓	✓	✓	✓
A5 Data Generation (Random Sampling)random sampling		✓	✓	✓
A1 Linearity linearity in parameters		✓	✓	✓
A3 Exogeneity of Independent Variables zero conditional mean		✓	✓	✓
A4 Homoskedasticity **H** homoskedasticity			✓	✓
A6 Normal Distribution normality of errors				✓

Usage of assumptions

- **Identification data description:** ensures the model is identifiable and coefficients can be estimated
- **Unbiasedness consistency:** guarantees that OLS estimates are unbiased and converge to the true parameter values as the sample size increases
- **Gauss-Markov (BLUE) and asymptotic inference:** requires additional assumptions (e.g., homoskedasticity) to ensure minimum variance of estimators and valid inference using large-sample tests (z and chi-squared)
- **Classical LM (BUE) small-sample inference:** builds on all previous assumptions and adds normality of errors for valid t- and F-tests in finite samples

2.1.2.2 Theorems

2.1.2.2.1 Frisch–Waugh–Lovell Theorem

The Frisch–Waugh–Lovell (FWL) theorem is a fundamental result in linear regression that allows for a deeper understanding of how coefficients are computed in a multiple regression setting (Lovell, 2008). Informally, it states:

> When estimating the effect of a subset of variables (X_1) on y in the presence of other variables (X_2), you can "partial out" the influence of X_2 from both y and X_1. Then, regressing the residuals of y on the residuals of X_1 produces coefficients for X_1 that are identical to those obtained from the full multiple regression.

Consider the multiple linear regression model:

$$\mathbf{y} = \mathbf{X}\boldsymbol{\beta} + \boldsymbol{\epsilon} = \mathbf{X}_1\boldsymbol{\beta}_1 + \mathbf{X}_2\boldsymbol{\beta}_2 + \boldsymbol{\epsilon}$$

where:

- y is an $n \times 1$ vector of the dependent variable
- X_1 is an $n \times k_1$ matrix of regressors of interest
- X_2 is an $n \times k_2$ matrix of additional regressors
- β_1 and β_2 are coefficient vectors of sizes $k_1 \times 1$ and $k_2 \times 1$, respectively
- ϵ is an $n \times 1$ error term vector

This can be equivalently represented in partitioned matrix form as:

$$\begin{pmatrix} X_1'X_1 & X_1'X_2 \\ X_2'X_1 & X_2'X_2 \end{pmatrix} \begin{pmatrix} \hat{\beta}_1 \\ \hat{\beta}_2 \end{pmatrix} = \begin{pmatrix} X_1'y \\ X_2'y. \end{pmatrix}$$

The OLS estimator for the vector $\begin{pmatrix} \beta_1 \\ \beta_2 \end{pmatrix}$ is:

$$\begin{pmatrix} \hat{\beta}_1 \\ \hat{\beta}_2 \end{pmatrix} = \begin{pmatrix} X_1'X_1 & X_1'X_2 \\ X_2'X_1 & X_2'X_2 \end{pmatrix}^{-1} \begin{pmatrix} X_1'y \\ X_2'y \end{pmatrix}.$$

If we only want the coefficients on X_1, a known result from partitioned-inversion gives:

$$\hat{\beta}_1 = \left(X_1' M_2 X_1\right)^{-1} X_1' M_2 y,$$

where

$$M_2 = I - X_2 \left(X_2' X_2\right)^{-1} X_2'.$$

The matrix M_2 is often called the **residual-maker** or **annihilator** matrix for X_2. It is an $n \times n$ symmetric, idempotent projection matrix that projects any vector in $\mathbb{R}^n$ onto the orthogonal complement of the column space of X_2. M_2 satisfies $M_2^2 = M_2$, and $M_2 = M_2'$.

Intuitively, M_2 captures the part of y (and any other vector) that is orthogonal to the columns of X_2. This "partialling out" of X_2 from both y and X_1 lets us isolate $\hat{\beta}_1$.

Equivalently, we can also represent $\hat{\beta}_1$ as:

$$\hat{\boldsymbol{\beta}}_1 = (\mathbf{X}_1'\mathbf{X}_1)^{-1}\mathbf{X}_1'\mathbf{y} - (\mathbf{X}_1'\mathbf{X}_1)^{-1}\mathbf{X}_1'\mathbf{X}_2\hat{\boldsymbol{\beta}}_2.$$

From this equation, we can see that:

1. **Betas from multiple versus simple regressions:**
   - The coefficients (β) from a multiple regression are generally **not the same** as the coefficients from separate individual simple regressions.

2. **Impact of additional variables** (X_2):
   - The inclusion of different sets of explanatory variables (X_2) affects all coefficient estimates, even for those in X_1.
3. **Special cases:**
   - If $X_1'X_2 = 0$ (orthogonality between X_1 and X_2) or $\hat{\beta}_2 = 0$, the above points (1 and 2) do not hold. In such cases, there is no interaction between the coefficients in X_1 and X_2, making the coefficients in X_1 unaffected by X_2.

Steps in FWL:

1. **Partial out** X_2 from y: Regress y on X_2 to obtain residuals:

$$\tilde{y} = M_2 y.$$

2. **Partial out** X_2 from X_1: For each column of X_1, regress it on X_2 to obtain residuals:

$$\tilde{X}_1 = M_2 X_1.$$

3. **Regression of residuals:** Regress $\tilde{y}$ on $\tilde{X}_1$:

$$\tilde{y} = \tilde{X}_1 \beta_1 + \text{error}.$$

The coefficients β_1 obtained here are identical to those from the full model regression:

$$y = X_1\beta_1 + X_2\beta_2 + \epsilon.$$

Why It Matters

1. **Interpretation of partial effects:** The FWL theorem provides a way to interpret β_1 as the effect of X_1 on y after removing any linear dependence on X_2.
2. **Computational simplicity:** It allows the decomposition of a large regression problem into smaller, computationally simpler pieces.

```
# Simulate data
set.seed(123)
n <- 100
X1 <- matrix(rnorm(n * 2), n, 2)  # Two regressors of interest
X2 <- matrix(rnorm(n * 2), n, 2)  # Two additional regressors
beta1 <- c(2,-1)  # Coefficients for X1
beta2 <- c(1, 0.5)  # Coefficients for X2undefined
u <- rnorm(n)  # Error term
```

```
y <- X1 %*% beta1 + X2 %*% beta2 + u  # Generate dependent
↪  variable

# Full regression
full_model <- lm(y ~ X1 + X2)
summary(full_model)
#>
#> Call:
#> lm(formula = y ~ X1 + X2)
#>
#> Residuals:
#>      Min       1Q   Median       3Q      Max
#> -2.47336 -0.58010  0.07461  0.68778  2.46552
#>
#> Coefficients:
#>             Estimate Std. Error t value Pr(>|t|)
#> (Intercept)  0.11614    0.10000   1.161    0.248
#> X11          1.77575    0.10899  16.293  < 2e-16 ***
#> X12         -1.14151    0.10204 -11.187  < 2e-16 ***
#> X21          0.94954    0.10468   9.071 1.60e-14 ***
#> X22          0.47667    0.09506   5.014 2.47e-06 ***
#> ---
#> Signif. codes:
#> 0 '***' 0.001 '**' 0.01 '*' 0.05 '.' 0.1 ' ' 1
#>
#> Residual standard error: 0.9794 on 95 degrees of freedom
#> Multiple R-squared:  0.8297, Adjusted R-squared:  0.8225
#> F-statistic: 115.7 on 4 and 95 DF,  p-value: < 2.2e-16

# Step 1: Partial out X2 from y
y_residual <- residuals(lm(y ~ X2))

# Step 2: Partial out X2 from X1
X1_residual <- residuals(lm(X1 ~ X2))

# Step 3: Regress residuals
fwl_model <- lm(y_residual ~ X1_residual - 1)
summary(fwl_model)
#>
#> Call:
#> lm(formula = y_residual ~ X1_residual - 1)
#>
#> Residuals:
#>      Min       1Q   Median       3Q      Max
```

```
#> -2.47336 -0.58010  0.07461  0.68778  2.46552
#>
#> Coefficients:
#>               Estimate Std. Error t value Pr(>|t|)
#> X1_residual1   1.7758     0.1073   16.55   <2e-16 ***
#> X1_residual2  -1.1415     0.1005  -11.36   <2e-16 ***
#> ---
#> Signif. codes:
#> 0 '***' 0.001 '**' 0.01 '*' 0.05 '.' 0.1 ' ' 1
#>
#> Residual standard error: 0.9643 on 98 degrees of freedom
#> Multiple R-squared:  0.8109, Adjusted R-squared:  0.807
#> F-statistic: 210.1 on 2 and 98 DF,  p-value: < 2.2e-16

# Comparison of coefficients
cat("Full model coefficients (X1):", coef(full_model)[2:3],
↪   "\n")
#> Full model coefficients (X1): 1.775754 -1.141514
cat("FWL model coefficients:", coef(fwl_model), "\n")
#> FWL model coefficients: 1.775754 -1.141514
```

2.1.2.2.2 Gauss-Markov Theorem

For a linear regression model:

$$\mathbf{y} = \mathbf{X}\boldsymbol{\beta} + \boldsymbol{\epsilon},$$

under the assumptions:

1. **A1**: Linearity of the model
2. **A2**: Full rank of $\mathbf{X}$
3. **A3**: Exogeneity of $\mathbf{X}$
4. **A4**: Homoskedasticity of ϵ

The ordinary least squares estimator:

$$\hat{\beta} = (\mathbf{X}'\mathbf{X})^{-1}\mathbf{X}'\mathbf{y},$$

is the **best linear unbiased estimator (BLUE)**. This means that $\hat{\beta}$ has the minimum variance among all linear unbiased estimators of β.

1. **Unbiasedness**

   Suppose we consider **any** linear estimator of β of the form:

$$\tilde{\beta} = \mathbf{C}\,\mathbf{y},$$

where

- $\mathbf{y}$ is the $n \times 1$ vector of observations
- $\mathbf{C}$ is a $k \times n$ matrix (with k the dimension of β) that depends **only** on the design matrix $\mathbf{X}$

Our regression model is

$$\mathbf{y} = \mathbf{X}\beta + \epsilon, \quad E[\epsilon \mid \mathbf{X}] = \mathbf{0}, \quad \mathrm{Var}(\epsilon \mid \mathbf{X}) = \sigma^2\mathbf{I}.$$

We say $\tilde{\beta}$ is **unbiased** if its conditional expectation (given $\mathbf{X}$) equals the true parameter β:

$$E(\tilde{\beta} \mid \mathbf{X}) = E(\mathbf{C}\,\mathbf{y} \mid \mathbf{X}) = \beta.$$

1. Substitute $\mathbf{y} = \mathbf{X}\beta + \epsilon$:

$$E(\mathbf{C}\,\mathbf{y} \mid \mathbf{X}) = E\left(\mathbf{C}(\mathbf{X}\beta + \epsilon) \mid \mathbf{X}\right) = \mathbf{C}\,\mathbf{X}\,\beta + \mathbf{C}\,E(\epsilon \mid \mathbf{X}) = \mathbf{C}\,\mathbf{X}\,\beta.$$

2. For this to hold for *all* β, we require

$$\mathbf{C}\,\mathbf{X} = \mathbf{I}.$$

In other words, $\mathbf{C}$ must be a "right-inverse" of $\mathbf{X}$.

On the other hand, the OLS estimator $\hat{\beta}$ is given by

$$\hat{\beta} = (\mathbf{X}'\mathbf{X})^{-1}\mathbf{X}'\,\mathbf{y}.$$

You can verify:

- Let $\mathbf{C}_{\mathrm{OLS}} = (\mathbf{X}'\mathbf{X})^{-1}\mathbf{X}'$.
- Then

$$\mathbf{C}_{\mathrm{OLS}}\,\mathbf{X} = (\mathbf{X}'\mathbf{X})^{-1}\mathbf{X}'\,\mathbf{X} = \mathbf{I}.$$

- By the argument above, this makes $\hat{\beta}$ unbiased.

Hence, any linear estimator $\tilde{\beta} = \mathbf{C}\,\mathbf{y}$ that is unbiased must satisfy $\mathbf{C}\,\mathbf{X} = \mathbf{I}$.

2. **Minimum Variance (Gauss–Markov Part)**
Among all estimators of the form $\tilde{\beta} = \mathbf{C}\,\mathbf{y}$ that are unbiased (so $\mathbf{C}\,\mathbf{X} = \mathbf{I}$), **OLS** achieves the smallest covariance matrix.

   - Variance of a general unbiased estimator

If $\tilde{\beta} = \mathbf{C}\,\mathbf{y}$ with $\mathbf{C}\,\mathbf{X} = \mathbf{I}$, then:

$$\mathrm{Var}(\tilde{\beta} \mid \mathbf{X}) = \mathrm{Var}(\mathbf{C}\,\mathbf{y} \mid \mathbf{X}) = \mathbf{C}\,\mathrm{Var}(\mathbf{y} \mid \mathbf{X})\,\mathbf{C}' = \mathbf{C}\left(\sigma^2 \mathbf{I}\right)\mathbf{C}' = \sigma^2\,\mathbf{C}\,\mathbf{C}'.$$

- Variance of the OLS estimator

  For OLS, $\hat{\beta} = (\mathbf{X}'\mathbf{X})^{-1}\mathbf{X}'\,\mathbf{y}$. Thus,

$$\mathrm{Var}(\hat{\beta} \mid \mathbf{X}) = \sigma^2\,(\mathbf{X}'\mathbf{X})^{-1}.$$

- Comparing $\mathrm{Var}(\tilde{\beta})$ to $\mathrm{Var}(\hat{\beta})$

We want to show:

$$\mathrm{Var}(\tilde{\beta} \mid \mathbf{X}) - \mathrm{Var}(\hat{\beta} \mid \mathbf{X}) \;\text{ is positive semi-definite.}$$

1. Since both $\tilde{\beta}$ and $\hat{\beta}$ are unbiased, we know:

$$\mathbf{C}\,\mathbf{X} = \mathbf{I}, \quad (\mathbf{X}'\mathbf{X})^{-1}\mathbf{X}'\,\mathbf{X} = \mathbf{I}.$$

2. One can show algebraically (as in the proof provided above) that

$$\mathrm{Var}(\tilde{\beta} \mid \mathbf{X}) - \mathrm{Var}(\hat{\beta} \mid \mathbf{X}) = \sigma^2\left[\mathbf{C}\mathbf{C}' - (\mathbf{X}'\mathbf{X})^{-1}\right].$$

   Under the condition $\mathbf{C}\,\mathbf{X} = \mathbf{I}$, the difference $\mathbf{C}\mathbf{C}' - (\mathbf{X}'\mathbf{X})^{-1}$ is positive semi-definite.
3. Positive semi-definite difference means

$$\mathbf{v}'\left(\mathbf{C}\mathbf{C}' - (\mathbf{X}'\mathbf{X})^{-1}\right)\mathbf{v} \geq 0 \quad \text{for all vectors } \mathbf{v}.$$

Hence, $\hat{\beta}$ has the smallest variance (in the sense of covariance matrices) among all linear unbiased estimators $\tilde{\beta} = \mathbf{C}\,\mathbf{y}$.

Summary of the Key Points

1. **Unbiasedness:**
   A linear estimator $\tilde{\beta} = \mathbf{C}\,\mathbf{y}$ is unbiased if $E(\tilde{\beta} \mid \mathbf{X}) = \beta$.
   This forces $\mathbf{C}\,\mathbf{X} = \mathbf{I}$.
2. **OLS is unbiased:**
   The OLS estimator $\hat{\beta} = (X'X)^{-1}X'\,y$ satisfies $(X'X)^{-1}X'\,X = I$, hence is unbiased.

3. **OLS has minimum variance:**
   Among all $\mathbf{C}$ that satisfy $\mathbf{C}\mathbf{X} = \mathbf{I}$, the matrix $(\mathbf{X}'\mathbf{X})^{-1}$ gives the smallest possible $\text{Var}(\tilde{\beta})$.
   In matrix form, $\text{Var}(\tilde{\beta}) - \text{Var}(\hat{\beta})$ is positive semi-definite, showing OLS is optimal (the **best linear unbiased estimator** [BLUE]).

2.1.2.3 Finite Sample Properties

The finite sample properties of an estimator are considered when the sample size n is fixed (not asymptotically large). Key properties include **bias**, **distribution**, and **standard deviation** of the estimator.
Bias measures how close an estimator is, on average, to the true parameter value β. It is defined as:

$$\text{Bias} = E(\hat{\beta}) - \beta$$

where:

- β: true parameter value
- $\hat{\beta}$: estimator for β

Unbiased estimator: An estimator is unbiased if:

$$\text{Bias} = E(\hat{\beta}) - \beta = 0 \quad \text{or equivalently} \quad E(\hat{\beta}) = \beta.$$

This means the estimator will produce estimates that are, on average, equal to the value it is trying to estimate.

An estimator is a function of random variables (data). Its **distribution** describes how the estimates vary across repeated samples. Key aspects include:

- **Center**: mean of the distribution, which relates to bias
- **Spread**: variability of the estimator, captured by its standard deviation or variance

The **standard deviation** of an estimator measures the spread of its sampling distribution. It indicates the variability of the estimator across different samples.

2.1.2.3.1 Ordinary Least Squares Properties

Under the standard assumptions for OLS:

1. A1: The relationship between Y and X is linear.
2. A2: The matrix $\mathbf{X}'\mathbf{X}$ is invertible.
3. A3: $E(\epsilon|X) = 0$ (errors are uncorrelated with predictors).

OLS is unbiased under these assumptions. The proof is as follows:

$$
\begin{aligned}
E(\hat{\beta}) &= E((\mathbf{X}'\mathbf{X})^{-1}\mathbf{X}'\mathbf{y}) && \text{A2} \\
&= E((\mathbf{X}'\mathbf{X})^{-1}\mathbf{X}'(\mathbf{X}\boldsymbol{\beta} + \boldsymbol{\epsilon})) && \text{A1} \\
&= E((\mathbf{X}'\mathbf{X})^{-1}\mathbf{X}'\mathbf{X}\boldsymbol{\beta} + (\mathbf{X}'\mathbf{X})^{-1}\mathbf{X}'\boldsymbol{\epsilon}) \\
&= E(\beta + (\mathbf{X}'\mathbf{X})^{-1}\mathbf{X}'\boldsymbol{\epsilon}) \\
&= \beta + E((\mathbf{X}'\mathbf{X})^{-1}\mathbf{X}'\boldsymbol{\epsilon}) \\
&= \beta + E(E((\mathbf{X}'\mathbf{X})^{-1}\mathbf{X}'\boldsymbol{\epsilon}|X)) && \text{LIE (law of iterated expectation)} \\
&= \beta + E((\mathbf{X}'\mathbf{X})^{-1}\mathbf{X}'E(\epsilon|X)) \\
&= \beta + E((\mathbf{X}'\mathbf{X})^{-1}\mathbf{X}' \cdot 0) && \text{A3} \\
&= \beta
\end{aligned}
$$

Key Points:

1. **Linearity of expectation**: used to separate terms involving β and ϵ
2. [Law of iterated expectation] **(LIE)**: simplifies nested expectations
3. **Exogeneity of errors (A3)**: ensures $E(\epsilon|X) = 0$, eliminating bias

Implications of unbiasedness

- OLS estimators are centered around the true value β across repeated samples.
- In small samples, OLS estimators may exhibit variability, but their expected value remains β.

If the assumption of exogeneity (A3) is violated, the OLS estimator becomes **biased**. Specifically, omitted variables or endogeneity can introduce systematic errors into the estimation.

From the Frisch–Waugh–Lovell theorem:

- If an omitted variable $\hat{\beta}_2 \neq 0$ (nonzero effect) and the omitted variable is correlated with the included predictors ($\mathbf{X}_1'\mathbf{X}_2 \neq 0$), then the OLS estimator will be biased.
- This bias arises because the omitted variable contributes to the variation in the dependent variable, but its effect is incorrectly attributed to other predictors.

2.1.2.3.2 Conditional Variance of OLS Estimator

Under assumptions A1, A2, A3, and A4, the **conditional variance of the OLS estimator** is:

$$
\begin{aligned}
Var(\hat{\beta}|\mathbf{X}) &= Var(\beta + (\mathbf{X}'\mathbf{X})^{-1}\mathbf{X}'\boldsymbol{\epsilon}|\mathbf{X}) && \text{A1-A2} \\
&= Var((\mathbf{X}'\mathbf{X})^{-1}\mathbf{X}'\boldsymbol{\epsilon}|\mathbf{X}) \\
&= (\mathbf{X}'\mathbf{X})^{-1}\mathbf{X}' Var(\epsilon|\mathbf{X})\mathbf{X}(\mathbf{X}'\mathbf{X})^{-1} \\
&= (\mathbf{X}'\mathbf{X})^{-1}\mathbf{X}\sigma^2 I\mathbf{X}(\mathbf{X}'\mathbf{X})^{-1} && \text{A4} \\
&= \sigma^2(\mathbf{X}'\mathbf{X})^{-1}.
\end{aligned}
$$

This result shows that the variance of $\hat{\beta}$ depends on:

1. σ^2: the variance of the errors
2. $\mathbf{X}'\mathbf{X}$: the information content in the design matrix $\mathbf{X}$

2.1.2.3.3 Sources of Variation in OLS Estimator

1. **Unexplained variation in the dependent variable**: $\sigma^2 = Var(\epsilon_i|\mathbf{X})$

- Large σ^2 indicates that the amount of unexplained variation (noise) is high relative to the explained variation ($\mathbf{x_i}\boldsymbol{\beta}$).
- This increases the variance of the OLS estimator.

2. **Small variation in predictor variables**

- If the variance of predictors ($Var(x_{i1}), Var(x_{i2}), \dots$) is small, the design matrix $\mathbf{X}$ lacks information, leading to:
  - High variability in $\hat{\beta}$
  - Potential issues in estimating coefficients accurately
- **Small sample size** exacerbates this issue, as fewer observations reduce the robustness of parameter estimates.

3. **Correlation between explanatory variables (collinearity)**

- **Strong correlation** among explanatory variables creates problems:
  - x_{i1} being highly correlated with a linear combination of $1, x_{i2}, x_{i3}, \dots$ contributes to inflated standard errors for $\hat{\beta}_1$.
  - Including many irrelevant variables exacerbates this issue.
- **Perfect collinearity**:
  - If x_1 is perfectly determined by a linear combination of other predictors, the matrix $\mathbf{X}'\mathbf{X}$ becomes singular.
  - This violates A2, making OLS impossible to compute.

- **Multicollinearity**:
  - If x_1 is highly correlated (but not perfectly) with a linear combination of other variables, the variance of $\hat{\beta}_1$ increases.
  - Multicollinearity does not violate OLS assumptions but weakens inference by inflating standard errors.

2.1.2.3.4 Standard Errors

Standard errors measure the variability of an estimator, specifically the standard deviation of $\hat{\beta}$. They are crucial for inference, as they quantify the uncertainty associated with parameter estimates.
The variance of the OLS estimator $\hat{\beta}$ is:

$$Var(\hat{\beta}|\mathbf{X}) = \sigma^2(\mathbf{X}'\mathbf{X})^{-1}$$

where:

- σ^2: variance of the error terms
- $(\mathbf{X}'\mathbf{X})^{-1}$: inverse of the design matrix product, capturing the geometry of the predictors

Estimation of σ^2
Under assumptions A1 through A5, we can estimate σ^2 as:

$$s^2 = \frac{1}{n-k}\sum_{i=1}^{n} e_i^2 = \frac{1}{n-k}SSR$$

where:

- n: number of observations
- k: number of predictors, including the intercept
- e_i: residuals from the regression model ($e_i = y_i - \hat{y}_i$)
- SSR: sum of squared residuals ($\sum e_i^2$)

The degrees of freedom adjustment $(n - k)$ accounts for the fact that residuals e_i are not true errors ϵ_i. Since the regression model uses k parameters, we lose k degrees of freedom in estimating variance.

The standard error for σ is:

$$s = \sqrt{s^2}$$

However, s is a biased estimator of σ due to [Jensen's inequality].
The standard error of each regression coefficient $\hat{\beta}_{j-1}$ is:

$$SE(\hat{\beta}_{j-1}) = s\sqrt{[(\mathbf{X}'\mathbf{X})^{-1}]_{jj}}.$$

Alternatively, it can be expressed in terms of SST_{j-1} and R_{j-1}^2:

$$SE(\hat{\beta}_{j-1}) = \frac{s}{\sqrt{SST_{j-1}(1 - R_{j-1}^2)}}$$

where:

- SST_{j-1}: total sum of squares for x_{j-1} from a regression of x_{j-1} on all other predictors
- R_{j-1}^2: coefficient of determination for the same regression

This formulation highlights the role of multicollinearity, as R_{j-1}^2 reflects the correlation between x_{j-1} and other predictors.

2.1.2.3.5 Summary of Finite Sample Properties of OLS Under Different Assumptions

1. **Under A1–A3**: OLS is unbiased.

$$E(\hat{\beta}) = \beta.$$

2. **Under A1–A4**: The variance of the OLS estimator is:

$$Var(\hat{\beta}|\mathbf{X}) = \sigma^2(\mathbf{X}'\mathbf{X})^{-\mathbf{1}}.$$

3. **Under A1–A4, A6**: The OLS estimator is normally distributed:

$$\hat{\beta} \sim N(\beta, \sigma^2(\mathbf{X}'\mathbf{X})^{-\mathbf{1}}).$$

4. **Under A1–A4, Gauss–Markov theorem**: OLS is BLUE (best linear unbiased estimator).
5. **Under A1–A5**: The standard errors for $\hat{\beta}$ are unbiased estimators of the standard deviation of $\hat{\beta}$.

2.1.2.4 Large Sample Properties

Large sample properties provide a framework to evaluate the quality of estimators when finite sample properties are either uninformative or computationally infeasible. This perspective becomes crucial in modern data analysis, especially for methods like GLS or MLE, where assumptions for finite sample analysis may not hold (Table 2.17).

Table 2.17 When to use finite versus large sample analysis

Aspect	Finite sample properties	Large sample properties
Applicability	Limited to fixed sample sizes	Relevant for $n \to \infty$
Exactness	Exact results (e.g., distributions, unbiasedness)	Approximate results
Assumptions	May require stronger assumptions (e.g., normality, independence)	Relies on asymptotic approximations (e.g., CLT, LLN)
Estimator behavior	Performance may vary significantly	Estimators stabilize and improve in accuracy
Ease of use	Often complex due to reliance on exact distributions	Simplifies analysis by leveraging asymptotic approximations
Real-world relevance	More realistic for small datasets	More relevant for large datasets

- **Finite sample analysis**:
  - Small sample sizes (e.g., $n < 30$)
  - Critical for studies where exact results are needed
  - Useful in experimental designs and case studies
- **Large sample analysis**:
  - Large datasets (e.g., $n > 100$)
  - Necessary when asymptotic approximations improve computational simplicity

Key Concepts:

1. **Consistency**:
   - Consistency ensures that an estimator converges in probability to the true parameter value as the sample size increases.
   - Mathematically, an estimator $\hat{\theta}$ is consistent for θ if:

$$\hat{\theta}_n \to^p \theta \quad \text{as } n \to \infty.$$

   - Consistency does not imply unbiasedness, and unbiasedness does not guarantee consistency.
2. **Asymptotic distribution**:
   - The limiting distribution describes the shape of the scaled estimator as $n \to \infty$.
   - Asymptotic distributions often follow normality due to the [central limit theorem], which underpins much of inferential statistics.

3. **Asymptotic variance**:
   - Represents the spread of the estimator with respect to its limiting distribution
   - Smaller asymptotic variance implies greater precision in large samples

Motivation
Finite sample properties, such as unbiasedness, rely on strong assumptions like:

- A1 Linearity
- A3 Exogeneity of Independent Variables
- A4 Homoskedasticity
- A6 Normal Distribution

When these assumptions are violated or impractical to verify, finite sample properties lose relevance. In such cases, large sample properties serve as an essential alternative for evaluating estimators.

For example, let the conditional expectation function (CEF) be:

$$\mu(\mathbf{X}) = E(y|\mathbf{X}),$$

which represents the **minimum mean squared predictor** over all possible functions $f(\mathbf{X})$:

$$\min_f E((y - f(\mathbf{X}))^2).$$

Under the assumptions A1 and A3, the CEF simplifies to:

$$\mu(\mathbf{X}) = \mathbf{X}\beta.$$

The **linear projection** is given by:

$$L(y|1, \mathbf{X}) = \gamma_0 + \mathbf{X}\text{Var}(\mathbf{X})^{-1}\text{Cov}(\mathbf{X}, y),$$

where:

$$\gamma = \mathbf{X}\text{Var}(\mathbf{X})^{-1}\text{Cov}(\mathbf{X}, y).$$

This linear projection minimizes the mean squared error:

$$(\gamma_0, \gamma) = \arg\min_{(a,b)} E\left[(E(y|\mathbf{X}) - (a + \mathbf{X}b))^2\right].$$

Implications for OLS

- **Consistency**: OLS is always consistent for the linear projection, ensuring convergence to the true parameter value as $n \to \infty$.

- **Causal interpretation**: The linear projection has no inherent causal interpretation—it approximates the conditional mean function.
- **Assumption independence**: Unlike the CEF, the linear projection does not depend on assumptions A1 and A3.

Evaluating Estimators via Large Sample Properties

1. **Consistency**:
   - Measures the estimator's centrality to the true value.
   - A consistent estimator ensures that with larger samples, estimates become arbitrarily close to the population parameter.
2. **Limiting distribution**:
   - Helps infer the sampling behavior of the estimator as n grows.
   - Often approximated by a normal distribution for practical use in hypothesis testing and confidence interval construction.
3. **Asymptotic variance**:
   - Quantifies the dispersion of the estimator around its limiting distribution.
   - Smaller variance is desirable for greater reliability.

An estimator $\hat{\theta}$ is **consistent** for a parameter θ if, as the sample size n increases, $\hat{\theta}$ converges in probability to θ:

$$\hat{\theta}_n \to^p \theta \quad \text{as } n \to \infty.$$

1. **Convergence in probability**:
   - The probability that $\hat{\theta}$ deviates from θ by more than a small margin (no matter how small) approaches zero as n increases.

   Formally:

$$\forall \epsilon > 0, \quad P(|\hat{\theta}_n - \theta| > \epsilon) \to 0 \quad \text{as } n \to \infty.$$

2. **Interpretation**: Consistency ensures that the estimator becomes arbitrarily close to the true population parameter θ as the sample size grows.
3. **Asymptotic behavior**: Large sample properties rely on consistency to provide valid approximations of an estimator's behavior in finite samples.

Relationship Between Consistency and Unbiasedness

- **Unbiasedness**:
  - An estimator $\hat{\theta}$ is unbiased if its expected value equals the true parameter:

$$E(\hat{\theta}) = \theta.$$

  - Unbiasedness is a finite-sample property and does not depend on the sample size.

- **Consistency**:
  - Consistency is a large-sample property and requires $\hat{\theta}$ to converge to θ as $n \to \infty$.

Important Distinctions

1. **Unbiasedness does not imply consistency**:
   - Example: Consider an unbiased estimator with extremely high variance that does not diminish as n increases. Such an estimator does not converge to θ in probability.
2. **Consistency does not imply unbiasedness**:
   - Example: $\hat{\theta}_n = \frac{n-1}{n}\theta$ is biased for all finite n, but as $n \to \infty$, $\hat{\theta}_n \to^p \theta$, making it consistent.

From the OLS formula:

$$\hat{\beta} = (\mathbf{X}'\mathbf{X})^{-1}\mathbf{X}'\mathbf{y},$$

we can expand as:

$$\hat{\beta} = (\sum_{\mathbf{i=1}}^{\mathbf{n}} \mathbf{x_i'x_i})^{-1} \sum_{\mathbf{i}} = \mathbf{1^n x_i' y_i},$$

or equivalently:

$$\hat{\beta} = (n^{-1} \sum_{\mathbf{i=1}}^{\mathbf{n}} \mathbf{x_i'x_i})^{-1} \mathbf{n}^{-1} \sum_{\mathbf{i=1}}^{\mathbf{n}} \mathbf{x_i'y_i}.$$

Taking the probability limit under the assumptions A2 and A5, we apply the [weak law of large numbers] (for a random sample, averages converge to expectations as $n \to \infty$):

$$plim(\hat{\beta}) = plim((n^{-1} \sum_{\mathbf{i=1}}^{\mathbf{n}} \mathbf{x_i'x_i})^{-1}) plim(n^{-1} \sum_{\mathbf{i=1}}^{\mathbf{n}} \mathbf{x_i'y_i}),$$

which simplifies to:

$$
\begin{aligned}
plim(\hat{\beta}) &= (E(\mathbf{x_i'x_i}))^{-1}E(\mathbf{x_i'y_i}) && \text{A1} \\
&= (E(\mathbf{x_i'x_i}))^{-1}\left(E(\mathbf{x_i'x_i}\,\beta) + E(\mathbf{x_i}\,\epsilon_\mathbf{i})\right) && (A3a) \\
&= (E(\mathbf{x_i'x_i}))^{-1}E(\mathbf{x_i'x_i})\,\beta && (A2) \\
&= \beta.
\end{aligned}
$$

Proof For a model of $y_i = x_i\beta + \epsilon_i$, where β is the true parameter vector, ϵ_i is the random error:

Expanding $E(x_i y_i)$:

$$E(x_i' y_i) = E(x_i'(x_i\beta + \epsilon_i)).$$

By the linearity of expectation:

$$E(x_i' y_i) = E(x_i' x_i \beta) + E(x_i \epsilon_i).$$

The second term $E(x_i\epsilon_i) = 0$ under assumption A3a.

Thus,

$$E(x_i' y_i) = E(x_i' x_i)\beta.$$

Consistency of OLS

Hence, in short, under the assumptions:

- A1 Linearity
- A2 Full Rank
- A3a: Weaker Exogeneity Assumption
- A5 Data Generation (Random Sampling)

the term $E(\mathbf{x_i'}\epsilon_\mathbf{i}) = 0$, and the OLS estimator is **consistent**:

$$plim(\hat{\beta}) = \beta.$$

However, OLS consistency does not guarantee unbiasedness in small samples.

2.1.2.4.1 Asymptotic Distribution of OLS

Under the same assumptions :

- A1 Linearity
- A2 Full Rank
- A3a: Weaker Exogeneity Assumption
- A5 Data Generation (Random Sampling)

and if $\mathbf{x_i'x_i}$ has finite first and second moments (required for the [central limit theorem]), we have:

1. Convergence of $n^{-1}\sum_{i=1}^{n}\mathbf{x_i'x_i}$:

$$n^{-1}\sum_{i=1}^{n}\mathbf{x_i'x_i} \to^{p} E(\mathbf{x_i'x_i}).$$

2. Asymptotic normality of $\sqrt{n}(n^{-1}\sum_{i=1}^{n}\mathbf{x_i'\epsilon_i})$:

$$\sqrt{n}(n^{-1}\sum_{i=1}^{n}\mathbf{x_i'\epsilon_i}) \to^{d} N(0, \mathbf{B}),$$

   where $\mathbf{B} = Var(\mathbf{x_i'\epsilon_i})$.

From these results, the scaled difference between $\hat{\beta}$ and β follows:

$$\sqrt{n}(\hat{\beta}-\beta) = (n^{-1}\sum_{i=1}^{n}\mathbf{x_i'x_i})^{-1}\sqrt{n}(n^{-1}\sum_{i=1}^{n}\mathbf{x_i'\epsilon_i}).$$

By the [central limit theorem]:

$$\sqrt{n}(\hat{\beta}-\beta) \to^{d} N(0, \Sigma),$$

where:

$$\Sigma = (E(\mathbf{x_i'x_i}))^{-1}\mathbf{B}(E(\mathbf{x_i'x_i}))^{-1}.$$

The sandwich form of Σ is standard.

Implications for Homoskedasticity (A4) Versus Heteroskedasticity

1. No homoskedasticity (A4 Homoskedasticity) needed:

   - The CLT and the large-sample distribution of $\hat{\beta}$ do *not* require homoskedasticity. The only place homoskedasticity would simplify things is that:

$$\mathbf{B} = Var(\mathbf{x_i'\epsilon_i}) = \sigma^2 E(\mathbf{x_i'x_i}),$$

   only if $Var(\epsilon_i|\mathbf{x}_i)\sigma^2$.
   Then:

$$\Sigma = \sigma^2(E(\mathbf{x_i'x_i}))^{-1}.$$

2. Adjusting for heteroskedasticity:
   - In practice, σ_i^2 can vary with $\mathbf{x}_i$, leading to heteroskedasticity.
   - The standard OLS formula for $Var(\hat{\beta})$ is inconsistent under heteroskedasticity, so one uses *robust (White) standard errors.*
   - Heteroskedasticity can arise from (but not limited to):
     - Limited dependent variables
     - Dependent variables with large or skewed ranges

2.1.2.4.2 Derivation of Asymptotic Variance

The asymptotic variance of the OLS estimator is derived as follows:

$$\Sigma = (E(\mathbf{x_i'x_i}))^{-1}\mathbf{B}(E(\mathbf{x_i'x_i}))^{-1}.$$

1. Substituting $\mathbf{B} = Var(\mathbf{x_i'}\epsilon_i)$:

$$\Sigma = (E(\mathbf{x_i'x_i}))^{-1}Var(\mathbf{x_i'}\epsilon_i)(E(\mathbf{x_i'x_i}))^{-1}.$$

2. Using the definition of variance:

$$\Sigma = (E(\mathbf{x_i'x_i}))^{-1}E[(\mathbf{x_i'}\epsilon_i - 0)(\mathbf{x_i'}\epsilon_i - 0)'](E(\mathbf{x_i'x_i}))^{-1}.$$

3. By the [law of iterated expectations] and A3a Weaker Exogeneity Assumption, we have:

$$E[(\mathbf{x_i'}\epsilon_i)(\mathbf{x_i'}\epsilon_i)'] = E[E(\epsilon_i^2|\mathbf{x_i})\mathbf{x_i'x_i}],$$

4. Assuming homoskedasticity (A4 Homoskedasticity), $E(\epsilon_i^2|\mathbf{x_i}) = \sigma^2$, so:

$$\Sigma = (E(\mathbf{x_i'x_i}))^{-1}\sigma^2E(\mathbf{x_i'x_i})(E(\mathbf{x_i'x_i}))^{-1}.$$

5. Simplifying:

$$\Sigma = \sigma^2(E(\mathbf{x_i'x_i}))^{-1}.$$

Hence, under the assumptions:

- A1 Linearity
- A2 Full Rank
- A3a: Weaker Exogeneity Assumption

- A4 Homoskedasticity
- A5 Data Generation (Random Sampling)

we have

$$\sqrt{n}(\hat{\beta} - \beta) \rightarrow^d N(0, \sigma^2(E(\mathbf{x_i'x_i}))^{-1}).$$

The asymptotic variance provides an approximation of the scaled estimator's variance for large n. This leads to:

$$Avar(\sqrt{n}(\hat{\beta} - \beta)) = \sigma^2(E(\mathbf{x_i'x_i}))^{-1}.$$

The finite sample variance of an estimator can be approximated using the asymptotic variance for large sample sizes:

$$Avar(\sqrt{n}(\hat{\beta} - \beta)) \approx Var(\sqrt{n}(\hat{\beta} - \beta))$$
$$Avar(\sqrt{n}(\hat{\beta} - \beta))/n \approx Var(\sqrt{n}(\hat{\beta} - \beta))/n = Var(\hat{\beta}).$$

However, it is critical to note that **asymptotic variance** ($Avar(.)$) does not behave in the same manner as finite sample variance ($Var(.)$). This distinction is evident in the following expressions:

$$Avar(\sqrt{n}(\hat{\beta} - \beta))/n \neq Avar(\sqrt{n}(\hat{\beta} - \beta)/\sqrt{n})$$
$$\neq Avar(\hat{\beta}).$$

Thus, while $Avar(.)$ provides a useful approximation for large samples, its conceptual properties differ from finite sample variance.
In finite sample properties, the standard errors are calculated as estimates of the conditional standard deviation:

$$SE_{fs}(\hat{\beta}_{j-1}) = \sqrt{\hat{Var}(\hat{\beta}_{j-1}|\mathbf{X})} = \sqrt{s^2 \cdot [(\mathbf{X'X})^{-1}]_{jj}},$$

where:

- s^2 is the estimator of the error variance
- $[(\mathbf{X'X})^{-1}]_{jj}$ represents the jth diagonal element of the inverse design matrix

In contrast, in large sample properties, the standard errors are calculated as estimates of the square root of the asymptotic variance:

$$SE_{ls}(\hat{\beta}_{j-1}) = \sqrt{\hat{Avar}(\sqrt{n}\hat{\beta}_{j-1})/n} = \sqrt{s^2 \cdot [(\mathbf{X'X})^{-1}]_{jj}}.$$

Interestingly, the standard error estimator is **identical** for both finite and large samples:

- The expressions for SE_{fs} and SE_{ls} are mathematically the same.
- However, they are conceptually estimating two **different quantities**:
  - **Finite sample standard error**: an estimate of the conditional standard deviation of $\hat{\beta}_{j-1}$ given $\mathbf{X}$
  - **Large sample standard error**: an estimate of the square root of the asymptotic variance of $\hat{\beta}_{j-1}$

The assumptions required for these estimators to be valid differ in their stringency:

1. **Finite sample variance (conditional variance):**
   - Requires stronger assumptions (A1–A5)
2. **Asymptotic variance:**
   - Valid under weaker assumptions (A1, A2, A3a, A4, A5)

This distinction highlights the utility of asymptotic properties in providing robust approximations when finite sample assumptions may not hold.

2.1.2.5 Diagnostics

2.1.2.5.1 Normality of Errors

Ensuring the normality of errors is a critical assumption in many regression models. Deviations from this assumption can impact inference and model interpretation. For diagnoses assessing normality, see [Normality Assessment].

Plots are invaluable for visual inspection of normality. One common approach is the Q-Q plot (Fig. 2.6), which compares the quantiles of the residuals against those of a standard normal distribution:

```
# Example Q-Q plot
set.seed(123) # For reproducibility
y <- 1:100
x <- rnorm(100) # Generating random normal data
qqplot(x,
       y,
       main = "Q-Q Plot",
       xlab = "Theoretical Quantiles",
       ylab = "Sample Quantiles")
```

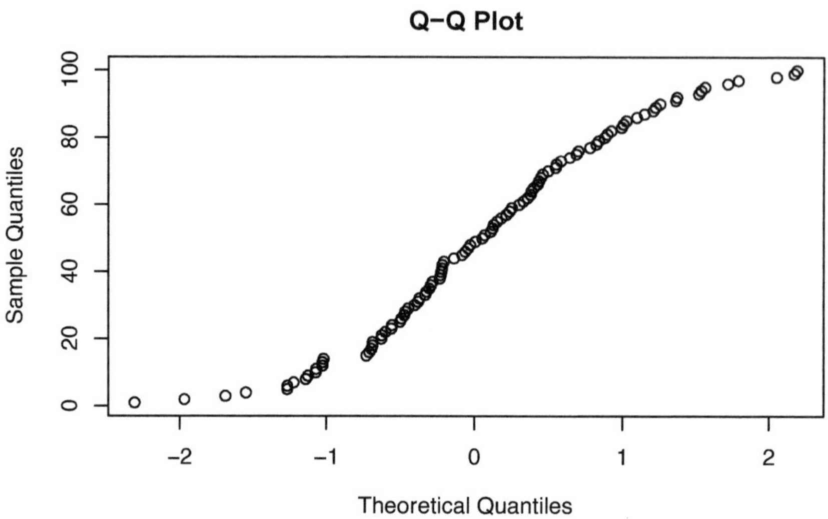

Fig. 2.6 Q-Q plot for assessing normality of sample data

2.1.2.5.2 Influential Observations and Outliers

Identifying influential observations or outliers is essential for robust regression modeling. The **hat matrix (H)** plays a key role in diagnosing influence.

2.1.2.5.3 Hat Matrix: Outliers in X-Space

The hat matrix is primarily concerned with **leverage**, which reflects how far an observation's predictor values (X-space) are from the centroid of the predictor space.

- **What it measures:** The diagonal elements of the hat matrix quantify **leverage**, not residual size or model fit. High leverage suggests that an observation has an unusual predictor configuration and might disproportionately influence the regression line, irrespective of the response variable.
- **What it does not measure:** It does not directly account for outliers in Y-space or residuals.

The hat matrix, defined as:

$$\mathbf{H} = \mathbf{X}(\mathbf{X}'\mathbf{X})^{-1}$$

has the following properties:

- **Fitted values**: $\hat{\mathbf{Y}} = \mathbf{HY}$.
- **Residuals**: $\mathbf{e} = (\mathbf{I} - \mathbf{H})\mathbf{Y}$.

- **Variance of residuals**: $\text{var}(\mathbf{e}) = \sigma^2(\mathbf{I} - \mathbf{H})$.

Diagonal elements of the hat matrix (h_{ii})

- h_{ii} is the ith element on the main diagonal of $\mathbf{H}$. It must satisfy $0 \le h_{ii} \le 1$.
- $\sum_{i=1}^{n} h_{ii} = p$, where p is the number of parameters (including the intercept).
- The variance of residuals for observation i is given by $\sigma^2(e_i) = \sigma^2(1 - h_{ii})$.
- The covariance between residuals for observations i and j ($i \neq j$) is $-h_{ij}\sigma^2$.

For large datasets, the off-diagonal elements (h_{ij}) tend to have small covariance if model assumptions hold.

Estimations Using MSE

- Variance of residuals: $s^2(e_i) = MSE(1 - h_{ii})$, where MSE is the mean squared error.
- Covariance of residuals: $\hat{\text{cov}}(e_i, e_j) = -h_{ij}(MSE)$.

Interpretation of h_{ii}

If $\mathbf{x}_i = [1, X_{i,1}, \dots, X_{i,p-1}]'$ represents the vector of predictor values for observation i, then:

$$h_{ii} = \mathbf{x}_i'(\mathbf{X}'\mathbf{X})^{-1}\mathbf{x}_i.$$

The value of h_{ii} depends on the relative positions of the design points $X_{i,1}, \dots, X_{i,p-1}$. Observations with high h_{ii} are more influential and warrant closer inspection for leverage or outlier behavior.

2.1.2.5.4 Studentized Residuals: Outliers in Y-Space

Residuals focus on discrepancies between observed (Y) and predicted ($\hat{Y}$) values, helping to identify outliers in Y-space.

- **What they measure:**
  - Standardized or studentized residuals assess how far an observation's response is from the regression line, adjusted for variance.
  - **Externally studentized residuals** are more robust because they exclude the i-th observation when estimating variance.
  - Large studentized residuals (e.g., > 2 or > 3) indicate observations that are unusual in Y-space.
- **What they do not measure:** They do not consider leverage or the X-space configuration. A point with a large residual could have low leverage, making it less influential overall.

Studentized residuals, also known as standardized residuals, adjust for the variance of residuals by dividing the residuals by their standard error:

$$r_i = \frac{e_i}{s(e_i)}$$
$$r_i \sim N(0, 1),$$

where $s(e_i) = \sqrt{MSE(1 - h_{ii})}$, and r_i accounts for the varying variances of residuals. These residuals allow for a better comparison of model fit across observations.

- **Semi-studentized residuals**: In contrast, the semi-studentized residuals are defined as:

$$e_i^* = \frac{e_i}{\sqrt{MSE}}.$$

This approach does not adjust for the heterogeneity in variances of residuals, as e_i^* assumes equal variance for all residuals.
To assess the influence of individual observations, we consider the model without a particular value. When the ith observation is removed, the **deleted residual** is defined as:

$$\begin{aligned} d_i &= Y_i - \hat{Y}_{i(i)} \\ &= \frac{e_i}{1 - h_{ii}}, \end{aligned}$$

where Y_i is the actual observation, and $\hat{Y}_{i(i)}$ is the predicted value for the ith observation, computed using the regression model fitted to the remaining $n - 1$ observations. Importantly, we do not need to refit the regression model for each observation to compute d_i.

- As h_{ii} (leverage) increases, d_i also increases, indicating higher influence of the observation.

The variance of the deleted residual is given by:

$$s^2(d_i) = \frac{MSE_{(i)}}{1 - h_{ii}},$$

where $MSE_{(i)}$ is the mean squared error when the ith case is omitted.
The **studentized deleted residual** accounts for variability and follows a t-distribution with $n - p - 1$ degrees of freedom:

$$t_i = \frac{d_i}{s(d_i)} = \frac{e_i}{\sqrt{MSE_{(i)}(1 - h_{ii})}},$$

where t_i helps identify outliers more effectively.

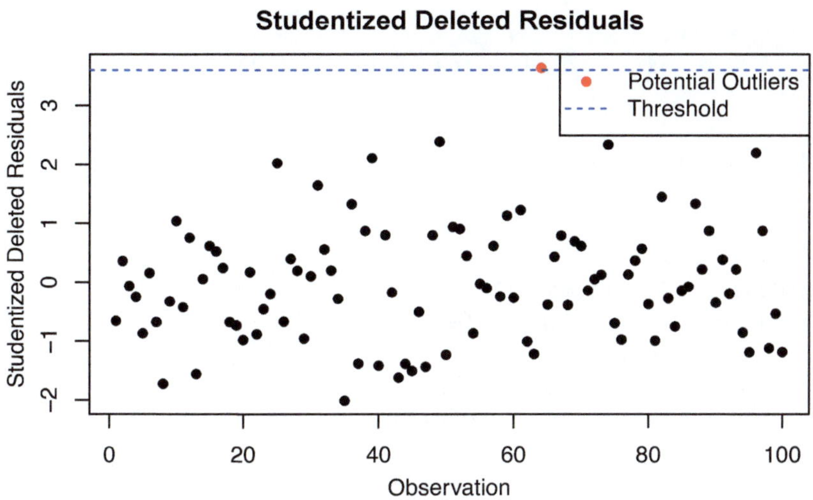

Fig. 2.7 Studentized deleted residuals for outlier detection

We can compute t_i without fitting the regression model multiple times. Using the relationship:

$$(n-p)MSE = (n-p-1)MSE_{(i)} + \frac{e_i^2}{1-h_{ii}},$$

we derive:

$$t_i = e_i\sqrt{\frac{n-p-1}{SSE(1-h_{ii})-e_i^2}}.$$

This formulation avoids the need for recalculating $MSE_{(i)}$ explicitly for each case.

Outlying Y-observations are those with large studentized deleted residuals in absolute value (Fig. 2.7). To handle multiple testing when there are many residuals, we use a Bonferroni-adjusted critical value:

$$t_{1-\alpha/2n;n-p-1},$$

where α is the desired significance level, n is the sample size, and p is the number of parameters in the model. Observations exceeding this threshold are flagged as potential outliers.

```
# Example R Code for Demonstrating Residual Diagnostics
set.seed(123) # For reproducibility

# Simulate some data
n <- 100
```

```r
x <- rnorm(n, mean = 10, sd = 3)
y <- 5 + 2 * x + rnorm(n, mean = 0, sd = 2)

# Fit a linear regression model
model <- lm(y ~ x)


# Extract residuals, fitted values, and hat values
residuals <- resid(model)
hat_values <- lm.influence(model)$hat
mse <- mean(residuals ^ 2)
n <- length(residuals)
p <- length(coefficients(model)) # Number of parameters

# Compute studentized residuals
studentized_residuals <- residuals / sqrt(mse * (1 -
↪  hat_values))

# Compute deleted residuals
deleted_residuals <- residuals / (1 - hat_values)

# Compute studentized deleted residuals
studentized_deleted_residuals <-
    residuals * sqrt((n - p - 1)
                    / (sum(residuals ^ 2) * (1 - hat_values)
                    ↪  - residuals ^ 2))

# Flag potential outliers using Bonferroni-adjusted critical
↪  value
alpha <- 0.05
bonferroni_threshold <- qt(1 - alpha / (2 * n), df = n - p -
↪  1)
outliers <-
    abs(studentized_deleted_residuals) > bonferroni_threshold

# Print results
results <- data.frame(
    Residuals                     = residuals,
    Hat_Values                    = hat_values,
    Studentized_Residuals         = studentized_residuals,
    Deleted_Residuals             = deleted_residuals,
    Studentized_Deleted_Residuals =
    ↪  studentized_deleted_residuals,
    Outlier                       = outliers
```

```
)

causalverse::nice_tab(head(results))
#> Warning: replacing previous import 'data.table::wday' by
#> 'lubridate::wday' when loading 'causalverse'
#> Warning: replacing previous import 'data.table::second' by
#> 'lubridate::second' when loading 'causalverse'
#> Warning: replacing previous import 'data.table::isoweek' by
#> 'lubridate::isoweek' when loading 'causalverse'
#> Warning: replacing previous import 'data.table::yday' by
#> 'lubridate::yday' when loading 'causalverse'
#> Warning: replacing previous import 'data.table::hour' by
#> 'lubridate::hour' when loading 'causalverse'
#> Warning: replacing previous import 'data.table::year' by
#> 'lubridate::year' when loading 'causalverse'
#> Warning: replacing previous import 'data.table::month' by
#> 'lubridate::month' when loading 'causalverse'
#> Warning: replacing previous import 'data.table::week' by
#> 'lubridate::week' when loading 'causalverse'
#> Warning: replacing previous import 'data.table::minute' by
#> 'lubridate::minute' when loading 'causalverse'
#> Warning: replacing previous import 'data.table::mday' by
#> 'lubridate::mday' when loading 'causalverse'
#> Warning: replacing previous import 'data.table::quarter' by
#> 'lubridate::quarter' when loading 'causalverse'
#> Warning: replacing previous import 'formattable::digits' by
#> 'xtable::digits' when loading 'causalverse'
#>   Residuals Hat_Values Studentized_Residuals
#> 1     -1.27       0.02                 -0.67
#> 2      0.70       0.01                  0.36
#> 3     -0.12       0.04                 -0.07
#> 4     -0.48       0.01                 -0.25
#> 5     -1.68       0.01                 -0.88
#> 6      0.30       0.04                  0.16
#>   Deleted_Residuals Studentized_Deleted_Residuals Outlier
#> 1             -1.29                         -0.66   FALSE
#> 2              0.70                          0.36   FALSE
#> 3             -0.13                         -0.06   FALSE
#> 4             -0.49                         -0.25   FALSE
#> 5             -1.70                         -0.87   FALSE
#> 6              0.31                          0.15   FALSE
```

```
# Plot studentized deleted residuals for visualization
plot(
    studentized_deleted_residuals,
    main = "Studentized Deleted Residuals",
    xlab = "Observation",
    ylab = "Studentized Deleted Residuals",
    pch = 16,
    col = ifelse(outliers, "red", "black")
)
abline(
    h = c(-bonferroni_threshold, bonferroni_threshold),
    col = "blue",
    lty = 2
)
legend(
    "topright",
    legend = c("Potential Outliers", "Threshold"),
    col = c("red", "blue"),
    pch = c(16, NA),
    lty = c(NA, 2)
)
```

2.1.2.5.5 Identifying Influential Cases

By **influential**, we refer to observations whose exclusion causes major changes in the fitted regression model. Note that not all outliers are influential.
Types of influence measures

- **Influence on single fitted values**: DFFITS
- **Influence on all fitted balues**: Cook's D
- **Influence on the regression coefficients**: DFBETAS

Measures like Cook's D, DFFITS, and DFBETAS combine leverage (from the hat matrix) and residual size (from studentized residuals) to assess the **influence** of an observation on the model as a whole. Hence, these effectively combine impact of X-space and Y-space.

2.1.2.5.6 DFFITS

DFFITS measures the **influence on single fitted values**. It is defined as:

$$
\begin{aligned}
(DFFITS)_i &= \frac{\hat{Y}_i - \hat{Y}_{i(i)}}{\sqrt{MSE_{(i)}h_{ii}}} \\
&= t_i \left(\frac{h_{ii}}{1-h_{ii}}\right)^{1/2}
\end{aligned}
$$

where:

- $\hat{Y}_i$ = fitted value for observation i using all data
- $\hat{Y}_{i(i)}$ = fitted value for observation i with the ith case removed
- $MSE_{(i)}$ = mean squared error with observation i excluded
- h_{ii} = leverage of the ith observation
- t_i = studentized deleted residual

High DFFITS values occur when leverage and residuals are jointly significant. DFFITS captures the **standardized difference between the fitted value for observation** i with and without the ith case in the model. It is a product of:

1. The **studentized deleted residual**
2. A scaling factor based on the leverage of the ith observation, h_{ii}

An observation is considered influential based on DFFITS if:

- **Small to medium data sets**: $|DFFITS| > 1$
- **Large data sets**: $|DFFITS| > 2\sqrt{p/n}$

where:

- p = number of predictors (including the intercept)
- n = total number of observations

This provides a practical threshold for detecting influential observations in different dataset sizes.

```
# Load necessary package
library(car)
#> Loading required package: carData

# Fit a linear model
model <- lm(mpg ~ hp + wt, data = mtcars)

# Compute DFFITS
dffits_values <- dffits(model)

# Display influential observations based on the threshold for
↪  a large dataset
threshold <- 2 * sqrt(length(coefficients(model)) /
↪  nrow(mtcars))
```

```
influential_obs <- which(abs(dffits_values) > threshold)

# Results
list(
    DFFITS                     = dffits_values,
    Threshold                  = threshold,
    Influential_Observations = influential_obs
)
#> $DFFITS
#>           Mazda RX4       Mazda RX4 Wag          Datsun 710
↪
#>        -0.218494101        -0.126664789        -0.249103400
↪
#>      Hornet 4 Drive   Hornet Sportabout             Valiant
↪
#>         0.011699160         0.028162679        -0.253806124
↪
#>          Duster 360           Merc 240D            Merc 230
↪
#>        -0.191618944         0.221917842         0.079763706
↪
#>            Merc 280           Merc 280C          Merc 450SE
↪
#>        -0.067222732        -0.190099538         0.064280875
↪
#>          Merc 450SL         Merc 450SLC  Cadillac Fleetwood
↪
#>         0.020560728        -0.135714533         0.008984366
↪
#> Lincoln Continental   Chrysler Imperial            Fiat 128
↪
#>         0.227919348         1.231668760         0.749153703
↪
#>         Honda Civic      Toyota Corolla       Toyota Corona
↪
#>         0.165329646         0.865985851        -0.292008465
↪
#>    Dodge Challenger         AMC Javelin          Camaro Z28
↪
#>        -0.253389811        -0.294709853        -0.170476763
↪
#>    Pontiac Firebird           Fiat X1-9       Porsche 914-2
↪
#>         0.207813200        -0.041423665        -0.004054382
↪
```

```
#>          Lotus Europa        Ford Pantera L          Ferrari Dino
↪
#>           0.471518032          -0.161026362          -0.129395315
↪
#>         Maserati Bora            Volvo 142E
#>           0.907521354          -0.128232538
#>
#> $Threshold
#> [1] 0.6123724
#>
#> $Influential_Observations
#> Chrysler Imperial             Fiat 128    Toyota Corolla
#>                17                   18                20
#>      Maserati Bora
#>                31
```

2.1.2.5.7 Cook's D

Cook's D measures the **influence of the** ith case on all fitted values in a regression model. It is defined as:

$$
\begin{aligned}
D_i &= \frac{\sum_{j=1}^{n}(\hat{Y}_j - \hat{Y}_{j(i)})^2}{p(MSE)} \\
&= \frac{e_i^2}{p(MSE)}\left(\frac{h_{ii}}{(1-h_{ii})^2}\right)
\end{aligned}
$$

where:

- $\hat{Y}_j$ = fitted value for observation j using all data
- $\hat{Y}_{j(i)}$ = fitted value for observation j with the ith case removed
- e_i = residual for observation i
- h_{ii} = leverage of the ith observation
- p = number of predictors (including the intercept)
- MSE = mean squared error of the model

Key Insights

- **Cook's D quantifies the overall influence** of the ith observation on the entire set of fitted values.
- If either the residual e_i increases or the leverage h_{ii} increases, then D_i also increases, indicating higher influence.
- Observations with both high leverage and large residuals are flagged as influential.

Threshold for Influence

- D_i can be interpreted as a percentile of an $F_{(p,n-p)}$ distribution.
- Practical thresholds:
  - **If** $D_i > 4/n$, the ith case has major influence.
  - Alternatively, cases where D_i exceeds the 50th percentile of the F-distribution may also be considered influential.

```
# Load necessary package
library(car)

# Fit a linear model
model <- lm(mpg ~ hp + wt, data = mtcars)

# Compute Cook's D
cooks_d_values <- cooks.distance(model)

# Display influential observations based on the threshold
threshold <- 4 / nrow(mtcars)
influential_obs <- which(cooks_d_values > threshold)

# Results
list(
  Cooks_D                 = cooks_d_values,
  Threshold               = threshold,
  Influential_Observations = influential_obs
)
#> $Cooks_D
#>           Mazda RX4       Mazda RX4 Wag          Datsun 710
↪
#>        1.589652e-02        5.464779e-03        2.070651e-02
↪
#>      Hornet 4 Drive   Hornet Sportabout             Valiant
↪
#>        4.724822e-05        2.736184e-04        2.155064e-02
↪
#>          Duster 360           Merc 240D            Merc 230
↪
#>        1.255218e-02        1.677650e-02        2.188702e-03
↪
#>            Merc 280           Merc 280C          Merc 450SE
↪
#>        1.554996e-03        1.215737e-02        1.423008e-03
↪
```

```
#>          Merc 450SL          Merc 450SLC  Cadillac Fleetwood
↪
#>        1.458960e-04        6.266049e-03        2.786686e-05
↪
#> Lincoln Continental   Chrysler Imperial            Fiat 128
↪
#>        1.780910e-02        4.236109e-01        1.574263e-01
↪
#>         Honda Civic      Toyota Corolla       Toyota Corona
↪
#>        9.371446e-03        2.083933e-01        2.791982e-02
↪
#>    Dodge Challenger         AMC Javelin          Camaro Z28
↪
#>        2.087419e-02        2.751510e-02        9.943527e-03
↪
#>    Pontiac Firebird           Fiat X1-9       Porsche 914-2
↪
#>        1.443199e-02        5.920440e-04        5.674986e-06
↪
#>        Lotus Europa      Ford Pantera L        Ferrari Dino
↪
#>        7.353985e-02        8.919701e-03        5.732672e-03
↪
#>       Maserati Bora          Volvo 142E
#>        2.720397e-01        5.600804e-03
#>
#> $Threshold
#> [1] 0.125
#>
#> $Influential_Observations
#> Chrysler Imperial            Fiat 128     Toyota Corolla
#>                17                  18                 20
#>     Maserati Bora
#>                31
```

2.1.2.5.8 *DFBETAS*

DFBETAS measures the **influence of the** ith observation on each regression coefficient in a regression model. It is defined as:

$$(DFBETAS)_{k(i)} = \frac{b_k - b_{k(i)}}{\sqrt{MSE_{(i)}c_{kk}}}$$

where:

- b_k = regression coefficient for the kth predictor using all observations
- $b_{k(i)}$ = regression coefficient for the kth predictor with the ith observation omitted
- $MSE_{(i)}$ = mean squared error with the ith observation excluded
- c_{kk} = kth diagonal element of $\mathbf{X}'\mathbf{X}^{-1}$, representing the variance of b_k

Key Insights

- **DFBETAS quantifies the change in each regression coefficient** (b_k) caused by omitting the ith observation.
- The sign of DFBETAS indicates whether the inclusion of an observation increases or decreases the regression coefficient.
  - Positive DFBETAS: Inclusion increases b_k.
  - Negative DFBETAS: Inclusion decreases b_k.
- High DFBETAS indicate that a single observation disproportionately affects one or more predictors.

The thresholds for identifying influential observations based on DFBETAS are:

- **Small data sets**: $|DFBETAS| > 1$
- **Large data sets**: $|DFBETAS| > 2/\sqrt{n}$

where n is the total number of observations.

```
# Load necessary package
library(car)

# Fit a linear model
model <- lm(mpg ~ hp + wt, data = mtcars)

# Compute DFBETAS
dfbetas_values <- dfbetas(model)

# Display influential observations based on the threshold for
↪  each predictor
threshold <- 2 / sqrt(nrow(mtcars))
influential_obs <-
    apply(dfbetas_values, 2, function(x)
        which(abs(x) > threshold))

# Results
list(
    DFBETAS                  = dfbetas_values,
    Threshold                = threshold,
    Influential_Observations = influential_obs
)
```

```
#> $DFBETAS
#>                      (Intercept)           hp            wt
#> Mazda RX4           -0.161347204  0.032966471  0.0639304305
#> Mazda RX4 Wag       -0.069324050  0.045785122 -0.0004066495
#> Datsun 710          -0.211199646  0.043374926  0.0972314374
#> Hornet 4 Drive       0.002672687 -0.006839301  0.0044886906
#> Hornet Sportabout    0.001784844  0.009208434 -0.0015536931
#> Valiant             -0.005985946  0.180374447 -0.1516565139
#> Duster 360           0.004705177 -0.159988770  0.0781031774
#> Merc 240D            0.034255292 -0.189552940  0.1224118752
#> Merc 230             0.019788247 -0.055075623  0.0332570461
#> Merc 280            -0.003198686  0.036709039 -0.0337297820
#> Merc 280C           -0.009045583  0.103809696 -0.0953846390
#> Merc 450SE          -0.026973686 -0.005712458  0.0356973740
#> Merc 450SL          -0.003961562  0.003399822  0.0049302300
#> Merc 450SLC          0.031572445 -0.016800308 -0.0400515832
#> Cadillac Fleetwood  -0.006420656 -0.002577897  0.0075499557
#> Lincoln Continental -0.168791258 -0.058242601  0.1903129995
#> Chrysler Imperial   -0.924056752 -0.148009806  0.9355996760
#> Fiat 128             0.605181396 -0.311246566 -0.1672758566
#> Honda Civic          0.156388333 -0.034026915 -0.0819144214
#> Toyota Corolla       0.804669969 -0.170934240 -0.4114605894
#> Toyota Corona       -0.231328587  0.066064464  0.0882138248
#> Dodge Challenger     0.003923967  0.049775308 -0.0888481611
#> AMC Javelin         -0.019610048  0.037837437 -0.0734203131
#> Camaro Z28           0.029920076 -0.128670440  0.0390740055
#> Pontiac Firebird    -0.058806962 -0.002278294  0.0868742949
#> Fiat X1-9           -0.037559007  0.010208853  0.0174261386
#> Porsche 914-2       -0.003655931  0.000316321  0.0020588013
#> Lotus Europa         0.423409344  0.188396749 -0.4072338373
#> Ford Pantera L      -0.022536462 -0.148176049  0.0999346699
#> Ferrari Dino        -0.065508308 -0.085182962  0.0869804902
#> Maserati Bora       -0.007482815  0.865763737 -0.4999048760
#> Volvo 142E          -0.080001907  0.038406565  0.0127537553
#>
#> $Threshold
#> [1] 0.3535534
#>
#> $Influential_Observations
#> $Influential_Observations$`(Intercept)`
#> Chrysler Imperial          Fiat 128    Toyota Corolla
#>                17                18                20
#>      Lotus Europa
#>                28
```

```
#>
#> $Influential_Observations$hp
#> Maserati Bora
#>            31
#>
#> $Influential_Observations$wt
#> Chrysler Imperial     Toyota Corolla       Lotus Europa
#>                17                 20                 28
#>     Maserati Bora
#>                31
```

2.1.2.5.9 Collinearity

Collinearity (or multicollinearity) refers to the correlation among explanatory variables in a regression model. It can lead to various issues, including:

- Large changes in the estimated regression coefficients when a predictor variable is added or removed, or when observations are altered
- Nonsignificant results for individual tests on regression coefficients of important predictor variables
- Regression coefficients with signs opposite to theoretical expectations or prior experience
- Large coefficients of simple correlation between pairs of predictor variables in the correlation matrix
- Wide confidence intervals for the regression coefficients representing important predictor variables

When some X variables are highly correlated, the inverse $(X'X)^{-1}$ either does not exist or is computationally unstable. This can result in:

1. **Non-interpretability of parameters**:

$$\mathbf{b} = (\mathbf{X}'\mathbf{X})^{-1}\mathbf{X}'\mathbf{y}$$

2. **Infinite sampling variability**:

$$\mathbf{s}^2(\mathbf{b}) = \mathbf{MSE}(\mathbf{X}'\mathbf{X})^{-1}$$

If some predictor variables (X) are “perfectly” correlated, the system becomes undetermined, leading to an infinite number of models that fit the data. Specifically:

- If $X'X$ is singular, then $(X'X)^{-1}$ does not exist.
- This results in poor parameter estimation and invalid statistical inference.

2.1.2.5.10 *Variance Inflation Factors (VIFs)*

The **variance inflation factor (VIF)** quantifies how much the variance of a regression coefficient is inflated due to multicollinearity. It is defined as:

$$VIF_k = \frac{1}{1 - R_k^2}$$

where:

- R_k^2 is the coefficient of multiple determination when X_k is regressed on the other $p - 2$ predictor variables in the model.

Interpretation of VIFs

- Large VIF_k values indicate that near collinearity is inflating the variance of b_k. The relationship is given by:

$$var(b_k) \propto \sigma^2(VIF_k).$$

- **Thresholds**:
  - $VIF > 4$: Investigate the cause of multicollinearity.
  - $VIF_k > 10$: Serious multicollinearity problem that can lead to poor parameter estimates.
- The **mean VIF** provides an estimate of the degree of multicollinearity:
  - If $avg(VIF) >> 1$, serious multicollinearity is present.
- **Multicollinearity and VIF:**
  - High VIFs with indicator variables are normal and not problematic.
  - VIF is generally not useful for detecting multicollinearity concerns in models with fixed effects.
- **Overemphasis on multicollinearity:**
  - Multicollinearity inflates standard errors and widens confidence intervals but does not bias results.
  - If key variables have narrow confidence intervals, multicollinearity is not an issue.
- **Goldberger's insight** (Goldberger, 1991)**:**
  - Multicollinearity is akin to small sample size ("micronumerosity").
  - Large standard errors are expected with highly correlated independent variables.
- **Practical implications:**
  - Evaluate whether confidence intervals for key variables are sufficiently narrow.

- If not, the study is inconclusive, and a larger dataset or redesigned study is needed.

```
# Load necessary packages
library(car)

# Fit a regression model
model <- lm(mpg ~ hp + wt + disp, data = mtcars)

# Compute Variance Inflation Factors
vif_values <- vif(model)

# Check for high multicollinearity
threshold <- 10
high_vif <- which(vif_values > threshold)

# Results
list(
  VIFs = vif_values,
  High_VIF_Threshold = threshold,
  High_VIF_Indices = high_vif
)
#> $VIFs
#>       hp       wt     disp
#> 2.736633 4.844618 7.324517
#>
#> $High_VIF_Threshold
#> [1] 10
#>
#> $High_VIF_Indices
#> named integer(0)
```

2.1.2.5.11 Condition Number

Condition number is a diagnostic measure for detecting multicollinearity, derived from the spectral decomposition of the matrix $\mathbf{X'X}$.

The spectral decomposition of $\mathbf{X'X}$ is:

$$\mathbf{X'X} = \sum_{i=1}^{p} \lambda_i \mathbf{u_i u_i'}$$

where:

- λ_i: eigenvalue associated with the ith eigenvector
- $\mathbf{u}_i$: eigenvector associated with λ_i
- $\lambda_1 > \lambda_2 > \cdots > \lambda_p$ (ordered eigenvalues)
- The eigenvectors are orthogonal:

$$\begin{cases} \mathbf{u_i'u_j} = 0 & \text{for } i \neq j \\ \mathbf{u_i'u_j} = 1 & \text{for } i = j \end{cases}$$

Definition of the Condition Number

The **condition number** is defined as:

$$k = \sqrt{\frac{\lambda_{\text{max}}}{\lambda_{\text{min}}}}$$

where:

- λ_{max}: largest eigenvalue
- λ_{min}: smallest eigenvalue

Interpretation

- $k > 30$: cause for concern
- $30 < k < 100$: moderate dependencies among predictors
- $k > 100$: strong collinearity, indicating serious multicollinearity

The **condition index** for the ith eigenvalue is defined as:

$$\delta_i = \sqrt{\frac{\lambda_{\text{max}}}{\lambda_i}}$$

where $i = 1, \ldots, p$.

Variance proportions
can be used to identify collinearity issues. The proportion of total variance associated with the kth regression coefficient and the ith eigen mode is given by:

$$\frac{u_{ik}^2/\lambda_i}{\sum_j \left(u_{jk}^2/\lambda_j\right)}.$$

Key Indicators:
- A **large condition index** δ_i suggests potential collinearity.
- Variance proportions > 0.5 for at least two regression coefficients indicate serious collinearity problems.

```
# Load necessary package
library(car)

# Fit a regression model
model <- lm(mpg ~ hp + wt + disp, data = mtcars)

# Compute eigenvalues and eigenvectors of the correlation
↪  matrix of predictors
cor_matrix <- cor(mtcars[, c("hp", "wt", "disp")])
eigen_decomp <- eigen(cor_matrix)

# Extract eigenvalues and compute the Condition Number
eigenvalues <- eigen_decomp$values
condition_number <- sqrt(max(eigenvalues) / min(eigenvalues))

# Compute Condition Indices
condition_indices <- sqrt(max(eigenvalues) / eigenvalues)

# Results
list(
  Condition_Number = condition_number,
  Condition_Indices = condition_indices
)
#> $Condition_Number
#> [1] 5.469549
#>
#> $Condition_Indices
#> [1] 1.000000 2.697266 5.469549
```

1. **Condition number**: a single value indicating the degree of multicollinearity
2. **Condition indices**: a vector showing the relative dependency associated with each eigenvalue

2.1.2.5.12 Constancy of Error Variance

Testing for the constancy of error variance (homoskedasticity) ensures that the assumptions of linear regression are met.

Below are two commonly used tests to assess error variance.

2.1.2.5.13 Brown–Forsythe Test (Modified Levene Test)

The **Brown–Forsythe test** does not depend on the normality of errors and is suitable when error variance increases or decreases with X.

Procedure

1. Split the residuals into two groups:

$$e_{i1}, i = 1, \ldots, n_1 \quad \text{and} \quad e_{j2}, j = 1, \ldots, n_2.$$

2. Compute absolute deviations from the group medians:

$$d_{i1} = |e_{i1} - \tilde{e}_1| \quad \text{and} \quad d_{j2} = |e_{j2} - \tilde{e}_2|$$

   where $\tilde{e}_1$ and $\tilde{e}_2$ are the medians of groups 1 and 2, respectively.
3. Perform a two-sample t-test on d_{i1} and d_{j2}:

$$t_L = \frac{\bar{d}_1 - \bar{d}_2}{s\sqrt{\frac{1}{n_1} + \frac{1}{n_2}}}$$

   where

$$s^2 = \frac{\sum_i (d_{i1} - \bar{d}_1)^2 + \sum_j (d_{j2} - \bar{d}_2)^2}{n - 2}.$$

4. Reject the null hypothesis of constant error variance if:

$$|t_L| > t_{1-\alpha/2;n-2}.$$

```
# Load necessary package
library(car)

# Fit a linear model
model <- lm(mpg ~ hp + wt, data = mtcars)

# Perform the Brown-Forsythe Test
levene_test <- leveneTest(model$residuals ~ cut(mtcars$hp, 2))
↪  # Split HP into 2 groups

# Results
levene_test
#> Levene's Test for Homogeneity of Variance (center = median)
#>       Df F value Pr(>F)
#> group  1  0.0851 0.7724
#>       30
```

The p-value determines whether to reject the null hypothesis of constant variance.

2.1.2.5.14 Breusch–Pagan Test (Cook–Weisberg Test)

The **Breusch–Pagan Test** assumes independent and normally distributed errors. It tests the hypothesis:

- $H_0 : \gamma_1 = 0$ (constant variance).
- $H_1 : \gamma_1 \neq 0$ (nonconstant variance).

Procedure

1. Assume the variance of the error terms is modeled as:

$$\sigma_i^2 = \gamma_0 + \gamma_1 X_i.$$

2. Regress the squared residuals (e_i^2) on X_i:
   - Obtain the **regression sum of squares** (SSR^*).
3. Compute the Breusch–Pagan statistic:

$$X_{BP}^2 = \frac{SSR^*/2}{\left(\frac{SSE}{n}\right)^2}$$

   where SSE is the error sum of squares from the regression of Y on X.
4. Compare X_{BP}^2 to the critical value from the χ^2 distribution with 1 degree of freedom:
   - Reject H_0 (homogeneous variance) if:

$$X_{BP}^2 > \chi_{1-\alpha;1}^2.$$

```
# Load necessary package
library(lmtest)
#> Loading required package: zoo
#>
#> Attaching package: 'zoo'
#> The following objects are masked from 'package:base':
#>
#>     as.Date, as.Date.numeric

# Perform the Breusch-Pagan Test
bp_test <- bptest(model)

# Results
bp_test
#>
```

```
#>  studentized Breusch-Pagan test
#>
#> data:  model
#> BP = 0.88072, df = 2, p-value = 0.6438
```

If the p-value is below the chosen significance level (e.g., 0.05), reject H_0 and conclude nonconstant variance.

2.1.2.5.15 Independence

Testing for independence ensures that the residuals of a regression model are uncorrelated. Violation of this assumption may lead to biased or inefficient parameter estimates. Below, we discuss three primary methods for diagnosing dependence: plots, the Durbin–Watson test, and specific methods for time-series and spatial data.

2.1.2.5.16 Plots

A residual plot can help detect dependence in the residuals (Fig. 2.8):

1. Plot residuals (e_i) versus the predicted values ($\hat{Y}_i$).
2. Patterns such as systematic waves, trends, or clusters indicate possible dependence.
3. Independence is suggested if residuals are randomly scattered without clear patterns.

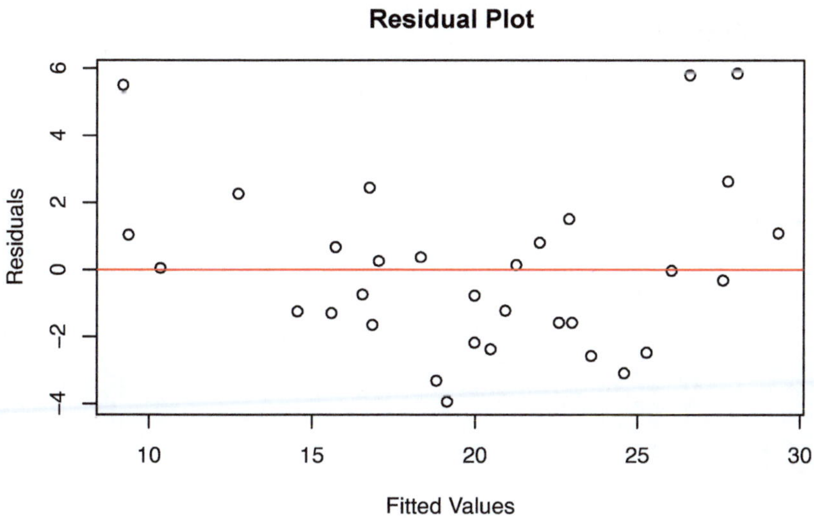

Fig. 2.8 Residual plot for linear model diagnostics

```r
# Fit a regression model
model <- lm(mpg ~ hp + wt, data = mtcars)

# Residual plot
plot(model$fitted.values, model$residuals,
     xlab = "Fitted Values",
     ylab = "Residuals",
     main = "Residual Plot")
abline(h = 0, col = "red")
```

2.1.2.5.17 Durbin–Watson Test

The **Durbin–Watson test** specifically detects autocorrelation in residuals, especially in time-series data.

- **Hypotheses**:
  - H_0: Residuals are uncorrelated.
  - H_1: Residuals are autocorrelated.
- The Durbin–Watson statistic (d) is calculated as: $d = \frac{\sum_{t=2}^{n}(e_t - e_{t-1})^2}{\sum_{t=1}^{n} e_t^2}$.
  - $d \approx 2$: no autocorrelation
  - $d < 2$: positive autocorrelation
  - $d > 2$: negative autocorrelation

```r
# Load necessary package
library(lmtest)

# Perform the Durbin-Watson test
dw_test <- dwtest(model)

# Results
dw_test
#> 
#>  Durbin-Watson test
#> 
#> data:  model
#> DW = 1.3624, p-value = 0.02061
#> alternative hypothesis: true autocorrelation is greater
↪  than 0
```

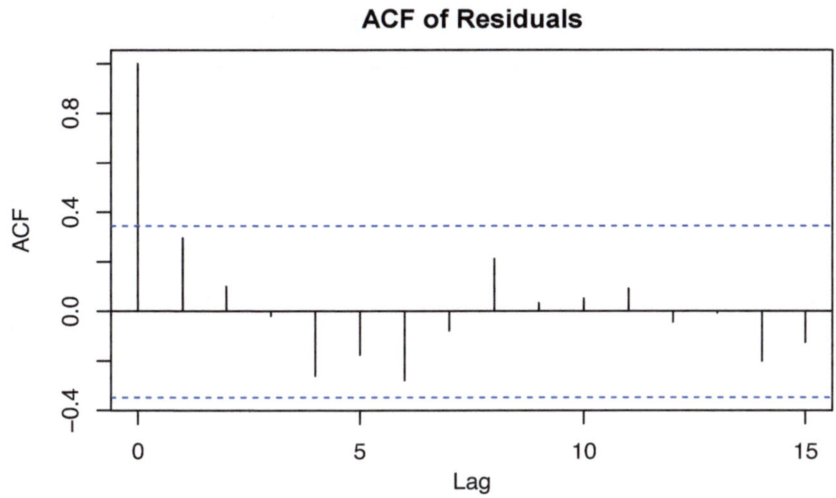

Fig. 2.9 ACF of residuals

2.1.2.5.18 Time-Series Autocorrelation

For time-series data, autocorrelation often occurs due to the temporal structure. Key diagnostics include:

1. **Autocorrelation function (ACF)** (Fig. 2.9):
   - Shows the correlation of residuals with their lagged values.
   - Significant spikes in the ACF plot indicate autocorrelation.
2. **Partial autocorrelation function (PACF)** (Fig. 2.10):
   - Identifies the correlation of residuals after removing the influence of intermediate lags.

```
# Load necessary package
library(forecast)
#> Registered S3 method overwritten by 'quantmod':
#>   method            from
#>   as.zoo.data.frame zoo

# Residuals from a time-series model
time_series_res <- ts(model$residuals)

# Plot ACF
acf(time_series_res, main = "ACF of Residuals")
```

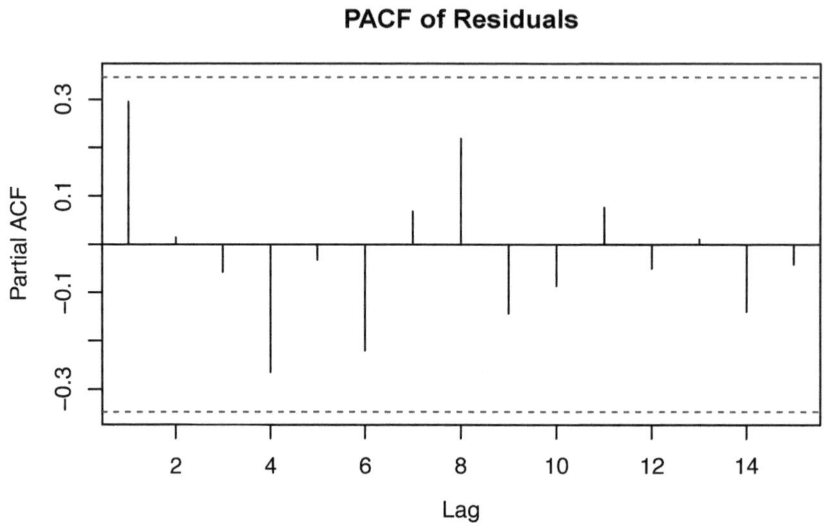

Fig. 2.10 PACF of residuals

```
# Plot PACF
pacf(time_series_res, main = "PACF of Residuals")
```

2.1.2.5.19 Spatial Statistics

Spatial dependence occurs when residuals are correlated across geographical or spatial locations. Two primary tests are used to diagnose spatial autocorrelation:

Moran's I
measures global spatial autocorrelation. It determines whether similar values cluster spatially. The statistic is defined as:

$$I = \frac{n}{W} \cdot \frac{\sum_i \sum_j w_{ij}(x_i - \bar{x})(x_j - \bar{x})}{\sum_i (x_i - \bar{x})^2}$$

where:

- n: number of observations
- W: sum of all spatial weights w_{ij}
- w_{ij}: spatial weight between locations i and j
- x_i: residual value at location i
- $\bar{x}$: mean of the residuals

Spatial weight matrix (W): W defines the spatial relationship between observations. It is often derived from:

- **Distance-based methods**: e.g., k-nearest neighbors or distance bands
- **Adjacency methods**: Based on shared boundaries in geographic data

Interpretation:

- $I > 0$: positive spatial autocorrelation (similar values cluster together)
- $I < 0$: negative spatial autocorrelation (dissimilar values are neighbors)
- $I \approx 0$: random spatial distribution

```
# Load necessary packages
library(sf)
#> Linking to GEOS 3.13.0, GDAL 3.10.1, PROJ 9.5.1;
#> sf_use_s2() is TRUE
library(spdep)
#> Loading required package: spData
#> To access larger datasets in this package, install
#> the spDataLarge package with:
#> `install.packages('spDataLarge',
#> repos='https://nowosad.github.io/drat/',
#> type='source')`

# Simulate spatial data (example with mtcars dataset)
coords <- cbind(mtcars$hp, mtcars$wt)  # Coordinates based on
↪  two variables

# Add small jitter to avoid duplicate coordinates
set.seed(123)  # For reproducibility
coords_jittered <-
    coords + matrix(runif(length(coords),-0.01, 0.01), ncol =
↪  2)

# Find nearest neighbors
neighbors <- knn2nb(knearneigh(coords_jittered, k = 3))
#> Warning in knn2nb(knearneigh(coords_jittered, k = 3)):
#> neighbour object has 4 sub-graphs

# Create spatial weights matrix
weights <- nb2listw(neighbors, style = "W")  #
↪  Row-standardized weights

# Fit the linear model
model <- lm(mpg ~ hp + wt, data = mtcars)
```

```r
# Check lengths of residuals and weights
length(model$residuals) # Should be 32
#> [1] 32
length(weights$neighbours) # Should also be 32
#> [1] 32

# Compute Moran's I for residuals
moran_test <- moran.test(model$residuals, weights)
print(moran_test)
#>
#>  Moran I test under randomisation
#>
#> data:  model$residuals
#> weights: weights
#>
#> Moran I statistic standard deviate = 1.7679, p-value
#> = 0.03854
#> alternative hypothesis: greater
#> sample estimates:
#> Moran I statistic       Expectation          Variance
#>        0.18544790       -0.03225806        0.01516371
```

```r
# Moran's scatterplot
moran.plot(model$residuals, weights, main = "Moran's
↪ Scatterplot")
```

Significant Moran's I: indicates global clustering of similar residuals, suggesting spatial dependence (Fig. 2.11).

Geary's C

measures spatial autocorrelation at a local level, emphasizing differences between neighboring observations. The statistic is defined as:

$$C = \frac{n-1}{2W} \cdot \frac{\sum_i \sum_j w_{ij}(x_i - x_j)^2}{\sum_i (x_i - \bar{x})^2}$$

where:

- C ranges from 0 to 2:
  - $C \approx 0$: high positive spatial autocorrelation
  - $C \approx 1$: spatial randomness
  - $C \approx 2$: high negative spatial autocorrelation

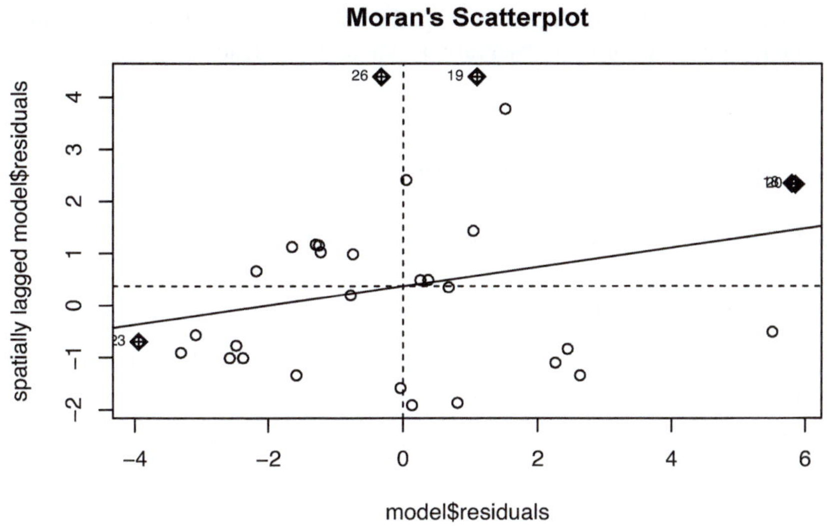

Fig. 2.11 Moran's scatterplot

Comparison of Moran's I and Geary's C:

- Moran's I is more global and measures the overall pattern of autocorrelation.
- Geary's C is more sensitive to local spatial autocorrelation, detecting small-scale patterns.

```
# Compute Geary's C for residuals
geary_test <- geary.test(model$residuals, weights)

# Results
geary_test
#>
#>  Geary C test under randomisation
#>
#> data:  model$residuals
#> weights: weights
#>
#> Geary C statistic standard deviate = 1.0606, p-value
#> = 0.1444
#> alternative hypothesis: Expectation greater than statistic
#> sample estimates:
#> Geary C statistic       Expectation          Variance
#>        0.84515868        1.00000000        0.02131399
```

Significant Geary's C: highlights local spatial autocorrelation, useful for identifying specific regions or groups of observations where dependence is strong.

2.2 Generalized Least Squares

2.2.1 *Infeasible Generalized Least Squares*

Motivation for a more efficient estimator

- The Gauss–Markov theorem guarantees that OLS is the **best linear unbiased estimator (BLUE)** under assumptions A1–A4:
  - A4: $Var(\epsilon|\mathbf{X}) = \sigma^2\mathbf{I}_n$ (homoskedasticity and no autocorrelation).
- When A4 does not hold:
  - **Heteroskedasticity:** $Var(\epsilon_i|\mathbf{X}) \neq \sigma^2$.
  - **Serial correlation:** $Cov(\epsilon_i, \epsilon_j|\mathbf{X}) \neq 0$ for $(i \neq j)$.

Without A4, OLS is **unbiased** but no longer **efficient**. This motivates the need for an alternative approach to identify the most efficient estimator.

The unweighted (standard) regression model is given by:

$$\mathbf{y} = \mathbf{X}\boldsymbol{\beta} + \boldsymbol{\epsilon}.$$

Assuming **A1–A3** hold (linearity, full rank, exogeneity), but A4 does not, the variance of the error term is no longer proportional to an identity:

$$Var(\boldsymbol{\epsilon}|\mathbf{X}) = \boldsymbol{\Omega} \neq \sigma^2\mathbf{I}_n.$$

To address the violation of A4 ($\boldsymbol{\Omega} \neq \sigma^2\mathbf{I}_n$), one can transform the model by premultiplying both sides by a full-rank matrix $\mathbf{w}$ to have a weighted (transformed) regression model:

$$\mathbf{wy} = \mathbf{wX}\boldsymbol{\beta} + \mathbf{w}\boldsymbol{\epsilon},$$

where $\mathbf{w}$ is a full-rank matrix chosen such that:

$$\mathbf{w}'\mathbf{w} = \boldsymbol{\Omega}^{-1}.$$

- $\mathbf{w}$ is the [Cholesky decomposition] of $\boldsymbol{\Omega}^{-1}$.
- The Cholesky decomposition ensures $\mathbf{w}$ satisfies $\mathbf{w}'\mathbf{w} = \boldsymbol{\Omega}^{-1}$, where $\mathbf{w}$ is the "square root" of $\boldsymbol{\Omega}^{-1}$ in matrix terms.

By transforming the original model, the variance of the transformed errors becomes:

$$\mathbf{\Omega} = Var(\boldsymbol{\epsilon}|\mathbf{X}),$$
$$\mathbf{\Omega}^{-1} = Var(\boldsymbol{\epsilon}|\mathbf{X})^{-1}.$$

The transformed equation allows us to compute a more efficient estimator.

Using the transformed model, the **infeasible generalized least squares (IGLS)** estimator is:

$$\begin{aligned}\boldsymbol{\beta}_{\hat{IGLS}} &= (\mathbf{X}'\mathbf{w}'\mathbf{w}\mathbf{X})^{-1}\mathbf{X}'\mathbf{w}'\mathbf{w}\mathbf{y} \\ &= (\mathbf{X}'\boldsymbol{\Omega}^{-1}\mathbf{X})^{-1}\mathbf{X}'\boldsymbol{\Omega}^{-1}\mathbf{y} \\ &= \boldsymbol{\beta} + (\mathbf{X}'\boldsymbol{\Omega}^{-1}\mathbf{X})^{-1}\mathbf{X}'\boldsymbol{\Omega}^{-1}\boldsymbol{\epsilon}.\end{aligned}$$

1. **Unbiasedness**

   Since assumptions **A1–A3** hold for the unweighted model:

$$\begin{aligned}\mathbf{E}(\hat{\boldsymbol{\beta}}_{\mathbf{IGLS}}|\mathbf{X}) &= \mathbf{E}(\boldsymbol{\beta} + (\mathbf{X}'\boldsymbol{\Omega}^{-1}\mathbf{X}'\boldsymbol{\Omega}^{-1}\boldsymbol{\epsilon})|\mathbf{X}) \\ &= \boldsymbol{\beta} + \mathbf{E}(\mathbf{X}'\boldsymbol{\Omega}^{-1}\mathbf{X}'\boldsymbol{\Omega}^{-1}\boldsymbol{\epsilon}|\mathbf{X}) \\ &= \boldsymbol{\beta} + \mathbf{X}'\boldsymbol{\Omega}^{-1}\mathbf{X}'\boldsymbol{\Omega}^{-1}\mathbf{E}(\boldsymbol{\epsilon}|\mathbf{X}) \qquad \text{since A3: } E(\epsilon|\mathbf{X}) = 0, \\ &= \boldsymbol{\beta}.\end{aligned}$$

Thus, the **IGLS estimator is unbiased**.

2. **Variance**

The variance of the transformed errors is given by:

$$\begin{aligned}\mathbf{Var}(\mathbf{w}\boldsymbol{\epsilon}|\mathbf{X}) &= \mathbf{w}\mathbf{Var}(\boldsymbol{\epsilon}|\mathbf{X})\mathbf{w}' \\ &= \mathbf{w}\boldsymbol{\Omega}\mathbf{w}' \\ &= \mathbf{w}(\mathbf{w}'\mathbf{w})^{-1}\mathbf{w}' \qquad \text{since } \mathbf{w} \text{ is full-rank,} \\ &= \mathbf{w}\mathbf{w}^{-1}(\mathbf{w}')^{-1}\mathbf{w}' \\ &= \mathbf{I_n}.\end{aligned}$$

Hence, A4 holds for the transformed (weighted) equation, satisfying the Gauss–Markov conditions.

The variance of the IGLS estimator is:

$$\begin{aligned}
\mathbf{Var}(\hat{\boldsymbol{\beta}}_{\mathbf{IGLS}}|\mathbf{X}) &= \mathbf{Var}(\boldsymbol{\beta} + (\mathbf{X}'\boldsymbol{\Omega}^{-1}\mathbf{X})^{-1}\mathbf{X}'\boldsymbol{\Omega}^{-1}\boldsymbol{\epsilon}|\mathbf{X}) \\
&= \mathbf{Var}((\mathbf{X}'\boldsymbol{\Omega}^{-1}\mathbf{X})^{-1}\mathbf{X}'\boldsymbol{\Omega}^{-1}\boldsymbol{\epsilon}|\mathbf{X}) \\
&= (\mathbf{X}'\boldsymbol{\Omega}^{-1}\mathbf{X})^{-1}\mathbf{X}'\boldsymbol{\Omega}^{-1}\mathbf{Var}(\boldsymbol{\epsilon}|\mathbf{X})\boldsymbol{\Omega}^{-1}\mathbf{X}(\mathbf{X}'\Omega^{-1}\mathbf{X})^{-1} \\
&\qquad \text{because A4 holds,} \\
&= (\mathbf{X}'\boldsymbol{\Omega}^{-1}\mathbf{X})^{-1}\mathbf{X}'\boldsymbol{\Omega}^{-1}\boldsymbol{\Omega}\boldsymbol{\Omega}^{-1}\boldsymbol{\Omega}^{-1}\mathbf{X}(\mathbf{X}'\boldsymbol{\Omega}^{-1}\mathbf{X})^{-1}, \\
&= (\mathbf{X}'\boldsymbol{\Omega}^{-1}\mathbf{X})^{-1}.
\end{aligned}$$

3. **Efficiency**

The difference in variances between OLS and IGLS is:

$$\mathbf{Var}(\hat{\beta}_{\mathbf{OLS}}|\mathbf{X}) - \mathbf{Var}(\hat{\boldsymbol{\beta}}_{\mathbf{IGLS}}|\mathbf{X}) = \mathbf{A}\Omega\mathbf{A}',$$

where:

$$\mathbf{A} = (\mathbf{X}'\mathbf{X})^{-1}\mathbf{X}' - (\mathbf{X}'\boldsymbol{\Omega}^{-1}\mathbf{X})^{-1}\mathbf{X}'\boldsymbol{\Omega}^{-1}.$$

Since $\boldsymbol{\Omega}$ is positive semi-definite, $\mathbf{A}\boldsymbol{\Omega}\mathbf{A}'$ is also positive semi-definite. Thus, the IGLS estimator is more efficient than OLS under heteroskedasticity or autocorrelation.

In short, properties of $\hat{\boldsymbol{\beta}}_{\mathbf{IGLS}}$:

1. **Unbiasedness**: $\hat{\boldsymbol{\beta}}_{\mathbf{IGLS}}$ remains unbiased as long as **A1–A3** hold.
2. **Efficiency**: $\hat{\boldsymbol{\beta}}_{\mathbf{IGLS}}$ is more efficient than OLS under heteroskedasticity or serial correlation since it accounts for the structure of $\boldsymbol{\Omega}$.

Why Is IGLS "Infeasible?"

The name **infeasible** arises because it is generally impossible to compute the estimator directly due to the structure of $\mathbf{w}$ (or equivalently $\boldsymbol{\Omega}^{-1}$). The matrix $\mathbf{w}$ is defined as:

$$\mathbf{w} = \begin{pmatrix} w_{11} & 0 & 0 & \cdots & 0 \\ w_{21} & w_{22} & 0 & \cdots & 0 \\ w_{31} & w_{32} & w_{33} & \cdots & \cdots \\ w_{n1} & w_{n2} & w_{n3} & \cdots & w_{nn} \end{pmatrix},$$

with $n(n+1)/2$ unique elements for n observations. This results in more parameters than data points, making direct estimation infeasible.

To make the estimation feasible, assumptions about the structure of $\boldsymbol{\Omega}$ are required. Common approaches include:

- Heteroskedasticity errors: Assume a multiplicative exponential model for the variance, such as $Var(\epsilon_i|\mathbf{X}) = \sigma_i^2$.
  - Assume no correlation between errors, but allow heterogeneous variances:

$$\mathbf{\Omega} = \begin{pmatrix} \sigma_1^2 & 0 & \cdots & 0 \\ 0 & \sigma_2^2 & \cdots & 0 \\ \vdots & \vdots & \ddots & \vdots \\ 0 & 0 & \cdots & \sigma_n^2 \end{pmatrix}.$$

  - Estimate σ_i^2 using methods such as:

    Modeling σ_i^2 as a function of predictors (e.g., $\sigma_i^2 = \exp(\mathbf{x}_i \gamma)$)

- Serial correlation: Assume serial correlation follows an autoregressive process AR(1) model, e.g., $\epsilon_t = \rho\epsilon_{t-1} + u_t$ and $Cov(\epsilon_t, \epsilon_{t-h}) = \rho^h \sigma^2$, where we have a variance–covariance matrix with off-diagonal elements decaying geometrically:

$$\mathbf{\Omega} = \frac{\sigma^2}{1-\rho^2} \begin{pmatrix} 1 & \rho & \rho^2 & \cdots & \rho^{n-1} \\ \rho & 1 & \rho & \cdots & \rho^{n-2} \\ \rho^2 & \rho & 1 & \cdots & \rho^{n-3} \\ \vdots & \vdots & \vdots & \ddots & \vdots \\ \rho^{n-1} & \rho^{n-2} & \rho^{n-3} & \cdots & 1 \end{pmatrix}.$$

- Cluster errors: Assume a block-diagonal structure for $\mathbf{\Omega}$ to account for grouped or panel data.

Each assumption simplifies the estimation of $\mathbf{\Omega}$ and thus $\mathbf{w}$, enabling feasible generalized least squares with fewer unknown parameters to estimate.

2.2.2 *Feasible Generalized Least Squares*

2.2.2.1 Heteroskedasticity Errors

Heteroskedasticity occurs when the variance of the error term is not constant across observations. Specifically:

$$Var(\epsilon_i|x_i) = E(\epsilon_i^2|x_i) \neq \sigma^2,$$

but instead depends on a function of x_i:

$$Var(\epsilon_i|x_i) = h(x_i) = \sigma_i^2.$$

This violates the assumption of homoskedasticity (constant variance), impacting the efficiency of OLS estimates.

For the model:

$$y_i = x_i\beta + \epsilon_i,$$

we apply a transformation to standardize the variance:

$$\frac{y_i}{\sigma_i} = \frac{x_i}{\sigma_i}\beta + \frac{\epsilon_i}{\sigma_i}.$$

By scaling each observation with $1/\sigma_i$, the variance of the transformed error term becomes:

$$\begin{aligned} Var\left(\frac{\epsilon_i}{\sigma_i}\middle| X\right) &= \frac{1}{\sigma_i^2}Var(\epsilon_i|X) \\ &= \frac{1}{\sigma_i^2}\sigma_i^2 \\ &= 1. \end{aligned}$$

Thus, the heteroskedasticity is corrected in the transformed model.

In matrix notation, the transformed model is:

$$\mathbf{wy} = \mathbf{wX}\boldsymbol{\beta} + \mathbf{w}\boldsymbol{\epsilon},$$

where $\mathbf{w}$ is the weight matrix used to standardize the variance. The weight matrix $\mathbf{w}$ is defined as:

$$\mathbf{w} = \begin{pmatrix} 1/\sigma_1 & 0 & 0 & \cdots & 0 \\ 0 & 1/\sigma_2 & 0 & \cdots & 0 \\ 0 & 0 & 1/\sigma_3 & \cdots & 0 \\ \vdots & \vdots & \vdots & \ddots & \vdots \\ 0 & 0 & 0 & \cdots & 1/\sigma_n \end{pmatrix}.$$

In the presence of heteroskedasticity, the variance of the error term, $Var(\epsilon_i|\mathbf{x}_i)$, is not constant across observations. This leads to inefficient OLS estimates.

Infeasible weighted least squares (IWLS) assumes that the variances $\sigma_i^2 = Var(\epsilon_i|\mathbf{x}_i)$ are **known**. This allows us to adjust the regression equation to correct for heteroskedasticity.

The model is transformed as follows:

$$y_i = \mathbf{x}_i\beta + \epsilon_i \quad \text{(original equation)},$$

where ϵ_i has variance σ_i^2. To make the errors homoskedastic, we divide through by σ_i:

$$\frac{y_i}{\sigma_i} = \frac{\mathbf{x}_i}{\sigma_i}\beta + \frac{\epsilon_i}{\sigma_i}.$$

Now, the transformed error term ϵ_i/σ_i has a constant variance of 1:

$$Var\left(\frac{\epsilon_i}{\sigma_i}|\mathbf{x}_i\right) = 1.$$

The IWLS estimator minimizes the **weighted sum of squared residuals** for the transformed model:

$$\text{Minimize: } \sum_{i=1}^{n}\left(\frac{y_i - \mathbf{x}_i\beta}{\sigma_i}\right)^2.$$

In matrix form, the IWLS estimator is:

$$\hat{\beta}_{IWLS} = (\mathbf{X}'\mathbf{W}\mathbf{X})^{-1}\mathbf{X}'\mathbf{W}\mathbf{y},$$

where $\mathbf{W}$ is a diagonal matrix of weights:

$$\mathbf{W} = \begin{pmatrix} 1/\sigma_1^2 & 0 & \cdots & 0 \\ 0 & 1/\sigma_2^2 & \cdots & 0 \\ \vdots & \vdots & \ddots & \vdots \\ 0 & 0 & \cdots & 1/\sigma_n^2 \end{pmatrix}.$$

Properties of IWLS

1. **Valid standard errors**:
   - If $Var(\epsilon_i|\mathbf{X}) = \sigma_i^2$, the usual standard errors from IWLS are valid.
2. **Robustness**:
   - If the variance assumption is incorrect ($Var(\epsilon_i|\mathbf{X}) \neq \sigma_i^2$), heteroskedasticity-robust standard errors must be used instead.

The primary issue with IWLS is that $\sigma_i^2 = Var(\epsilon_i|\mathbf{x}_i)$ is generally unknown. Specifically, we do not know:

$$\sigma_i^2 = Var(\epsilon_i|\mathbf{x}_i) = E(\epsilon_i^2|\mathbf{x}_i).$$

The challenges are:

1. **Single observation**:
   - For each observation i, there is only one ϵ_i, which is insufficient to estimate the variance σ_i^2 directly.

2. **Dependence on assumptions**:
   - To estimate σ_i^2, we must impose assumptions about its relationship to $\mathbf{x}_i$.

To make IWLS feasible, we model σ_i^2 as a function of the predictors $\mathbf{x}_i$. A common approach is:

$$\epsilon_i^2 = \upsilon_i \exp(\mathbf{x}_i \gamma),$$

where:

- υ_i is an independent error term with strictly positive values, representing random noise
- $\exp(\mathbf{x}_i \gamma)$ is a deterministic function of the predictors $\mathbf{x}_i$

Taking the natural logarithm of both sides linearizes the model:

$$\ln(\epsilon_i^2) = \mathbf{x}_i \gamma + \ln(\upsilon_i),$$

where $\ln(\upsilon_i)$ is independent of $\mathbf{x}_i$. This transformation enables us to estimate γ using standard OLS techniques.

Estimation Procedure for Feasible GLS (FGLS)
Since we do not observe the true errors ϵ_i, we approximate them using the OLS residuals e_i. Here is the step-by-step process:

1. **Compute OLS residuals**: First, fit the original model using OLS and calculate the residuals:

$$e_i = y_i - \mathbf{x}_i \hat{\beta}_{OLS}.$$

2. **Approximate** ϵ_i^2 **with** e_i^2: Use the squared residuals as a proxy for the squared errors:

$$e_i^2 \approx \epsilon_i^2.$$

3. **Log-linear model**: Fit the log-transformed model to estimate γ:

$$\ln(e_i^2) = \mathbf{x}_i \gamma + \ln(\upsilon_i).$$

   Estimate γ using OLS, where $\ln(\upsilon_i)$ is treated as the error term.
4. **Estimate variances**: Use the fitted values $\hat{\gamma}$ to estimate σ_i^2 for each observation:

$$\hat{\sigma}_i^2 = \exp(\mathbf{x}_i \hat{\gamma}).$$

5. **Perform weighted least squares**: Use the estimated variances $\hat{\sigma}_i^2$ to construct the weight matrix $\hat{\mathbf{W}}$:

$$\hat{\mathbf{W}} = \begin{pmatrix} 1/\hat{\sigma}_1^2 & 0 & \cdots & 0 \\ 0 & 1/\hat{\sigma}_2^2 & \cdots & 0 \\ \vdots & \vdots & \ddots & \vdots \\ 0 & 0 & \cdots & 1/\hat{\sigma}_n^2 \end{pmatrix}.$$

Then, compute the **feasible GLS (FGLS)** estimator:

$$\hat{\beta}_{FGLS} = (\mathbf{X}'\hat{\mathbf{W}}\mathbf{X})^{-1}\mathbf{X}'\hat{\mathbf{W}}\mathbf{y}.$$

2.2.2.2 Serial Correlation

Serial correlation (also called autocorrelation) occurs when the error terms in a regression model are correlated across observations. Formally:

$$Cov(\epsilon_i, \epsilon_j|\mathbf{X}) \neq 0 \quad \text{for } i \neq j.$$

This violates the Gauss–Markov assumption that $Cov(\epsilon_i, \epsilon_j|\mathbf{X}) = 0$, leading to inefficiencies in OLS estimates.

2.2.2.2.1 Covariance Stationarity

If the errors are **covariance stationary**, the covariance between errors depends only on their relative time or positional difference (h), not their absolute position:

$$Cov(\epsilon_i, \epsilon_j|\mathbf{X}) = Cov(\epsilon_i, \epsilon_{i+h}|\mathbf{x}_i, \mathbf{x}_{i+h}) = \gamma_h,$$

where γ_h represents the covariance at lag h.

Under covariance stationarity, the variance–covariance matrix of the error term $\boldsymbol{\epsilon}$ takes the following form:

$$Var(\boldsymbol{\epsilon}|\mathbf{X}) = \boldsymbol{\Omega} = \begin{pmatrix} \sigma^2 & \gamma_1 & \gamma_2 & \cdots & \gamma_{n-1} \\ \gamma_1 & \sigma^2 & \gamma_1 & \cdots & \gamma_{n-2} \\ \gamma_2 & \gamma_1 & \sigma^2 & \cdots & \vdots \\ \vdots & \vdots & \vdots & \ddots & \gamma_1 \\ \gamma_{n-1} & \gamma_{n-2} & \cdots & \gamma_1 & \sigma^2 \end{pmatrix}.$$

Key Points:

1. The diagonal elements represent the variance of the error term: σ^2.
2. The off-diagonal elements γ_h represent covariances at different lags h.

Why Serial Correlation Is a Problem?
The matrix $\mathbf{\Omega}$ introduces n parameters to estimate (e.g., $\sigma^2, \gamma_1, \gamma_2, \ldots, \gamma_{n-1}$). Estimating such a large number of parameters becomes impractical, especially for large datasets. To address this, we impose additional structure to reduce the number of parameters.

2.2.2.2.2 AR(1) Model

In the **AR(1)** process, the errors follow a first-order autoregressive process:

$$y_t = \beta_0 + x_t\beta_1 + \epsilon_t,$$
$$\epsilon_t = \rho\epsilon_{t-1} + u_t,$$

where:

- ρ is the first-order autocorrelation coefficient, capturing the relationship between consecutive errors
- u_t is white noise, satisfying $Var(u_t) = \sigma_u^2$ and $Cov(u_t, u_{t-h}) = 0$ for $h \neq 0$

Under the AR(1) assumption, the variance–covariance matrix of the error term $\boldsymbol{\epsilon}$ becomes:

$$Var(\boldsymbol{\epsilon}|\mathbf{X}) = \frac{\sigma_u^2}{1-\rho^2}\begin{pmatrix} 1 & \rho & \rho^2 & \cdots & \rho^{n-1} \\ \rho & 1 & \rho & \cdots & \rho^{n-2} \\ \rho^2 & \rho & 1 & \cdots & \rho^{n-3} \\ \vdots & \vdots & \vdots & \ddots & \vdots \\ \rho^{n-1} & \rho^{n-2} & \cdots & \rho & 1 \end{pmatrix}.$$

Key Features:

1. The diagonal elements represent the variance: $Var(\epsilon_t|\mathbf{X}) = \sigma_u^2/(1-\rho^2)$.
2. The off-diagonal elements decay exponentially with lag h: $Cov(\epsilon_t, \epsilon_{t-h}|\mathbf{X}) = \rho^h \cdot Var(\epsilon_t|\mathbf{X})$.

Under AR(1), only **one parameter** ρ needs to be estimated (in addition to σ_u^2), greatly simplifying the structure of $\mathbf{\Omega}$.

OLS Properties Under AR(1)
1. **Consistency**: If assumptions A1, A2, A3a, and A5a hold, OLS remains consistent.
2. **Asymptotic normality**: OLS estimates are asymptotically normal.

3. **Inference with serial correlation**:
   - Standard OLS errors are invalid.
   - Use **Newey–West standard errors** to obtain robust inference.

2.2.2.2.3 Infeasible Cochrane–Orcutt

The **infeasible Cochrane–Orcutt** procedure addresses serial correlation in the error terms by assuming an **AR(1)** process for the errors:

$$\epsilon_t = \rho\epsilon_{t-1} + u_t,$$

where u_t is white noise and ρ is the autocorrelation coefficient.

By transforming the original regression equation:

$$y_t = \beta_0 + x_t\beta_1 + \epsilon_t,$$

we subtract ρ times the lagged equation:

$$\rho y_{t-1} = \rho(\beta_0 + x_{t-1}\beta_1 + \epsilon_{t-1}),$$

to obtain the weighted first-difference equation:

$$y_t - \rho y_{t-1} = (1-\rho)\beta_0 + (x_t - \rho x_{t-1})\beta_1 + u_t.$$

Key Points:

1. **Dependent variable**: $y_t - \rho y_{t-1}$.
2. **Independent variable**: $x_t - \rho x_{t-1}$.
3. **Error term**: u_t, which satisfies the Gauss–Markov assumptions (A3, A4, A5).

The ICO estimator minimizes the sum of squared residuals for this transformed equation.

1. **Standard errors**:
   - If the errors truly follow an AR(1) process, the standard errors for the transformed equation are valid.
   - For more complex error structures, **Newey–West HAC standard errors** are required.

2. **Loss of observations**:
   - The transformation involves first differences, which means the first observation (y_1) cannot be used. This reduces the effective sample size by one.

The Problem: ρ Is Unknown
The ICO procedure is **infeasible** because it requires knowledge of ρ, the autocorrelation coefficient. In practice, we estimate ρ from the data.

To estimate ρ, we use the OLS residuals (e_t) as a proxy for the errors (ϵ_t). The estimate $\hat{\rho}$ is given by:

$$\hat{\rho} = \frac{\sum_{t=2}^{T} e_t e_{t-1}}{\sum_{t=2}^{T} e_t^2}.$$

Estimation via OLS:

1. Regress the OLS residuals e_t on their lagged values e_{t-1}, without an intercept:

$$e_t = \rho e_{t-1} + u_t.$$

2. The slope of this regression is the estimate $\hat{\rho}$.

This estimation is efficient under the AR(1) assumption and provides a practical approximation for ρ.

2.2.2.2.4 Feasible Prais–Winsten

The **Feasible Prais–Winsten (FPW)** method addresses **AR(1) serial correlation** in regression models by transforming the data to eliminate serial dependence in the errors. Unlike the infeasible Cochrane–Orcutt procedure, which discards the first observation, the Prais–Winsten method retains it using a weighted transformation.

The FPW transformation uses the following weighting matrix **w**:

$$\mathbf{w} = \begin{pmatrix} \sqrt{1-\hat{\rho}^2} & 0 & 0 & \cdots & 0 \\ -\hat{\rho} & 1 & 0 & \cdots & 0 \\ 0 & -\hat{\rho} & 1 & \cdots & 0 \\ \vdots & \vdots & \vdots & \ddots & \vdots \\ 0 & 0 & 0 & -\hat{\rho} & 1 \end{pmatrix}.$$

where:

- The first row accounts for the transformation of the first observation, using $\sqrt{1-\hat{\rho}^2}$
- Subsequent rows represent the AR(1) transformation for the remaining observations

Step-by-Step Procedure
Step 1: Initial OLS Estimation
Estimate the regression model using OLS:

$$y_t = \mathbf{x}_t\beta + \epsilon_t,$$

and compute the residuals:

$$e_t = y_t - \mathbf{x}_t\hat{\beta}.$$

Step 2: Estimate the AR(1) Correlation Coefficient
Estimate the AR(1) correlation coefficient ρ by regressing e_t on e_{t-1} without an intercept:

$$e_t = \rho e_{t-1} + u_t.$$

The slope of this regression is the estimated $\hat{\rho}$.

Step 3: Transform the Data
Apply the transformation using the weighting matrix $\mathbf{w}$ to transform both the dependent variable $\mathbf{y}$ and the independent variables $\mathbf{X}$:

$$\mathbf{wy} = \mathbf{wX}\beta + \mathbf{w}\epsilon.$$

Specifically: 1. For $t = 1$, the transformed dependent and independent variables are:

$$\tilde{y}_1 = \sqrt{1-\hat{\rho}^2} \cdot y_1, \quad \tilde{\mathbf{x}}_1 = \sqrt{1-\hat{\rho}^2} \cdot \mathbf{x}_1.$$

2. For $t = 2, \ldots, T$, the transformed variables are:

$$\tilde{y}_t = y_t - \hat{\rho} \cdot y_{t-1}, \quad \tilde{\mathbf{x}}_t = \mathbf{x}_t - \hat{\rho} \cdot \mathbf{x}_{t-1}.$$

Step 4: Feasible Prais–Winsten Estimation
Run OLS on the transformed equation:

$$\mathbf{wy} = \mathbf{wX}\beta + \mathbf{w}\epsilon.$$

The resulting estimator is the **feasible Prais–Winsten (FPW) estimator**:

$$\hat{\beta}_{FPW} = (\mathbf{X}'\mathbf{w}'\mathbf{wX})^{-1}\mathbf{X}'\mathbf{w}'\mathbf{wy}.$$

Properties of feasible Prais–Winsten estimator

1. **Infeasible Prais–Winsten estimator**:
   - The infeasible Prais-Winsten (PW) estimator assumes the AR(1) parameter ρ is known.
   - Under assumptions A1, A2, and A3 for the unweighted equation, the infeasible PW estimator is **unbiased** and efficient.

2. **Feasible Prais–Winsten (FPW) estimator**: The FPW estimator replaces the unknown ρ with an estimate $\hat{\rho}$ derived from the OLS residuals, introducing bias in small samples.

   1. **Bias**:

      - The FPW estimator is **biased** due to the estimation of $\hat{\rho}$, which introduces an additional layer of approximation.

   2. **Consistency**:

      1. The FPW estimator is **consistent** under the following assumptions:

         - A1: The model is linear in parameters.
         - A2: The independent variables are linearly independent.
         - A5: The data is generated through random sampling.
         - Additionally:

$$E\big((\mathbf{x_t} - \rho\mathbf{x_{t-1}})'(\epsilon_t - \rho\epsilon_{t-1})\big) = 0.$$

         This condition ensures the transformed error term $\epsilon_t - \rho\epsilon_{t-1}$ is uncorrelated with the transformed regressors $\mathbf{x_t} - \rho\mathbf{x_{t-1}}$.

      3. **Note**: A3a (zero conditional mean of the error term, $E(\epsilon_t|\mathbf{x}_t) = 0$) is **not sufficient** for the above condition. Full exogeneity of the independent variables (A3) is required.

   4. **Efficiency**

      1. **Asymptotic efficiency**: The FPW estimator is asymptotically **more efficient than OLS** if the errors are truly generated by an AR(1) process:

$$\epsilon_t = \rho\epsilon_{t-1} + u_t, \quad Var(u_t) = \sigma^2.$$

      2. **Standard errors**:

         1. **Usual standard errors**: If the errors are correctly specified as an AR(1) process, the usual standard errors from FPW are valid.
         2. **Robust standard errors**: If there is concern about a more complex dependence structure (e.g., higher-order autocorrelation or heteroskedasticity), use **Newey–West standard errors** for inference. These are robust to both serial correlation and heteroskedasticity.

2.2.2.3 Cluster Errors

Consider the regression model with clustered errors:

$$y_{gi} = \mathbf{x}_{gi}\beta + \epsilon_{gi},$$

where:

- g indexes the group (e.g., households, firms, schools)
- i indexes the individual within the group

The covariance structure for the errors ϵ_{gi} is defined as:

$$Cov(\epsilon_{gi}, \epsilon_{hj}) \begin{cases} = 0 & \text{if } g \neq h \text{ (independent across groups),} \\ \neq 0 & \text{for any pair } (i, j) \text{ within group } g. \end{cases}$$

Within each group, individuals' errors may be correlated (i.e., intra-group correlation), while errors are independent across groups. This violates A4 (constant variance and no correlation of errors).

Suppose there are three groups with varying sizes. The variance–covariance matrix $\mathbf{\Omega}$ for the errors $\boldsymbol{\epsilon}$ is:

$$Var(\boldsymbol{\epsilon}|\mathbf{X}) = \mathbf{\Omega} = \begin{pmatrix} \sigma^2 & \delta_{12}^1 & \delta_{13}^1 & 0 & 0 & 0 \\ \delta_{12}^1 & \sigma^2 & \delta_{23}^1 & 0 & 0 & 0 \\ \delta_{13}^1 & \delta_{23}^1 & \sigma^2 & 0 & 0 & 0 \\ 0 & 0 & 0 & \sigma^2 & \delta_{12}^2 & 0 \\ 0 & 0 & 0 & \delta_{12}^2 & \sigma^2 & 0 \\ 0 & 0 & 0 & 0 & 0 & \sigma^2 \end{pmatrix}$$

where:

- $\delta_{ij}^g = Cov(\epsilon_{gi}, \epsilon_{gj})$ is the covariance between errors for individuals i and j in group g
- $Cov(\epsilon_{gi}, \epsilon_{hj}) = 0$ for $g \neq h$ (independent groups)

Infeasible Generalized Least Squares (Cluster)

1. **Assume known variance–covariance matrix**: If σ^2 and δ_{ij}^g are known, construct $\mathbf{\Omega}$ and compute its inverse $\mathbf{\Omega}^{-1}$.
2. **Infeasible GLS estimator**: The infeasible generalized least squares (IGLS) estimator is:

$$\hat{\beta}_{IGLS} = (\mathbf{X}'\mathbf{\Omega}^{-1}\mathbf{X})^{-1}\mathbf{X}'\mathbf{\Omega}^{-1}\mathbf{y}.$$

Problem:

- We do not know σ^2 and δ_{ij}^g, making this approach **infeasible**.
- Even if $\mathbf{\Omega}$ is estimated, incorrect assumptions about its structure may lead to invalid inference.

To make the estimation feasible, we assume a **group-level random effects** specification for the error:

$$\begin{aligned} y_{gi} &= \mathbf{x}_{gi}\beta + c_g + u_{gi}, \\ Var(c_g|\mathbf{x}_i) &= \sigma_c^2, \\ Var(u_{gi}|\mathbf{x}_i) &= \sigma_u^2, \end{aligned}$$

where:

- c_g represents the **group-level random effect** (common shocks within each group, independent across groups)
- u_{gi} represents the **individual-level error** (idiosyncratic shocks within each group, independent across individuals and groups)
- $\epsilon_{gi} = c_g + u_{gi}$

Independence assumptions:

- c_g and u_{gi} are independent of each other.
- Both are mean-independent of $\mathbf{x}_i$.

Under this specification, the variance–covariance matrix $\mathbf{\Omega}$ becomes block diagonal, where each block corresponds to a group:

$$Var(\boldsymbol{\epsilon}|\mathbf{X}) = \mathbf{\Omega} = \begin{pmatrix} \sigma_c^2 + \sigma_u^2 & \sigma_c^2 & \sigma_c^2 & 0 & 0 & 0 \\ \sigma_c^2 & \sigma_c^2 + \sigma_u^2 & \sigma_c^2 & 0 & 0 & 0 \\ \sigma_c^2 & \sigma_c^2 & \sigma_c^2 + \sigma_u^2 & 0 & 0 & 0 \\ 0 & 0 & 0 & \sigma_c^2 + \sigma_u^2 & \sigma_c^2 & 0 \\ 0 & 0 & 0 & \sigma_c^2 & \sigma_c^2 + \sigma_u^2 & 0 \\ 0 & 0 & 0 & 0 & 0 & \sigma_c^2 + \sigma_u^2 \end{pmatrix}.$$

When the variance components σ_c^2 and σ_u^2 are unknown, we can use the **feasible group-level random effect (RE)** estimator to simultaneously estimate these variances and the regression coefficients β. This practical approach allows us to account for intra-group correlation in the errors and still obtain consistent and efficient estimates of the parameters.

Step-by-Step Procedure

Step 1: Initial OLS Estimation

Estimate the regression model using OLS:

$$y_{gi} = \mathbf{x}_{gi}\beta + \epsilon_{gi},$$

and compute the residuals:

$$e_{gi} = y_{gi} - \mathbf{x}_{gi}\hat{\beta}.$$

Step 2: Estimate Variance Components

Use the standard OLS variance estimator s^2 to estimate the **total variance**:

$$s^2 = \frac{1}{n-k}\sum_{i=1}^{n} e_i^2,$$

where n is the total number of observations and k is the number of regressors (including the intercept).
Estimate the **between-group variance** $\hat{\sigma}_c^2$ using:

$$\hat{\sigma}_c^2 = \frac{1}{G}\sum_{g=1}^{G}\left(\frac{1}{\sum_{i=1}^{n_g-1} i}\sum_{i\neq j}\sum_{j=1}^{n_g} e_{gi}e_{gj}\right),$$

where:

- G is the total number of groups
- n_g is the size of group g
- The term $\sum_{i\neq j} e_{gi}e_{gj}$ accounts for within-group covariance

Estimate the **within-group variance** as:

$$\hat{\sigma}_u^2 = s^2 - \hat{\sigma}_c^2.$$

Step 3: Construct the Variance–Covariance Matrix
Use the estimated variances $\hat{\sigma}_c^2$ and $\hat{\sigma}_u^2$ to construct the variance–covariance matrix $\hat{\Omega}$ for the error term:

$$\hat{\Omega}_{gi,gj} = \begin{cases} \hat{\sigma}_c^2 + \hat{\sigma}_u^2 & \text{if } i = j \text{ (diagonal elements),} \\ \hat{\sigma}_c^2 & \text{if } i \neq j \text{ (off-diagonal elements within group),} \\ 0 & \text{if } g \neq h \text{ (across groups).} \end{cases}$$

Step 4: Feasible GLS Estimation
With $\hat{\Omega}$ in hand, perform feasible generalized least squares (FGLS) to estimate β:

$$\hat{\beta}_{RE} = (\mathbf{X}'\hat{\Omega}^{-1}\mathbf{X})^{-1}\mathbf{X}'\hat{\Omega}^{-1}\mathbf{y}.$$

If the assumptions about $\boldsymbol{\Omega}$ are incorrect or infeasible, use **cluster-robust standard errors** to account for intra-group correlation without explicitly modeling the variance–covariance structure. These standard errors remain valid under arbitrary within-cluster dependence, provided clusters are independent.

Properties of the Feasible Group-Level Random Effects Estimator
1. **Infeasible group RE estimator**

- The infeasible RE estimator (assuming known variances) is **unbiased** under assumptions A1, A2, and A3 for the unweighted equation.

- A3 requires:

$$E(\epsilon_{gi}|\mathbf{x}_i) = E(c_g|\mathbf{x}_i) + E(u_{gi}|\mathbf{x}_i) = 0.$$

  This assumes:

  - $E(c_g|\mathbf{x}_i) = 0$: the **random effect assumption** (group-level effects are uncorrelated with the regressors)
  - $E(u_{gi}|\mathbf{x}_i) = 0$: no endogeneity at the individual level

2. **Feasible group RE estimator**

- The feasible RE estimator is **biased** because the variances σ_c^2 and σ_u^2 are estimated, introducing approximation errors.
- However, the estimator is **consistent** under A1, A2, A3a ($E(\mathbf{x}_i'\epsilon_{gi}) = E(\mathbf{x}_i'c_g) + E(\mathbf{x}_i'u_{gi}) = 0$), A5a.
- **Efficiency**
  - **Asymptotic efficiency**:

    The feasible RE estimator is asymptotically more efficient than OLS if the errors follow the random effect specification.

  - **Standard errors**:
    * If the random effect specification is correct, the usual standard errors are consistent.
    * If there is concern about more complex dependence structures or heteroskedasticity, use **cluster-robust standard errors**.

2.2.3 *Weighted Least Squares*

In the presence of heteroskedasticity, the errors ϵ_i have nonconstant variance $Var(\epsilon_i|\mathbf{x}_i) = \sigma_i^2$. This violates the Gauss–Markov assumption of homoskedasticity, leading to inefficient OLS estimates.

Weighted least squares (WLS) addresses this by applying weights inversely proportional to the variance of the errors, ensuring that observations with larger variances have less influence on the estimation.

- Weighted least squares is essentially generalized least squares in the special case that $\mathbf{\Omega}$ is a *diagonal* matrix with variances σ_i^2 on the diagonal (i.e., errors are uncorrelated but have nonconstant variance).

- That is, assume the errors are uncorrelated but **heteroskedastic**: $\boldsymbol{\Omega} = \text{diag}\left(\sigma_1^2, \ldots, \sigma_n^2\right)$.
- Then $\boldsymbol{\Omega}^{-1} = \text{diag}\left(1/\sigma_1^2, \ldots, 1/\sigma_n^2\right)$.

Steps for Feasible Weighted Least Squares (FWLS)

1. Initial OLS Estimation

   First, estimate the model using OLS:

$$y_i = \mathbf{x}_i \beta + \epsilon_i,$$

and compute the residuals:

$$e_i = y_i - \mathbf{x}_i \hat{\beta}.$$

2. Model the Error Variance

   Transform the residuals to model the variance as a function of the predictors:

$$\ln(e_i^2) = \mathbf{x}_i \gamma + \ln(v_i),$$

where:

- e_i^2 approximates ϵ_i^2
- $\ln(v_i)$ is the error term in this auxiliary regression, assumed independent of $\mathbf{x}_i$

Estimate this equation using OLS to obtain the predicted values:

$$\hat{g}_i = \mathbf{x}_i \hat{\gamma}.$$

3. Estimate Weights

Use the predicted values from the auxiliary regression to compute the weights:

$$\hat{\sigma}_i = \sqrt{\exp(\hat{g}_i)}.$$

   These weights approximate the standard deviation of the errors.

4. Weighted Regression

Transform the original equation by dividing through by $\hat{\sigma}_i$:

$$\frac{y_i}{\hat{\sigma}_i} = \frac{\mathbf{x}_i}{\hat{\sigma}_i} \beta + \frac{\epsilon_i}{\hat{\sigma}_i}.$$

Estimate the transformed equation using OLS. The resulting estimator is the **feasible weighted least squares (FWLS)** estimator:

$$\hat{\beta}_{FWLS} = (\mathbf{X}'\hat{\mathbf{W}}\mathbf{X})^{-1}\mathbf{X}'\hat{\mathbf{W}}\mathbf{y},$$

where $\hat{\mathbf{W}}$ is a diagonal weight matrix with elements $1/\hat{\sigma}_i^2$.

Properties of the FWLS Estimator

1. **Unbiasedness**:
   - The infeasible WLS estimator (where σ_i is known) is unbiased under assumptions A1–A3 for the unweighted model.
   - The **FWLS estimator** is **not unbiased** due to the approximation of σ_i using $\hat{\sigma}_i$.
2. **Consistency**:
   - The FWLS estimator is **consistent** under the following assumptions:
     - A1 (for the unweighted equation): The model is linear in parameters.
     - A2 (for the unweighted equation): The independent variables are linearly independent.
     - A5: The data is randomly sampled.
     - $E(\mathbf{x}_i'\epsilon_i/\sigma_i^2) = 0$. A3a: Weaker Exogeneity Assumption is not sufficient, but A3 is.
3. **Efficiency**:
   - The FWLS estimator is asymptotically more efficient than OLS if the errors have multiplicative exponential heteroskedasticity:

$$Var(\epsilon_i|\mathbf{x}_i) = \sigma_i^2 = \exp(\mathbf{x}_i\gamma).$$

The FWLS estimator is asymptotically more efficient than OLS if the errors have multiplicative exponential heteroskedasticity.

1. **Usual standard errors**:
   - If the errors are truly multiplicative exponential heteroskedastic, the usual standard errors for FWLS are valid.
2. **Heteroskedastic robust standard errors**:
   - If there is potential mis-specification of the multiplicative exponential model for σ_i^2, heteroskedastic-robust standard errors should be reported to ensure valid inference.

2.3 Maximum Likelihood

The **maximum likelihood estimation (MLE)** is a statistical method used to estimate the parameters of a model by maximizing the likelihood of observing the given data. The premise is to find the parameter values that maximize the probability (or likelihood) of the observed data.

The likelihood function, denoted as $L(\theta)$, is expressed as:

$$L(\theta) = \prod_{i=1}^{n} f(y_i|\theta)$$

where:

- $f(y|\theta)$ is the probability density or mass function of observing a single value of Y given the parameter θ
- The product runs over all n observations

For different types of data, $f(y|\theta)$ can take different forms. For example, if y is dichotomous (e.g., success/failure), then the likelihood function becomes:

$$L(\theta) = \prod_{i=1}^{n} \theta^{y_i}(1-\theta)^{1-y_i}$$

Here, $\hat{\theta}$ is the maximum likelihood estimator (MLE) if:

$$L(\hat{\theta}) > L(\theta_0), \quad \forall\theta_0 \text{ in the parameter space.}$$

See [Distributions] for a review on variable distributions.

2.3.1 *Motivation for MLE*

Suppose we know the **conditional distribution** of Y given X, denoted as:

$$f_{Y|X}(y, x; \theta)$$

where θ is an unknown parameter of the distribution. Sometimes, we are only concerned with the unconditional distribution $f_Y(y; \theta)$.

For a sample of independent and identically distributed (i.i.d.) data, the joint probability of the sample is:

$$f_{Y_1,\ldots,Y_n|X_1,\ldots,X_n}(y_1, \ldots, y_n, x_1, \ldots, x_n; \theta) = \prod_{i=1}^{n} f_{Y|X}(y_i, x_i; \theta).$$

The **joint distribution**, evaluated at the observed data, defines the likelihood function. The goal of MLE is to find the parameter θ that maximizes this likelihood.

To estimate θ, we maximize the likelihood function:

$$\max_{\theta} \prod_{i=1}^{n} f_{Y|X}(y_i, x_i; \theta).$$

In practice, it is easier to work with the natural logarithm of the likelihood (log-likelihood), as it transforms the product into a sum:

$$\max_{\theta} \sum_{i=1}^{n} \ln(f_{Y|X}(y_i, x_i; \theta)).$$

Solving for the Maximum Likelihood Estimator

1. **First-order condition**: Solve the first derivative of the log-likelihood function with respect to θ:

$$\frac{\partial}{\partial\theta}\ell(\theta) = \frac{\partial}{\partial\theta}\ln L(\theta) = \frac{\partial}{\partial\theta}\sum_{i=1}^{n} \ln(f_{Y|X}(y_i, x_i; \hat{\theta}_{MLE})) = 0.$$

   This yields the critical points where the likelihood is maximized. This derivative, sometimes written as $U(\theta)$, is called the **score**. Intuitively, the log-likelihood's "peak" indicates the parameter value(s) that make the observed data "most likely."
2. **Second-order condition**: Verify that the second derivative of the log-likelihood function is negative at the critical point:

$$\frac{\partial^2}{\partial\theta^2}\sum_{i=1}^{n} \ln(f_{Y|X}(y_i, x_i; \hat{\theta}_{MLE})) < 0.$$

   This ensures that the solution corresponds to a maximum.

Examples of Likelihood Functions

1. Unconditional [Poisson distribution]

The Poisson distribution models count data, such as the number of website visits in a day or product orders per hour. Its likelihood function is:

$$L(\theta) = \prod_{i=1}^{n} \frac{\theta^{y_i} e^{-\theta}}{y_i!}.$$

2. [Exponential distribution]

The exponential distribution is often used to model the time between events, such as the time until a machine fails. Its probability density function (PDF) is:

$$f_{Y|X}(y, x; \theta) = \frac{\exp(-y/(x\theta))}{x\theta}.$$

The joint likelihood for n observations is:

$$L(\theta) = \prod_{i=1}^{n} \frac{\exp(-y_i/(x_i\theta))}{x_i\theta}$$

By taking the logarithm, we obtain the log-likelihood for ease of maximization.

2.3.2 Key Quantities for Inference

1. **Score function**
   The **score** is given by:

$$U(\theta) \;=\; \frac{d}{d\theta}\ell(\theta).$$

   Setting $U(\hat{\theta}_{\mathrm{MLE}}) = 0$ yields the critical points of the log-likelihood, from which we can find $\hat{\theta}_{\mathrm{MLE}}$.
2. **Observed information**
   The second derivative of the log-likelihood, taken at the MLE, is called the **observed information**:

$$I_O(\theta) \;=\; -\frac{d^2}{d\theta^2}\ell(\theta).$$

   (The negative sign is often included so that $I_O(\theta)$ is *positive* if $\ell(\theta)$ is concave near its maximum. In some texts, you will see it defined without the negative sign, but the idea is the same: It measures the "pointedness" or curvature of $\ell(\theta)$ at its maximum.)
3. **Fisher information**
   The **Fisher information** (or **expected information**) is the expectation of the observed information over the distribution of the data:

$$I(\theta) \;=\; \mathbb{E}\left[I_O(\theta)\right].$$

It quantifies how much information the data carry about the parameter θ. A larger Fisher information suggests that you can estimate θ more precisely.

4. **Approximate variance of $\hat{\theta}_{\text{MLE}}$**
   One of the key results from standard asymptotic theory is that, for large n, the variance of $\hat{\theta}_{\text{MLE}}$ can be approximated by the inverse of the Fisher information:

$$\text{Var}\left(\hat{\theta}_{\text{MLE}}\right) \approx I(\theta)^{-1}.$$

   This also lays the groundwork for constructing confidence intervals for θ in large samples.

2.3.3 Assumptions of MLE

MLE has desirable properties—*consistency*, *asymptotic normality*, and *efficiency*—but these do not come "for free." Instead, they rely on certain assumptions. Below is a breakdown of the main **regularity conditions**. These conditions are typically mild in many practical settings (for example, in exponential families, such as the normal distribution), but need to be checked in more complex models.

High-Level Regulatory Assumptions

1. **Independence and identical distribution (iid)**
   The sample $\{(x_i, y_i)\}$ is usually assumed to be composed of independent and identically distributed observations. This independence assumption simplifies the likelihood to a product of individual densities:

$$L(\theta) = \prod_{i=1}^{n} f_{Y|X}(y_i, x_i; \theta).$$

   In practice, if you have dependent data (e.g., time series, spatial data), modifications are required in the likelihood function.
2. **Same density function**
   All observations must come from the *same* conditional probability density function $f_{Y|X}(\cdot, \cdot; \theta)$. If the model changes across observations, you cannot simply multiply all of them together in one unified likelihood.
3. **Multivariate normality (for certain models)**
   In many practical cases—especially for continuous outcomes—you might assume (multivariate) normal distributions with finite second or fourth moments (Little, 1988). Under these assumptions, the MLE for the mean vector and covariance matrix is consistent and (under further conditions) asymptotically normal. This assumption is quite common in regression, ANOVA, and other classical statistical frameworks.

2.3.3.1 Large Sample Properties of MLE

2.3.3.1.1 Consistency of MLE

Definition An estimator $\hat{\theta}_n$ is *consistent* if it converges in probability to the true parameter value θ_0 as the sample size $n \to \infty$:

$$\hat{\theta}_n \to^p \theta_0.$$

For the MLE, a set of regularity conditions $R1$–$R4$ is commonly used to ensure consistency:

1. **R1**
   If $\theta \neq \theta_0$, then:

   $$f_{Y|X}(y_i, x_i; \theta) \neq f_{Y|X}(y_i, x_i; \theta_0).$$

   In simpler terms, the model is identifiable: No two distinct parameter values generate the *exact* same distribution for the data.
2. **R2**
   The parameter space Θ is compact (closed and bounded), and it contains the true parameter θ_0. This ensures that θ lies in a "nice" region (no parameter going to infinity, etc.), making it easier to prove that a maximum in that space indeed exists.
3. **R3**
   The log-likelihood function $\ln(f_{Y|X}(y_i, x_i; \theta))$ is continuous in θ with probability 1. Continuity is important so that we can apply theorems (like the continuous mapping theorem or the extreme value theorem) to find maxima.
4. **R4**
   The expected supremum of the absolute value of the log-likelihood is finite.

   $$\mathbb{E}\left(\sup_{\theta \in \Theta} \left|\ln(f_{Y|X}(y_i, x_i; \theta))\right|\right) < \infty.$$

   This is a technical condition that helps ensure we can "exchange" expectations and suprema, a step needed in many consistency proofs.

When these conditions are satisfied, you can show via standard arguments (e.g., the [law of large numbers], uniform convergence of the log-likelihood) that:

$$\hat{\theta}_{\text{MLE}} \to^p \theta_0 \quad \text{(consistency)}.$$

2.3.3.1.2 Asymptotic Normality of MLE

Definition An estimator $\hat{\theta}_n$ is *asymptotically normal* if

$$\sqrt{n}\,(\hat{\theta}_n - \theta_0) \;\to^d\; \mathcal{N}(0, \Sigma)\,,$$

where $\to^d$ denotes convergence in distribution and Σ is some covariance matrix. For the MLE, Σ is typically $I(\theta_0)^{-1}$, where $I(\theta_0)$ is the Fisher information evaluated at the true parameter.

Beyond $R1$–$R4$, we need the following additional assumptions:

1. **R5**
   The true parameter θ_0 is in the *interior* of the parameter space Θ. If θ_0 sits on the boundary, different arguments are required to handle edge effects.
2. **R6**
   The pdf $f_{Y|X}(y_i, x_i; \theta)$ is *twice* continuously differentiable (in θ) and strictly positive in a neighborhood N of θ_0. This allows us to use second-order Taylor expansions around θ_0 to get the approximate distribution of $\hat{\theta}_{\text{MLE}}$.
3. **R7**
   The following integrals are finite in some neighborhood N of θ_0:
   - $\int \sup_{\theta\in N} \left\| \frac{\partial f_{Y|X}(y_i,x_i;\theta)}{\partial\theta} \right\| d(y,x) < \infty.$
   - $\int \sup_{\theta\in N} \left\| \frac{\partial^2 f_{Y|X}(y_i,x_i;\theta)}{\partial\theta\partial\theta'} \right\| d(y,x) < \infty.$
   - $\mathbb{E}\left(\sup_{\theta\in N} \left\| \frac{\partial^2 \ln(f_{Y|X}(y_i,x_i;\theta))}{\partial\theta\partial\theta'} \right\|\right) < \infty.$

   These conditions ensure that differentiating inside integrals is justified (via the dominated convergence theorem) and that we can expand the log-likelihood in a Taylor series safely.
4. **R8**
   The information matrix $I(\theta_0)$ exists and is nonsingular:

$$I(\theta_0) = \text{Var}\left(\frac{\partial}{\partial\theta}\ln\left(f_{Y|X}(y_i, x_i; \theta_0)\right)\right) \neq 0.$$

   Nonsingularity implies there is enough information in the data to estimate θ uniquely.

Under $R1$–$R8$, you can show that

$$\sqrt{n}\,(\hat{\theta}_{\text{MLE}} - \theta_0) \to^d \mathcal{N}\left(0,\; I(\theta_0)^{-1}\right).$$

This result is central to frequentist inference, allowing you to construct approximate confidence intervals and hypothesis tests using the normal approximation for large n.

2.3.4 Properties of MLE

Having established in earlier sections that maximum likelihood estimators (MLEs) are **consistent** (consistency of MLE) and **asymptotically normal** (asymptotic normality of MLE) under standard regularity conditions, we now highlight additional properties that make MLE a powerful estimation technique.

1. **Asymptotic efficiency**

- **Definition**: An estimator is *asymptotically efficient* if it attains the smallest possible asymptotic variance among all consistent estimators (i.e., it achieves the *Cramér–Rao lower bound*).
- **Interpretation**: In large samples, MLE typically has smaller standard errors than other consistent estimators that do not fully use the assumed distributional form.
- **Implication**: When the true model is correctly specified, MLE is the *most efficient* among a broad class of estimators, leading to more precise inference for θ.
  - **Cramér–Rao lower bound (CRLB)**: A theoretical lower limit on the variance of any unbiased (or asymptotically unbiased) estimator (Rao, 1992; Cramér, 1999).
  - **When MLE meets CRLB**: Under correct specification and standard regularity conditions, the asymptotic variance of the MLE matches the CRLB, making it *asymptotically efficient.*
  - **Interpretation**: Achieving CRLB means no other unbiased estimator can consistently outperform MLE in terms of variance for large n.

2. **Invariance**

- **Core idea**: If $\hat{\theta}$ is the MLE for θ, then for any *smooth* transformation $g(\theta)$, the MLE for $g(\theta)$ is simply $g(\hat{\theta})$.
- **Example**: If θ is a mean parameter and you want the MLE for the *variance* θ^2, you can just square the MLE for θ.
- **Key point**: This *invariance property* saves considerable effort—there is no need to re-derive a new likelihood for the transformed parameter.

3. Explicit versus implicit MLE

- **Explicit MLE**:
  Occurs when the score equation can be solved in *closed form*. A classic example is the MLE for the mean and variance in a normal distribution.
- **Implicit MLE**:
  Happens when no closed-form solution exists. Iterative numerical methods, such as **Newton–Raphson, expectation–maximization (EM)**, or other optimization algorithms, are used to find $\hat{\theta}$.

Distributional Mis-Specification

- **Definition**: If you assume a distribution for $f_{Y|X}(\cdot;\theta)$ that does *not* reflect the true data-generating process, the MLE may become *inconsistent* or biased in finite samples.
- **Quasi-MLE**:
  - A strategy to handle certain forms of mis-specification.
  - If the chosen distribution belongs to a flexible class or meets certain conditions (e.g., generalized linear models with a robust link), the resulting parameter estimates can remain consistent *for some parameters of interest.*
- **Nonparametric and semiparametric approaches**:
  - Require minimal or no distributional assumptions
  - More robust to mis-specification but can be *harder to implement* and may exhibit higher variance or require larger sample sizes to achieve comparable precision

2.3.5 Practical Considerations

1. **Use cases**
   - MLE is extremely popular for:
     - **Binary outcomes** (logistic regression)
     - **Count data** (Poisson regression)
     - **Strictly positive outcomes** (gamma regression)
     - **Heteroskedastic settings** (models with variance related to mean, e.g., GLM)
2. **Distributional assumptions**
   - The efficiency gains of MLE stem from using a specific probability model.
   - If the assumed model closely reflects the data-generating process, MLE gives accurate parameter estimates and reliable standard errors.
   - MLE assumes knowledge of the conditional distribution of the outcome variable. This assumption parallels the normality assumption in linear regression models (e.g., A6 normal distribution).
   - If severely mis-specified, consider robust or semi-/nonparametric methods.
3. **Comparison with OLS**: See Comparison of MLE and OLS for more details.
   - Ordinary least squares is a special case of MLE when errors are normally distributed and homoskedastic.
   - In more general settings (e.g., non-Gaussian or heteroskedastic data), MLE can outperform OLS in terms of smaller standard errors and better inference.

Table 2.18 Comparative summary of MLE and OLS across estimation, assumptions, and interpretation

Aspect	MLE	OLS
Estimator efficiency	Efficient for correctly specified distributions	Efficient under Gauss–Markov assumptions
Assumptions about errors	Requires specifying a distribution (e.g., normal, binomial)	Requires only mean-zero errors and homoskedasticity
Use of likelihood	Based on maximizing the likelihood function for parameter estimation	Based on minimizing the sum of squared residuals
Model flexibility	More flexible (supports various distributions, nonlinear models)	Primarily linear models (extensions for nonlinear exist)
Interpretation	Log-likelihood values guide model comparison (AIC/BIC)	R-Squared and adjusted R-squared measure fit

4. **Numerical stability and computation**

   - For complex likelihoods, iterative methods can fail to converge or converge to local maxima.
   - Proper initialization and diagnostics (e.g., checking multiple start points) are crucial.

2.3.6 *Comparison of MLE and OLS*

While maximum likelihood estimation is a powerful estimation method, it does not solve all of the challenges associated with ordinary least squares (Table 2.18). Below is a detailed comparison highlighting similarities, differences, and limitations.

Key Points of Comparison

1. **Inference methods**:

   - MLE:

     – Joint inference is typically conducted using **log-likelihood calculations**, such as likelihood ratio tests or information criteria (e.g., AIC, BIC).
     – These methods replace the use of F-statistics commonly associated with OLS.

   - OLS:

     – Relies on the **F-statistic** for hypothesis testing and joint inference.

2. **Sensitivity to functional form**:
   - Both MLE and OLS are sensitive to the **functional form** of the model. Incorrect specification (e.g., linear vs. nonlinear relationships) can lead to biased or inefficient estimates in both cases.
3. **Perfect collinearity and multicollinearity**:
   - Both methods are affected by collinearity:
     - **Perfect collinearity** (e.g., two identical predictors) makes parameter estimation impossible.
     - **Multicollinearity** (highly correlated predictors) inflates standard errors, reducing the precision of estimates.
   - Neither MLE nor OLS directly resolves these issues without additional measures, such as regularization or variable selection.
4. **Endogeneity**:
   - Problems like **omitted variable bias** or **simultaneous equations** affect both MLE and OLS:
     - If relevant predictors are omitted, estimates from both methods are likely to be biased and inconsistent.
     - Similarly, in systems of simultaneous equations, both methods yield biased results unless endogeneity is addressed through instrumental variables or other approaches.
   - MLE, while efficient under correct model specification, does not inherently address endogeneity.

Situations Where MLE and OLS Differ
Practical Considerations

1. **When to use MLE**:
   - Situations where the dependent variable is:
     - Binary (e.g., logistic regression)
     - Count data (e.g., Poisson regression)
     - Skewed or bounded (e.g., survival models)
   - When the model naturally arises from a probabilistic framework
2. **When to use OLS**:
   - Suitable for continuous dependent variables with approximately linear relationships between predictors and outcomes
   - Simpler to implement and interpret when the assumptions of linear regression are reasonably met

2.3.7 Applications of MLE

MLE is widely used across various applications to estimate parameters in models tailored for specific data structures. Below are key applications of MLE, categorized by problem type and estimation method (Table 2.19).

2.3.7.1 Binary Response Models

A binary response variable (y_i) follows a Bernoulli distribution:

$$f_Y(y_i; p) = p^{y_i}(1-p)^{(1-y_i)}$$

where p is the probability of success. For conditional models, the likelihood becomes:

$$f_{Y|X}(y_i, x_i; p(.)) = p(x_i)^{y_i}(1-p(x_i))^{(1-y_i)}.$$

To model $p(x_i)$, we use a function of x_i and unknown parameters θ. A common approach involves a **latent variable model**:

$$\begin{aligned} y_i &= 1\{y_i^* > 0\}, \\ y_i^* &= x_i\beta - \epsilon_i, \end{aligned}$$

where:

- y_i^* is an unobserved (latent) variable
- ϵ_i is a random variable with mean 0, representing unobserved noise

Rewriting in terms of observed data:

$$y_i = 1\{x_i\beta > \epsilon_i\}.$$

The probability function becomes:

$$\begin{aligned} p(x_i) &= P(y_i = 1|x_i) \\ &= P(x_i\beta > \epsilon_i|x_i) \\ &= F_{\epsilon|X}(x_i\beta|x_i), \end{aligned}$$

where $F_{\epsilon|X}(.)$ is the cumulative distribution function (CDF) of ϵ_i. Assuming independence of ϵ_i and x_i, the probability function simplifies to:

$$p(x_i) = F_\epsilon(x_i\beta).$$

Table 2.19 Applications of maximum likelihood estimation in nonlinear and limited dependent variable models

Model type	Examples	Key characteristics	Common estimation methods	Additional notes
Corner solution models	Hours worked Donations to charity Household consumption of a good	Dependent variable is often bfcensored at zero (or another threshold). Large fraction of observations at the corner (e.g., 0 hours, 0 donations)	Tobit regression (latent variable approach with censoring)	Useful when a continuous outcome has a bfmass point at zero but also positive values (e.g., 30% of individuals donate $0, the rest donate greater $0)
Nonnegative count models	Number of arrests Number of cigarettes smoked Doctor visits per year	Dependent variable consists of bfnonnegative integer counts. Possible bfoverdispersion (variance greater mean)	Poisson regression, negative binomial regression	Poisson assumes mean = variance, so often negative binomial is preferred for real data. Zero-inflated models (ZIP/ZINB) may be used for data with bfexcess zeros
Multinomial choice models	Demand for different car brands Votes in a primary election Choice of travel mode	Dependent variable is a bfcategorical choice among bf3+ alternatives. Each category is distinct, with no inherent ordering (e.g., brand A, B, or C)	Multinomial logit, multinomial probit	Extension of binary choice (logit/probit) to multiple categories. bfIndependence of irrelevant alternatives (IIAs) can be a concern for the multinomial logit
Ordinal choice models	Self-reported happiness (low/medium/high) Income level brackets Likert-scale surveys	Dependent variable is bfordered (e.g., bflow less medium less high). Distances between categories are not necessarily equal	Ordered logit, ordered probit	Probit/logit framework adapted to preserve bfordinal information. Interprets latent continuous variable mapped to discrete ordered categories

The conditional expectation function is equivalent:

$$E(y_i|x_i) = P(y_i = 1|x_i) = F_\epsilon(x_i\beta).$$

Common Distributional Assumptions

1. **Probit model**:
   - Assumes ϵ_i follows a standard normal distribution.
   - $F_\epsilon(.) = \Phi(.)$, where $\Phi(.)$ is the standard normal CDF.
2. **Logit model**:
   - Assumes ϵ_i follows a standard logistic distribution.
   - $F_\epsilon(.) = \Lambda(.)$, where $\Lambda(.)$ is the logistic CDF.

Steps to Derive MLE for Binary Models

1. **Specify the log-likelihood**:
   - For a chosen distribution (e.g., normal for probit or logistic for logit), the log-likelihood is:

$$\ln(f_{Y|X}(y_i, x_i; \beta)) = y_i \ln(F_\epsilon(x_i\beta)) + (1 - y_i)\ln(1 - F_\epsilon(x_i\beta)).$$

2. **Maximize the log-likelihood**:
   - Find the parameter estimates that maximize the log-likelihood:

$$\hat{\beta}_{MLE} = \underset{\beta}{\operatorname{argmax}} \sum_{i=1}^{n} \ln(f_{Y|X}(y_i, x_i; \beta)).$$

Properties of Probit and Logit Estimators

- **Consistency and asymptotic normality**:
  - Probit and logit estimators are consistent and asymptotically normal if:

    A2 Full Rank: $E(x_i'x_i)$ exists and is nonsingular.
    A5 Data Generation (Random Sampling): $\{y_i, x_i\}$ are iid (or stationary and weakly dependent).
    Distributional assumptions on ϵ_i hold (e.g., normal or logistic, independent of x_i).

- **Asymptotic efficiency**:
  - Under these assumptions, probit and logit estimators are asymptotically efficient with variance:

$$I(\beta_0)^{-1} = \left[E\left(\frac{(f_\epsilon(x_i\beta_0))^2}{F_\epsilon(x_i\beta_0)(1 - F_\epsilon(x_i\beta_0))} x_i' x_i \right) \right]^{-1},$$

where $f_\epsilon(x_i\beta_0)$ is the PDF (derivative of the CDF).

Interpretation of Binary Response Models

Binary response models, such as probit and logit, estimate the probability of an event occurring ($y_i = 1$) given predictor variables x_i. However, interpreting the estimated coefficients (β) in these models differs significantly from linear models. Below, we explore how to interpret these coefficients and the concept of **partial effects**.

1. Interpreting β in binary response models

In binary response models, the coefficient β_j represents the average change in the **latent variable** y_i^* (an unobserved variable) for a one-unit change in x_{ij}. While this provides insight into the direction of the relationship:

- **Magnitudes** of β_j do not have a direct, meaningful interpretation in terms of y_i.
- **Direction** of β_j is meaningful:
  - $\beta_j > 0$: a positive association between x_{ij} and the probability of $y_i = 1$
  - $\beta_j < 0$: a negative association between x_{ij} and the probability of $y_i = 1$

2. Partial effects in nonlinear binary models

To interpret the effect of a change in a predictor x_{ij} on the probability of an event occurring ($P(y_i = 1|x_i)$), we use the **partial effect**:

$$E(y_i|x_i) = F_\epsilon(x_i\beta),$$

where $F_\epsilon(.)$ is the cumulative distribution function (CDF) of the error term ϵ_i (e.g., standard normal for probit, logistic for logit). The **partial effect** is the derivative of the expected probability with respect to x_{ij}:

$$PE(x_{ij}) = \frac{\partial E(y_i|x_i)}{\partial x_{ij}} = f_\epsilon(x_i\beta)\beta_j,$$

where:

- $f_\epsilon(.)$ is the probability density function (PDF) of the error term ϵ_i
- β_j is the coefficient associated with x_{ij}

3. Key characteristics of partial effects

- **Scaling factor**:
  - The partial effect depends on a **scaling factor**, $f_\epsilon(x_i\beta)$, which is derived from the density function $f_\epsilon(.)$.

  - The scaling factor varies depending on the values of x_i, making the partial effect nonlinear and **context-dependent**.

- **Nonconstant partial effects**:
  - Unlike linear models where coefficients directly represent constant marginal effects, the partial effect in binary models changes based on x_i.
  - For example, in a logit model, the partial effect is largest when $P(y_i = 1|x_i)$ is around 0.5 (the midpoint of the S-shaped logistic curve) and smaller at the extremes (close to 0 or 1).

4. Single values for partial effects

   In practice, researchers often summarize partial effects using either:

- **Partial effect at the average (PEA)**:
  - The partial effect is calculated for an "average individual," where $x_i = \bar{x}$ (the sample mean of predictors):

$$PEA = f_\epsilon(\bar{x}\hat{\beta})\hat{\beta}_j.$$

  - This provides a single, interpretable value but assumes the average effect applies to all individuals.

- **Average partial effect (APE)**:
  - The average of all individual-level partial effects across the sample:

$$APE = \frac{1}{n}\sum_{i=1}^{n} f_\epsilon(x_i\hat{\beta})\hat{\beta}_j.$$

  - This accounts for the nonlinearity of the partial effects and provides a more accurate summary of the marginal effect in the population.

5. Comparing partial effects in linear and nonlinear models

- **Linear models**:
  - Partial effects are constant: $APE = PEA$.
  - The coefficients directly represent the marginal effects on $E(y_i|x_i)$.

- **Nonlinear models**:
  - Partial effects are **not constant** due to the dependence on $f_\epsilon(x_i\beta)$.
  - As a result, $APE \neq PEA$ in general.

2.4 Penalized (Regularized) Estimators

Penalized or regularized estimators are extensions of ordinary least squares designed to address its limitations, particularly in high-dimensional settings. Regularization methods introduce a penalty term to the loss function to prevent overfitting, handle multicollinearity, and improve model interpretability.

There are three popular regularization techniques (but not limited to):

1. Ridge regression
2. Lasso regression
3. Elastic net

2.4.1 Motivation for Penalized Estimators

OLS minimizes the residual sum of squares (RSS):

$$RSS = \sum_{i=1}^{n} \left(y_i - \hat{y}_i\right)^2 = \sum_{i=1}^{n} \left(y_i - x_i'\beta\right)^2,$$

where:

- y_i is the observed outcome
- x_i is the vector of predictors for observation i
- β is the vector of coefficients

While OLS works well under ideal conditions (e.g., low dimensionality, no multicollinearity), it struggles when:

- **Multicollinearity**: Predictors are highly correlated, leading to large variances in β estimates.
- **High dtimensionality**: The number of predictors (p) exceeds or approaches the sample size (n), making OLS inapplicable or unstable.
- **Overfitting**: When p is large, OLS fits noise in the data, reducing generalizability.

To address these issues, **penalized regression** modifies the OLS loss function by adding a **penalty term** that shrinks the coefficients toward zero. This discourages overfitting and improves predictive performance. The general form of the penalized loss function is:

$$L(\beta) = \sum_{i=1}^{n} \left(y_i - x_i'\beta\right)^2 + \lambda P(\beta),$$

where:

- $\lambda \geq 0$: tuning parameter controlling the strength of regularization
- $P(\beta)$: penalty term that quantifies model complexity

Different choices of $P(\beta)$ lead to ridge regression, lasso regression, or elastic net.

2.4.2 *Ridge Regression*

Ridge regression, also known as **L2 regularization**, penalizes the sum of squared coefficients:

$$P(\beta) = \sum_{j=1}^{p} \beta_j^2.$$

The ridge objective function becomes:

$$L_{ridge}(\beta) = \sum_{i=1}^{n} \left(y_i - x_i'\beta\right)^2 + \lambda \sum_{j=1}^{p} \beta_j^2,$$

where:

- $\lambda \geq 0$ controls the degree of shrinkage. Larger λ leads to greater shrinkage.

Ridge regression has a closed-form solution:

$$\hat{\beta}_{ridge} = \left(X'X + \lambda I\right)^{-1} X'y,$$

where I is the $p \times p$ identity matrix.

Key Features

- Shrinks coefficients but does **not set them exactly to zero**
- Handles multicollinearity effectively by stabilizing the coefficient estimates (Hoerl & Kennard, 1970)
- Works well when all predictors contribute to the response

Example Use Case

Ridge regression is ideal for applications with many correlated predictors, such as:

- Predicting housing prices based on a large set of features (e.g., size, location, age of the house)

2.4.3 Lasso Regression

Lasso regression, or **L1 regularization**, penalizes the sum of absolute coefficients:

$$P(\beta) = \sum_{j=1}^{p} |\beta_j|.$$

The lasso objective function is:

$$L_{lasso}(\beta) = \sum_{i=1}^{n} \left(y_i - x_i'\beta\right)^2 + \lambda \sum_{j=1}^{p} |\beta_j|.$$

Key Features

- Unlike ridge regression, lasso can set coefficients to **exactly zero**, performing automatic feature selection.
- Encourages sparse models, making it suitable for high-dimensional data (Tibshirani, 1996).

Optimization

Lasso does not have a closed-form solution due to the non-differentiability of $|\beta_j|$ at $\beta_j = 0$. It requires iterative algorithms, such as:

- **Coordinate descent**
- **Least angle regression (LARS)**

Example Use Case

Lasso regression is useful when many predictors are irrelevant, such as:

- Genomics, where only a subset of genes are associated with a disease outcome

2.4.4 Elastic Net

Elastic net combines the penalties of ridge and lasso regression:

$$P(\beta) = \alpha \sum_{j=1}^{p} |\beta_j| + \frac{1-\alpha}{2} \sum_{j=1}^{p} \beta_j^2,$$

where:

- $0 \leq \alpha \leq 1$ determines the balance between lasso (L1) and ridge (L2) penalties
- λ controls the overall strength of regularization

The elastic net objective function is:

$$L_{elastic\ net}(\beta) = \sum_{i=1}^{n} \left(y_i - x_i'\beta\right)^2 + \lambda \left(\alpha \sum_{j=1}^{p} |\beta_j| + \frac{1-\alpha}{2} \sum_{j=1}^{p} \beta_j^2 \right).$$

Key Features

- Combines the strengths of lasso (sparse models) and ridge (stability with correlated predictors) (Zou & Hastie, 2005)
- Effective when predictors are highly correlated or when $p > n$

Example Use Case

Elastic net is ideal for high-dimensional datasets with correlated predictors, such as:

- Predicting customer churn using demographic and behavioral features

2.4.5 Tuning Parameter Selection

Choosing the regularization parameter λ (and α for elastic net) is critical for balancing model complexity (fit) and regularization (parsimony). If λ is too large, coefficients are overly shrunk (or even set to zero in the case of L1 penalty), leading to underfitting. If λ is too small, the model might overfit because coefficients are not penalized sufficiently. Hence, a systematic approach is needed to determine the optimal λ. For elastic net, we also choose an appropriate α to balance the L1 and L2 penalties.

2.4.5.1 Cross-Validation

A common approach to selecting λ (and α) is K-fold cross-validation:

1. **Partition the data** into K roughly equal-sized "folds."
2. **Train the model** on $K - 1$ folds and **validate** on the remaining fold, computing a validation error.
3. Repeat this process **for all folds**, and compute the average validation error across the K folds.
4. **Select** the value of λ (and α if tuning it) that **minimizes** the cross-validated error.

This method helps us maintain a good bias-variance trade-off because every point is used for both training and validation exactly once.

2.4.5.2 Information Criteria

Alternatively, one can use **information criteria**—like the Akaike information criterion (AIC) or the Bayesian information criterion (BIC)—to guide model selection. These criteria reward goodness-of-fit while penalizing model complexity, thereby helping in selecting an appropriately regularized model.

2.4.6 Properties of Penalized Estimators

1. **Bias–variance trade-off**:
   - Regularization introduces some bias in exchange for reducing variance, often resulting in better predictive performance on new data.
2. **Shrinkage**:
   - Ridge shrinks coefficients toward zero but usually retains all predictors.
   - Lasso shrinks some coefficients exactly to zero, performing inherent feature selection.
3. **Flexibility**:
   - Elastic net allows for a continuum between ridge and lasso, so it can adapt to different data structures (e.g., many correlated features or very-high-dimensional feature spaces).

```
# Load required libraries
library(glmnet)
#> Loading required package: Matrix
#> Loaded glmnet 4.1-8

# Simulate data
set.seed(123)
n <- 100    # Number of observations
p <- 20     # Number of predictors
X <- matrix(rnorm(n * p), nrow = n, ncol = p)  # Predictor
↪  matrix
y <- rnorm(n)                                   # Response
↪  vector

# Ridge regression (alpha = 0)
ridge_fit <- glmnet(X, y, alpha = 0)
plot(ridge_fit, xvar = "lambda", label = TRUE)
title("Coefficient Paths for Ridge Regression")
```

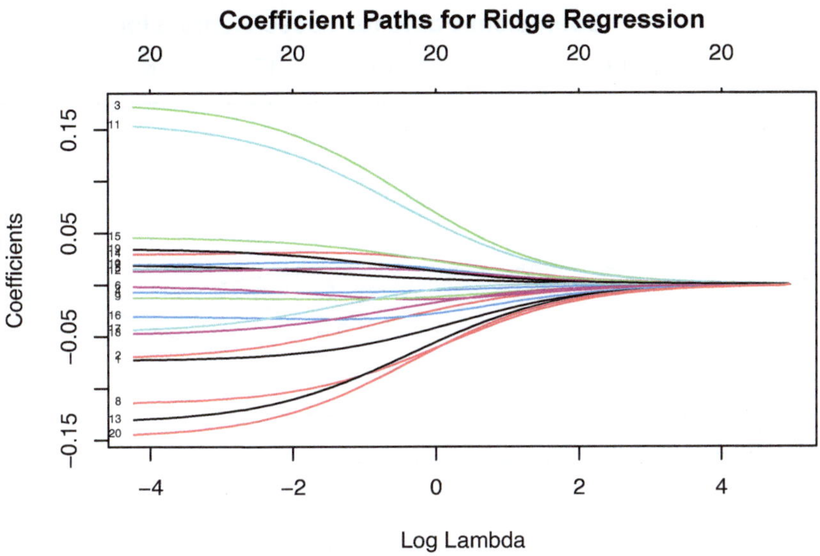

Fig. 2.12 Coefficient paths for ridge regression

- In Fig. 2.12, each curve represents a coefficient's value as a function of λ.
- As λ increases (moving from left to right on a log scale by default), coefficients shrink toward zero but typically stay nonzero.
- Ridge regression tends to shrink coefficients but does not force them to be exactly zero.

```
# Lasso regression (alpha = 1)
lasso_fit <- glmnet(X, y, alpha = 1)
plot(lasso_fit, xvar = "lambda", label = TRUE)
title("Coefficient Paths for Lasso Regression")
```

Here, as λ grows, several coefficient paths **hit zero exactly**, illustrating the variable selection property of lasso (Fig. 2.13).

```
# Elastic net (alpha = 0.5)
elastic_net_fit <- glmnet(X, y, alpha = 0.5)
plot(elastic_net_fit, xvar = "lambda", label = TRUE)
title("Coefficient Paths for Elastic Net (alpha = 0.5)")
```

- Elastic net combines ridge and lasso penalties. At $\lambda = 0.5$, we see partial shrinkage and some coefficients going to zero (Fig. 2.14).
- This model is often helpful when you suspect both group-wise shrinkage (like ridge) and sparse solutions (like lasso) might be beneficial.

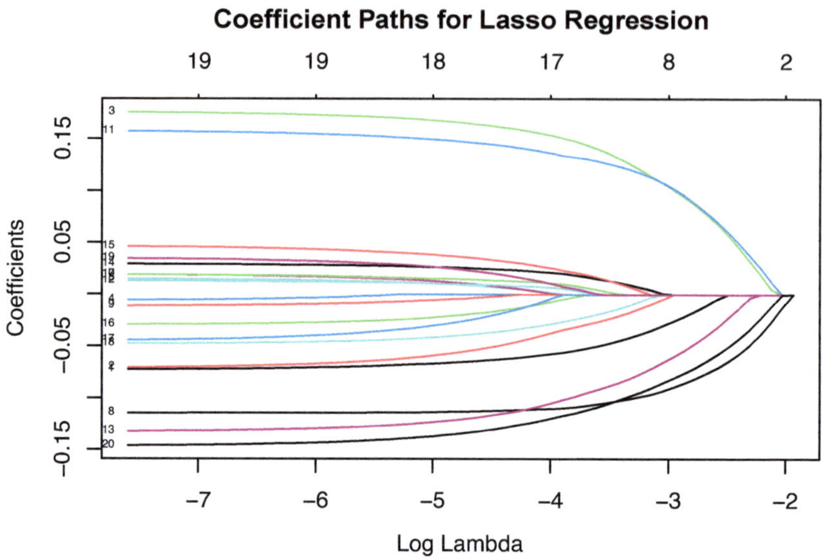

Fig. 2.13 Coefficient Paths for Lasso Regression

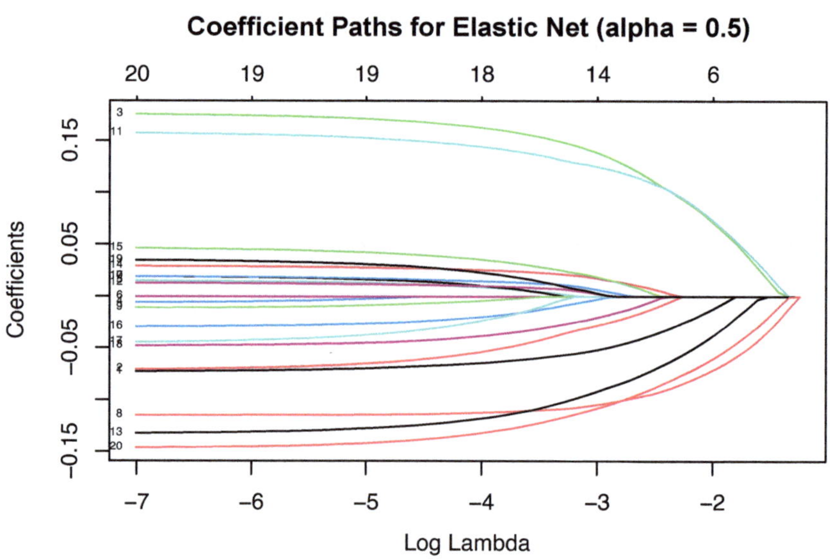

Fig. 2.14 Coefficient paths for elastic net

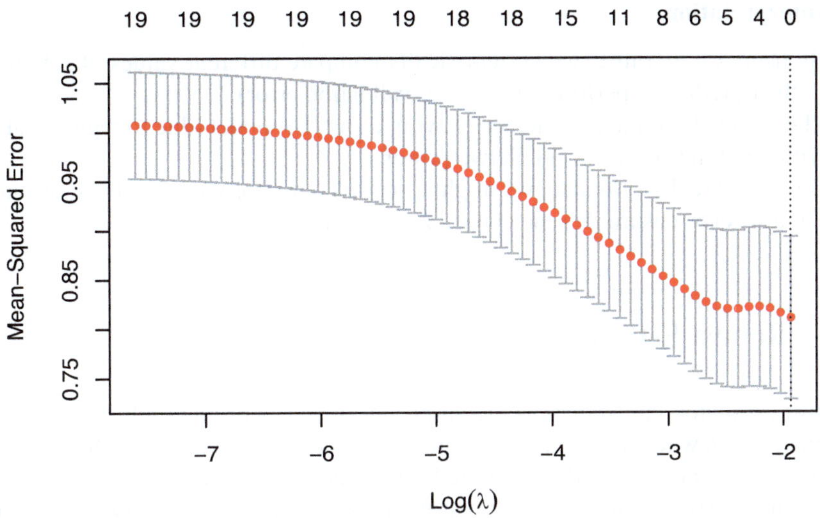

Fig. 2.15 Cross-validation curve for lasso regression: mean squared error by log lambda

We can further refine our choice of λ by performing cross-validation on the lasso model:

```
cv_lasso <- cv.glmnet(X, y, alpha = 1)
plot(cv_lasso)
```

```
best_lambda <- cv_lasso$lambda.min
best_lambda
#> [1] 0.1449586
```

- Figure 2.15 displays the cross-validated error (often mean squared error or deviance) on the y-axis versus log(λ) on the x-axis.
- Two vertical dotted lines typically appear:

  1. $\lambda.min$: The λ that achieves the minimum cross-validated error.
  2. $\lambda.1se$: The largest λ such that the cross-validated error is still within one standard error of the minimum. This is a more conservative choice that favors higher regularization (simpler models).

- `best_lambda` above prints the numeric value of $\lambda.min$. This is the λ that gave the lowest cross-validation error for the lasso model.

Interpretation:

- By using `cv.glmnet`, we systematically compare different values of λ in terms of their predictive performance (cross-validation error).
- The selected λ typically balances having a smaller model (due to regularization) with retaining sufficient predictive power.
- If we used real-world data, we might also look at performance metrics on a hold-out test set to ensure that the chosen λ generalizes well.

2.5 Robust Estimators

Robust estimators are statistical techniques designed to provide reliable parameter estimates even when the assumptions underlying classical methods, such as ordinary least squares, are violated. Specifically, they address issues caused by outliers, nonnormal errors, or heavy-tailed distributions, which can render OLS inefficient or biased.

The goal of robust estimation is to reduce the sensitivity of the estimator to extreme or aberrant data points, thereby ensuring a more reliable and accurate fit to the majority of the data.

We will cover the key robust estimation techniques, their properties, and applications, along with practical examples and mathematical derivations. The focus will include M-estimators, R-estimators, L-estimators, LTS, S-estimators, MM-estimators, and more.

2.5.1 Motivation for Robust Estimation

OLS seeks to minimize the residual sum of squares (RSS):

$$RSS = \sum_{i=1}^{n} (y_i - x_i'\beta)^2,$$

where:

- y_i is the observed response for the ith observation
- x_i is the vector of predictors for the ith observation
- β is the vector of coefficients

OLS assumes:

1. Errors are normally distributed and no outliers in the data (A6 Normal Distribution).
2. Homoskedasticity (constant variance of errors) (A4 Homoskedasticity).

In real-world scenarios:

- **Outliers** in y or x can disproportionately affect the estimates, leading to biased or inefficient results.
- **Heavy-tailed distributions** (e.g., Cauchy) violate the normality assumption, making OLS inappropriate.

For example, Huber (1964) demonstrates that a single extreme observation can arbitrarily distort OLS estimates, while Hampel et al. (2005) define the breakdown point as a measure of robustness. Robust estimators aim to mitigate these problems by limiting the influence of problematic observations.

OLS inherently squares the residuals $e_i = y_i - x_i'\beta$, amplifying the influence of large residuals. For example, if a single residual is much larger than the others, its squared value can dominate the RSS, distorting the estimated coefficients.

Consider a simple case where $y_i = \beta_0 + \beta_1 x_i + e_i$, with $e_i \sim N(0, \sigma^2)$ under the classical assumptions. Now introduce an outlier: a single observation with an unusually large e_i. The squared residual for this point will dominate the RSS and pull the estimated regression line toward it, leading to biased estimates of β_0 and β_1.

The **breakdown point** of an estimator is the proportion of contamination (e.g., outliers) that the estimator can tolerate before yielding arbitrarily large or incorrect results. For OLS, the breakdown point is $1/n$, meaning even one outlier can cause substantial distortion in the estimates.

2.5.2 *M-Estimators*

To address the sensitivity of OLS, robust estimators minimize a different objective function:

$$\sum_{i=1}^{n} \rho\left(\frac{y_i - x_i'\beta}{\sigma}\right),$$

where:

- $\rho(\cdot)$ is a robust loss function that grows slower than the quadratic function used in OLS
- σ is a scale parameter to normalize residuals

In OLS, the quadratic loss function $\rho(z) = z^2$ penalizes large residuals disproportionately. Robust estimators replace this with alternative ρ functions that limit the penalty for large residuals, thus reducing their influence on the parameter estimates.

A robust ρ function should satisfy the following properties:

1. **Bounded influence**: Large residuals contribute a finite amount to the objective function.

2. **Symmetry**: $\rho(z) = \rho(-z)$ ensures that positive and negative residuals are treated equally.
3. **Differentiability**: For computational tractability, ρ should be smooth and differentiable.

2.5.2.1 Examples of Robust ρ Functions

1. **Huber's Loss Function** (Huber, 1964)

Huber's loss function transitions between quadratic and linear growth:

$$\rho(z) = \begin{cases} \frac{z^2}{2} & \text{if } |z| \leq c, \\ c|z| - \frac{c^2}{2} & \text{if } |z| > c. \end{cases}$$

Key features:

- For small residuals ($|z| \leq c$), the loss is quadratic, mimicking OLS.
- For large residuals ($|z| > c$), the loss grows linearly, limiting their influence.

The parameter c controls the threshold at which the loss function transitions from quadratic to linear. Smaller values of c make the estimator more robust but potentially less efficient under normality.

2. **Tukey's Bisquare Function** (Beaton & Tukey, 1974)

   Tukey's bisquare function completely bounds the influence of large residuals:

$$\rho(z) = \begin{cases} c^2 \left(1 - \left(1 - \left(\frac{z}{c}\right)^2\right)^3\right)/6 & \text{if } |z| \leq c, \\ c^2/6 & \text{if } |z| > c. \end{cases}$$

Key features:

- Residuals larger than c contribute a constant value to the objective function, effectively excluding them from the estimation process.
- This approach achieves high robustness at the cost of lower efficiency for small residuals.

3. **Andrews' Sine Function** (Andrews, 1974):

   - Smoothly downweights extreme residuals:

$$\rho(z) = \begin{cases} c^2 \left(1 - \cos\left(\frac{z}{c}\right)\right)/2 & \text{if } |z| \leq \pi c, \\ c^2/2 & \text{if } |z| > \pi c. \end{cases}$$

2.5.2.2 Weighting Scheme: Influence Functions

A critical concept in robust estimation is the **influence function**, which describes the sensitivity of the estimator to individual observations. For M-estimators, the influence function is derived as the derivative of the loss function $\rho(z)$ with respect to z:

$$\psi(z) = \frac{d}{dz}\rho(z).$$

This function plays a crucial role in downweighting large residuals. The weight assigned to each residual is proportional to $\psi(z)/z$, which decreases as $|z|$ increases for robust estimators.

For Huber's loss function, the influence function is:

$$\psi(z) = \begin{cases} z & \text{if } |z| \leq c, \\ c \cdot \text{sign}(z) & \text{if } |z| > c. \end{cases}$$

- For small residuals, $\psi(z) = z$, matching OLS.
- For large residuals, $\psi(z)$ is constant, ensuring bounded influence.

A key consideration when selecting a robust estimator is the trade-off between **robustness** (resistance to outliers) and **efficiency** (performance under ideal conditions). The tuning parameters in ρ functions (e.g., c in Huber's loss) directly affect this balance:

- Smaller c increases robustness but reduces efficiency under normality.
- Larger c improves efficiency under normality but decreases robustness to outliers.

This trade-off reflects the fundamental goal of robust estimation: to achieve a balance between reliability and precision across a wide range of data scenarios.

2.5.2.3 Properties of M-Estimators

Robust estimators, particularly M-estimators, possess the following mathematical properties:

1. **Asymptotic normality**: Under mild regularity conditions, M-estimators are asymptotically normal:

   $$\sqrt{n}(\hat{\beta} - \beta) \xrightarrow{d} N(0, \Sigma),$$

   where Σ depends on the choice of ρ and the distribution of residuals.

2. **Consistency**: As $n \to \infty$, $\hat{\beta} \to \beta$ in probability, provided the majority of the data satisfies the model assumptions.
3. **Breakdown point**: M-estimators typically have a moderate breakdown point, sufficient to handle a reasonable proportion of contamination.

2.5.3 *R-Estimators*

R-Estimators are a class of robust estimators that rely on the ranks of residuals rather than their raw magnitudes. This approach makes them naturally resistant to the influence of outliers and highly effective in scenarios involving ordinal data or heavy-tailed error distributions. By leveraging rank-based methods, R-estimators are particularly useful in situations where classical assumptions about the data, such as normality or homoskedasticity, do not hold.

The general form of an R-estimator can be expressed as:

$$\hat{\beta}_R = \arg\min_{\beta} \sum_{i=1}^{n} w_i R_i \left(y_i - x_i' \beta \right),$$

where:

- R_i are the ranks of residuals $e_i = y_i - x_i'\beta$
- w_i are rank-based weights determined by a chosen scoring function
- y_i are observed responses, x_i are predictor values, and β is the vector of coefficients

This formulation differs from M-estimators, which directly minimize a loss function ρ, by instead using the **ordering of residuals** to drive the estimation.

2.5.3.1 Ranks and Scoring Function

2.5.3.1.1 Definition of Ranks

The rank R_i of a residual e_i is its position in the sorted sequence of all residuals:

$$R_i = \text{rank}(e_i) = \sum_{j=1}^{n} \mathbb{I}(e_j \le e_i),$$

where $\mathbb{I}(\cdot)$ is the indicator function, equal to 1 if the condition is true and 0 otherwise. This step transforms the residuals into an ordinal scale, eliminating their dependency on magnitude.

2.5.3.1.2 Scoring Function

The weights w_i are derived from a scoring function $S(R_i)$, which assigns importance to each rank. A common choice is the **Wilcoxon scoring function**, defined as:

$$S(R_i) = \frac{R_i}{n + 1},$$

which gives equal weight to all ranks, scaled by their position relative to the total number of observations n.

Other scoring functions can emphasize different parts of the rank distribution:

- **Normal scores**: Derived from the quantiles of a standard normal distribution.
- **Logarithmic scores**: Weight lower ranks more heavily.

The flexibility of the scoring function allows R-estimators to adapt to various data structures and assumptions.

2.5.3.2 Properties of R-Estimators

2.5.3.2.1 Influence Function and Robustness

A key feature of R-estimators is their **bounded influence function**, which ensures robustness. Because the estimator depends only on the ranks of the residuals, extreme values in y or x do not disproportionately affect the results.

For R-estimators, the influence function $\psi(e_i)$ is proportional to the derivative of the rank-based objective function:

$$\psi(e_i) = S'(R_i),$$

where $S'(R_i)$ is the derivative of the scoring function. Since R_i depends only on the ordering of residuals, outliers in the data cannot produce excessive changes in R_i, resulting in bounded influence.

2.5.3.2.2 Breakdown Point

The **breakdown point** of R-estimators is higher than that of OLS and comparable to other robust methods. This means they can tolerate a larger proportion of contaminated data without yielding unreliable results.

2.5.3.2.3 Asymptotic Efficiency

Under specific scoring functions, R-estimators achieve high asymptotic efficiency. For example, the Wilcoxon R-estimator performs nearly as well as OLS under normality while retaining robustness to nonnormality.

2.5.3.3 Derivation of R-Estimators for Simple Linear Regression

Consider the simple linear regression model:

$$y_i = \beta_0 + \beta_1 x_i + e_i,$$

where $e_i = y_i - (\beta_0 + \beta_1 x_i)$ are the residuals.

1. **Rank the residuals**: Compute the residuals e_i for all observations and rank them from smallest to largest.
2. **Assign weights**: Compute weights w_i for each residual rank based on the scoring function $S(R_i)$.
3. **Minimize the rank-based objective**: Solve the following optimization problem:

$$\hat{\beta}_R = \arg\min_{\beta_0, \beta_1} \sum_{i=1}^{n} w_i R_i \left(y_i - (\beta_0 + \beta_1 x_i)\right).$$

   This minimization can be performed iteratively using numerical methods, as the rank-based nature of the function makes direct analytic solutions challenging.

2.5.3.4 Comparison to M-Estimators

While M-estimators downweight large residuals using robust loss functions, R-estimators completely avoid reliance on the magnitude of residuals by using their ranks. This distinction has important implications:

- R-Estimators are naturally robust to leverage points and extreme outliers.
- The performance of R-estimators is less sensitive to the choice of scale parameter compared to M-estimators.
- However, R-estimators may be less efficient than M-estimators under normality because they do not use the full information contained in the residual magnitudes.

2.5.4 L-Estimators

L-Estimators are a class of robust estimators constructed as **linear combinations of order statistics**, where order statistics are simply the sorted values of a dataset. These estimators are particularly appealing due to their intuitive nature and computational simplicity. By using the relative ranks of observations, L-estimators offer robustness against outliers and heavy-tailed distributions.

Order statistics are denoted as $y_{(1)}, y_{(2)}, \ldots, y_{(n)}$, where $y_{(i)}$ is the ith smallest observation in the sample.

The general form of an L-estimator is:

$$\hat{\theta}_L = \sum_{i=1}^{n} c_i \, y_{(i)},$$

where:

- $y_{(i)}$ are the order statistics (sorted observations)
- c_i are coefficients (weights) that determine the contribution of each order statistic to the estimator

By appropriately choosing the weights c_i, different types of L-estimators can be constructed to suit specific needs, such as handling outliers or capturing central tendencies robustly.

Examples of L-Estimators

1. **Sample median**: The sample median is a simple L-estimator where only the middle-order statistic contributes (for odd n) or the average of the two middle-order statistics contributes (for even n):

$$\hat{\mu}_{\text{median}} = \begin{cases} y_{\left(\frac{n+1}{2}\right)} & \text{if } n \text{ is odd,} \\ \frac{1}{2}\left(y_{(\frac{n}{2})} + y_{(\frac{n}{2}+1)}\right) & \text{if } n \text{ is even.} \end{cases}$$

   - **Robustness**: The median has a breakdown point of 50%, meaning it remains unaffected unless more than half the data are corrupted.
   - **Efficiency**: Under normality, the efficiency of the median is lower than that of the mean (about 64%).

2. **Trimmed mean**: The trimmed mean excludes the smallest and largest $k\%$ of observations before averaging the remaining values:

$$\hat{\mu}_T = \frac{1}{n-2k} \sum_{i=k+1}^{n-k} y_{(i)},$$

where:

- k is the number of observations trimmed from each tail
- n is the sample size
- **Robustness**: The trimmed mean is less sensitive to extreme values than the sample mean.
- **Efficiency**: By retaining most observations, the trimmed mean achieves a good balance between robustness and efficiency.

3. **Winsorized mean**: Similar to the trimmed mean, but instead of excluding extreme values, it replaces them with the nearest remaining observations:

$$\hat{\mu}_W = \frac{1}{n}\sum_{i=1}^{n} y^*_{(i)},$$

where $y^*_{(i)}$ are "Winsorized" values:

$$y^*_{(i)} = \begin{cases} y_{(k+1)} & \text{if } i \le k, \\ y_{(i)} & \text{if } k+1 \le i \le n-k, \\ y_{(n-k)} & \text{if } i > n-k. \end{cases}$$

- **Robustness**: The Winsorized mean reduces the influence of outliers without discarding data.
- **Efficiency**: Slightly less efficient than the trimmed mean under normality.

4. **Midrange**: The midrange is the average of the smallest and largest observations:

$$\hat{\mu}_{\text{midrange}} = \frac{y_{(1)} + y_{(n)}}{2}.$$

- **Robustness**: Poor robustness, as it depends entirely on the extreme observations
- **Simplicity**: Highly intuitive and computationally trivial

2.5.4.1 Properties of L-Estimators

1. **Robustness to outliers**: L-Estimators gain robustness by downweighting or excluding extreme observations. For instance:
   - The trimmed mean completely removes outliers from the estimation process.
   - The Winsorized mean limits the influence of outliers by bounding their values.

2. **Breakdown point**:
   - The breakdown point of an L-estimator depends on how many extreme observations are excluded or replaced.

- The median has the highest possible breakdown point (50%), while the trimmed and Winsorized means have breakdown points proportional to the trimming percentage.

3. **Efficiency**:
   - The efficiency of L-estimators varies depending on the underlying data distribution and the specific estimator.
   - For symmetric distributions, the trimmed mean and Winsorized mean approach the efficiency of the sample mean while being much more robust.

4. **Computational simplicity**:
   - L-Estimators involve simple operations like sorting and averaging, making them computationally efficient even for large datasets.

2.5.4.2 Derivation of the Trimmed Mean

To understand the robustness of the trimmed mean, consider a dataset with n observations. Sorting the data gives $y_{(1)} \leq y_{(2)} \leq \cdots \leq y_{(n)}$. After trimming the smallest k and largest k observations, the remaining $n - 2k$ observations are used to compute the mean:

$$\hat{\mu}_T = \frac{1}{n - 2k} \sum_{i=k+1}^{n-k} y_{(i)}.$$

Key observations:

- **Impact of** k: Larger k increases robustness by removing more extreme values but reduces efficiency by discarding more data.
- **Choosing** k: In practice, k is often chosen as a percentage of the total sample size, such as 10% trimming ($k = 0.1n$).

2.5.5 Least Trimmed Squares (LTS)

Least trimmed squares (LTS) is a robust regression method that minimizes the sum of the smallest h squared residuals, rather than using all residuals as in ordinary least squares. This approach ensures that large residuals, often caused by outliers or leverage points, have no influence on the parameter estimation.

The LTS estimator is defined as:

$$\hat{\beta}_{LTS} = \arg\min_{\beta} \sum_{i=1}^{h} r_{[i]}^2,$$

where:

- $r_{[i]}^2$ are the ordered squared residuals, ranked from smallest to largest
- h is the subset size of residuals to include in the minimization, typically chosen as $h = \lfloor n/2 \rfloor + 1$ (where n is the sample size)

This trimming process ensures robustness by focusing on the best-fitting h observations and ignoring the most extreme residuals.

2.5.5.1 Motivation for LTS

In OLS regression, the objective is to minimize the residual sum of squares (RSS):

$$RSS = \sum_{i=1}^{n} r_i^2,$$

where $r_i = y_i - x_i'\beta$ are the residuals. However, this method is highly sensitive to outliers because even one large residual (r_i^2) can dominate the RSS, distorting the parameter estimates β.

LTS addresses this issue by trimming the largest residuals and focusing only on the h smallest ones, thus preventing extreme values from affecting the fit. This approach provides a more robust estimate of the regression coefficients β.

2.5.5.2 Properties of LTS

1. **Objective function**: The LTS objective function is non-differentiable because it involves ordering the squared residuals. Formally, the ordered residuals are denoted as:

$$r_{[1]}^2 \leq r_{[2]}^2 \leq \cdots \leq r_{[n]}^2,$$

   and the objective is to minimize:

$$\sum_{i=1}^{h} r_{[i]}^2.$$

   This requires sorting the squared residuals, making the computation more complex than OLS.
2. **Choice of h**: The parameter h determines the number of residuals included in the minimization. A common choice is:

$$h = \lfloor n/2 \rfloor + 1,$$

which ensures a high breakdown point (discussed below). Smaller values of h increase robustness but reduce efficiency, while larger h values improve efficiency but decrease robustness.

3. **Breakdown point**: LTS has a **breakdown point** of approximately 50%, the highest possible for a regression estimator. This means that LTS can handle up to 50% of contaminated data (e.g., outliers) without yielding unreliable estimates.
4. **Robustness**: By focusing only on the h best-fitting observations, LTS naturally excludes outliers from the estimation process, making it highly robust to both vertical outliers (extreme values in y) and leverage points (extreme values in x).

2.5.5.3 Algorithm for LTS

Computing the LTS estimator involves the following steps:

1. **Initialization**: Select an initial subset of h observations to compute a preliminary fit for β.
2. **Residual calculation**: For each observation, compute the squared residuals:

$$r_i^2 = \left(y_i - x_i'\beta\right)^2 .$$

3. **Trimming**: Rank the residuals from smallest to largest and retain only the h smallest residuals.
4. **Refitting**: Use the h retained observations to recompute the regression coefficients β.
5. **Iterative refinement**: Repeat the process (residual calculation, trimming, refitting) until convergence, typically when β stabilizes.

Efficient algorithms, such as the **Fast-LTS** algorithm, are used in practice to reduce computational complexity.

2.5.5.4 Comparison of LTS with OLS

Table 2.20 shows comparison of LTS with OLS.

Table 2.20 Comparison of LTS with OLS

Property	OLS	LTS
Objective	Minimize $\sum_{i=1}^{n} r_i^2$	Minimize $\sum_{i=1}^{h} r_{[i]}^2$
Sensitivity to outliers	High	Low
Breakdown point	$1/n$	$\approx 50\%$
Computational cost	Low	Moderate (requires sorting and iterations)

2.5.6 S-Estimators

S-Estimators are a class of robust estimators that focus on minimizing a robust measure of the **dispersion** of residuals. Unlike methods such as M-estimators, which directly minimize a loss function based on residuals, S-estimators aim to find the parameter values β that produce residuals with the smallest robust scale. These estimators are particularly useful in handling datasets with outliers, heavy-tailed distributions, or other violations of classical assumptions.

The scale σ is estimated by solving the following minimization problem:

$$\hat{\sigma}_S = \arg\min_{\sigma} \frac{1}{n} \sum_{i=1}^{n} \rho \left(\frac{y_i - x_i' \beta}{\sigma} \right),$$

where:

- ρ is a robust loss function that controls the influence of residuals
- y_i are observed responses x_i are predictors β is the vector of regression coefficients
- σ represents the robust scale of the residuals

Once σ is estimated, the S-estimator of β is obtained by solving:

$$\hat{\beta}_S = \arg\min_{\beta} \hat{\sigma}_S.$$

2.5.6.1 Motivation for S-Estimators

In regression analysis, classical methods such as ordinary least squares rely on minimizing the residual sum of squares (RSS). However, OLS is highly sensitive to outliers because even a single extreme residual can dominate the sum of squared residuals, leading to biased estimates of β.

S-Estimators address this limitation by using a robust scale σ to evaluate the dispersion of residuals. By minimizing this scale, S-estimators effectively downweight the influence of outliers, resulting in parameter estimates that are more resistant to contamination in the data.

2.5.6.2 Key Concepts in S-Estimators

1. **Robust scale function**: The key idea of S-estimators is to minimize a robust measure of scale. The scale σ is computed such that the residuals normalized by σ produce a value close to the expected contribution of well-behaved observations.

Formally, σ satisfies:

$$\frac{1}{n}\sum_{i=1}^{n}\rho\left(\frac{y_i - x_i'\beta}{\sigma}\right) = \delta,$$

where δ is a constant that depends on the choice of ρ and ensures consistency under normality. This equation balances the residuals and controls their influence on the scale estimate.

2. **Choice of** ρ-function: The choice of the robust ρ-function is critical in determining the behavior of S-estimators. Common ρ-functions include:

   - **Huber's** ρ-function:

$$\rho(z) = \begin{cases} z^2/2 & \text{if } |z| \leq c, \\ c|z| - c^2/2 & \text{if } |z| > c. \end{cases}$$

   - **Tukey's bisquare**:

$$\rho(z) = \begin{cases} c^2\left(1 - \left(1 - \left(\frac{z}{c}\right)^2\right)^3\right)/6 & \text{if } |z| \leq c, \\ c^2/6 & \text{if } |z| > c. \end{cases}$$

   - **Andrews' sine**:

$$\rho(z) = \begin{cases} c^2\left(1 - \cos\left(\frac{z}{c}\right)\right)/2 & \text{if } |z| \leq \pi c, \\ c^2/2 & \text{if } |z| > \pi c. \end{cases}$$

Robust ρ-functions grow more slowly than the quadratic function used in OLS, limiting the impact of large residuals.

2.5.6.3 Properties of S-Estimators

1. **Breakdown point**: S-Estimators have a breakdown point of up to 50%, meaning they can tolerate up to half the data being contaminated (e.g., outliers) without yielding unreliable estimates.
2. **Efficiency**: The efficiency of S-estimators depends on the choice of ρ. While they are highly robust, their efficiency under ideal conditions (e.g., normality) may be lower than that of OLS. Proper tuning of ρ can balance robustness and efficiency.
3. **Influence function**: The **influence function** measures the sensitivity of the estimator to a small perturbation in the data. For S-estimators, the influence function is bounded, ensuring robustness to outliers.

4. **Consistency**: Under mild regularity conditions, S-estimators are consistent, meaning $\hat{\beta}_S \to \beta$ as the sample size $n \to \infty$.
5. **Asymptotic normality**: S-Estimators are asymptotically normal, with:

$$\sqrt{n}(\hat{\beta}_S - \beta) \xrightarrow{d} N(0, \Sigma),$$

where Σ depends on the choice of ρ and the distribution of residuals.

2.5.6.4 Algorithm for Computing S-Estimators

1. **Initial guess**: Compute an initial estimate of β using a robust method (e.g., LTS or an M-estimator).
2. **Scale estimation**: Compute a robust estimate of scale $\hat{\sigma}$ by solving:

$$\frac{1}{n}\sum_{i=1}^{n} \rho\left(\frac{y_i - x_i'\beta}{\sigma}\right) = \delta.$$

3. **Iterative refinement**:
   - Recalculate residuals $r_i = y_i - x_i'\beta$.
   - Update β and σ iteratively until convergence, typically using numerical optimization techniques.

2.5.7 MM-Estimators

MM-Estimators are a robust regression method that combines the strengths of two powerful techniques: S-estimators and M-estimators. They are designed to achieve both a **high breakdown point** (up to 50%) and **high efficiency** under ideal conditions (e.g., normality). This combination makes MM-estimators one of the most versatile and widely used robust regression methods.

The process of computing MM-estimators involves three main steps:

1. Compute an initial robust estimate of scale using an S-estimator.
2. Use this robust scale to define weights for an M-estimator.
3. Estimate regression coefficients by solving the weighted M-estimation problem.

This stepwise approach ensures robustness in the initial scale estimation while leveraging the efficiency of M-estimators for the final parameter estimates.

Step 1: Robust Scale Estimation
The first step is to estimate the robust scale σ using an S-estimator. This involves solving:

$$\hat{\sigma}_S = \arg\min_{\sigma} \frac{1}{n} \sum_{i=1}^{n} \rho_S \left(\frac{y_i - x_i'\beta}{\sigma} \right),$$

where ρ_S is a robust loss function chosen to control the influence of extreme residuals. Common choices for ρ_S include Huber's or Tukey's bisquare functions. This scale estimation provides a robust baseline for weighting residuals in the subsequent M-estimation step.

Step 2: Weight Definition for M-Estimation
Using the robust scale $\hat{\sigma}_S$ obtained in Step 1, the weights for the M-estimator are defined based on a second loss function, ρ_M. The weights downweight residuals proportional to their deviation relative to $\hat{\sigma}_S$. For each residual $r_i = y_i - x_i'\beta$, the weight is computed as:

$$w_i = \psi_M \left(\frac{r_i}{\hat{\sigma}_S} \right) / \frac{r_i}{\hat{\sigma}_S},$$

where:

- ψ_M is the derivative of the robust ρ_M function, known as the **influence function**
- ρ_M is often chosen to provide high efficiency under normality, such as Huber's or Hampel's function

These weights reduce the impact of large residuals while preserving the influence of small, well-behaved residuals.

Step 3: Final M-Estimation
The final step involves solving the M-estimation problem using the weights defined in Step 2. The coefficients $\hat{\beta}_{MM}$ are estimated by minimizing the weighted residuals:

$$\hat{\beta}_{MM} = \arg\min_{\beta} \sum_{i=1}^{n} w_i \rho_M \left(\frac{y_i - x_i'\beta}{\hat{\sigma}_S} \right).$$

This ensures that the final estimates combine the robustness of the initial S-estimator with the efficiency of the M-estimator.

2.5.7.1 Properties of MM-Estimators

1. **High breakdown point**:
   - The S-estimator in the first step ensures a breakdown point of up to 50%, meaning the estimator can handle up to half the data being contaminated without producing unreliable results.

2. **Asymptotic efficiency**:
   - The use of an efficient ρ_M function in the final M-estimation step ensures that MM-estimators achieve high asymptotic efficiency under normality, often close to that of OLS.
3. **Robustness**:
   - The combination of robust scale estimation and downweighting of large residuals makes MM-estimators highly robust to outliers and leverage points.
4. **Influence function**:
   - The influence function of MM-estimators is bounded, ensuring that no single observation can exert disproportionate influence on the parameter estimates.
5. **Consistency**:
   - MM-estimators are consistent, converging to the true parameter values as the sample size increases, provided the majority of the data satisfies the model assumptions.
6. **Asymptotic normality**:
   - MM-estimators are asymptotically normal, with:

$$\sqrt{n}(\hat{\beta}_{MM} - \beta) \xrightarrow{d} N(0, \Sigma),$$

   where Σ depends on the choice of ρ_M and the distribution of residuals.

2.5.7.2 Choice of ρ-Functions for MM-Estimators

The robustness and efficiency of MM-estimators depend on the choice of ρ_S (for scale) and ρ_M (for final estimation). Common choices include:

1. **Huber's** ρ-function. Combines quadratic and linear growth to balance robustness and efficiency:

$$\rho(z) = \begin{cases} \frac{z^2}{2} & \text{if } |z| \leq c, \\ c|z| - \frac{c^2}{2} & \text{if } |z| > c. \end{cases}$$

2. **Tukey's bisquare function**. Provides high robustness by completely bounding large residuals:

$$\rho(z) = \begin{cases} c^2\left(1 - \left(1 - \left(\frac{z}{c}\right)^2\right)^3\right)/6 & \text{if } |z| \leq c, \\ c^2/6 & \text{if } |z| > c. \end{cases}$$

3. **Hampel's three-part redescending function**. Further limits the influence of large residuals by assigning a constant penalty beyond a certain threshold.

$$\rho(z) = \begin{cases} z^2/2 & \text{if } |z| \leq a, \\ a|z| - a^2/2 & \text{if } a < |z| \leq b, \\ \text{constant} & \text{if } |z| > b. \end{cases}$$

2.5.8 Practical Considerations

The following table summarizes the key properties, advantages, and limitations of the robust estimators discussed (Table 2.21):
Notes on Choosing an Estimator

- M-Estimators: best suited for general-purpose robust regression, offering a balance between robustness and efficiency with moderate contamination
- R-Estimators: ideal for rank-based data or ordinal data, especially when outliers are present in both predictors and responses
- L-Estimators: simple and effective for descriptive statistics or data cleaning with limited computational resources
- **LTS**: recommended for datasets with significant contamination or leverage points due to its high breakdown point
- S-Estimators: focus on robust scale estimation, suitable for identifying and mitigating the influence of extreme residuals
- MM-Estimators: combines the robustness of S-estimators with the efficiency of M-estimators, making it the most versatile choice for heavily contaminated data

```
# Load necessary libraries
library(MASS)        # For robust regression functions like rlm
library(robustbase)  # For LTS regression and MM-estimators
library(dplyr)       # For data manipulation
#>
#> Attaching package: 'dplyr'
#> The following object is masked from 'package:MASS':
#>
#>     select
#> The following objects are masked from 'package:stats':
#>
#>     filter, lag
#> The following objects are masked from 'package:base':
#>
```

Table 2.21 Comparison of robust regression estimators

Estimator	Key features	Breakdown point	Efficiency (under normality)	Applications	Advantages
M-Estimators	Generalization of OLS; robust ρ reduces large residual influence; flexible tuning via ρ-function	Moderate (up to 0.29)	High with proper tuning	Wide applicability in regression with moderate robustness	Balances robustness and efficiency
R-Estimators	Rank-based method; immune to outliers in x and y; suitable for ordinal or rank-based data	High (depends on ranks)	Moderate	Ordinal data or heavily skewed distributions	Handles both predictor and response outliers
L-Estimators	Linear combination of order statistics; easy to compute even for large datasets	High (up to 50%)	Moderate	Descriptive statistics, robust averages	Simple and intuitive
LTS	Minimizes smallest h squared residuals; resistant to leverage points	High (up to 50%)	Moderate	Data with high contamination, fault detection	High robustness to outliers
S-Estimators	Minimizes robust scale of residuals; effective at detecting extreme outliers	High (up to 50%)	Low to moderate	Outlier detection, data with heavy-tailed distributions	Focus on robust scale estimation
MM-Estimators	High robustness (scale) + high efficiency (coefficients); versatile and flexible	High (up to 50%)	High	Mixed contamination and heavy-tailed errors	Combines robustness and efficiency effectively

```
#>     intersect, setdiff, setequal, union
library(ggplot2)     # For visualization

# Simulate dataset
set.seed(123)
n <- 100
x <- rnorm(n, mean = 5, sd = 2)    # Predictor
y <- 3 + 2 * x + rnorm(n, sd = 1) # Response

# Introduce outliers
y[95:100] <- y[95:100] + 20  # Vertical outliers
x[90:95] <- x[90:95] + 10    # Leverage points

data <- data.frame(x, y)
```

```
# Visualize the data
ggplot(data, aes(x, y)) +
    geom_point() +
    labs(title = "Scatterplot of Simulated Data with
    ↪ Outliers",
         x = "Predictor (x)",
         y = "Response (y)") +
    theme_minimal()
```

```
# Ordinary Least Squares
ols_model <- lm(y ~ x, data = data)
summary(ols_model)
#>
#> Call:
#> lm(formula = y ~ x, data = data)
#>
#> Residuals:
#>      Min       1Q   Median       3Q      Max
#> -12.6023  -2.4590  -0.5717   0.9247  24.4024
#>
#> Coefficients:
#>             Estimate Std. Error t value Pr(>|t|)
#> (Intercept)   8.8346     1.1550   7.649 1.41e-11 ***
#> x             0.9721     0.1749   5.558 2.36e-07 ***
#> ---
#> Signif. codes:
```

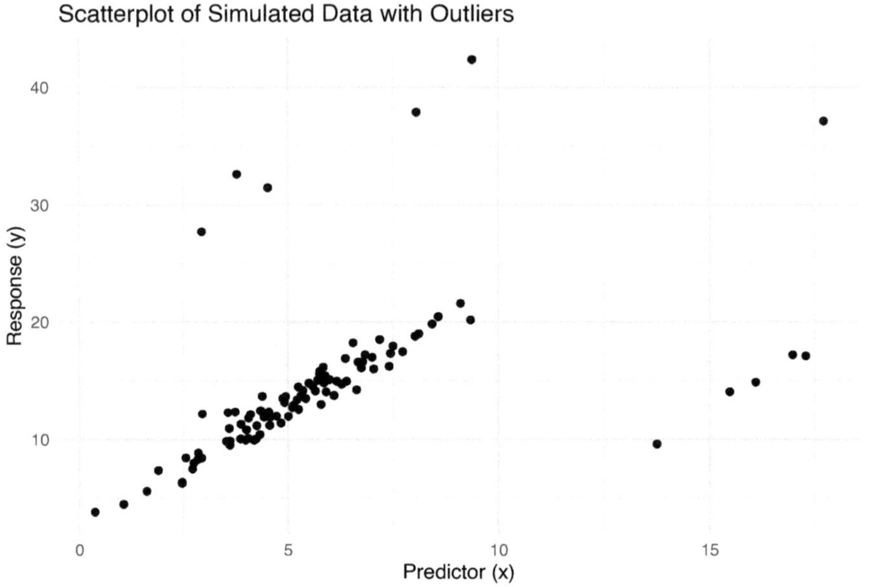

Fig. 2.16 Scatterplot of simulated data with outliers

```
#> 0 '***' 0.001 '**' 0.01 '*' 0.05 '.' 0.1 ' ' 1
#>
#> Residual standard error: 5.583 on 98 degrees of freedom
#> Multiple R-squared:  0.2396, Adjusted R-squared:  0.2319
#> F-statistic: 30.89 on 1 and 98 DF,  p-value: 2.358e-07
```

OLS coefficients are highly influenced by the presence of outliers. For example, the slope (x coefficient) and intercept are shifted to fit the outliers, resulting in a poor fit to the majority of the data (Fig. 2.16).

```
# $M$-Estimators
m_model <- rlm(y ~ x, data = data, psi = psi.huber)
summary(m_model)
#>
#> Call: rlm(formula = y ~ x, data = data, psi = psi.huber)
#> Residuals:
#>       Min        1Q    Median       3Q      Max
#> -18.43919  -0.97575  -0.03297  0.76967 21.85546
#>
#> Coefficients:
#>             Value  Std. Error t value
#> (Intercept) 4.3229 0.2764     15.6421
```

```
#> x              1.7250  0.0419    41.2186
#>
#> Residual standard error: 1.349 on 98 degrees of freedom
```

The M-estimator reduces the influence of large residuals using Huber's psi function. This results in coefficients that are less affected by outliers compared to OLS.

```
# Least Trimmed Squares (LTS)
lts_model <- ltsReg(y ~ x, data = data)
lts_coefficients <- coef(lts_model)
```

LTS minimizes the smallest squared residuals, ignoring extreme residuals. This results in a more robust fit, particularly in the presence of both vertical outliers and leverage points.

```
# $MM$-Estimators
mm_model <- lmrob(y ~ x, data = data, setting = "KS2014")
summary(mm_model)
#>
#> Call:
#> lmrob(formula = y ~ x, data = data, setting = "KS2014")
#>  \--> method = "SMDM"
#> Residuals:
#>       Min        1Q    Median        3Q       Max
#> -20.45989  -0.69436  -0.01455   0.73614  22.10173
#>
#> Coefficients:
#>             Estimate Std. Error t value Pr(>|t|)
#> (Intercept)  3.02192    0.25850   11.69   <2e-16 ***
#> x            1.96672    0.04538   43.34   <2e-16 ***
#> ---
#> Signif. codes:
#> 0 '***' 0.001 '**' 0.01 '*' 0.05 '.' 0.1 ' ' 1
#>
#> Robust residual standard error: 0.9458
#> Multiple R-squared:  0.9562, Adjusted R-squared:  0.9558
#> Convergence in 7 IRWLS iterations
#>
#> Robustness weights:
#>  10 observations c(90,91,92,93,94,96,97,98,99,100)
#>   are outliers with |weight| = 0 ( < 0.001);
#>  67 weights are ~= 1. The remaining 23 ones are summarized
↪  as
#>    Min. 1st Qu.  Median    Mean 3rd Qu.    Max.
#>  0.2496  0.7969  0.9216  0.8428  0.9548  0.9943
```

```
#> Algorithmic parameters:
#>         tuning.chi1          tuning.chi2          tuning.chi3
#>          -5.000e-01           1.500e+00                   NA
#>         tuning.chi4                   bb          tuning.psi1
#>           5.000e-01           5.000e-01           -5.000e-01
#>         tuning.psi2          tuning.psi3          tuning.psi4
#>           1.500e+00           9.500e-01                   NA
#>          refine.tol              rel.tol            scale.tol
#>           1.000e-07           1.000e-07            1.000e-10
#>           solve.tol             zero.tol          eps.outlier
#>           1.000e-07           1.000e-10            1.000e-03
#>               eps.x warn.limit.reject warn.limit.meanrw
#>           3.223e-11           5.000e-01            5.000e-01
#>      nResample          max.it       best.r.s       k.fast.s
↪
#>           1000             500             20              2
↪
#>          k.max     maxit.scale      trace.lev            mts
↪
#>           2000             200              0           1000
↪
#>     compute.rd       numpoints fast.s.large.n
#>              0              10           2000
#>                  setting                       psi
#>                 "KS2014"                     "lqq"
#>              subsampling                       cov
#>            "nonsingular"                 ".vcov.w"
#> compute.outlier.stats
#>                "SMDM"
#> seed : int(0)
```

MM-Estimators combine robust scale estimation (from S-estimators) with efficient coefficient estimation (from M-estimators). This achieves both high robustness and high efficiency under normal conditions.

```
# Visualizing results
data <- data %>%
    mutate(
        ols_fit = predict(ols_model, newdata = data),
        m_fit = predict(m_model, newdata = data),
        lts_fit = fitted(lts_model),
        # Use `fitted()` for ltsReg objects
        mm_fit = predict(mm_model, newdata = data)
    )
```

```
ggplot(data, aes(x, y)) +
    geom_point() +
    geom_line(
        aes(y = ols_fit),
        color = "red",
        linetype = "dashed",
        size = 1,
        label = "OLS"
    ) +
    geom_line(
        aes(y = m_fit),
        color = "blue",
        linetype = "dashed",
        size = 1,
        label = "$M$-Estimator"
    ) +
    geom_line(
        aes(y = lts_fit),
        color = "green",
        linetype = "dashed",
        size = 1,
        label = "LTS"
    ) +
    geom_line(
        aes(y = mm_fit),
        color = "purple",
        linetype = "dashed",
        size = 1,
        label = "$MM$-Estimator"
    ) +
    labs(title = "Comparison of Regression Fits",
         x = "Predictor (x)",
         y = "Response (y)") +
    theme_minimal()
#> Warning: Using `size` aesthetic for lines was deprecated in
↪   ggplot2
#> 3.4.0.
#> i Please use `linewidth` instead.
#> This warning is displayed once every 8 hours.
#> Call `lifecycle::last_lifecycle_warnings()` to see where
#> this warning was generated.
#> Warning in geom_line(aes(y = ols_fit), color = "red",
#> linetype = "dashed", : Ignoring unknown parameters: `label`
#> Warning in geom_line(aes(y = m_fit), color = "blue",
```

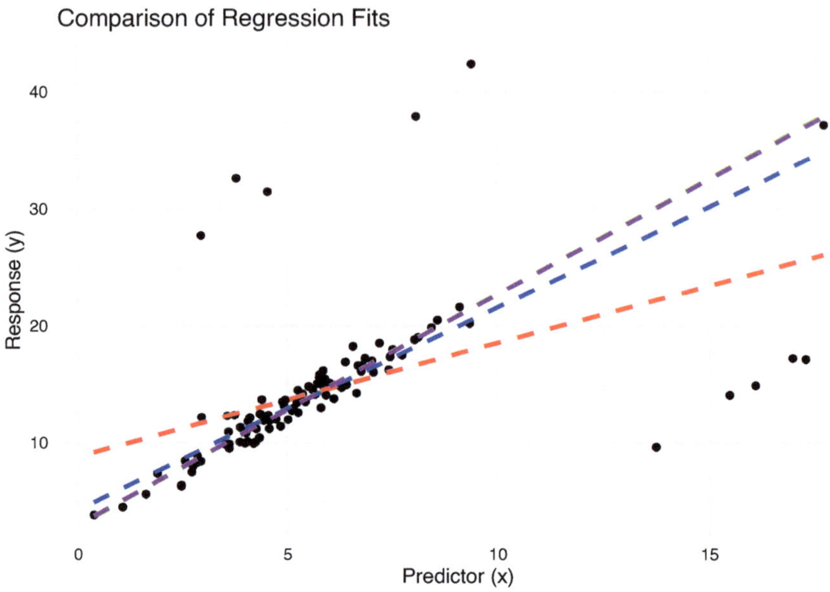

Fig. 2.17 Comparison of regression fits

```
#> linetype = "dashed", : Ignoring unknown parameters: `label`
#> Warning in geom_line(aes(y = lts_fit), color = "green",
#> linetype = "dashed", : Ignoring unknown parameters: `label`
#> Warning in geom_line(aes(y = mm_fit), color = "purple",
#> linetype = "dashed", : Ignoring unknown parameters: `label`
```

Visualization shows the differences in regression fits (Fig. 2.17):

- OLS is heavily influenced by outliers and provides a poor fit to the majority of the data.
- The M-estimator downweights large residuals, resulting in a better fit.
- LTS regression ignores the extreme residuals entirely, providing the most robust fit.
- MM-Estimators balance robustness and efficiency, producing coefficients close to the LTS but with improved efficiency under normality.

```
# Comparing Coefficients
comparison <- data.frame(
    Method = c("OLS", "$M$-Estimator", "LTS",
    ↪  "$MM$-Estimator"),
    Intercept = c(
        coef(ols_model)[1],
        coef(m_model)[1],
```

```
        lts_coefficients[1],
        coef(mm_model)[1]
    ),
    Slope = c(
        coef(ols_model)[2],
        coef(m_model)[2],
        lts_coefficients[2],
        coef(mm_model)[2]
    )
)

print(comparison)
#>          Method Intercept     Slope
#> 1           OLS  8.834553 0.9720994
#> 2  $M$-Estimator  4.322869 1.7250441
#> 3           LTS  2.954960 1.9777635
#> 4 $MM$-Estimator  3.021923 1.9667208
```

The table above shows how the coefficients vary across methods:

- OLS coefficients are the most distorted by outliers.
- M-Estimators and MM-estimators provide coefficients that are less influenced by extreme values.
- LTS regression, with its trimming mechanism, produces the most robust coefficients by excluding the largest residuals.

2.6 Partial Least Squares

Partial least squares (PLS) is a dimensionality reduction technique used for regression and predictive modeling. It is particularly useful when predictors are highly collinear or when the number of predictors (p) exceeds the number of observations (n). Unlike methods such as principal component regression (PCR), PLS simultaneously considers the relationship between predictors and the response variable.

2.6.1 Motivation for PLS

Limitations of Classical Methods

1. **Multicollinearity**:
    - OLS fails when predictors are highly correlated because the design matrix $X'X$ becomes nearly singular, leading to unstable estimates.

2. **High-dimensional data**:
   - When $p > n$, OLS cannot be directly applied as $X'X$ is not invertible.
3. **Principal component regression (PCR)**:
   - While PCR addresses multicollinearity by using principal components of X, it does not account for the relationship between predictors and the response variable y when constructing components.

PLS overcomes these limitations by constructing components that maximize the covariance between predictors X and the response y. It finds a compromise between explaining the variance in X and predicting y, making it particularly suited for regression in high-dimensional or collinear datasets.
Let:

- X be the $n \times p$ matrix of predictors
- y be the $n \times 1$ response vector
- t_k be the kth latent component derived from X
- p_k and q_k be the loadings for X and y, respectively

PLS aims to construct latent components $t_1, t_2, \ldots, t_K$ such that:

1. Each t_k is a linear combination of the predictors: $t_k = Xw_k$, where w_k is a weight vector. 2
2. The covariance between t_k and y is maximized:

$$\text{Maximize } Cov(t_k, y) = w_k' X' y.$$

2.6.2 Steps to Construct PLS Components

1. **Compute Weights**:
   - The weights w_k for the kth component are obtained by solving:

$$w_k = \frac{X'y}{\|X'y\|}.$$

2. **Construct Latent Component**:
   - Form the kth latent component:

$$t_k = Xw_k.$$

3. **Deflate the Predictors**:
   - After extracting t_k, the predictors are deflated to remove the information explained by t_k:

$$X \leftarrow X - t_k p_k',$$

   where $p_k = \frac{X' t_k}{t_k' t_k}$ are the loadings for X.

4. **Deflate the Response**:
   - Similarly, deflate y to remove the variance explained by t_k:

$$y \leftarrow y - t_k q_k,$$

   where $q_k = \frac{t_k' y}{t_k' t_k}$.

5. **Repeat for All Components**:
   - Repeat the steps above until K components are extracted.

After constructing K components, the response y is modeled as:

$$y = TC + \epsilon,$$

where:

- $T = [t_1, t_2, \ldots, t_K]$ is the matrix of latent components
- C is the vector of regression coefficients

The estimated coefficients for the original predictors are then:

$$\hat{\beta} = W(P'W)^{-1}C,$$

where $W = [w_1, w_2, \ldots, w_K]$ and $P = [p_1, p_2, \ldots, p_K]$.

2.6.3 Properties of PLS

1. **Dimensionality reduction**:
   - PLS reduces X to K components, where $K \leq \min(n, p)$.
2. **Handles multicollinearity**:
   - By constructing uncorrelated components, PLS avoids the instability caused by multicollinearity in OLS.

3. **Supervised dimensionality reduction**:
   - Unlike PCR, PLS considers the relationship between X and y when constructing components.
4. **Efficiency**:
   - PLS requires fewer components than PCR to achieve a similar level of predictive accuracy.

Practical Considerations

1. **Number of components**:
   - The optimal number of components K can be determined using cross-validation.
2. **Preprocessing**:
   - Standardizing predictors is essential for PLS, as it ensures that all variables are on the same scale.
3. **Comparison with other methods**:
   - PLS outperforms OLS and PCR in cases of multicollinearity or when $p > n$, but it may be less interpretable than sparse methods like Lasso.

```
# Load required library
library(pls)
#>
#> Attaching package: 'pls'
#> The following object is masked from 'package:stats':
#>
#>     loadings

# Step 1: Simulate data
set.seed(123)  # Ensure reproducibility
n <- 100       # Number of observations
p <- 10        # Number of predictors
X <- matrix(rnorm(n * p), nrow = n, ncol = p)  # Design matrix
↪  (predictors)
beta <- runif(p)                               # True
↪  coefficients
y <- X %*% beta + rnorm(n)                     # Response
↪  variable with noise

# Step 2: Fit Partial Least Squares (PLS) Regression
pls_fit <- plsr(y ~ X, ncomp = 5, validation = "CV")
```

```
# Step 3: Summarize the PLS Model
summary(pls_fit)
#> Data:    X dimension: 100 10
#>  Y dimension: 100 1
#> Fit method: kernelpls
#> Number of components considered: 5
#>
#> VALIDATION: RMSEP
#> Cross-validated using 10 random segments.
#>        (Intercept)  1 comps  2 comps  3 comps  4 comps
#> CV           1.339    1.123    1.086    1.090    1.088
#> adjCV        1.339    1.112    1.078    1.082    1.080
#>        5 comps
#> CV       1.087
#> adjCV    1.080
#>
#> TRAINING: % variance explained
#>    1 comps  2 comps  3 comps  4 comps  5 comps
#> X    10.88    20.06    30.80    42.19    51.61
#> y    44.80    48.44    48.76    48.78    48.78
```

```
# Step 4: Perform Cross-Validation and Select Optimal
↪  Components
validationplot(pls_fit, val.type = "MSEP")
```

```
# Step 5: Extract Coefficients for Predictors
pls_coefficients <- coef(pls_fit)
print(pls_coefficients)
#> , , 5 comps
#>
#>               y
#> X1   0.30192935
#> X2  -0.03161151
#> X3   0.22392538
#> X4   0.42315637
#> X5   0.33000198
#> X6   0.66228763
#> X7   0.40452691
#> X8  -0.05704037
#> X9  -0.02699757
#> X10  0.05944765
```

```
# Step 6: Evaluate Model Performance
predicted_y <- predict(pls_fit, X)
actual_vs_predicted <- data.frame(
  Actual = y,
  Predicted = predicted_y[, , 5]  # Predicted values using 5
   ↪  components
)
```

```
# Plot Actual vs Predicted
library(ggplot2)
ggplot(actual_vs_predicted, aes(x = Actual, y = Predicted)) +
    geom_point() +
    geom_abline(
        intercept = 0,
        slope = 1,
        color = "red",
        linetype = "dashed"
    ) +
    labs(title = "Actual vs Predicted Values (PLS
    ↪  Regression)",
         x = "Actual Values",
         y = "Predicted Values") +
    theme_minimal()
```

```
# Step 7: Extract and Interpret Variable Importance (Loadings)
loadings_matrix <- as.matrix(unclass(loadings(pls_fit)))
variable_importance <- as.data.frame(loadings_matrix)
colnames(variable_importance) <-
    paste0("Component_", 1:ncol(variable_importance))
rownames(variable_importance) <-
    paste0("X", 1:nrow(variable_importance))

# Print variable importance
print(variable_importance)
#>     Component_1 Component_2 Component_3 Component_4
#> X1  -0.04991097   0.5774569  0.24349681 -0.41550345
#> X2   0.08913192  -0.1139342 -0.17582957 -0.05709948
#> X3   0.13773357   0.1633338  0.07622919 -0.07248620
#> X4   0.40369572  -0.2730457  0.69994206 -0.07949013
#> X5   0.50562681  -0.1788131 -0.27936562  0.36197480
```

```
#> X6    0.57044281   0.3358522 -0.38683260  0.17656349
#> X7    0.36258623   0.1202109 -0.01753715 -0.12980483
#> X8    0.12975452  -0.1164935 -0.30479310 -0.65654861
#> X9   -0.29521786   0.6170234 -0.32082508 -0.01041860
#> X10   0.23930055  -0.3259554  0.20006888 -0.53547258
#>      Component_5
#> X1   -0.02098351
#> X2   -0.06707863
#> X3   -0.61962875
#> X4    0.35239113
#> X5   -0.41919645
#> X6    0.31154275
#> X7   -0.06919411
#> X8    0.49948167
#> X9    0.04904396
#> X10  -0.17963372
```

The loadings provide the contribution of each predictor to the PLS components. Higher absolute values indicate stronger contributions to the corresponding component.

- **Summary of the model**:
  - The **proportion of variance explained** indicates how much of the variability in both the predictors and response is captured by each PLS component.
  - The goal is to retain enough components to explain most of the variance while avoiding overfitting.
- **Validation plot** (Fig. 2.18):
  - The **mean squared error of prediction (MSEP)** curve is used to select the optimal number of components.
  - Adding too many components can lead to overfitting, while too few may underfit the data.
- **Coefficients**:
  - The extracted coefficients are the weights applied to the predictors in the final PLS model.
  - These coefficients are derived from the PLS components and may differ from OLS regression coefficients due to dimensionality reduction.
- **Actual versus predicted plot** (Fig. 2.19):
  - This visualization evaluates how well the PLS model predicts the response variable.
  - Points tightly clustered around the diagonal indicate good performance.

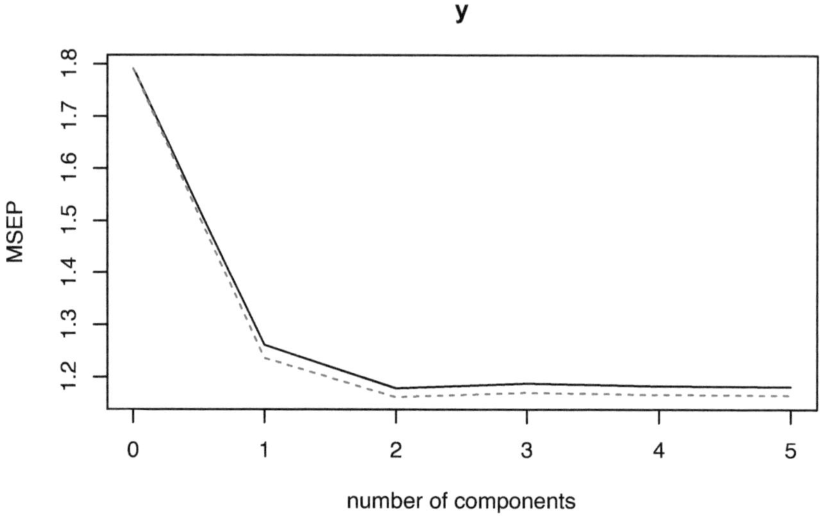

Fig. 2.18 Validation plot of PLS model: mean squared error of prediction (MSEP)

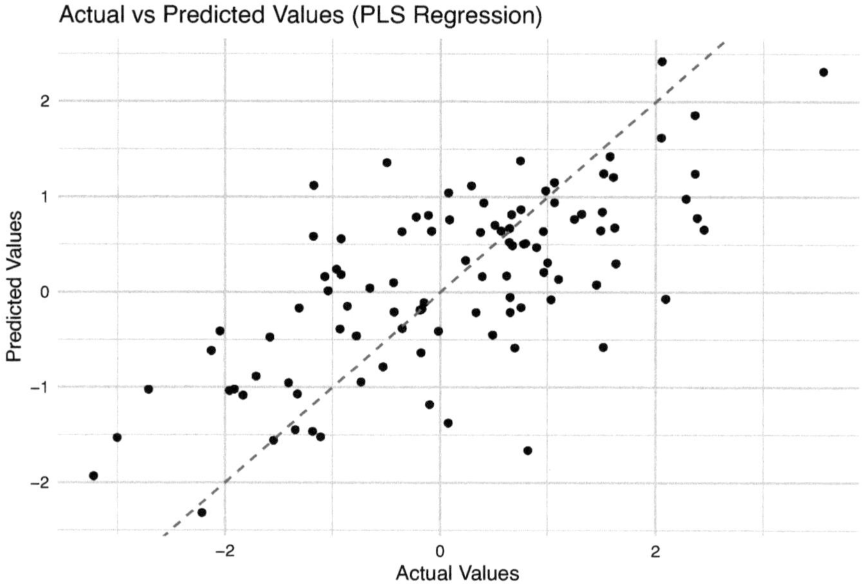

Fig. 2.19 Actual versus predicted values (PLS regression)

- **VIP scores**:
  - VIP scores help identify the most important predictors in the PLS model.
  - Predictors with higher VIP scores contribute more to explaining the response variable.

2.6.4 Comparison with Related Methods

Table 2.22 shows comparison between methods for high-dimensional data.

Table 2.22 Comparison of regression methods for high-dimensional data

Method	Handles multicollinearity	Supervised dimensionality reduction	Sparse solution	Interpretability
OLS	No	No	No	High
Ridge regression	Yes	No	No	Moderate
Lasso regression	Yes	No	Yes	High
PCR	Yes	No	No	Low
PLS	Yes	Yes	No	Moderate

Chapter 3
Nonlinear Regression

Moving beyond linearity, this chapter examines deterministic and stochastic nonlinear models. We formalize inference under nonlinear least squares, emphasize the distinction between intrinsic and extrinsic curvature, and explore identifiability issues. Practical considerations—starting-value sensitivity, convergence diagnostics, and interval estimation via the delta method.

Nonlinear regression models differ fundamentally from linear regression models in that the derivatives of the mean function with respect to parameters depend on one or more of the parameters (Table 3.1). This dependence not only adds complexity but also provides greater flexibility to model intricate relationships.

Linear Regression

- **Model Form Example:** A typical linear regression model looks like $y = \beta_0 + \beta_1 x$, where β_0 and β_1 are the parameters.
- **Parameter Effect:** The influence of each parameter on y is constant. For example, if β_1 increases by 1, the change in y is always x, regardless of the current value of β_1.
- **Derivatives:** The partial derivatives of y with respect to each parameter (e.g., $\frac{\partial y}{\partial \beta_1} = x$) do not depend on the parameters β_0 or β_1 themselves—they only depend on the data x. This makes the mathematics of finding the best-fit line straightforward.
- Straightforward estimation via closed-form solutions like Ordinary Least Squares.

Nonlinear Regression

- **Model Form Example:** Consider $y = \alpha \cdot e^{\beta x}$. Here, α and β arc parameters, but the relationship is not a straight line.
- **Parameter Effect:** The effect of changing α or β on y is not constant. For instance, if you change β, the impact on y depends on both x and the current value of β. This makes predictions and adjustments more complex.

M. Nguyen, *Regression Techniques for Data Analysis*,
https://doi.org/10.1007/978-3-032-01835-9_3

Table 3.1 Summary table: linear vs. nonlinear regression

Feature	Linear regression	Nonlinear regression
Relationship	Linear in parameters	Nonlinear in parameters
Interpretability	High	Often challenging
Estimation	Closed-form solutions	Iterative algorithms
Computational cost	Low	Higher

- **Derivatives:** Taking the partial derivative with respect to β gives $\frac{\partial y}{\partial \beta} = \alpha x e^{\beta x}$. Notice this derivative depends on α, β, and x. Unlike linear regression, the sensitivity of y to changes in β changes as β itself changes.
- Estimation requires iterative algorithms like the Gauss-Newton Algorithm, as closed-form solutions are not feasible.

Key Features of Nonlinear Regression

- **Complex Functional Forms**: Nonlinear regression allows for relationships that are not straight lines or planes.
- **Interpretability Challenges**: Nonlinear models can be difficult to interpret, especially if the functional forms are complex.
- **Practical Use Cases**:
  - Growth curves
  - High-order polynomials
  - Linear approximations (e.g., Taylor expansions)
  - Collections of locally linear models or basis functions (e.g., splines)

While these approaches can approximate data, they may suffer from interpretability issues or may not generalize well when data is sparse. Hence, intrinsically nonlinear models are often preferred.

Intrinsically Nonlinear Models

The general form of an intrinsically nonlinear regression model is:

$$Y_i = f(\mathbf{x}_i; \boldsymbol{\theta}) + \epsilon_i$$

Where:

- $f(\mathbf{x}_i; \boldsymbol{\theta})$: A **nonlinear function** that relates $E(Y_i)$ to the independent variables $\mathbf{x}_i$.
- $\mathbf{x}_i$: A $k \times 1$ vector of independent variables (fixed).
- $\boldsymbol{\theta}$: A $p \times 1$ vector of parameters.
- ϵ_i: Independent and identically distributed random errors, often assumed to have a mean of 0 and a constant variance σ^2. In some cases, $\epsilon_i \sim \mathcal{N}(0, \sigma^2)$.

Example: Exponential Growth Model

A common nonlinear model is the exponential growth function:

$$y = \theta_1 e^{\theta_2 x} + \epsilon$$

Where:

- θ_1: Initial value.
- θ_2: Growth rate.
- x: Independent variable (e.g., time).
- ϵ: Random error.

3.1 Inference

Since $Y_i = f(\mathbf{x}_i, \theta) + \epsilon_i$, where $\epsilon_i \sim \text{iid}(0, \sigma^2)$, we can estimate parameters $(\hat{\theta})$ by minimizing the sum of squared errors:

$$\sum_{i=1}^{n} \left(Y_i - f(\mathbf{x}_i, \theta)\right)^2$$

Let $\hat{\theta}$ be the minimizer, and the variance of residuals is estimated as:

$$s^2 = \hat{\sigma}_\epsilon^2 = \frac{\sum_{i=1}^{n} \left(Y_i - f(\mathbf{x}_i, \hat{\theta})\right)^2}{n - p}$$

where p is the number of parameters in $\boldsymbol{\theta}$ and n is the number of observations.

Asymptotic Distribution of $\hat{\theta}$

Under regularity conditions—most notably that $\epsilon_i \sim N(0, \sigma^2)$ or that n is sufficiently large for a central-limit-type argument—the parameter estimates $\hat{\theta}$ have the following asymptotic normal distribution:

$$\hat{\theta} \sim AN(\boldsymbol{\theta}, \sigma^2[\mathbf{F}(\theta)'\mathbf{F}(\theta)]^{-1})$$

Where:

- AN stands for "asymptotic normality."
- $\mathbf{F}(\theta)$ is the $n \times p$ Jacobian matrix of partial derivatives of $f(\mathbf{x}_i, \theta)$ with respect to $\boldsymbol{\theta}$, evaluated at $\hat{\theta}$. Specifically,

$$\mathbf{F}(\theta) = \begin{pmatrix} \frac{\partial f(\mathbf{x}_1, \boldsymbol{\theta})}{\partial \theta_1} & \cdots & \frac{\partial f(\mathbf{x}_1, \boldsymbol{\theta})}{\partial \theta_p} \\ \vdots & \ddots & \vdots \\ \frac{\partial f(\mathbf{x}_n, \boldsymbol{\theta})}{\partial \theta_1} & \cdots & \frac{\partial f(\mathbf{x}_n, \boldsymbol{\theta})}{\partial \theta_p} \end{pmatrix}$$

Asymptotic normality means that as the sample size n becomes large, the sampling distribution of $\hat{\theta}$ approaches a normal distribution, which enables inference on the parameters.

3.1.1 Linear Functions of the Parameters

A "linear function of the parameters" refers to a quantity that can be written as $\mathbf{a}'\boldsymbol{\theta}$, where $\mathbf{a}$ is some (constant) contrast vector. Common examples include:

- A single parameter θ_j (using a vector $\mathbf{a}$ with 1 in the jth position and 0 elsewhere).
- Differences, sums, or other contrasts, e.g., $\theta_1 - \theta_2$.

Suppose we are interested in a linear combination of the parameters, such as $\theta_1 - \theta_2$. Define the contrast vector $\mathbf{a}$ as:

$$\mathbf{a} = (0, 1, -1)'$$

We then consider inference for $\mathbf{a}'\boldsymbol{\theta}$ ($\mathbf{a}$ can be p-dimensional vector). Using rules for the expectation and variance of a linear combination of a random vector $\mathbf{Z}$:

$$E(\mathbf{a}'\mathbf{Z}) = \mathbf{a}'E(\mathbf{Z})$$
$$\text{Var}(\mathbf{a}'\mathbf{Z}) = \mathbf{a}'\text{Var}(\mathbf{Z})\mathbf{a}$$

We have:

$$E(\mathbf{a}'\hat{\boldsymbol{\theta}}) = \mathbf{a}'E(\hat{\theta}) \approx \mathbf{a}'\theta$$
$$\text{Var}(\mathbf{a}'\hat{\theta}) = \mathbf{a}'\text{Var}(\hat{\theta})\mathbf{a} \approx \sigma^2\mathbf{a}'[\mathbf{F}(\theta)'\mathbf{F}(\theta)]^{-1}\mathbf{a}$$

Hence,

$$\mathbf{a}'\hat{\boldsymbol{\theta}} \sim AN\big(\mathbf{a}'\boldsymbol{\theta}, \sigma^2\mathbf{a}'[\mathbf{F}(\theta)'\mathbf{F}(\theta)]^{-1}\mathbf{a}\big)$$

Confidence Intervals for Linear Contrasts

Since $\mathbf{a}'\hat{\boldsymbol{\theta}}$ is asymptotically independent of s^2 (up to order $O1/n$), a two-sided $100(1-\alpha)\%$ confidence interval for $\mathbf{a}'\boldsymbol{\theta}$ is given by:

$$\mathbf{a}'\boldsymbol{\theta} \pm t_{(1-\alpha/2,n-p)}s\sqrt{\mathbf{a}'[\mathbf{F}(\hat{\boldsymbol{\theta}})'\mathbf{F}(\hat{\boldsymbol{\theta}})]^{-1}\mathbf{a}}$$

Where:

- $t_{(1-\alpha/2,n-p)}$ is the critical value of the t-distribution with $n - p$ degrees of freedom.

- $s = \sqrt{\hat{\sigma^2}_\epsilon}$ is the estimated standard deviation of residuals.

Special Case: A Single Parameter θ_j
If we focus on a single parameter θ_j, let $\mathbf{a}' = (0, \ldots, 1, \ldots, 0)$ (with 1 at the jth position). Then, the confidence interval for θ_j becomes:

$$\hat{\theta}_j \pm t_{(1-\alpha/2, n-p)} s \sqrt{\hat{c}^j}$$

where $\hat{c}^j$ is the jth diagonal element of $[\mathbf{F}(\hat{\theta})'\mathbf{F}(\hat{\theta})]^{-1}$.

3.1.2 Nonlinear Functions of Parameters

In many cases, we are interested in nonlinear functions of $\boldsymbol{\theta}$. Let $h(\boldsymbol{\theta})$ be such a function (e.g., a ratio of parameters, a difference of exponentials, etc.).

When $h(\theta)$ is a nonlinear function of the parameters, we can use a **Taylor series expansion** about θ to approximate $h(\hat{\theta})$:

$$h(\hat{\theta}) \approx h(\theta) + \mathbf{h}'[\hat{\theta} - \theta]$$

Where:

- $\mathbf{h} = \left(\frac{\partial h}{\partial \theta_1}, \frac{\partial h}{\partial \theta_2}, \ldots, \frac{\partial h}{\partial \theta_p}\right)'$ is the gradient vector of partial derivatives.

Key Approximations

1. **Expectation and variance of $\hat{\theta}$** (using the asymptotic normality of $\hat{\theta}$):

$$E(\hat{\theta}) \approx \theta,$$
$$\text{Var}(\hat{\theta}) \approx \sigma^2[\mathbf{F}(\theta)'\mathbf{F}(\theta)]^{-1}.$$

2. **Expectation and variance of $h(\hat{\theta})$** (properties of expectation and variance of (approximately) linear transformations):

$$E(h(\hat{\theta})) \approx h(\theta),$$
$$\text{Var}(h(\hat{\theta})) \approx \sigma^2\mathbf{h}'[\mathbf{F}(\theta)'\mathbf{F}(\theta)]^{-1}\mathbf{h}.$$

Combining these results, we find:

$$h(\hat{\theta}) \sim AN(h(\theta), \sigma^2\mathbf{h}'[\mathbf{F}(\theta)'\mathbf{F}(\theta)]^{-1}\mathbf{h}),$$

where AN represents asymptotic normality.

Confidence Interval for $h(\theta)$
An approximate $100(1-\alpha)\%$ confidence interval for $h(\theta)$ is:

$$h(\hat{\theta}) \pm t_{(1-\alpha/2,n-p)} s \sqrt{\mathbf{h}'[\mathbf{F}(\theta)'\mathbf{F}(\theta)]^{-1}\mathbf{h}},$$

where $\mathbf{h}$ and $\mathbf{F}(\theta)$ are evaluated at $\hat{\theta}$.

To compute a **prediction interval** for a new observation Y_0 at $x = x_0$:

1. **Model Definition**:

$$Y_0 = f(x_0; \theta) + \epsilon_0, \quad \epsilon_0 \sim N(0, \sigma^2),$$

with the predicted value:

$$\hat{Y}_0 = f(x_0, \hat{\theta}).$$

2. **Approximation for** $\hat{Y}_0$: As $n \to \infty$, $\hat{\theta} \to \theta$, so we have:

$$f(x_0, \hat{\theta}) \approx f(x_0, \theta) + \mathbf{f}_0(\theta)'[\hat{\theta} - \theta],$$

where:

$$\mathbf{f}_0(\theta) = \left(\frac{\partial f(x_0, \theta)}{\partial \theta_1}, \ldots, \frac{\partial f(x_0, \theta)}{\partial \theta_p}\right)'.$$

3. **Error Approximation**:

$$\begin{aligned} Y_0 - \hat{Y}_0 &\approx Y_0 - f(x_0, \theta) - f_0(\theta)'[\hat{\theta} - \theta] \\ &= \epsilon_0 - f_0(\theta)'[\hat{\theta} - \theta] \end{aligned}$$

4. **Variance of** $Y_0 - \hat{Y}_0$:

$$\begin{aligned} \text{Var}(Y_0 - \hat{Y}_0) &\approx \text{Var}(\epsilon_0 - \mathbf{f}_0(\theta)'[\hat{\theta} - \theta]) \\ &= \sigma^2 + \sigma^2 \mathbf{f}_0(\theta)'[\mathbf{F}(\theta)'\mathbf{F}(\theta)]^{-1}\mathbf{f}_0(\theta) \\ &= \sigma^2\big(1 + \mathbf{f}_0(\theta)'[\mathbf{F}(\theta)'\mathbf{F}(\theta)]^{-1}\mathbf{f}_0(\theta)\big). \end{aligned}$$

Hence, the prediction error $Y_0 - \hat{Y}_0$ follows an asymptotic normal distribution:

$$Y_0 - \hat{Y}_0 \sim AN\big(0, \sigma^2\big(1 + \mathbf{f}_0(\theta)'[\mathbf{F}(\theta)'\mathbf{F}(\theta)]^{-1}\mathbf{f}_0(\theta)\big)\big).$$

A $100(1-\alpha)\%$ prediction interval for Y_0 is:

$$\hat{Y}_0 \pm t_{(1-\alpha/2,n-p)} s \sqrt{1 + \mathbf{f}_0(\hat{\theta})'[\mathbf{F}(\hat{\theta})'\mathbf{F}(\hat{\theta})]^{-1}\mathbf{f}_0(\hat{\theta})}.$$

where we substitute $\hat{\theta}$ into $\mathbf{f}_0$ and $\mathbf{F}$. Recall s is the estimate of σ.
Sometimes we want a confidence interval for $E(Y_i)$ (i.e., the mean response at x_0), rather than a prediction interval for an individual future observation. In that case, the variance term for the random error ϵ_0 is not included. Hence, the formula is the same but without the "+1":

$$E(Y_0) \approx f(x_0; \theta),$$

and the confidence interval is:

$$f(x_0, \hat{\theta}) \pm t_{(1-\alpha/2,n-p)} s \sqrt{\mathbf{f}_0(\hat{\theta})'[\mathbf{F}(\hat{\theta})'\mathbf{F}(\hat{\theta})]^{-1}\mathbf{f}_0(\hat{\theta})}.$$

Summary

- Linear Functions of the Parameters: A function $f(x, \theta)$ is linear in θ if it can be written in the form

$$f(x, \theta) = \theta_1 g_1(x) + \theta_2 g_2(x) + \cdots + \theta_p g_p(x)$$

  where $g_j(x)$ do not depend on θ. In this case, the Jacobian $\mathbf{F}(\theta)$ does not depend on θ itself (only on x_i), and exact formulas often match what is familiar from linear regression.
- Nonlinear Functions of Parameters: If $f(x, \theta)$ depends on θ in a nonlinear way (e.g., $\theta_1 e^{\theta_2 x}$, β_1/β_2 or more complicated expressions), $\mathbf{F}(\theta)$ depends on θ. Estimation generally requires iterative numerical methods (e.g., Gauss-Newton, Levenberg-Marquardt), and the asymptotic results rely on evaluating partial derivatives at $\hat{\theta}$. Nevertheless, the final inference formulas—confidence intervals and prediction intervals—have a similar form, thanks to the asymptotic normality arguments.

3.2 Nonlinear Least Squares Estimation

The least squares (LS) estimate of θ, denoted as $\hat{\theta}$, minimizes the residual sum of squares:

$$S(\hat{\theta}) = SSE(\hat{\theta}) = \sum_{i=1}^{n} \{Y_i - f(\mathbf{x}_i; \hat{\theta})\}^2$$

To solve this, we consider the partial derivatives of $S(\theta)$ with respect to each θ_j and set them to 0, leading to the normal equations:

$$\frac{\partial S(\theta)}{\partial \theta_j} = -2\sum_{i=1}^{n}\{Y_i - f(\mathbf{x}_i;\theta)\}\frac{\partial f(\mathbf{x}_i;\theta)}{\partial \theta_j} = 0$$

However, these equations are inherently nonlinear and, in most cases, cannot be solved analytically. As a result, various estimation techniques are employed to approximate solutions efficiently (Table 3.2). These approaches include:

- Iterative Optimization—Methods that refine estimates through successive iterations to minimize error.
- Derivative-Free Methods—Techniques that do not rely on gradient information, useful for complex or non-smooth functions.
- Stochastic Heuristic—Algorithms that incorporate randomness, such as Genetic Algorithms or Simulated Annealing, to explore solution spaces.
- Linearization—Approximating nonlinear models with linear ones to enable analytical or numerical solutions.
- Hybrid Approaches—Combining multiple methods to leverage their respective strengths for improved estimation.

3.2.1 *Iterative Optimization*

3.2.1.1 Gauss-Newton Algorithm

The Gauss-Newton Algorithm is an iterative optimization method used to estimate parameters in nonlinear least squares problems. It refines parameter estimates by approximating the Hessian matrix using first-order derivatives, making it computationally efficient for many practical applications (e.g., regression models in finance and marketing analytics). The objective is to minimize the sum of squared errors (SSE):

$$SSE(\theta) = \sum_{i=1}^{n}[Y_i - f_i(\theta)]^2,$$

where $\mathbf{Y} = [Y_1, \dots, Y_n]'$ are the observed responses and $f_i(\theta)$ are the model-predicted values.

Iterative Refinement via Taylor Expansion

The Gauss-Newton Algorithm iteratively refines an initial estimate $\hat{\theta}^{(0)}$ using the **Taylor series expansion** of $f(\mathbf{x}_i;\theta)$ about $\hat{\theta}^{(0)}$. We start with the observation model:

$$Y_i = f(\mathbf{x}_i;\theta) + \epsilon_i.$$

Table 3.2 Comparison of optimization methods for nonlinear estimation

Category	Method	Best for	Derivative?	Optimization	Comp. cost
Iterative optimization	Steepest Descent (Gradient Descent)	Simple problems, slow convergence	Yes	Local	Low
Iterative optimization	Gauss-Newton Algorithm	Faster than GD, ignores exact second-order info	Yes	Local	Medium
Iterative optimization	Levenberg-Marquardt Algorithm	Balances GN and GD, robust	Yes	Local	Medium
Iterative optimization	Newton-Raphson Method	Quadratic convergence, needs Hessian	Yes	Local	High
Iterative optimization	Quasi-Newton Method	Approximates Hessian for large problems	Yes	Local	Medium
Iterative optimization	Trust-Region Reflective Algorithm	Handles constraints, robust	Yes	Local	High
Derivative-free	Secant Method	Approximates derivative from function evaluations	No	Local	Medium
Derivative-free	Nelder-Mead (Simplex)	No derivatives, heuristic	No	Local	Medium
Derivative-free	Powell's Method	Line search, no explicit gradient	No	Local	Medium
Derivative-free	Grid Search	Exhaustive search (best in low dims)	No	Global	Very High
Derivative-free	Hooke-Jeeves Pattern Search	Pattern-based, black-box optimization	No	Local	Medium
Derivative-free	Bisection Method	Simple root/interval-based approach	No	Local	Low
Stochastic heuristic	Simulated Annealing	Escapes local minima, non-smooth problems	No	Global	High
Stochastic heuristic	Genetic Algorithm	Large search spaces, evolving parameters	No	Global	High
Stochastic heuristic	Particle Swarm Optimization	Swarm-based, often fast global convergence	No	Global	Medium

(continued)

Table 3.2 (continued)

Category	Method	Best for	Derivative?	Optimization	Comp. cost
Stochastic heuristic	Simulated Annealing	Escapes local minima, non-smooth problems	No	Global	High
Stochastic heuristic	Genetic Algorithm	Large search spaces, evolving parameters	No	Global	High
Stochastic heuristic	Particle Swarm Optimization	Swarm-based, often fast global convergence	No	Global	Medium
Stochastic heuristic	Evolutionary Strategies	High-dimensional, adaptive step size	No	Global	High
Stochastic heuristic	Differential Evolution Algorithm	Robust global optimizer, population-based	No	Global	High
Stochastic heuristic	Ant Colony Optimization (ACO)	Discrete/combinatorial problems	No	Global	High
Linearization	Taylor Series Approximation	Local approximation of nonlinearity	Yes	Local	Low
Linearization	Log-Linearization	Transforms nonlinear equations	Yes	Local	Low
Hybrid	Adaptive Levenberg-Marquardt	Dynamically adjusts damping in LM	Yes	Local	Medium
Hybrid	Hybrid Genetic Algorithm and LM (GA-LM)	GA for coarse search, LM for fine-tuning	No	Hybrid	High
Hybrid	Neural Network-Based NLLS	Deep learning for complex nonlinear least squares	No	Hybrid	Very High

By expanding $f(\mathbf{x}_i; \theta)$ around $\hat{\theta}^{(0)}$ and ignoring higher-order terms (assuming the remainder is small), we get:

$$Y_i \approx f(\mathbf{x}_i; \hat{\theta}^{(0)}) + \sum_{j=1}^{p} \left. \frac{\partial f(\mathbf{x}_i; \theta)}{\partial \theta_j} \right|_{\theta=\hat{\theta}^{(0)}} \left(\theta_j - \hat{\theta}_j^{(0)} \right) + \epsilon_i .$$

In **matrix form**, let

$$\mathbf{Y} = \begin{bmatrix} Y_1 \\ \vdots \\ Y_n \end{bmatrix}, \quad \mathbf{f}(\hat{\theta}^{(0)}) = \begin{bmatrix} f(\mathbf{x}_1, \hat{\theta}^{(0)}) \\ \vdots \\ f(\mathbf{x}_n, \hat{\theta}^{(0)}) \end{bmatrix},$$

and define the **Jacobian matrix** of partial derivatives:

$$\mathbf{F}(\hat{\theta}^{(0)}) = \begin{bmatrix} \frac{\partial f(\mathbf{x}_1,\theta)}{\partial \theta_1} & \cdots & \frac{\partial f(\mathbf{x}_1,\theta)}{\partial \theta_p} \\ \vdots & \ddots & \vdots \\ \frac{\partial f(\mathbf{x}_n,\theta)}{\partial \theta_1} & \cdots & \frac{\partial f(\mathbf{x}_n,\theta)}{\partial \theta_p} \end{bmatrix}_{\theta=\hat{\theta}^{(0)}} .$$

Then,

$$\mathbf{Y} \approx \mathbf{f}(\hat{\theta}^{(0)}) + \mathbf{F}(\hat{\theta}^{(0)}) \, (\theta - \hat{\theta}^{(0)}) + \epsilon,$$

with $\epsilon = [\epsilon_1, \ldots, \epsilon_n]'$ assumed i.i.d. with mean 0 and variance σ^2.

From this linear approximation,

$$\mathbf{Y} - \mathbf{f}(\hat{\theta}^{(0)}) \approx \mathbf{F}(\hat{\theta}^{(0)}) \, (\theta - \hat{\theta}^{(0)}).$$

Solving for $\theta - \hat{\theta}^{(0)}$ in the **least squares** sense gives the Gauss increment $\hat{\delta}^{(1)}$, so we can update:

$$\hat{\theta}^{(1)} = \hat{\theta}^{(0)} + \hat{\delta}^{(1)}.$$

Step-by-Step Procedure

1. **Initialize**: Start with an initial estimate $\hat{\theta}^{(0)}$ and set $j = 0$.
2. **Compute Taylor Expansion**: Calculate $\mathbf{f}(\hat{\theta}^{(j)})$ and $\mathbf{F}(\hat{\theta}^{(j)})$.
3. **Solve for Increment**: Treating $\mathbf{Y} - \mathbf{f}(\hat{\theta}^{(j)}) \approx \mathbf{F}(\hat{\theta}^{(j)}) \, (\theta - \hat{\theta}^{(j)})$ as a linear model, use Ordinary Least Squares to compute $\hat{\delta}^{(j+1)}$.
4. **Update Parameters**: Set $\hat{\theta}^{(j+1)} = \hat{\theta}^{(j)} + \hat{\delta}^{(j+1)}$.
5. **Check for Convergence**: If the convergence criteria are met (see below), stop; otherwise, return to Step 2.
6. **Estimate Variance**: After convergence, we assume $\epsilon \sim (\mathbf{0}, \sigma^2 \mathbf{I})$. The variance σ^2 can be estimated by:

$$\hat{\sigma}^2 = \frac{1}{n-p}\left(\mathbf{Y} - \mathbf{f}(\mathbf{x}; \hat{\theta})\right)'\left(\mathbf{Y} - \mathbf{f}(\mathbf{x}; \hat{\theta})\right).$$

Convergence Criteria
Common criteria for deciding when to stop iterating include:

1. **Objective Function Change**:

$$\frac{\left|SSE(\hat{\theta}^{(j+1)}) - SSE(\hat{\theta}^{(j)})\right|}{SSE(\hat{\theta}^{(j)})} < \gamma_1.$$

2. **Parameter Change**:

$$\left|\hat{\theta}^{(j+1)} - \hat{\theta}^{(j)}\right| < \gamma_2.$$

3. **Residual Projection Criterion**: The residuals satisfy convergence as defined in (Bates and Watts, 1981).

Another way to see the update step is by viewing the **necessary condition for a minimum**: the gradient of $SSE(\theta)$ with respect to θ should be 0. For

$$SSE(\theta) = \sum_{i=1}^{n}[Y_i - f_i(\theta)]^2,$$

the gradient is

$$\frac{\partial SSE(\theta)}{\partial \theta} = 2\,\mathbf{F}(\theta)'\,[\mathbf{Y} - \mathbf{f}(\theta)]\,.$$

Using the Gauss-Newton update rule from iteration j to $j+1$:

$$\begin{aligned}
\hat{\theta}^{(j+1)} &= \hat{\theta}^{(j)} + \hat{\delta}^{(j+1)} \\
&= \hat{\theta}^{(j)} + \left[\mathbf{F}(\hat{\theta}^{(j)})'\,\mathbf{F}(\hat{\theta}^{(j)})\right]^{-1}\,\mathbf{F}(\hat{\theta}^{(j)})'\left[\mathbf{Y} - \mathbf{f}(\hat{\theta}^{(j)})\right] \\
&= \hat{\theta}^{(j)} - \frac{1}{2}\left[\mathbf{F}(\hat{\theta}^{(j)})'\,\mathbf{F}(\hat{\theta}^{(j)})\right]^{-1}\frac{\partial SSE(\hat{\theta}^{(j)})}{\partial \theta},
\end{aligned}$$

Where:

- $\frac{\partial SSE(\hat{\theta}^{(j)})}{\partial \theta}$ is the **gradient vector**, pointing in the direction of **steepest ascent** of SSE.

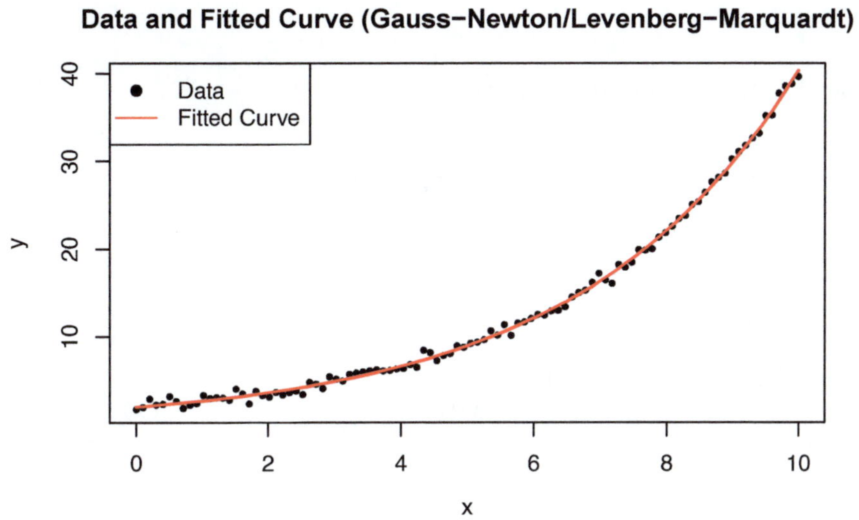

Fig. 3.1 Data and fitted curve (Gauss-Newton/Levenberg-Marquardt)

- $\left[\mathbf{F}(\hat{\theta}^{(j)})'\mathbf{F}(\hat{\theta}^{(j)})\right]^{-1}$ determines the **step size**, controlling how far to move in the direction of improvement.
- The factor $-\frac{1}{2}$ ensures movement in the **direction of steepest descent**, helping **minimize** the SSE.

The Gauss-Newton method works well when the nonlinear model can be approximated accurately by a **first-order Taylor expansion** near the solution (Fig. 3.1). If the assumption of near-linearity in the residual function $\mathbf{r}(\theta) = \mathbf{Y} - \mathbf{f}(\theta)$ is violated, convergence may be slow or fail altogether. In such cases, more robust methods like Levenberg-Marquardt Algorithm (which modifies Gauss-Newton with a damping parameter) are often preferred.

```
# Load necessary libraries
library(minpack.lm)  # Provides nonlinear least squares
↪  functions

# Define a nonlinear function (exponential model)
nonlinear_model <- function(theta, x) {
  # theta is a vector of parameters: theta[1] = A, theta[2]
  ↪  = B
  # x is the independent variable
  # The model is A * exp(B * x)
  theta[1] * exp(theta[2] * x)
}
```

```
# Define SSE function for clarity
sse <- function(theta, x, y) {
    # SSE = sum of squared errors between actual y and model
    ↪ predictions
    sum((y - nonlinear_model(theta, x)) ^ 2)
}

# Generate synthetic data
set.seed(123)                          # for reproducibility
x <- seq(0, 10, length.out = 100) # 100 points from 0 to 10
true_theta <- c(2, 0.3)                # true parameter values
y <- nonlinear_model(true_theta, x) + rnorm(length(x), sd =
↪ 0.5)

# Display the first few data points
head(data.frame(x, y))
#>           x        y
#> 1 0.0000000 1.719762
#> 2 0.1010101 1.946445
#> 3 0.2020202 2.904315
#> 4 0.3030303 2.225593
#> 5 0.4040404 2.322373
#> 6 0.5050505 3.184724

# Gauss-Newton optimization using nls.lm (Levenberg-Marquardt
↪ as extension).
# Initial guess for theta: c(1, 0.1)
fit <- nls.lm(
    par = c(1, 0.1),
    fn = function(theta)
        y - nonlinear_model(theta, x)
)

# Display estimated parameters
cat("Estimated parameters (A, B):\n")
#> Estimated parameters (A, B):
print(fit$par)
#> [1] 1.9934188 0.3008742
```

1. We defined the model "nonlinear_model(theta, x)" which returns A*exp*(*B*x).
2. We generated synthetic data using the "true_theta" values and added random noise.
3. We used `nls.lm(...)` from the `minpack.lm` package to fit the data:

- `par = c(1, 0.1)` is our initial parameter guess.
- `fn = function(theta) y - nonlinear_model(theta, x)` is the residual function, i.e., observed minus predicted.

4. The `fit$par` provides the estimated parameters after the algorithm converges.

```
# Visualize the data and the fitted model
plot(
  x,
  y,
  main = "Data and Fitted Curve
  ↪ (Gauss-Newton/Levenberg-Marquardt)",
  xlab = "x",
  ylab = "y",
  pch = 19,
  cex = 0.5
)
curve(
  nonlinear_model(fit$par, x),
  from = 0,
  to = 10,
  add = TRUE,
  col = "red",
  lwd = 2
)
legend(
  "topleft",
  legend = c("Data", "Fitted Curve"),
  pch = c(19, NA),
  lty = c(NA, 1),
  col = c("black", "red")
)
```

3.2.1.2 Modified Gauss-Newton Algorithm

The Modified Gauss-Newton Algorithm introduces a learning rate α_j to control step size and prevent overshooting the local minimum. The standard Gauss-Newton Algorithm assumes that the full step direction $\hat{\delta}^{(j+1)}$ is optimal, but in practice, especially for highly nonlinear problems, it can overstep the minimum or cause numerical instability (Fig. 3.2). The modification introduces a step size reduction, making it more robust.

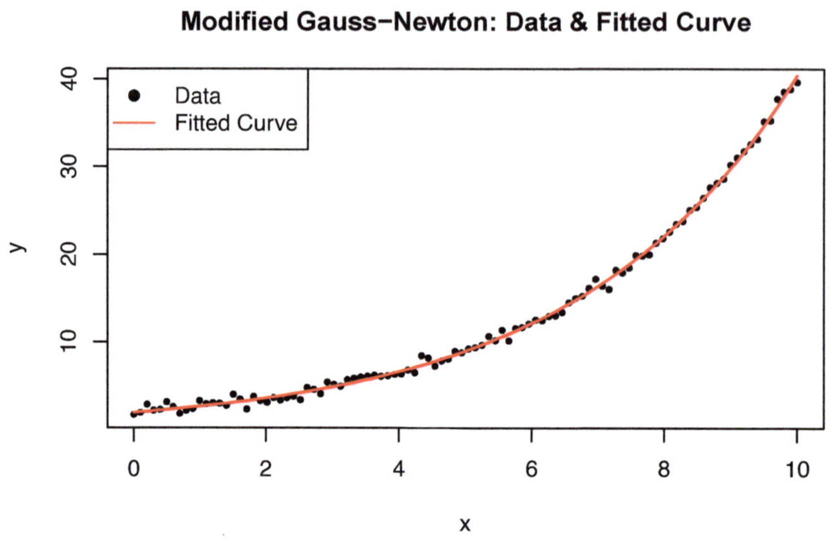

Fig. 3.2 Modified Gauss-Newton

We redefine the update step as:

$$\hat{\theta}^{(j+1)} = \hat{\theta}^{(j)} + \alpha_j \hat{\delta}^{(j+1)}, \quad 0 < \alpha_j < 1,$$

Where:

- α_j is a **learning rate**, controlling how much of the step $\hat{\delta}^{(j+1)}$ is taken.
- If $\alpha_j = 1$, we recover the standard Gauss-Newton method.
- If α_j is too small, convergence is slow; if too large, the algorithm may diverge.

This learning rate α_j allows for **adaptive step size adjustments**, helping prevent excessive parameter jumps and ensuring that SSE decreases at each iteration.
A common approach to determine α_j is **step halving**, ensuring that each iteration moves in a direction that reduces SSE. Instead of using a fixed α_j, we iteratively reduce the step size until SSE decreases:

$$\hat{\theta}^{(j+1)} = \hat{\theta}^{(j)} + \frac{1}{2^k} \hat{\delta}^{(j+1)},$$

Where:

- k is the smallest non-negative integer such that

$$SSE(\hat{\theta}^{(j)} + \frac{1}{2^k} \hat{\delta}^{(j+1)}) < SSE(\hat{\theta}^{(j)}).$$

This means we start with the full step $\hat{\delta}^{(j+1)}$, then try $\hat{\delta}^{(j+1)}/2$, then $\hat{\delta}^{(j+1)}/4$, and so on, until SSE is reduced.

Algorithm for Step Halving:

1. Compute the Gauss-Newton step $\hat{\delta}^{(j+1)}$.
2. Set an initial $\alpha_j = 1$.
3. If the updated parameters $\hat{\theta}^{(j)} + \alpha_j \hat{\delta}^{(j+1)}$ increase SSE, divide α_j by 2.
4. Repeat until SSE decreases.

This **ensures monotonic SSE reduction**, preventing divergence due to an overly aggressive step.

Generalized Form of the Modified Algorithm

A more general form of the update rule, incorporating step size control and a matrix $\mathbf{A}_j$, is:

$$\hat{\theta}^{(j+1)} = \hat{\theta}^{(j)} - \alpha_j \mathbf{A}_j \frac{\partial Q(\hat{\theta}^{(j)})}{\partial \theta},$$

Where:

- $\mathbf{A}_j$ is a **positive definite matrix** that preconditions the update direction.
- α_j is the **learning rate**.
- $\frac{\partial Q(\hat{\theta}^{(j)})}{\partial \theta}$ is the **gradient** of the objective function $Q(\theta)$, typically SSE in nonlinear regression.

Connection to the Modified Gauss-Newton Algorithm

The **Modified Gauss-Newton Algorithm** fits this framework:

$$\hat{\theta}^{(j+1)} = \hat{\theta}^{(j)} - \alpha_j [\mathbf{F}(\hat{\theta}^{(j)})' \mathbf{F}(\hat{\theta}^{(j)})]^{-1} \frac{\partial SSE(\hat{\theta}^{(j)})}{\partial \theta}.$$

Here, we recognize:

- **Objective function**: $Q = SSE$.
- **Preconditioner matrix**: $[\mathbf{F}(\hat{\theta}^{(j)})' \mathbf{F}(\hat{\theta}^{(j)})]^{-1} = \mathbf{A}$.

Thus, the standard Gauss-Newton method can be interpreted as a special case of this broader optimization framework, with a preconditioned Gradient Descent approach.

```
# Load required library
library(minpack.lm)

# Define a nonlinear function (exponential model)
nonlinear_model <- function(theta, x) {
  theta[1] * exp(theta[2] * x)
}
```

```
# Define the Sum of Squared Errors function
sse <- function(theta, x, y) {
  sum((y - nonlinear_model(theta, x)) ^ 2)
}

# Generate synthetic data
set.seed(123)
x <- seq(0, 10, length.out = 100)
true_theta <- c(2, 0.3)
y <- nonlinear_model(true_theta, x) + rnorm(length(x), sd =
↪  0.5)

# Gauss-Newton with Step Halving
gauss_newton_modified <-
  function(theta_init,
           x,
           y,
           tol = 1e-6,
           max_iter = 100) {
    theta <- theta_init
    for (j in 1:max_iter) {
      # Compute Jacobian matrix numerically
      epsilon <- 1e-6
      F_matrix <-
        matrix(0, nrow = length(y), ncol = length(theta))
      for (p in 1:length(theta)) {
        theta_perturb <- theta
        theta_perturb[p] <- theta[p] + epsilon
        F_matrix[, p] <-
          (nonlinear_model(theta_perturb,
↪  x)-nonlinear_model(theta, x))/epsilon
      }

      # Compute residuals
      residuals <- y - nonlinear_model(theta, x)

      # Compute Gauss-Newton step
      delta <-
        solve(t(F_matrix) %*% F_matrix) %*% t(F_matrix) %*%
        ↪  residuals

      # Step Halving Implementation
      alpha <- 1
      k <- 0
```

```
      while (sse(theta + alpha * delta, x, y) >= sse(theta, x,
       ↪  y) &&
              k < 10) {
        alpha <- alpha / 2
        k <- k + 1
      }

      # Update theta
      theta_new <- theta + alpha * delta

      # Check for convergence
      if (sum(abs(theta_new - theta)) < tol) {
        break
      }

      theta <- theta_new
    }
    return(theta)
  }

# Run Modified Gauss-Newton Algorithm
theta_init <- c(1, 0.1)  # Initial parameter guess
estimated_theta <- gauss_newton_modified(theta_init, x, y)

# Display estimated parameters
cat("Estimated parameters (A, B) with Modified
 ↪   Gauss-Newton:\n")
#> Estimated parameters (A, B) with Modified Gauss-Newton:
print(estimated_theta)
#>           [,1]
#> [1,] 1.9934188
#> [2,] 0.3008742
```

```
# Plot data and fitted curve
plot(
  x,
  y,
  main = "Modified Gauss-Newton: Data & Fitted Curve",
  pch = 19,
  cex = 0.5,
  xlab = "x",
  ylab = "y"
)
```

```
curve(
  nonlinear_model(estimated_theta, x),
  from = 0,
  to = 10,
  add = TRUE,
  col = "red",
  lwd = 2
)
legend(
  "topleft",
  legend = c("Data", "Fitted Curve"),
  pch = c(19, NA),
  lty = c(NA, 1),
  col = c("black", "red")
)
```

3.2.1.3 Steepest Descent (Gradient Descent)

The Steepest Descent Method, commonly known as Gradient Descent, is a fundamental iterative optimization technique used for finding parameter estimates that minimize an objective function $\mathbf{Q}(\theta)$. It is a special case of the Modified Gauss-Newton Algorithm, where the preconditioning matrix $\mathbf{A}_j$ is replaced by the identity matrix (Table 3.3).

The update rule is given by:

$$\hat{\theta}^{(j+1)} = \hat{\theta}^{(j)} - \alpha_j \mathbf{I}_{p\times p} \frac{\partial \mathbf{Q}(\hat{\theta}^{(j)})}{\partial \theta},$$

Where:

- α_j is the **learning rate**, determining the step size.
- $\mathbf{I}_{p\times p}$ is the **identity matrix**, meaning updates occur in the direction of the **negative gradient**.
- $\frac{\partial \mathbf{Q}(\hat{\theta}^{(j)})}{\partial \theta}$ is the **gradient vector**, which provides the direction of **steepest ascent**; its negation ensures movement toward the minimum.

Table 3.3 Comparison of steepest descent and Gauss-Newton optimization methods

Method	Update rule	Advantages	Disadvantages
Steepest Descent	$\hat{\theta}^{(j+1)} = \hat{\theta}^{(j)} - \alpha_j \mathbf{I} \nabla Q(\theta)$	Simple, requires only first derivatives	Slow convergence, sensitive to α_j
Gauss-Newton	$\hat{\theta}^{(j+1)} = \hat{\theta}^{(j)} - \mathbf{H}^{-1} \nabla Q(\theta)$	Uses curvature, faster near solution	Requires Jacobian computation, may diverge

Characteristics of Steepest Descent

- **Slow to converge:** The algorithm moves in the direction of the gradient but does not take into account curvature, which may result in slow convergence, especially in ill-conditioned problems.
- **Moves rapidly initially:** The method can exhibit **fast initial progress**, but as it approaches the minimum, step sizes become small, leading to slow convergence.
- **Useful for initialization:** Due to its simplicity and ease of implementation, it is often used to obtain **starting values** for more advanced methods like **Newton's Method** or Gauss-Newton Algorithm.

The **key difference** is that steepest descent **only** considers the gradient direction, while Gauss-Newton and Newton's Method incorporate curvature information.

Choosing the Learning Rate α_j
A well-chosen **learning rate** is crucial for the success of Gradient Descent:

- **Too large**: The algorithm may overshoot the minimum and diverge.
- **Too small**: Convergence is **very slow**.
- **Adaptive strategies**:
  - **Fixed step size**: α_j is constant.
  - **Step size decay**: α_j decreases over iterations (e.g., $\alpha_j = \frac{1}{j}$).
  - **Line search**: Choose α_j by minimizing $\mathbf{Q}(\theta^{(j+1)})$ along the gradient direction.

A common approach is **backtracking line search**, where α_j is reduced iteratively until a decrease in $\mathbf{Q}(\theta)$ is observed.

```
# Load necessary libraries
library(ggplot2)


# Define the nonlinear function (exponential model)
nonlinear_model <- function(theta, x) {
    theta[1] * exp(theta[2] * x)
}

# Define the Sum of Squared Errors function
sse <- function(theta, x, y) {
    sum((y - nonlinear_model(theta, x)) ^ 2)
}

# Define Gradient of SSE w.r.t theta (computed numerically)
gradient_sse <- function(theta, x, y) {
    n <- length(y)
```

```
    residuals <- y - nonlinear_model(theta, x)

    # Partial derivative w.r.t theta_1
    grad_1 <- -2 * sum(residuals * exp(theta[2] * x))

    # Partial derivative w.r.t theta_2
    grad_2 <- -2 * sum(residuals * theta[1] * x * exp(theta[2]
↪   * x))

    return(c(grad_1, grad_2))
}

# Generate synthetic data
set.seed(123)
x <- seq(0, 10, length.out = 100)
true_theta <- c(2, 0.3)
y <- nonlinear_model(true_theta, x) + rnorm(length(x), sd =
↪  0.5)

# Safe Gradient Descent Implementation
gradient_descent <-
    function(theta_init,
             x,
             y,
             alpha = 0.01,
             tol = 1e-6,
             max_iter = 500) {
        theta <- theta_init
        sse_values <- numeric(max_iter)

        for (j in 1:max_iter) {
            grad <- gradient_sse(theta, x, y)

            # Check for NaN or Inf values in the gradient
            ↪  (prevents divergence)
            if (any(is.na(grad)) || any(is.infinite(grad))) {
                cat("Numerical instability detected at
                ↪  iteration",
                    j,
                    "\n")
                break
            }

            # Update step
```

```
            theta_new <- theta - alpha * grad
            sse_values[j] <- sse(theta_new, x, y)

            # Check for convergence
            if (!is.finite(sse_values[j])) {
                cat("Divergence detected at iteration", j,
                ↪  "\n")
                break
            }

            if (sum(abs(theta_new - theta)) < tol) {
                cat("Converged in", j, "iterations.\n")
                return(list(theta = theta_new, sse_values =
                ↪  sse_values[1:j]))
            }

            theta <- theta_new
        }

        return(list(theta = theta, sse_values = sse_values))
    }

# Run Gradient Descent with a Safe Implementation
theta_init <- c(1, 0.1)  # Initial guess
alpha <- 0.001            # Learning rate
result <- gradient_descent(theta_init, x, y, alpha)
#> Divergence detected at iteration 1

# Extract results
estimated_theta <- result$theta
sse_values <- result$sse_values

# Display estimated parameters
cat("Estimated parameters (A, B) using Gradient Descent:\n")
#> Estimated parameters (A, B) using Gradient Descent:
print(estimated_theta)
#> [1] 1.0 0.1
```

```
# Plot convergence of SSE over iterations
# Ensure sse_values has valid data
sse_df <- data.frame(
  Iteration = seq_along(sse_values),
```

```
  SSE = sse_values
)

# Generate improved plot using ggplot()
ggplot(sse_df, aes(x = Iteration, y = SSE)) +
  geom_line(color = "blue", linewidth = 1) +
  labs(
    title = "Gradient Descent Convergence",
    x = "Iteration",
    y = "SSE"
  ) +
  theme_minimal()
```

1. Steepest Descent (Gradient Descent) moves in the direction of steepest descent, which can lead to zigzagging behavior (Fig. 3.3).
2. Slow convergence occurs when the curvature of the function varies significantly across dimensions.

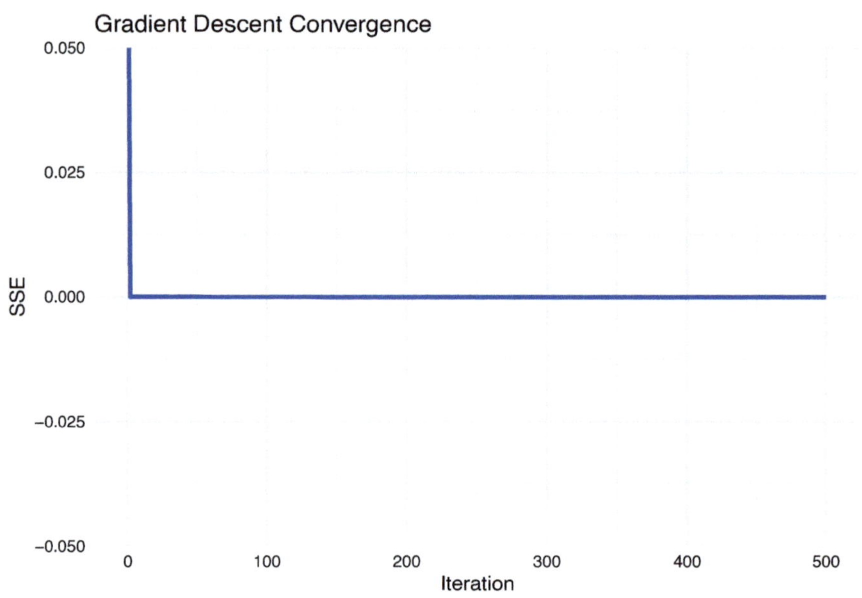

Fig. 3.3 Gradient Descent Convergence

3. Learning rate tuning is critical:
   - If too large, the algorithm diverges.
   - If too small, progress is very slow.
4. Useful for initialization: It is often used to get close to the optimal solution before switching to more advanced methods like Gauss-Newton Algorithm or Newton's Method.

Several advanced techniques improve the performance of steepest descent:

- **Momentum Gradient Descent**: Adds a momentum term to smooth updates, reducing oscillations.
- **Adaptive Learning Rates**:
  - AdaGrad: Adjusts α_j per parameter based on historical gradients.
  - RMSprop: Uses a moving average of past gradients to scale updates.
  - Adam (Adaptive Moment Estimation): Combines momentum and adaptive learning rates.

In practice, Adam is widely used in machine learning and deep learning, while Newton-based methods (including Gauss-Newton) are preferred for nonlinear regression.

3.2.1.4 Levenberg-Marquardt Algorithm

The Levenberg-Marquardt Algorithm is a widely used optimization method for solving nonlinear least squares problems. It is an adaptive technique that blends the Gauss-Newton Algorithm and Steepest Descent (Gradient Descent), dynamically switching between them based on problem conditions.

The update rule is:

$$\hat{\theta}^{(j+1)} = \hat{\theta}^{(j)} - \alpha_j [\mathbf{F}(\hat{\theta}^{(j)})'\mathbf{F}(\hat{\theta}^{(j)}) + \tau \mathbf{I}_{p\times p}] \frac{\partial \mathbf{Q}(\hat{\theta}^{(j)})}{\partial \theta}$$

Where:

- τ is the **damping parameter**, controlling whether the step behaves like Gauss-Newton Algorithm or Steepest Descent (Gradient Descent).
- $\mathbf{I}_{p\times p}$ is the **identity matrix**, ensuring numerical stability.
- $\mathbf{F}(\hat{\theta}^{(j)})$ is the **Jacobian matrix** of partial derivatives.
- $\frac{\partial \mathbf{Q}(\hat{\theta}^{(j)})}{\partial \theta}$ is the **gradient vector**.
- α_j is the **learning rate**, determining step size.

The Levenberg-Marquardt Algorithm is particularly useful when the Jacobian matrix $\mathbf{F}(\hat{\theta}^{(j)})$ is nearly singular, meaning that Gauss-Newton Algorithm alone may fail:

- When τ is large, the method behaves like steepest descent, ensuring stability.
- When τ is small, it behaves like Gauss-Newton, accelerating convergence.
- Adaptive control of τ:
  - If $SSE(\hat{\theta}^{(j+1)}) < SSE(\hat{\theta}^{(j)})$, reduce τ:

$$\tau \leftarrow \tau/10$$

  - Otherwise, increase τ to stabilize:

$$\tau \leftarrow 10\tau$$

This adjustment ensures that the algorithm moves efficiently while avoiding instability.

```
# Load required libraries
library(minpack.lm)
library(ggplot2)

# Define a nonlinear function (exponential model)
nonlinear_model <- function(theta, x) {
    theta[1] * exp(theta[2] * x)
}

# Define SSE function
sse <- function(theta, x, y) {
    sum((y - nonlinear_model(theta, x)) ^ 2)
}

# Generate synthetic data
set.seed(123)
x <- seq(0, 10, length.out = 100)
true_theta <- c(2, 0.3)
y <- nonlinear_model(true_theta, x) + rnorm(length(x), sd =
↪ 0.5)

# Robust Levenberg-Marquardt Optimization Implementation
levenberg_marquardt <-
    function(theta_init,
             x,
             y,
             tol = 1e-6,
             max_iter = 500,
             tau_init = 1) {
```

```
    theta <- theta_init
    tau <- tau_init
    lambda <- 1e-8  # Small regularization term
    sse_values <- numeric(max_iter)

    for (j in 1:max_iter) {
        # Compute Jacobian matrix numerically
        epsilon <- 1e-6
        F_matrix <-
            matrix(0, nrow = length(y), ncol =
            ↪  length(theta))
        for (p in 1:length(theta)) {
            theta_perturb <- theta
            theta_perturb[p] <- theta[p] + epsilon
            F_matrix[, p] <-
                (nonlinear_model(theta_perturb, x)
                 - nonlinear_model(theta, x)) / epsilon
        }

        # Compute residuals
        residuals <- y - nonlinear_model(theta, x)

        # Compute Levenberg-Marquardt update
        A <-
            t(F_matrix) %*% F_matrix + tau *
            ↪  diag(length(theta))
        + lambda * diag(length(theta))  # Regularized A
        delta <- tryCatch(
            solve(A) %*% t(F_matrix) %*% residuals,
            error = function(e) {
                cat("Singular matrix detected at
                ↪  iteration",
                    j,
                    "- Increasing tau\n")
                tau <<- tau * 10  # Increase tau to
↪  stabilize
                # Return zero delta to avoid NaN updates
                return(rep(0, length(theta)))
            }
        )

        theta_new <- theta + delta

        # Compute new SSE
```

```
            sse_values[j] <- sse(theta_new, x, y)

            # Adjust tau dynamically
            if (sse_values[j] < sse(theta, x, y)) {
                # Reduce tau but prevent it from going too low
                tau <-
                    max(tau / 10, 1e-8)
            } else {
                tau <- tau * 10  # Increase tau if SSE
↪   increases
            }

            # Check for convergence
            if (sum(abs(delta)) < tol) {
                cat("Converged in", j, "iterations.\n")
                return(list(theta = theta_new, sse_values =
                 ↪  sse_values[1:j]))
            }

            theta <- theta_new
        }

        return(list(theta = theta, sse_values = sse_values))
    }

# Run Levenberg-Marquardt
theta_init <- c(1, 0.1)  # Initial guess
result <- levenberg_marquardt(theta_init, x, y)
#> Singular matrix detected at iteration 11 - Increasing tau
#> Converged in 11 iterations.

# Extract results
estimated_theta <- result$theta
sse_values <- result$sse_values

# Display estimated parameters
cat("Estimated parameters (A, B) using
↪  Levenberg-Marquardt:\n")
#> Estimated parameters (A, B) using Levenberg-Marquardt:
print(estimated_theta)
#>               [,1]
#> [1,] -6.473440e-09
#> [2,]  1.120637e+01
```

```r
# Plot convergence of SSE over iterations
sse_df <-
    data.frame(Iteration = seq_along(sse_values), SSE =
    ↪  sse_values)

ggplot(sse_df, aes(x = Iteration, y = SSE)) +
    geom_line(color = "blue", linewidth = 1) +
    labs(title = "Levenberg-Marquardt Convergence",
         x = "Iteration",
         y = "SSE") +
    theme_minimal()
```

Figure 3.4 shows convergence of SSE over iterations:

- Early Stability (Flat SSE)
  - The SSE remains near zero for the first few iterations, which suggests that the algorithm is initially behaving stably.
  - This might indicate that the initial parameter guess is reasonable or that the updates are too small to significantly affect SSE.

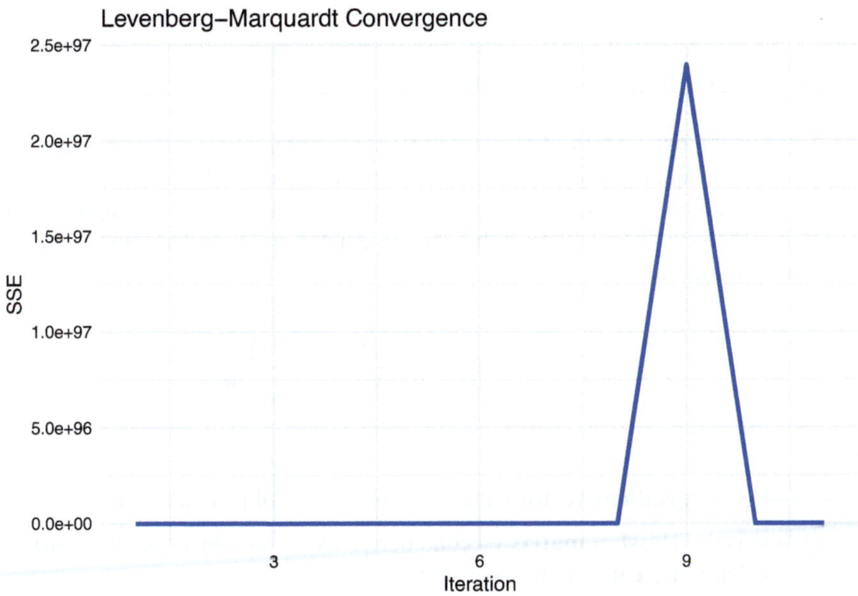

Fig. 3.4 Levenberg-Marquardt Convergence

- Sudden Explosion in SSE (Iteration ~8–9)
  - The sharp spike in SSE at iteration 9 indicates a numerical instability or divergence in the optimization process.
  - This could be due to:
    * An ill-conditioned Jacobian matrix: The step direction is poorly estimated, leading to an unstable jump.
    * A sudden large update (`delta`): The damping parameter (`tau`) might have been reduced too aggressively, causing an uncontrolled step.
    * Floating-point issues: If `A` becomes nearly singular, solving `A \ delta = residuals` may produce excessively large values.

- Return to Stability (After Iteration 9)
  - The SSE immediately returns to a low value after the spike, which suggests that the damping parameter (`tau`) might have been increased after detecting instability.
  - This is consistent with the adaptive nature of Levenberg-Marquardt:
    * If a step leads to a bad SSE increase, the algorithm increases `tau` to make the next step more conservative.
    * If the next step stabilizes, `tau` may be reduced again.

3.2.1.5 Newton-Raphson Algorithm

The Newton-Raphson Method is a second-order optimization technique used for nonlinear least squares problems. Unlike first-order methods (such as Steepest Descent (Gradient Descent) and Gauss-Newton Algorithm), Newton-Raphson uses both first and second derivatives of the objective function for faster convergence.

The update rule is:

$$\hat{\theta}^{(j+1)} = \hat{\theta}^{(j)} - \alpha_j \left[\frac{\partial^2 Q(\hat{\theta}^{(j)})}{\partial\theta\partial\theta'} \right]^{-1} \frac{\partial Q(\hat{\theta}^{(j)})}{\partial\theta}$$

Where:

- $\frac{\partial Q(\hat{\theta}^{(j)})}{\partial\theta}$ is the **gradient vector** (first derivative of the objective function).
- $\frac{\partial^2 Q(\hat{\theta}^{(j)})}{\partial\theta\partial\theta'}$ is the **Hessian matrix** (second derivative of the objective function).
- α_j is the **learning rate**, controlling step size.

The **Hessian matrix** in nonlinear least squares problems is:

$$\frac{\partial^2 Q(\hat{\theta}^{(j)})}{\partial\theta\partial\theta'} = 2\mathbf{F}(\hat{\theta}^{(j)})'\mathbf{F}(\hat{\theta}^{(j)}) - 2\sum_{i=1}^{n}[Y_i - f(x_i;\theta)]\frac{\partial^2 f(x_i;\theta)}{\partial\theta\partial\theta'}$$

Where:

- The first term $2\mathbf{F}(\hat{\theta}^{(j)})'\mathbf{F}(\hat{\theta}^{(j)})$ is the same as in the Gauss-Newton Algorithm.
- The second term $-2\sum_{i=1}^{n}[Y_i - f(x_i;\theta)]\frac{\partial^2 f(x_i;\theta)}{\partial\theta\partial\theta'}$ contains second-order derivatives.

Key Observations

- Gauss-Newton vs. Newton-Raphson:
  - Gauss-Newton approximates the Hessian by ignoring the second term.
  - Newton-Raphson explicitly incorporates second-order derivatives, making it more precise but computationally expensive.
- Challenges:
  - The Hessian matrix may be singular, making it impossible to invert.
  - Computing second derivatives is often difficult for complex functions.

```
# Load required libraries
library(ggplot2)

# Define a nonlinear function (exponential model)
nonlinear_model <- function(theta, x) {
    theta[1] * exp(theta[2] * x)
}

# Define SSE function
sse <- function(theta, x, y) {
    sum((y - nonlinear_model(theta, x)) ^ 2)
}

# Compute Gradient (First Derivative) of SSE
gradient_sse <- function(theta, x, y) {
    residuals <- y - nonlinear_model(theta, x)

    # Partial derivative w.r.t theta_1
    grad_1 <- -2 * sum(residuals * exp(theta[2] * x))

    # Partial derivative w.r.t theta_2
    grad_2 <- -2 * sum(residuals * theta[1] * x * exp(theta[2]
 ↪   * x))

    return(c(grad_1, grad_2))
}

# Compute Hessian (Second Derivative) of SSE
```

```
hessian_sse <- function(theta, x, y) {
    residuals <- y - nonlinear_model(theta, x)

    # Compute second derivatives
    H_11 <- 2 * sum(exp(2 * theta[2] * x))
    H_12 <- 2 * sum(x * exp(2 * theta[2] * x) * theta[1])
    H_21 <- H_12

    term1 <- 2 * sum((x ^ 2) * exp(2 * theta[2] * x) *
↪   theta[1] ^ 2)
    term2 <- 2 * sum(residuals * (x ^ 2) * exp(theta[2] * x))

    H_22 <- term1 - term2

    return(matrix(
        c(H_11, H_12, H_21, H_22),
        nrow = 2,
        byrow = TRUE
    ))
}

# Generate synthetic data
set.seed(123)
x <- seq(0, 10, length.out = 100)
true_theta <- c(2, 0.3)
y <- nonlinear_model(true_theta, x) + rnorm(length(x), sd =
↪   0.5)

# Newton-Raphson Optimization Implementation
newton_raphson <-
    function(theta_init,
             x,
             y,
             tol = 1e-6,
             max_iter = 500) {
        theta <- theta_init
        sse_values <- numeric(max_iter)

        for (j in 1:max_iter) {
            grad <- gradient_sse(theta, x, y)
            hessian <- hessian_sse(theta, x, y)

            # Check if Hessian is invertible
            if (det(hessian) == 0) {
```

```
            cat("Hessian is singular at iteration",
                j,
                "- Using identity matrix instead.\n")
            # Replace with identity matrix if singular
            hessian <-
                diag(length(theta))
        }

        # Compute Newton update
        delta <- solve(hessian) %*% grad
        theta_new <- theta - delta
        sse_values[j] <- sse(theta_new, x, y)

        # Check for convergence
        if (sum(abs(delta)) < tol) {
            cat("Converged in", j, "iterations.\n")
            return(list(theta = theta_new, sse_values =
             ↪  sse_values[1:j]))
        }

        theta <- theta_new
    }

    return(list(theta = theta, sse_values = sse_values))
}

# Run Newton-Raphson
theta_init <- c(1, 0.1)  # Initial guess
result <- newton_raphson(theta_init, x, y)
#> Converged in 222 iterations.

# Extract results
estimated_theta <- result$theta
sse_values <- result$sse_values

# Display estimated parameters
cat("Estimated parameters (A, B) using Newton-Raphson:\n")
#> Estimated parameters (A, B) using Newton-Raphson:
print(estimated_theta)
#>           [,1]
#> [1,] 1.9934188
#> [2,] 0.3008742
```

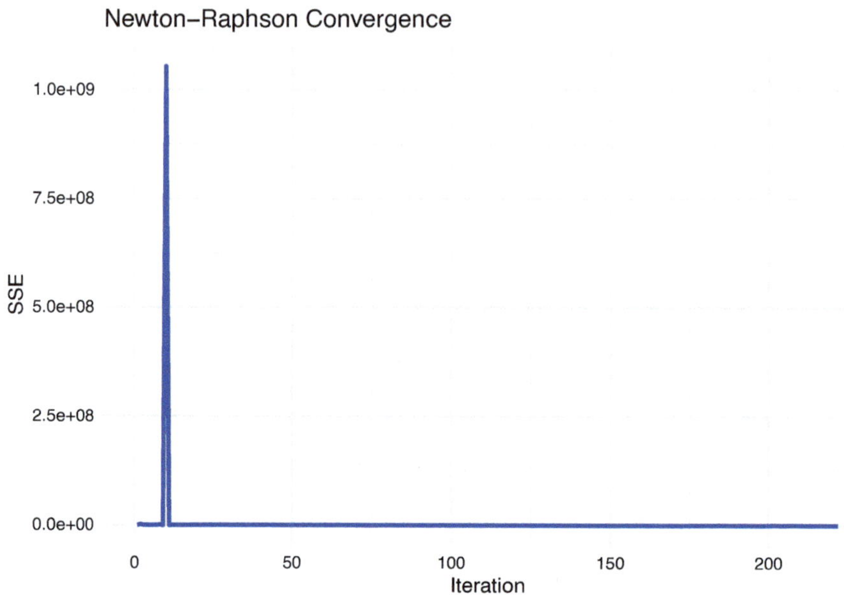

Fig. 3.5 Newton-Raphson Convergence

Figure 3.5 shows convergence plot of SSE over multiple iterations

```
# Plot convergence of SSE over iterations
sse_df <-
    data.frame(Iteration = seq_along(sse_values), SSE =
    ↪  sse_values)

ggplot(sse_df, aes(x = Iteration, y = SSE)) +
    geom_line(color = "blue", size = 1) +
    labs(title = "Newton-Raphson Convergence",
         x = "Iteration",
         y = "SSE") +
    theme_minimal()
#> Warning: Using `size` aesthetic for lines was deprecated in
↪  ggplot2
#> 3.4.0.
#> i Please use `linewidth` instead.
#> This warning is displayed once every 8 hours.
#> Call `lifecycle::last_lifecycle_warnings()` to see where
#> this warning was generated.
```

3.2.1.6 Quasi-Newton Method

The Quasi-Newton Method is an optimization technique that approximates Newton's Method without requiring explicit computation of the Hessian matrix. Instead, it iteratively constructs an approximation $\mathbf{H}_j$ to the Hessian based on the first derivative information.

The update rule is:

$$\hat{\theta}^{(j+1)} = \hat{\theta}^{(j)} - \alpha_j \mathbf{H}_j^{-1} \frac{\partial \mathbf{Q}(\hat{\theta}^{(j)})}{\partial \theta}$$

Where:

- $\mathbf{H}_j$ is a **symmetric positive definite approximation** to the **Hessian matrix**.
- As $j \to \infty$, $\mathbf{H}_j$ gets closer to the **true Hessian**.
- $\frac{\partial Q(\hat{\theta}^{(j)})}{\partial \theta}$ is the **gradient vector**.
- α_j is the **learning rate**, controlling step size.

Why Use Quasi-Newton Instead of Newton-Raphson Method?

- Newton-Raphson requires computing the Hessian matrix explicitly, which is computationally expensive and may be singular.
- Quasi-Newton avoids computing the Hessian directly by approximating it iteratively.
- Among first-order methods (which only require gradients, not Hessians), Quasi-Newton Methods perform best.

Hessian Approximation

Instead of directly computing the Hessian $\mathbf{H}_j$, Quasi-Newton Methods update an approximation $\mathbf{H}_j$ iteratively.

One of the most widely used formulas is the **Broyden-Fletcher-Goldfarb-Shanno (BFGS) update**:

$$\mathbf{H}_{j+1} = \mathbf{H}_j + \frac{(\mathbf{s}_j \mathbf{s}_j')}{\mathbf{s}_j' \mathbf{y}_j} - \frac{\mathbf{H}_j \mathbf{y}_j \mathbf{y}_j' \mathbf{H}_j}{\mathbf{y}_j' \mathbf{H}_j \mathbf{y}_j}$$

Where:

- $\mathbf{s}_j = \hat{\theta}^{(j+1)} - \hat{\theta}^{(j)}$ (change in parameters).
- $\mathbf{y}_j = \nabla Q(\hat{\theta}^{(j+1)}) - \nabla Q(\hat{\theta}^{(j)})$ (change in gradient).
- $\mathbf{H}_j$ is the current **inverse Hessian approximation**.

```
# Load required libraries
library(ggplot2)

# Define a nonlinear function (exponential model)
nonlinear_model <- function(theta, x) {
    theta[1] * exp(theta[2] * x)
}

# Define SSE function
sse <- function(theta, x, y) {
    sum((y - nonlinear_model(theta, x))^2)
}

# Generate synthetic data
set.seed(123)
x <- seq(0, 10, length.out = 100)
true_theta <- c(2, 0.3)
y <- nonlinear_model(true_theta, x) + rnorm(length(x), sd =
↪  0.5)

# Run BFGS Optimization using `optim()`
theta_init <- c(1, 0.1)  # Initial guess
result <- optim(
    par = theta_init,
    fn = function(theta) sse(theta, x, y),  # Minimize SSE
    method = "BFGS",
    control = list(trace = 0)  # Suppress optimization
     ↪  progress
    # control = list(trace = 1, REPORT = 1)  # Print
     ↪  optimization progress
)

# Extract results
estimated_theta <- result$par
sse_final <- result$value
convergence_status <- result$convergence  # 0 means successful
↪  convergence

# Display estimated parameters
cat("\n=== Optimization Results ===\n")
#>
#> === Optimization Results ===
cat("Estimated parameters (A, B) using Quasi-Newton BFGS:\n")
#> Estimated parameters (A, B) using Quasi-Newton BFGS:
```

```
print(estimated_theta)
#> [1] 1.9954216 0.3007569

# Display final SSE
cat("\nFinal SSE:", sse_final, "\n")
#>
#> Final SSE: 20.3227
```

3.2.1.7 Trust-Region Reflective Algorithm

The Trust-Region Reflective (TRR) algorithm is an optimization technique used for nonlinear least squares problems. Unlike Newton's method and gradient-based approaches, TRR dynamically restricts updates to a trust region, ensuring stability and preventing overshooting.

The goal is to minimize the objective function $Q(\theta)$ (e.g., SSE):

$$\hat{\theta} = \arg\min_{\theta} Q(\theta)$$

Instead of taking a full Newton step, TRR solves the following **quadratic subproblem**:

$$\min_{\delta} m_j(\delta) = Q(\hat{\theta}^{(j)}) + \nabla Q(\hat{\theta}^{(j)})'\delta + \frac{1}{2}\delta'\mathbf{H}_j\delta$$

subject to:

$$\|\delta\| \leq \Delta_j$$

Where:

- $\mathbf{H}_j$ is an **approximation of the Hessian matrix**.
- $\nabla Q(\hat{\theta}^{(j)})$ is the **gradient vector**.
- Δ_j is the **trust-region radius**, which is adjusted dynamically.

Trust-Region Adjustments
The algorithm **modifies the step size dynamically** based on the ratio ρ_j:

$$\rho_j = \frac{Q(\hat{\theta}^{(j)}) - Q(\hat{\theta}^{(j)} + \delta)}{m_j(0) - m_j(\delta)}$$

- **If** $\rho_j > 0.75$ and $\|\delta\| = \Delta_j$, then expand the trust region:

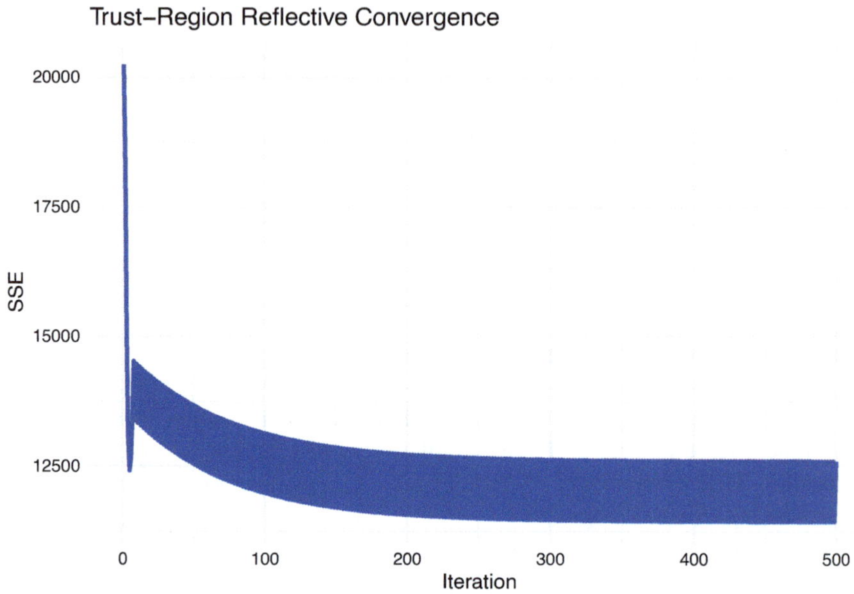

Fig. 3.6 Trust-region reflective convergence

$$\Delta_{j+1} = 2\Delta_j$$

- **If** $\rho_j < 0.25$, shrink the trust region:

$$\Delta_{j+1} = \frac{1}{2}\Delta_j$$

- **If** $\rho_j > 0$, accept the step; otherwise, reject it.

If a step **violates a constraint**, it is **reflected back** into the feasible region:

$$\hat{\theta}^{(j+1)} = \max(\hat{\theta}^{(j)} + \delta, \theta_{\min})$$

This ensures that the optimization **respects parameter bounds** (Fig. 3.6).

```
# Load required libraries
library(ggplot2)


# Define a nonlinear function (exponential model)
nonlinear_model <- function(theta, x) {
    theta[1] * exp(theta[2] * x)
}
```

```
# Define SSE function
sse <- function(theta, x, y) {
    sum((y - nonlinear_model(theta, x)) ^ 2)
}

# Compute Gradient (First Derivative) of SSE
gradient_sse <- function(theta, x, y) {
    residuals <- y - nonlinear_model(theta, x)

    # Partial derivative w.r.t theta_1
    grad_1 <- -2 * sum(residuals * exp(theta[2] * x))

    # Partial derivative w.r.t theta_2
    grad_2 <- -2 * sum(residuals * theta[1] * x * exp(theta[2]
↪    * x))

    return(c(grad_1, grad_2))
}

# Compute Hessian Approximation of SSE
hessian_sse <- function(theta, x, y) {
    residuals <- y - nonlinear_model(theta, x)

    # Compute second derivatives
    H_11 <- 2 * sum(exp(2 * theta[2] * x))
    H_12 <- 2 * sum(x * exp(2 * theta[2] * x) * theta[1])
    H_21 <- H_12

    term1 <- 2 * sum((x ^ 2) * exp(2 * theta[2] * x) *
↪   theta[1] ^ 2)
    term2 <- 2 * sum(residuals * (x ^ 2) * exp(theta[2] * x))

    H_22 <- term1 - term2

    return(matrix(
        c(H_11, H_12, H_21, H_22),
        nrow = 2,
        byrow = TRUE
    ))
}

# Generate synthetic data
set.seed(123)
```

```
x <- seq(0, 10, length.out = 100)
true_theta <- c(2, 0.3)
y <- nonlinear_model(true_theta, x) + rnorm(length(x), sd =
↪  0.5)

# Manual Trust-Region Reflective Optimization Implementation
trust_region_reflective <-
    function(theta_init,
             x,
             y,
             tol = 1e-6,
             max_iter = 500,
             delta_max = 1.0) {
        theta <- theta_init
        delta_j <- 0.5  # Initial trust-region size
        n <- length(theta)
        sse_values <- numeric(max_iter)

        for (j in 1:max_iter) {
            grad <- gradient_sse(theta, x, y)
            hessian <- hessian_sse(theta, x, y)

            # Check if Hessian is invertible
            if (det(hessian) == 0) {
                cat("Hessian is singular at iteration",
                    j,
                    "- Using identity matrix instead.\n")
                hessian <-
                    diag(n)  # Replace with identity matrix if
                     ↪  singular
            }

            # Compute Newton step
            delta_full <- -solve(hessian) %*% grad

            # Apply trust-region constraint
            if (sqrt(sum(delta_full ^ 2)) > delta_j) {
                # Scale step
                delta <-
                    (delta_j / sqrt(sum(delta_full ^ 2))) *
↪  delta_full
            } else {
                delta <- delta_full
            }
```

```
        # Compute new theta and ensure it respects
        ↪  constraints
        theta_new <-
            pmax(theta + delta, c(0,-Inf))  # Reflect to
            ↪  lower bound
        sse_new <- sse(theta_new, x, y)

        # Compute agreement ratio (rho_j)
        predicted_reduction <-
            -t(grad) %*% delta - 0.5 * t(delta) %*%
            ↪  hessian %*% delta
        actual_reduction <- sse(theta, x, y) - sse_new
        rho_j <- actual_reduction / predicted_reduction

        # Adjust trust region size
        if (rho_j < 0.25) {
            delta_j <- max(delta_j / 2, 1e-4)  # Shrink
        } else if (rho_j > 0.75 &&
                   sqrt(sum(delta ^ 2)) == delta_j) {
            delta_j <- min(2 * delta_j, delta_max)  #
↪  Expand
        }

        # Accept or reject step
        if (rho_j > 0) {
            theta <- theta_new  # Accept step
        } else {
            cat("Step rejected at iteration", j, "\n")
        }

        sse_values[j] <- sse(theta, x, y)

        # Check for convergence
        if (sum(abs(delta)) < tol) {
            cat("Converged in", j, "iterations.\n")
            return(list(theta = theta, sse_values =
            ↪  sse_values[1:j]))
        }
    }

    return(list(theta = theta, sse_values = sse_values))
}
```

```
# Run Manual Trust-Region Reflective Algorithm
theta_init <- c(1, 0.1) # Initial guess
result <- trust_region_reflective(theta_init, x, y)

# Extract results
estimated_theta <- result$theta
sse_values <- result$sse_values
```

```
# Plot convergence of SSE over iterations
sse_df <-
    data.frame(Iteration = seq_along(sse_values), SSE =
    ↪ sse_values)

ggplot(sse_df, aes(x = Iteration, y = SSE)) +
    geom_line(color = "blue", size = 1) +
    labs(title = "Trust-Region Reflective Convergence",
         x = "Iteration",
         y = "SSE") +
    theme_minimal()
```

3.2.2 Derivative-Free

3.2.2.1 Secant Method

The Secant Method is a root-finding algorithm that approximates the derivative using finite differences, making it a derivative-free alternative to Newton's method. It is particularly useful when the exact gradient (or Jacobian in the case of optimization problems) is unavailable or expensive to compute.

In nonlinear optimization, we apply the Secant Method to iteratively refine parameter estimates without explicitly computing second-order derivatives.
In one dimension, the Secant Method approximates the derivative as:

$$f'(\theta) \approx \frac{f(\theta_j) - f(\theta_{j-1})}{\theta_j - \theta_{j-1}}.$$

Using this approximation, the **update step** in the Secant Method follows:

$$\theta_{j+1} = \theta_j - f(\theta_j)\frac{\theta_j - \theta_{j-1}}{f(\theta_j) - f(\theta_{j-1})}.$$

Instead of computing the exact derivative (as in Newton's method), we use the difference between the last two iterates to approximate it. This makes the Secant Method **more efficient** in cases where gradient computation is expensive or infeasible.
In **higher dimensions**, the Secant Method extends to an approximate Quasi-Newton Method, often referred to as **Broyden's Method**. We iteratively approximate the inverse Hessian matrix using past updates.

The update formula for a vector-valued function $F(\theta)$ is:

$$\theta^{(j+1)} = \theta^{(j)} - \mathbf{B}^{(j)} F(\theta^{(j)}),$$

where $\mathbf{B}^{(j)}$ is an approximation of the inverse Jacobian matrix, updated at each step using:

$$\mathbf{B}^{(j+1)} = \mathbf{B}^{(j)} + \frac{(\Delta\theta^{(j)} - \mathbf{B}^{(j)}\Delta F^{(j)})(\Delta\theta^{(j)})'}{(\Delta\theta^{(j)})'\Delta F^{(j)}},$$

Where:

- $\Delta\theta^{(j)} = \theta^{(j+1)} - \theta^{(j)}$,
- $\Delta F^{(j)} = F(\theta^{(j+1)}) - F(\theta^{(j)})$.

This **secant-based update** approximates the behavior of the **true Jacobian inverse**, reducing computational cost compared to full Newton's method.

Algorithm: Secant Method for Nonlinear Optimization
The **Secant Method** for nonlinear optimization follows these steps:

1. Initialize parameters $\theta^{(0)}$ and $\theta^{(1)}$ (two starting points).
2. Compute the function values $F(\theta^{(0)})$ and $F(\theta^{(1)})$.
3. Use the Secant update formula to compute the next iterate $\theta^{(j+1)}$.
4. Update the approximate inverse Jacobian $\mathbf{B}^{(j)}$.
5. Repeat until convergence (Fig. 3.7).

```r
# Load required libraries
library(numDeriv)


# Define a nonlinear function (logistic model)
nonlinear_model <- function(theta, x) {
  return(theta[1] / (1 + exp(-theta[2] * (x - theta[3]))))
}

# Define the Sum of Squared Errors (SSE) function
sse <- function(theta, x, y) {
  return(sum((y - nonlinear_model(theta, x)) ^ 2))
}
```

```
# Generate synthetic data
set.seed(123)
x <- seq(-5, 5, length.out = 100)
true_theta <- c(4, 1.5, 0)  # True parameters (A, B, C)
y <- nonlinear_model(true_theta, x) + rnorm(length(x), sd =
↪  0.3)

# Improved Secant Method with Line Search
secant_method_improved <-
    function(theta0,
             theta1,
             x,
             y,
             tol = 1e-6,
             max_iter = 100) {
        theta_prev <- as.matrix(theta0)  # Convert to column
↪   vector
        theta_curr <- as.matrix(theta1)

        alpha <- 1  # Initial step size
        step_reduction_factor <- 0.5  # Reduce step if SSE
↪   increases
        B_inv <- diag(length(theta0))  # Identity matrix
↪   initialization

        for (j in 1:max_iter) {
            # Compute function values using numerical gradient
            F_prev <-
                as.matrix(grad(function(theta)
                    sse(theta, x, y), theta_prev))
            F_curr <-
                as.matrix(grad(function(theta)
                    sse(theta, x, y), theta_curr))

            # Compute secant step update (convert to column
            ↪   vectors)
            delta_theta <- as.matrix(theta_curr - theta_prev)
            delta_F <- as.matrix(F_curr - F_prev)

            # Prevent division by zero
            if (sum(abs(delta_F)) < 1e-8) {
                cat("Gradient diff is too small, stopping
                ↪   optimization.\n")
                break
```

```
        }

        # Ensure correct dimensions for Broyden update
        numerator <-
            (delta_theta - B_inv %*% delta_F) %*%
↪   t(delta_theta)
        # Convert scalar to numeric
        denominator <-
            as.numeric(t(delta_theta) %*% delta_F)

        # Updated inverse Jacobian approximation
        B_inv <-
            B_inv + numerator / denominator

        # Compute next theta using secant update
        theta_next <- theta_curr - alpha * B_inv %*%
↪   F_curr

        # Line search: Reduce step size if SSE increases
        while (sse(as.vector(theta_next), x, y) >
         ↪  sse(as.vector(theta_curr),
                                                      x,
↪   y)) {
            alpha <- alpha * step_reduction_factor
            theta_next <- theta_curr - alpha * B_inv %*%
↪   F_curr

            # Print progress
            # cat("Reducing step size to", alpha, "\n")
        }

        # Print intermediate results for debugging
        cat(
            sprintf(
                "Iteration %d: Theta = (%.4f, %.4f, %.4f),
                 ↪  Alpha = %.4f\n",
                j,
                theta_next[1],
                theta_next[2],
                theta_next[3],
                alpha
            )
        )
```

```
            # Check convergence
            if (sum(abs(theta_next - theta_curr)) < tol) {
                cat("Converged successfully.\n")
                break
            }

            # Update iterates
            theta_prev <- theta_curr
            theta_curr <- theta_next
        }

        return(as.vector(theta_curr))  # Convert back to
        ↪  numeric vector
    }

# Adjusted initial parameter guesses
theta0 <- c(2, 0.8,-1)   # Closer to true parameters
theta1 <- c(4, 1.2,-0.5)  # Slightly refined

# Run Improved Secant Method
estimated_theta <- secant_method_improved(theta0, theta1, x,
↪  y)
#> Iteration 1: Theta = (3.8521, 1.3054, 0.0057), Alpha =
↪  0.0156
#> Iteration 2: Theta = (3.8521, 1.3054, 0.0057), Alpha =
↪  0.0000
#> Converged successfully.

# Display estimated parameters
cat("Estimated parameters (A, B, C) using Secant Method:\n")
#> Estimated parameters (A, B, C) using Secant Method:
print(estimated_theta)
#> [1] 3.85213912 1.30538435 0.00566417
```

```
# Plot data and fitted curve
plot(
    x,
    y,
    main = "Secant Method: Data & Fitted Curve",
    pch = 19,
    cex = 0.5,
    xlab = "x",
    ylab = "y"
```

```
)
curve(
    nonlinear_model(estimated_theta, x),
    from = -5,
    to = 5,
    add = TRUE,
    col = "red",
    lwd = 2
)
legend(
    "topleft",
    legend = c("Data", "Fitted Curve"),
    pch = c(19, NA),
    lty = c(NA, 1),
    col = c("black", "red")
)
```

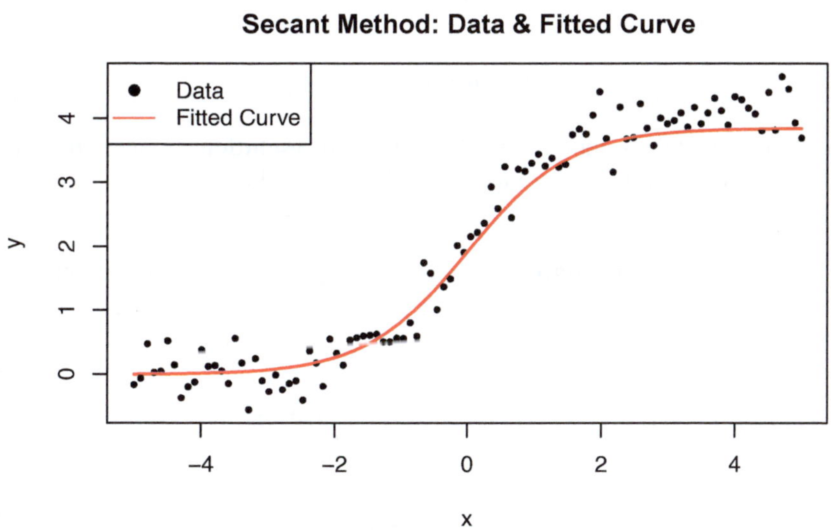

Fig. 3.7 Secant Method

3.2.2.2 Grid Search

Grid search (GS) is a brute-force optimization method that systematically explores a grid of possible parameter values to identify the combination that minimizes the residual sum of squares (RSS). Unlike gradient-based optimization, which moves iteratively toward a minimum, grid search evaluates all predefined parameter combinations, making it robust but computationally expensive.

Grid search is particularly useful when:

- **The function is highly nonlinear**, making gradient methods unreliable.
- **The parameter space is small**, allowing exhaustive search.
- **A global minimum is needed**, and prior knowledge about parameter ranges exists.

The goal of grid search is to find an optimal parameter set $\hat{\theta}$ that minimizes the SSE:

$$\hat{\theta} = \arg\min_{\theta \in \Theta} SSE(\theta),$$

Where:

$$SSE(\theta) = \sum_{i=1}^{n} (y_i - f(x_i; \theta))^2.$$

Grid search **discretizes** the search space Θ into a finite set of candidate values for each parameter:

$$\Theta = \theta_1 \times \theta_2 \times \cdots \times \theta_p.$$

The **accuracy** of the solution depends on the **grid resolution**—a finer grid leads to **better accuracy** but **higher computational cost** (Fig. 3.8).

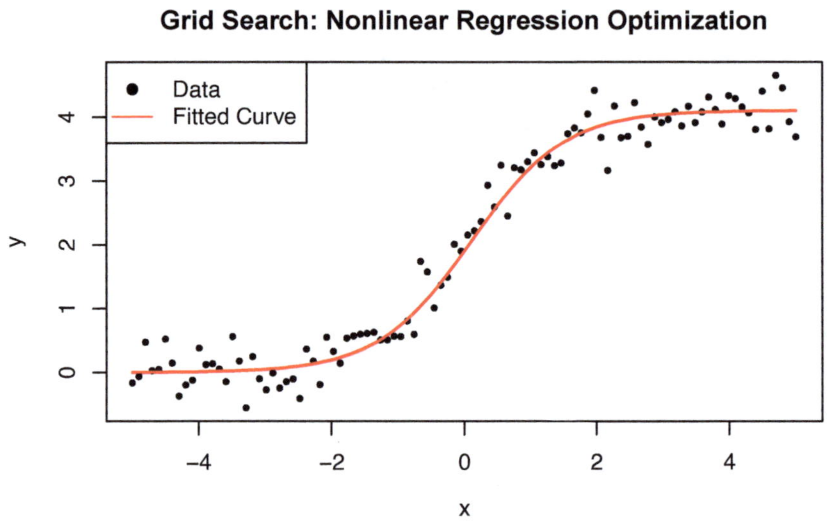

Fig. 3.8 Grid search: nonlinear regression optimization

Grid Search Algorithm

1. Define a grid of possible values for each parameter.
2. Evaluate SSE for each combination of parameters.
3. Select the parameter set that minimizes SSE.

```
# Define a numerically stable logistic function
safe_exp <- function(x) {
    return(ifelse(x > 700, Inf, exp(pmin(x, 700))))  # Prevent
    ↪ overflow
}

# Define the logistic growth model
nonlinear_model <- function(theta, x) {
    return(theta[1] / (1 + safe_exp(-theta[2] * (x -
    ↪ theta[3]))))
}

# Define the Sum of Squared Errors (SSE) function
sse <- function(theta, x, y) {
    predictions <- nonlinear_model(theta, x)
    return(sum((y - predictions)^2))
}

# Grid Search Optimization
grid_search_optimization <- function(x, y, grid_size = 10) {
    # Define grid of parameter values
    A_values <- seq(2, 6, length.out = grid_size)
    B_values <- seq(0.5, 3, length.out = grid_size)
    C_values <- seq(-2, 2, length.out = grid_size)

    # Generate all combinations of parameters
    param_grid <- expand.grid(A = A_values, B = B_values, C =
↪   C_values)

    # Evaluate SSE for each parameter combination
    param_grid$SSE <- apply(param_grid, 1, function(theta) {
        sse(as.numeric(theta[1:3]), x, y)
    })

    # Select the best parameter set
    best_params <- param_grid[which.min(param_grid$SSE), 1:3]
    return(as.numeric(best_params))
}

# Generate synthetic data
```

```
set.seed(123)
x <- seq(-5, 5, length.out = 100)
true_theta <- c(4, 1.5, 0)  # True parameters (A, B, C)
y <- nonlinear_model(true_theta, x) + rnorm(length(x), sd =
↪  0.3)

# Run Grid Search
estimated_theta <- grid_search_optimization(x, y, grid_size =
↪  20)

# Display results
cat("Estimated parameters (A, B, C) using Grid Search:\n")
#> Estimated parameters (A, B, C) using Grid Search:
print(estimated_theta)
#> [1] 4.1052632 1.4210526 0.1052632
```

```
# Plot data and fitted curve
plot(
    x, y,
    main = "Grid Search: Nonlinear Regression Optimization",
    pch = 19,
    cex = 0.5,
    xlab = "x",
    ylab = "y"
)
curve(
    nonlinear_model(estimated_theta, x),
    from = min(x),
    to = max(x),
    add = TRUE,
    col = "red",
    lwd = 2
)
legend(
    "topleft",
    legend = c("Data", "Fitted Curve"),
    pch = c(19, NA),
    lty = c(NA, 1),
    col = c("black", "red")
)
```

3.2.2.3 Nelder-Mead (Simplex)

The Nelder-Mead Algorithm, also known as the Simplex method, is a derivative-free optimization algorithm that iteratively adjusts a geometric shape (simplex) to find the minimum of an objective function. It is particularly useful for nonlinear regression when gradient-based methods fail due to non-differentiability or noisy function evaluations.

Nelder-Mead is particularly useful when:

- The function is non-differentiable or noisy.
- Gradient-based methods are unreliable.
- The parameter space is low-dimensional.

The goal of Nelder-Mead is to find an optimal parameter set $\hat{\theta}$ that minimizes the SSE:

$$\hat{\theta} = \arg\min_{\theta} SSE(\theta),$$

Where:

$$SSE(\theta) = \sum_{i=1}^{n} (y_i - f(x_i; \theta))^2.$$

1. **Simplex Representation**
   The algorithm maintains a simplex, a geometric shape with $p + 1$ vertices for a p-dimensional parameter space.
   Each vertex represents a parameter vector:

$$S = \{\theta_1, \theta_2, \dots, \theta_{p+1}\}.$$

2. **Simplex Operations**
   At each iteration, the algorithm **updates the simplex** based on the objective function values at each vertex:

   1. **Reflection**: Reflect the worst point θ_h across the centroid.

$$\theta_r = \theta_c + \alpha(\theta_c - \theta_h)$$

   2. **Expansion**: If reflection improves the objective, expand the step.

$$\theta_e = \theta_c + \gamma(\theta_r - \theta_c)$$

   3. **Contraction**: If reflection worsens the objective, contract toward the centroid.

$$\theta_c = \theta_c + \rho(\theta_h - \theta_c)$$

4. **Shrink**: If contraction fails, shrink the simplex.

$$\theta_i = \theta_1 + \sigma(\theta_i - \theta_1), \quad \forall i > 1$$

Where:

- $\alpha = 1$ (reflection coefficient)
- $\gamma = 2$ (expansion coefficient)
- $\rho = 0.5$ (contraction coefficient)
- $\sigma = 0.5$ (shrink coefficient)

The process continues until convergence (Fig. 3.9).

Nelder-Mead Algorithm

1. Initialize a simplex with $p + 1$ vertices.
2. Evaluate SSE at each vertex.
3. Perform reflection, expansion, contraction, or shrink operations.
4. Repeat until convergence.

```
# Load required library
library(stats)


# Define a numerically stable logistic function
safe_exp <- function(x) {
    return(ifelse(x > 700, Inf, exp(pmin(x, 700)))) # Prevent
    ↪ overflow
}

# Define the logistic growth model
nonlinear_model <- function(theta, x) {
    return(theta[1] / (1 + safe_exp(-theta[2] * (x -
    ↪ theta[3]))))
}

# Define the Sum of Squared Errors (SSE) function
sse <- function(theta, x, y) {
    predictions <- nonlinear_model(theta, x)
    return(sum((y - predictions) ^ 2))
}

# Nelder-Mead Optimization
nelder_mead_optimization <- function(x, y) {
```

```
    # Initial guess for parameters
    initial_guess <- c(2, 1, 0)

    # Run Nelder-Mead optimization
    result <- optim(
        par = initial_guess,
        fn = sse,
        x = x,
        y = y,
        method = "Nelder-Mead",
        control = list(maxit = 500)
    )

    return(result$par)  # Return optimized parameters
}

# Generate synthetic data
set.seed(123)
x <- seq(-5, 5, length.out = 100)
true_theta <- c(4, 1.5, 0)  # True parameters (A, B, C)
y <- nonlinear_model(true_theta, x) + rnorm(length(x), sd =
↪  0.3)

# Run Nelder-Mead Optimization
estimated_theta <- nelder_mead_optimization(x, y)

# Display results
cat("Estimated parameters (A, B, C) using Nelder-Mead:\n")
#> Estimated parameters (A, B, C) using Nelder-Mead:
print(estimated_theta)
#> [1] 4.06873116 1.42759898 0.01119379
```

```
# Plot data and fitted curve
plot(
    x,
    y,
    main = "Nelder-Mead: Nonlinear Regression Optimization",
    pch = 19,
    cex = 0.5,
    xlab = "x",
    ylab = "y"
)
curve(
```

```
    nonlinear_model(estimated_theta, x),
    from = min(x),
    to = max(x),
    add = TRUE,
    col = "red",
    lwd = 2
)
legend(
    "topleft",
    legend = c("Data", "Fitted Curve"),
    pch = c(19, NA),
    lty = c(NA, 1),
    col = c("black", "red")
)
```

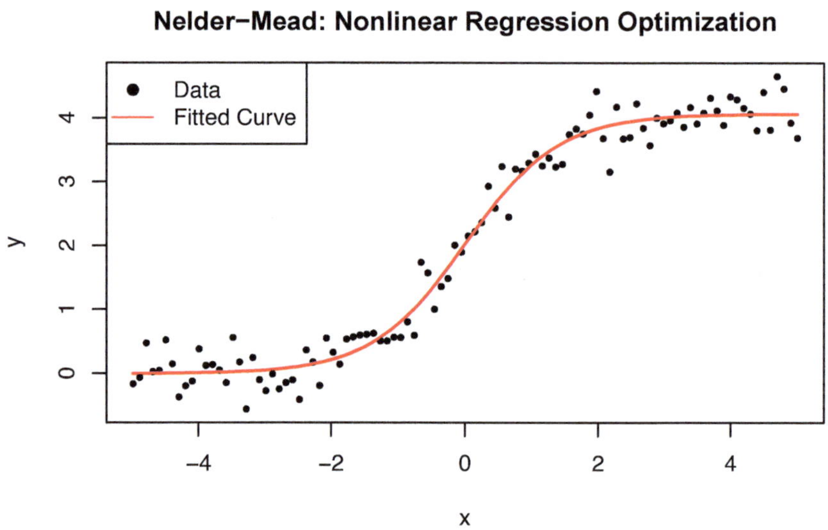

Fig. 3.9 Nelder-Mead: nonlinear regression optimization

3.2.2.4 Powell's Method

Powell's Method is a derivative-free optimization algorithm that minimizes a function without using gradients. It works by iteratively refining a set of search directions to efficiently navigate the parameter space. Unlike Nelder-Mead (Simplex), which adapts a simplex, Powell's Method builds an orthogonal basis of search directions.

Powell's Method is particularly useful when:

- The function is non-differentiable or noisy.
- Gradient-based methods are unreliable.
- The parameter space is low-to-moderate dimensional.

The goal of Powell's Method is to find an optimal parameter set $\hat{\theta}$ that minimizes the SSE:

$$\hat{\theta} = \arg\min_{\theta} SSE(\theta),$$

where:

$$SSE(\theta) = \sum_{i=1}^{n} (y_i - f(x_i; \theta))^2.$$

1. **Search Directions**
   Powell's Method **maintains a set of search directions $\mathbf{d}_1, \mathbf{d}_2, \ldots, \mathbf{d}_p$**:
   $$D = \{\mathbf{d}_1, \mathbf{d}_2, \ldots, \mathbf{d}_p\}.$$
   Initially, these directions are chosen as **unit basis vectors** (each optimizing a single parameter independently).
2. **Line Minimization**
   For each direction $\mathbf{d}_j$, Powell's Method **performs a 1D optimization**:
   $$\theta' = \theta + \lambda \mathbf{d}_j,$$
   where λ is chosen to minimize $SSE(\theta')$.
3. **Updating Search Directions**
   After optimizing along all directions:
   1. The **largest improvement** direction $\mathbf{d}_{\max}$ is replaced with:
   $$\mathbf{d}_{\text{new}} = \theta_{\text{final}} - \theta_{\text{initial}}.$$
   2. The **new direction set is orthogonalized** using Gram-Schmidt.

   This ensures **efficient exploration** of the parameter space.
4. **Convergence Criteria**
   Powell's Method **stops when function improvement is below a tolerance level**:
   $$|SSE(\theta_{t+1}) - SSE(\theta_t)| < \epsilon.$$

Powell's Method Algorithm

1. Initialize search directions (standard basis vectors).
2. Perform 1D line searches along each direction.
3. Update the search directions based on the largest improvement.
4. Repeat until convergence (Fig. 3.10).

```r
# Load required library
library(stats)

# Define a numerically stable logistic function
safe_exp <- function(x) {
    return(ifelse(x > 700, Inf, exp(pmin(x, 700)))) # Prevent
     ↪ overflow
}

# Define the logistic growth model
nonlinear_model <- function(theta, x) {
    return(theta[1] / (1 + safe_exp(-theta[2] * (x -
     ↪ theta[3]))))
}

# Define the Sum of Squared Errors (SSE) function
sse <- function(theta, x, y) {
    predictions <- nonlinear_model(theta, x)
    return(sum((y - predictions) ^ 2))
}

# Powell's Method Optimization
powell_optimization <- function(x, y) {
    # Initial guess for parameters
    initial_guess <- c(2, 1, 0)

    # Run Powell's optimization (via BFGS without gradient)
    result <- optim(
        par = initial_guess,
        fn = sse,
        x = x,
        y = y,
        method = "BFGS",
        control = list(maxit = 500),
        gr = NULL  # No gradient used
    )

    return(result$par)  # Return optimized parameters
```

```
}

# Generate synthetic data
set.seed(123)
x <- seq(-5, 5, length.out = 100)
true_theta <- c(4, 1.5, 0)  # True parameters (A, B, C)
y <- nonlinear_model(true_theta, x) + rnorm(length(x), sd =
↪  0.3)

# Run Powell's Method Optimization
estimated_theta <- powell_optimization(x, y)

# Display results
cat("Estimated parameters (A, B, C) using Powell's Method:\n")
#> Estimated parameters (A, B, C) using Powell's Method:
print(estimated_theta)
#> [1] 4.06876538 1.42765687 0.01128753
```

```
# Plot data and fitted curve
plot(
    x,
    y,
    main = "Powell's Method: Nonlinear Regression
     ↪  Optimization",
    pch = 19,
    cex = 0.5,
    xlab = "x",
    ylab = "y"
)
curve(
    nonlinear_model(estimated_theta, x),
    from = min(x),
    to = max(x),
    add = TRUE,
    col = "red",
    lwd = 2
)
legend(
    "topleft",
    legend = c("Data", "Fitted Curve"),
    pch = c(19, NA),
    lty = c(NA, 1),
    col = c("black", "red")
)
```

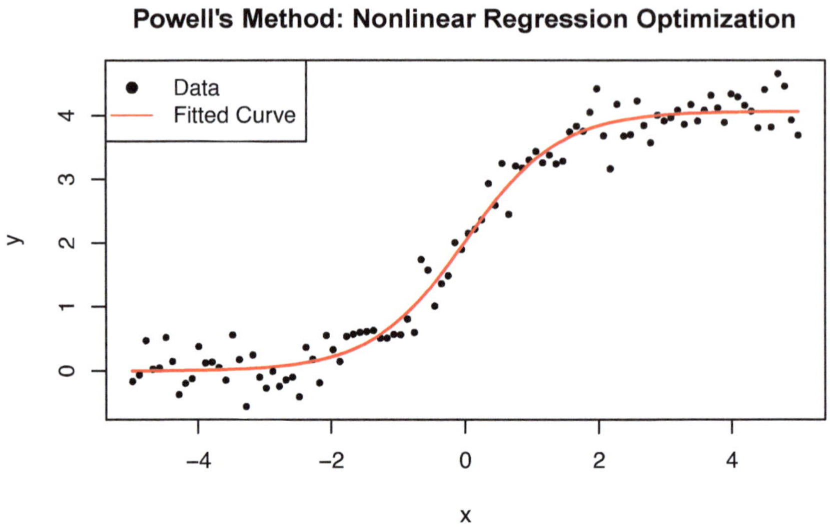

Fig. 3.10 Powell's Method: nonlinear regression optimization

3.2.2.5 Random Search

Random search (RS) is a simple yet effective optimization algorithm that explores the search space by randomly sampling candidate solutions. Unlike grid search, which evaluates all predefined parameter combinations, random search selects a random subset, reducing computational cost.

Random search is particularly useful when:

- The search space is large, making grid search impractical.
- Gradient-based methods fail due to non-differentiability or noisy data.
- The optimal region is unknown, making global exploration essential.

The goal of random search is to find an optimal parameter set $\hat{\theta}$ that minimizes the SSE:

$$\hat{\theta} = \arg\min_{\theta \in \Theta} SSE(\theta),$$

where:

$$SSE(\theta) = \sum_{i=1}^{n} (y_i - f(x_i; \theta))^2.$$

Unlike grid search, **random search does not evaluate all parameter combinations** but instead **randomly samples a subset**:

$$\Theta_{\text{sampled}} \subset \Theta.$$

The **accuracy** of the solution depends on the **number of random samples**—a higher number increases **the likelihood of finding a good solution**.

Random Search Algorithm

1. Define a random sampling range for each parameter.
2. Randomly sample N parameter sets.
3. Evaluate SSE for each sampled set.
4. Select the parameter set that minimizes SSE (Fig. 3.11).

```
# Load required library
library(stats)

# Define a numerically stable logistic function
safe_exp <- function(x) {
  return(ifelse(x > 700, Inf, exp(pmin(x, 700)))) # Prevent
   ↪ overflow
}

# Define the logistic growth model
nonlinear_model <- function(theta, x) {
  return(theta[1] / (1 + safe_exp(-theta[2] * (x -
   ↪ theta[3]))))
}

# Define the Sum of Squared Errors (SSE) function
sse <- function(theta, x, y) {
  predictions <- nonlinear_model(theta, x)
  return(sum((y - predictions) ^ 2))
}

# Random Search Optimization
random_search_optimization <- function(x, y, num_samples =
 ↪ 100) {
  # Define parameter ranges
  A_range <-
    runif(num_samples, 2, 6)   # Random values between 2 and 6
  B_range <-
    runif(num_samples, 0.5, 3) # Random values between 0.5 and
     ↪ 3
  C_range <-
    runif(num_samples,-2, 2)  # Random values between -2 and 2

  # Initialize best parameters
```

```
  best_theta <- NULL
  best_sse <- Inf

  # Evaluate randomly sampled parameter sets
  for (i in 1:num_samples) {
    theta <- c(A_range[i], B_range[i], C_range[i])
    current_sse <- sse(theta, x, y)

    if (current_sse < best_sse) {
      best_sse <- current_sse
      best_theta <- theta
    }
  }

  return(best_theta)
}

# Generate synthetic data
set.seed(123)
x <- seq(-5, 5, length.out = 100)
true_theta <- c(4, 1.5, 0)  # True parameters (A, B, C)
y <- nonlinear_model(true_theta, x) + rnorm(length(x), sd =
 ↪  0.3)

# Run Random Search
estimated_theta <-
  random_search_optimization(x, y, num_samples = 500)

# Display results
cat("Estimated parameters (A, B, C) using Random Search:\n")
#> Estimated parameters (A, B, C) using Random Search:
print(estimated_theta)
#> [1] 4.0893431 1.4687456 0.1024474
```

```
# Plot data and fitted curve
plot(
  x,
  y,
  main = "Random Search: Nonlinear Regression Optimization",
  pch = 19,
  cex = 0.5,
  xlab = "x",
  ylab = "y"
```

```
)
curve(
  nonlinear_model(estimated_theta, x),
  from = min(x),
  to = max(x),
  add = TRUE,
  col = "red",
  lwd = 2
)
legend(
  "topleft",
  legend = c("Data", "Fitted Curve"),
  pch = c(19, NA),
  lty = c(NA, 1),
  col = c("black", "red")
)
```

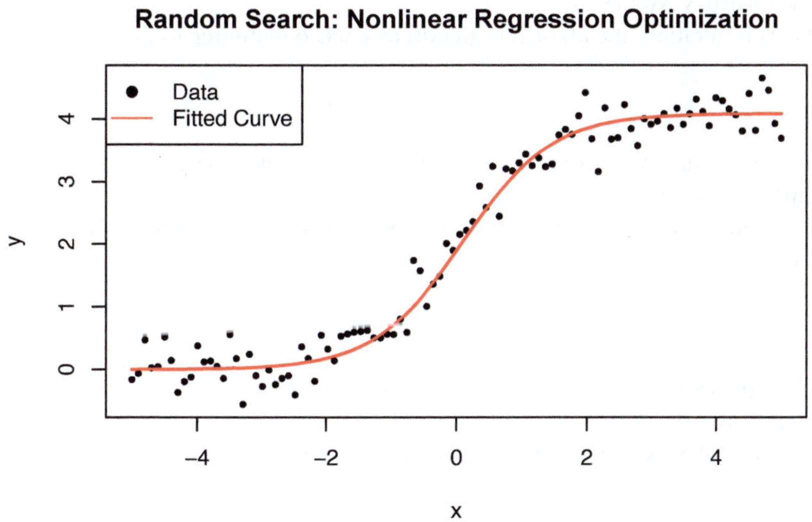

Fig. 3.11 Random search: nonlinear regression optimization

3.2.2.6 Hooke-Jeeves Pattern Search

Hooke-Jeeves Pattern Search is a derivative-free optimization algorithm that searches for an optimal solution by exploring and adjusting parameter values iteratively. Unlike gradient-based methods, it does not require differentiability, making it effective for non-smooth and noisy functions.

Hooke-Jeeves is particularly useful when:

- The function is non-differentiable or noisy.
- Gradient-based methods are unreliable.
- The parameter space is low-to-moderate dimensional.

The goal of Hooke-Jeeves Pattern Search is to find an optimal parameter set $\hat{\theta}$ that minimizes the SSE:

$$\hat{\theta} = \arg\min_{\theta} SSE(\theta),$$

where:

$$SSE(\theta) = \sum_{i=1}^{n} (y_i - f(x_i; \theta))^2.$$

1. **Exploratory Moves**
   At each iteration, the algorithm **perturbs each parameter** to find a lower SSE:

   $$\theta' = \theta \pm \delta.$$

   If a new parameter θ' reduces SSE, it becomes the new base point.
2. **Pattern Moves**
   After an improvement, the algorithm **accelerates movement** in the promising direction:

   $$\theta_{\text{new}} = \theta_{\text{old}} + (\theta_{\text{old}} - \theta_{\text{prev}}).$$

   This **speeds up convergence** toward an optimum.
3. **Step Size Adaptation**
   If no improvement is found, the step size δ is reduced:

   $$\delta_{\text{new}} = \beta \cdot \delta.$$

   where $\beta < 1$ is a reduction factor.
4. **Convergence Criteria**
   The algorithm **stops when step size is sufficiently small**:

   $$\delta < \epsilon.$$

Hooke-Jeeves Algorithm

1. Initialize a starting point θ and step size δ.

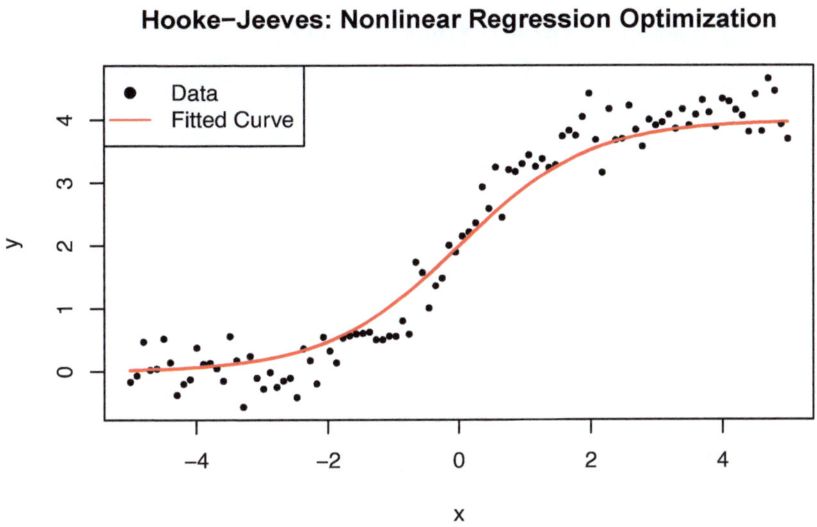

Fig. 3.12 Hooke-Jeeves: nonlinear regression optimization

2. Perform exploratory moves in each parameter direction.
3. If improvement is found, perform pattern moves.
4. Reduce step size if no improvement is found.
5. Repeat until convergence (Fig. 3.12).

```
# Load required library
library(stats)

# Define a numerically stable logistic function
safe_exp <- function(x) {
    return(ifelse(x > 700, Inf, exp(pmin(x, 700))))  # Prevent
     ↪  overflow
}

# Define the logistic growth model
nonlinear_model <- function(theta, x) {
    return(theta[1] / (1 + safe_exp(-theta[2] * (x -
     ↪  theta[3]))))
}

# Define the Sum of Squared Errors (SSE) function
sse <- function(theta, x, y) {
    predictions <- nonlinear_model(theta, x)
```

```
    return(sum((y - predictions) ^ 2))
}

# Hooke-Jeeves Pattern Search Optimization
hooke_jeeves_optimization <-
    function(x,
             y,
             step_size = 0.5,
             tol = 1e-6,
             max_iter = 500) {
        # Initial guess for parameters
        theta <- c(2, 1, 0)
        best_sse <- sse(theta, x, y)
        step <- step_size
        iter <- 0

        while (step > tol & iter < max_iter) {
            iter <- iter + 1
            improved <- FALSE
            new_theta <- theta

            # Exploratory move in each parameter direction
            for (i in 1:length(theta)) {
                for (delta in c(step,-step)) {
                    theta_test <- new_theta
                    theta_test[i] <- theta_test[i] + delta
                    sse_test <- sse(theta_test, x, y)

                    if (sse_test < best_sse) {
                        best_sse <- sse_test
                        new_theta <- theta_test
                        improved <- TRUE
                    }
                }
            }

            # Pattern move if improvement found
            if (improved) {
                theta <- 2 * new_theta - theta
                best_sse <- sse(theta, x, y)
            } else {
                step <- step / 2  # Reduce step size
            }
        }
```

```
        return(theta)
    }

# Generate synthetic data
set.seed(123)
x <- seq(-5, 5, length.out = 100)
true_theta <- c(4, 1.5, 0)  # True parameters (A, B, C)
y <- nonlinear_model(true_theta, x) + rnorm(length(x), sd =
↪  0.3)

# Run Hooke-Jeeves Optimization
estimated_theta <- hooke_jeeves_optimization(x, y)

# Display results
cat("Estimated parameters (A, B, C) using Hooke-Jeeves:\n")
#> Estimated parameters (A, B, C) using Hooke-Jeeves:
print(estimated_theta)
#> [1] 4 1 0
```

```
# Plot data and fitted curve
plot(
    x,
    y,
    main = "Hooke-Jeeves: Nonlinear Regression Optimization",
    pch = 19,
    cex = 0.5,
    xlab = "x",
    ylab = "y"
)
curve(
    nonlinear_model(estimated_theta, x),
    from = min(x),
    to = max(x),
    add = TRUE,
    col = "red",
    lwd = 2
)
legend(
    "topleft",
    legend = c("Data", "Fitted Curve"),
    pch = c(19, NA),
```

```
    lty = c(NA, 1),
    col = c("black", "red")
)
```

3.2.2.7 Bisection Method

The **Bisection Method** is a fundamental numerical technique primarily used for **root finding**, but it can also be adapted for **optimization problems** where the goal is to minimize or maximize a nonlinear function.

In nonlinear regression, optimization often involves **finding the parameter values that minimize the SSE**:

$$\hat{\theta} = \arg\min_{\theta} SSE(\theta)$$

where:

$$SSE(\theta) = \sum_{i=1}^{n} (y_i - f(x_i; \theta))^2 .$$

Since the minimum of a function occurs where the derivative equals zero, we apply the **Bisection Method** to the **derivative of the SSE function**:

$$\frac{d}{d\theta} SSE(\theta) = 0.$$

This transforms the **optimization problem** into a **root-finding problem**.
The **Bisection Method** is based on the **Intermediate Value Theorem**, which states:

> If a continuous function $f(\theta)$ satisfies $f(\theta_a) \cdot f(\theta_b) < 0$,
> then there exists at least one root in the interval (θ_a, θ_b).

For **optimization**, we apply this principle to the **first derivative** of the objective function $Q(\theta)$:

$$Q'(\theta) = 0.$$

Step-by-Step Algorithm for Optimization
1. **Choose an interval** $[\theta_a, \theta_b]$ such that:

$$Q'(\theta_a) \cdot Q'(\theta_b) < 0.$$

2. **Compute the midpoint**:

$$\theta_m = \frac{\theta_a + \theta_b}{2}.$$

3. **Evaluate** $Q'(\theta_m)$:

   - If $Q'(\theta_m) = 0$, then θ_m is the optimal point.
   - If $Q'(\theta_a) \cdot Q'(\theta_m) < 0$, set $\theta_b = \theta_m$.
   - Otherwise, set $\theta_a = \theta_m$.

4. **Repeat** until convergence:

$$|\theta_b - \theta_a| < \epsilon.$$

Determining the Nature of the Critical Point
Since the Bisection Method finds a **stationary point**, we use the **second derivative test** to classify it:

- If $Q''(\theta) > 0$, the point is a **local minimum**.
- If $Q''(\theta) < 0$, the point is a **local maximum**.

For **nonlinear regression**, we expect $Q(\theta) = SSE(\theta)$, so the solution found by Bisection should be a **minimum** (Fig. 3.13).

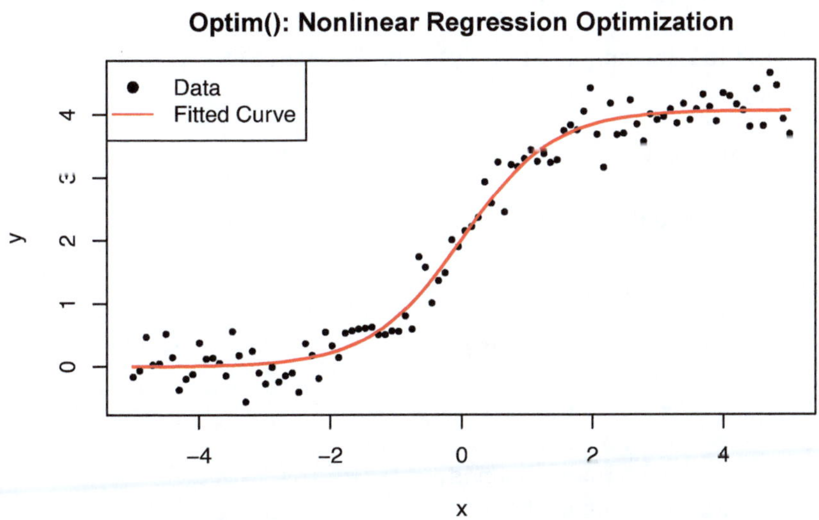

Fig. 3.13 Optim(): nonlinear regression optimization

```
# Load required library
library(stats)

# Define a numerically stable logistic function
safe_exp <- function(x) {
    return(ifelse(x > 700, Inf, exp(pmin(x, 700))))  # Prevent
     ↪  overflow
}

# Define the logistic growth model
nonlinear_model <- function(theta, x) {
    return(theta[1] / (1 + safe_exp(-theta[2] * (x -
     ↪  theta[3]))))
}

# Define the Sum of Squared Errors (SSE) function for
↪  optimization
sse <- function(theta, x, y) {
    predictions <- nonlinear_model(theta, x)
    return(sum((y - predictions)^2))
}

# Optimize all three parameters simultaneously
find_optimal_parameters <- function(x, y) {
    # Initial guess for parameters (based on data)
    initial_guess <- c(max(y), 1, median(x))

    # Bounds for parameters
    lower_bounds <- c(0.1, 0.01, min(x))  # Ensure positive
↪   scaling
    upper_bounds <- c(max(y) * 2, 10, max(x))

    # Run optim() with L-BFGS-B (bounded optimization)
    result <- optim(
        par = initial_guess,
        fn = sse,
        x = x, y = y,
        method = "L-BFGS-B",
        lower = lower_bounds,
        upper = upper_bounds
    )

    return(result$par)  # Extract optimized parameters
}
```

```r
# Generate synthetic data
set.seed(123)
x <- seq(-5, 5, length.out = 100)
true_theta <- c(4, 1.5, 0)  # True parameters (A, B, C)
y <- nonlinear_model(true_theta, x) + rnorm(length(x), sd =
↪  0.3)

# Find optimal parameters using optim()
estimated_theta <- find_optimal_parameters(x, y)

# Display results
cat("Estimated parameters (A, B, C) using optim():\n")
#> Estimated parameters (A, B, C) using optim():
print(estimated_theta)
#> [1] 4.06876536 1.42765688 0.01128756
```

```r
# Plot data and fitted curve
plot(
    x, y,
    main = "Optim(): Nonlinear Regression Optimization",
    pch = 19,
    cex = 0.5,
    xlab = "x",
    ylab = "y"
)
curve(
    nonlinear_model(estimated_theta, x),
    from = min(x),
    to = max(x),
    add = TRUE,
    col = "red",
    lwd = 2
)
legend(
    "topleft",
    legend = c("Data", "Fitted Curve"),
    pch = c(19, NA),
    lty = c(NA, 1),
    col = c("black", "red")
)
```

3.2.3 Stochastic Heuristic

3.2.3.1 Differential Evolution Algorithm

The **Differential Evolution (DE) Algorithm** is a **stochastic, population-based optimization algorithm** that is widely used for solving complex **global optimization problems**. Unlike gradient-based methods such as Newton's method or the Secant Method, DE does not require derivatives and is well-suited for optimizing **non-differentiable, nonlinear, and multimodal functions**.

Key Features of Differential Evolution

- **Population-based approach:** Maintains a population of candidate solutions instead of a single point.
- **Mutation and crossover:** Introduces variations to explore the search space.
- **Selection mechanism:** Retains the best candidates for the next generation.
- **Global optimization:** Avoids local minima by using **stochastic search strategies**.

Mathematical Formulation of Differential Evolution
Differential Evolution operates on a **population** of candidate solutions $\{\theta_i\}$, where each θ_i is a **vector of parameters**. The algorithm iteratively updates the population using three main operations:

1. **Mutation**
   For each candidate solution θ_i, a **mutant vector** $\mathbf{v}_i^{(j)}$ is generated as:

$$\mathbf{v}_i^{(j)} = \boldsymbol{\theta}_{r1}^{(j)} + F \cdot (\boldsymbol{\theta}_{r2}^{(j)} - \boldsymbol{\theta}_{r3}^{(j)})$$

   Where:

   - $\boldsymbol{\theta}_{r_1}, \boldsymbol{\theta}_{r_2}, \boldsymbol{\theta}_{r_3}$ are **randomly selected distinct vectors** from the population.
   - $F \in (0, 2)$ is the **mutation factor** controlling the step size.

2. **Crossover**
   A trial vector $\mathbf{u}_i^{(j)}$ is generated by combining the mutant vector $\mathbf{v}_i^{(j)}$ with the original solution $\boldsymbol{\theta}_i^{(j)}$:

$$u_{i,k}^{(j)} = \begin{cases} v_{i,k}^{(j)} & \text{if } rand_k \leq C_r \text{ or } k = k_{\text{rand}}, \\ \theta_{i,k}^{(j)} & \text{otherwise.} \end{cases}$$

   Where:

   - $C_r \in (0, 1)$ is the **crossover probability**.
   - $rand_k$ is a **random value between 0 and 1**.

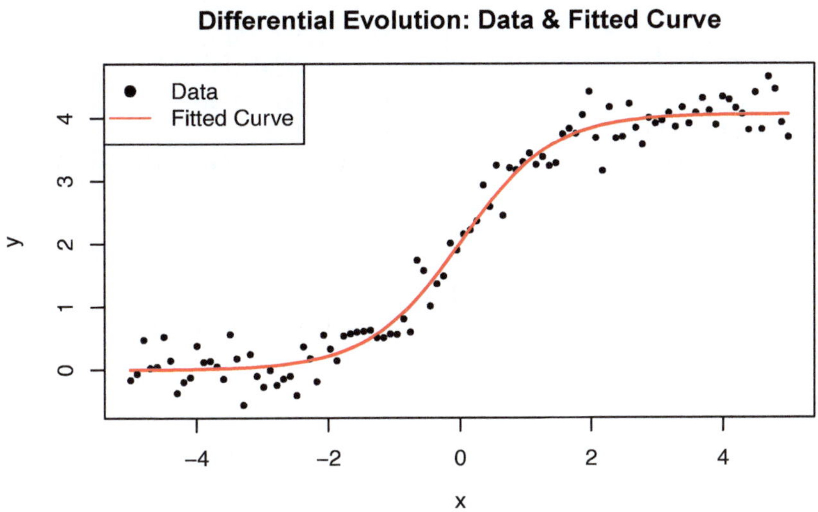

Fig. 3.14 Differential Evolution

- k_{rand} ensures at least **one parameter is mutated**.

3. **Selection**
The new candidate solution is accepted only if it improves the objective function:

$$\boldsymbol{\theta}_i^{(j+1)} = \begin{cases} \mathbf{u}_i^{(j)} & \text{if } Q(\mathbf{u}_i^{(j)}) < Q(\boldsymbol{\theta}_i^{(j)}), \\ \boldsymbol{\theta}_i^{(j)} & \text{otherwise.} \end{cases}$$

where $Q(\theta)$ is the **objective function** (e.g., sum of squared errors in regression problems).

Algorithm: Differential Evolution for Nonlinear Optimization
1. Initialize a population of candidate solutions.
2. Evaluate the objective function for each candidate.
3. Mutate individuals using a difference strategy.
4. Apply crossover to create trial solutions.
5. Select individuals based on their fitness (objective function value).
6. Repeat until convergence (or a stopping criterion is met) (Fig. 3.14).

```
# Load required library
library(DEoptim)
#> Loading required package: parallel
#>
#> DEoptim package
#> Differential Evolution algorithm in R
```

```
#> Authors: D. Ardia, K. Mullen, B. Peterson and J. Ulrich

# Define a nonlinear function (logistic model)
nonlinear_model <- function(theta, x) {
    return(theta[1] / (1 + exp(-theta[2] * (x - theta[3]))))
}

# Define the Sum of Squared Errors (SSE) function
sse <- function(theta, x, y) {
    return(sum((y - nonlinear_model(theta, x))^2))
}

# Generate synthetic data
set.seed(123)
x <- seq(-5, 5, length.out = 100)
true_theta <- c(4, 1.5, 0)  # True parameters (A, B, C)
y <- nonlinear_model(true_theta, x) + rnorm(length(x), sd =
↪  0.3)

# Define the objective function for DEoptim
objective_function <- function(theta) {
    return(sse(theta, x, y))
}

# Define parameter bounds
lower_bounds <- c(0, 0, -5)
upper_bounds <- c(10, 5, 5)

# Run Differential Evolution Algorithm
de_result <-
    DEoptim(
        objective_function,
        lower_bounds,
        upper_bounds,
        DEoptim.control(
            NP = 50,
            itermax = 100,
            F = 0.8,
            CR = 0.9,
            trace = F
        )
    )

# Extract optimized parameters
```

```r
estimated_theta <- de_result$optim$bestmem

# Display estimated parameters
cat("Estimated parameters (A, B, C) using Differential
↪ Evolution:\n")
#> Estimated parameters (A, B, C) using Differential
↪ Evolution:
print(estimated_theta)
#>       par1       par2       par3
#> 4.06876562 1.42765614 0.01128768
```

```r
# Plot data and fitted curve
plot(
    x,
    y,
    main = "Differential Evolution: Data & Fitted Curve",
    pch = 19,
    cex = 0.5,
    xlab = "x",
    ylab = "y"
)
curve(
    nonlinear_model(estimated_theta, x),
    from = -5,
    to = 5,
    add = TRUE,
    col = "red",
    lwd = 2
)
legend(
    "topleft",
    legend = c("Data", "Fitted Curve"),
    pch = c(19, NA),
    lty = c(NA, 1),
    col = c("black", "red")
)
```

3.2.3.2 Simulated Annealing

Simulated Annealing (SA) is a probabilistic global optimization algorithm inspired by annealing in metallurgy, where a material is heated and slowly cooled to remove defects. In optimization, SA gradually refines a solution by exploring the search space, allowing occasional jumps to escape local minima, before converging to an optimal solution.

Simulated Annealing is particularly useful when:

- The function is highly nonlinear and multimodal.
- Gradient-based methods struggle due to non-differentiability or poor initialization.
- A global minimum is needed, rather than a local one.

1. **Energy Function (Objective Function)**
   The goal of SA is to minimize an objective function $Q(\theta)$. For nonlinear regression, this is the SSE:

$$Q(\theta) = SSE(\theta) = \sum_{i=1}^{n} (y_i - f(x_i; \theta))^2.$$

2. **Probability of Acceptance**
   At each step, SA randomly perturbs the parameters θ to create a **new candidate solution** θ' and evaluates the change in SSE:

$$\Delta Q = Q(\theta') - Q(\theta).$$

   The **Metropolis Criterion** determines whether to accept the new solution:

$$P(\text{accept}) = \begin{cases} 1, & \text{if } \Delta Q < 0 \quad \text{(new solution improves fit)} \\ \exp\left(-\frac{\Delta Q}{T}\right), & \text{if } \Delta Q \geq 0 \quad \text{(accept with probability).} \end{cases}$$

   Where:

   - T is the **temperature** that **gradually decreases** over iterations.
   - **Worse solutions** are accepted **with small probability** to escape local minima.

3. **Cooling Schedule**
   The **temperature** follows a cooling schedule:

$$T_k = \alpha T_{k-1},$$

   where $\alpha \in (0, 1)$ is a decay factor that controls cooling speed.

Simulated Annealing Algorithm

1. Initialize parameters θ randomly.
2. Set an initial temperature T_0 and cooling rate α.

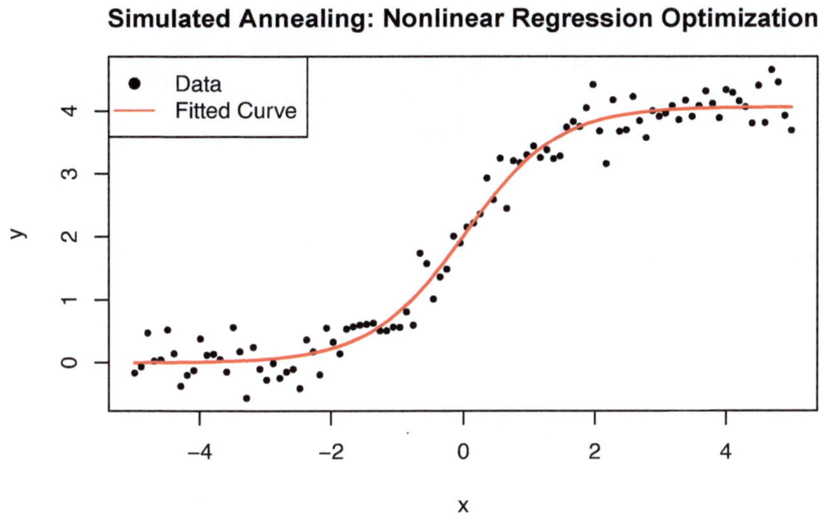

Fig. 3.15 Simulated Annealing: nonlinear regression optimization

3. Repeat for max iterations:
   - Generate perturbed candidate θ'.
   - Compute $\Delta Q = Q(\theta') - Q(\theta)$.
   - Accept if $\Delta Q < 0$ or with probability $\exp(-\Delta Q/T)$.
   - Reduce temperature: $T \leftarrow \alpha T$.
4. Return the best solution found (Fig. 3.15).

```
# Load required library
library(stats)

# Define a numerically stable logistic function
safe_exp <- function(x) {
    return(ifelse(x > 700, Inf, exp(pmin(x, 700)))) # Prevent
    ↪ overflow
}

# Define the logistic growth model
nonlinear_model <- function(theta, x) {
    return(theta[1] / (1 + safe_exp(-theta[2] * (x -
    ↪ theta[3]))))
}
```

```
# Define the Sum of Squared Errors (SSE) function
sse <- function(theta, x, y) {
    predictions <- nonlinear_model(theta, x)
    return(sum((y - predictions) ^ 2))
}

# Simulated Annealing Algorithm
simulated_annealing <-
    function(x,
             y,
             initial_theta,
             T_init = 1.0,
             alpha = 0.99,
             max_iter = 5000) {
        # Initialize parameters
        theta <- initial_theta
        best_theta <- theta
        best_sse <- sse(theta, x, y)
        T <- T_init  # Initial temperature

        for (iter in 1:max_iter) {
            # Generate new candidate solution (small random
            ↪  perturbation)
            theta_new <- theta + rnorm(length(theta), mean =
↪  0, sd = T)

            # Compute new SSE
            sse_new <- sse(theta_new, x, y)

            # Compute change in SSE
            delta_Q <- sse_new - best_sse

            # Acceptance criteria
            if (delta_Q < 0 || runif(1) < exp(-delta_Q / T)) {
                theta <- theta_new
                best_sse <- sse_new
                best_theta <- theta_new
            }

            # Reduce temperature
            T <- alpha * T

            # Stopping condition (very low temperature)
            if (T < 1e-6)
```

```
                break
        }

        return(best_theta)
    }

# Generate synthetic data
set.seed(123)
x <- seq(-5, 5, length.out = 100)
true_theta <- c(4, 1.5, 0)  # True parameters (A, B, C)
y <- nonlinear_model(true_theta, x) + rnorm(length(x), sd =
↪  0.3)

# Initial guess
initial_theta <-
    c(runif(1, 1, 5), runif(1, 0.1, 3), runif(1,-2, 2))

# Run Simulated Annealing
estimated_theta <- simulated_annealing(x, y, initial_theta)

# Display results
cat("Estimated parameters (A, B, C) using Simulated
↪  Annealing:\n")
#> Estimated parameters (A, B, C) using Simulated Annealing:
print(estimated_theta)
#> [1] 4.07180419 1.41457906 0.01422147
```

```
# Plot data and fitted curve
plot(
    x,
    y,
    main = "Simulated Annealing: Nonlinear Regression
    ↪  Optimization",
    pch = 19,
    cex = 0.5,
    xlab = "x",
    ylab = "y"
)
curve(
    nonlinear_model(estimated_theta, x),
    from = min(x),
    to = max(x),
    add = TRUE,
```

```
    col = "red",
    lwd = 2
)
legend(
    "topleft",
    legend = c("Data", "Fitted Curve"),
    pch = c(19, NA),
    lty = c(NA, 1),
    col = c("black", "red")
)
```

3.2.3.3 Genetic Algorithm

Genetic Algorithms (GA) are a class of evolutionary algorithms inspired by the principles of natural selection and genetics. Unlike deterministic optimization techniques, GA evolves a population of candidate solutions over multiple generations, using genetic operators such as selection, crossover, and mutation.

GA is particularly useful when:

- The function is nonlinear, non-differentiable, or highly multimodal.
- Gradient-based methods fail due to rugged function landscapes.
- A global minimum is required, rather than a local one.

The goal of a Genetic Algorithm is to find an optimal solution $\hat{\theta}$ that minimizes an objective function:

$$\hat{\theta} = \arg\min_{\theta} SSE(\theta),$$

where:

$$SSE(\theta) = \sum_{i=1}^{n} (y_i - f(x_i; \theta))^2.$$

1. **Population Representation**
   Each candidate solution (individual) is represented as a **chromosome**, which is simply a vector of parameters:

   $$\theta = (\theta_1, \theta_2, \theta_3)$$

   An entire **population** consists of multiple such solutions.
2. **Selection**
   Each individual's fitness is evaluated using:

   $$\text{Fitness}(\theta) = -SSE(\theta)$$

We use **Tournament Selection** or **Roulette Wheel Selection** to choose parents for reproduction.

3. **Crossover (Recombination)**
   A new solution θ' is generated by **combining** two parents:

$$\theta' = \alpha\theta_{\text{parent1}} + (1-\alpha)\theta_{\text{parent2}}, \quad \alpha \sim U(0, 1).$$

4. **Mutation**
   Random small changes are introduced to **increase diversity**:

$$\theta_i' = \theta_i + \mathcal{N}(0, \sigma),$$

   where $\mathcal{N}(0, \sigma)$ is a **small Gaussian perturbation**.
5. **Evolutionary Cycle**
   The **algorithm iterates** through:

1. Selection
2. Crossover
3. Mutation
4. Survival of the fittest
5. Termination when convergence is reached (Fig. 3.16).

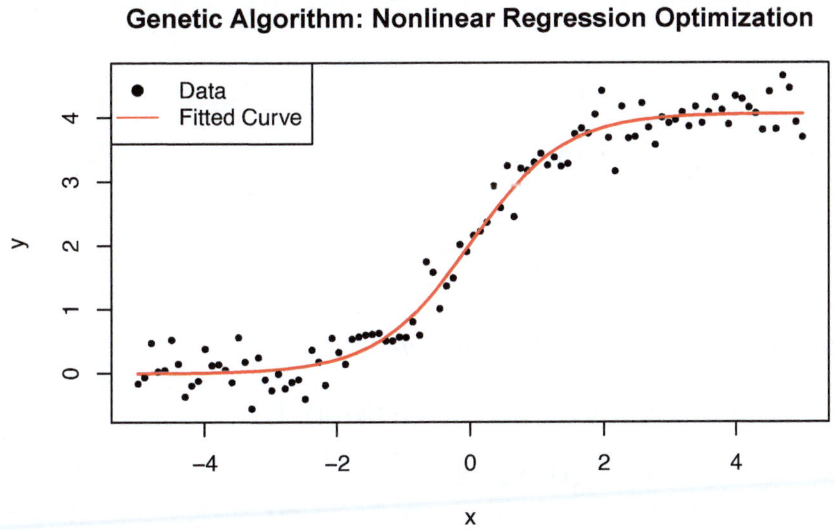

Fig. 3.16 Genetic Algorithm: nonlinear regression optimization

```
# Load required library
library(GA)
#> Loading required package: foreach
#> Loading required package: iterators
#> Package 'GA' version 3.2.4
#> Type 'citation("GA")' for citing this R package in
↪  publications.
#>
#> Attaching package: 'GA'
#> The following object is masked from 'package:utils':
#>
#>     de

# Define a numerically stable logistic function
safe_exp <- function(x) {
    return(ifelse(x > 700, Inf, exp(pmin(x, 700))))  # Prevent
     ↪  overflow
}

# Define the logistic growth model
nonlinear_model <- function(theta, x) {
    return(theta[1] / (1 + safe_exp(-theta[2] * (x -
     ↪  theta[3]))))
}

# Define the Sum of Squared Errors (SSE) function for
↪  optimization
sse <- function(theta, x, y) {
    predictions <- nonlinear_model(theta, x)
    return(sum((y - predictions) ^ 2))
}

# Genetic Algorithm for Optimization
ga_optimization <- function(x, y) {
    # Define fitness function (negative SSE for maximization)
    fitness_function <- function(theta) {
        # GA maximizes fitness, so we use negative SSE
        return(-sse(theta, x, y))
    }

    # Set parameter bounds
    lower_bounds <- c(0.1, 0.01, min(x))  # Ensure positive
↪  scaling
    upper_bounds <- c(max(y) * 2, 10, max(x))
```

```
    # Run GA optimization
    ga_result <- ga(
        type = "real-valued",
        fitness = fitness_function,
        lower = lower_bounds,
        upper = upper_bounds,
        popSize = 50,
        # Population size
        maxiter = 200,
        # Max generations
        pmutation = 0.1,
        # Mutation probability
        monitor = FALSE
    )

    return(ga_result@solution)  # Return optimized parameters
}

# Generate synthetic data
set.seed(123)
x <- seq(-5, 5, length.out = 100)
true_theta <- c(4, 1.5, 0)  # True parameters (A, B, C)
y <- nonlinear_model(true_theta, x) + rnorm(length(x), sd =
↪  0.3)

# Run Genetic Algorithm
estimated_theta <- ga_optimization(x, y)

# Display results
cat("Estimated parameters (A, B, C) using Genetic
↪  Algorithm:\n")
#> Estimated parameters (A, B, C) using Genetic Algorithm:
print(estimated_theta)
#>            x1       x2         x3
#> [1,] 4.066144 1.433886 0.00824126
```

```
# Plot data and fitted curve
plot(
    x,
    y,
    main = "Genetic Algorithm: Nonlinear Regression
     ↪ Optimization",
    pch = 19,
    cex = 0.5,
    xlab = "x",
    ylab = "y"
)
curve(
    nonlinear_model(estimated_theta, x),
    from = min(x),
    to = max(x),
    add = TRUE,
    col = "red",
    lwd = 2
)
legend(
    "topleft",
    legend = c("Data", "Fitted Curve"),
    pch = c(19, NA),
    lty = c(NA, 1),
    col = c("black", "red")
)
```

3.2.3.4 Particle Swarm Optimization

Particle Swarm Optimization (PSO) is a population-based global optimization algorithm inspired by the social behavior of birds and fish schools. Instead of using genetic operators (like in Genetic Algorithms), PSO models particles (solutions) flying through the search space, adjusting their position based on their own experience and the experience of their neighbors.

PSO is particularly useful when:

- The function is nonlinear and noisy or lacks smooth gradients.
- Gradient-based methods struggle due to non-differentiability.
- A global minimum is needed, rather than a local one.

The goal of PSO is to find an optimal solution $\hat{\theta}$ that minimizes an objective function:

$$\hat{\theta} = \arg\min_{\theta} SSE(\theta),$$

where:

$$SSE(\theta) = \sum_{i=1}^{n} (y_i - f(x_i; \theta))^2.$$

1. **Particle Representation**
   Each **particle** represents a candidate solution:

$$\theta_i = (\theta_{i1}, \theta_{i2}, \theta_{i3})$$

   where θ_{ij} is the jth parameter of particle i.
2. **Particle Velocity and Position Updates**
   Each particle moves in the search space with **velocity** v_i, which is updated as:

$$v_i^{(t+1)} = \omega v_i^{(t)} + c_1 r_1 (p_i - \theta_i^{(t)}) + c_2 r_2 (g - \theta_i^{(t)})$$

   Where:

   - ω is the **inertia weight** (controls exploration vs. exploitation).
   - c_1, c_2 are **acceleration coefficients.**
   - $r_1, r_2 \sim U(0, 1)$ are random numbers.
   - p_i is the **particle's personal best position.**
   - g is the **global best position.**

   Then, the **position update** is:

$$\theta_i^{(t+1)} = \theta_i^{(t)} + v_i^{(t+1)}$$

   This process continues until **convergence criteria** (like a max number of iterations or minimum error) is met.

Particle Swarm Optimization Algorithm

1. Initialize particles randomly within search bounds.
2. Set random initial velocities.
3. Evaluate SSE for each particle.
4. Update the personal and global best solutions.
5. Update velocities and positions using the update equations.
6. Repeat until convergence (Fig. 3.17).

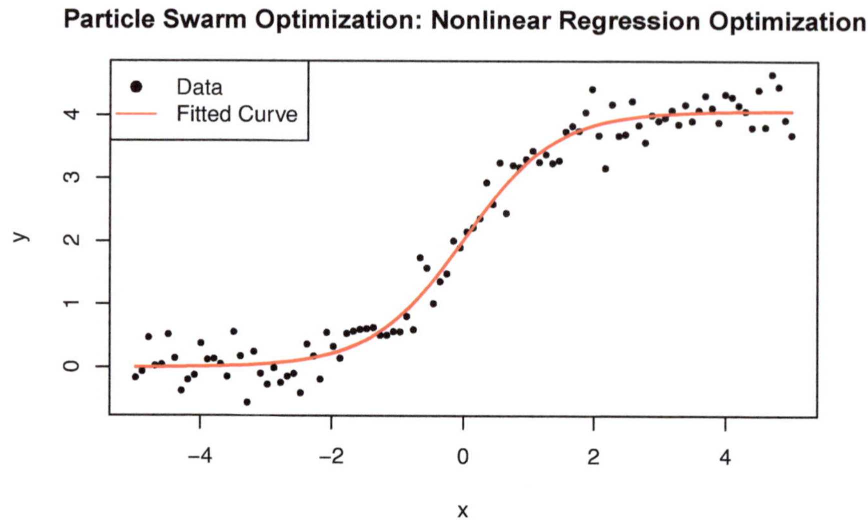

Fig. 3.17 Particle Swarm Optimization: nonlinear regression optimization

```
# Load required library
library(pso)

# Define a numerically stable logistic function
safe_exp <- function(x) {
    return(ifelse(x > 700, Inf, exp(pmin(x, 700))))  # Prevent
    ↪  overflow
}

# Define the logistic growth model
nonlinear_model <- function(theta, x) {
    return(theta[1] / (1 + safe_exp(-theta[2] * (x -
    ↪  theta[3]))))
}

# Define the Sum of Squared Errors (SSE) function for
↪  optimization
sse <- function(theta, x, y) {
    predictions <- nonlinear_model(theta, x)
    return(sum((y - predictions) ^ 2))
}

# Particle Swarm Optimization (PSO) for Nonlinear Regression
pso_optimization <- function(x, y) {
```

```
  # Define fitness function (minimize SSE)
  fitness_function <- function(theta) {
    return(sse(theta, x, y))
  }

  # Set parameter bounds
  lower_bounds <- c(0.1, 0.01, min(x))  # Ensure positive
↪ scaling
  upper_bounds <- c(max(y) * 2, 10, max(x))

  # Run PSO optimization
  pso_result <- psoptim(
    par = c(1, 1, 0),
    # Initial guess
    fn = fitness_function,
    lower = lower_bounds,
    upper = upper_bounds,
    control = list(maxit = 200, s = 50)  # 200 iterations,
     ↪ 50 particles
  )

  return(pso_result$par)  # Return optimized parameters
}

# Generate synthetic data
set.seed(123)
x <- seq(-5, 5, length.out = 100)
true_theta <- c(4, 1.5, 0)  # True parameters (A, B, C)
y <- nonlinear_model(true_theta, x) + rnorm(length(x), sd =
↪ 0.3)

# Run Particle Swarm Optimization
estimated_theta <- pso_optimization(x, y)

# Display results
cat("Estimated parameters (A, B, C) using Particle Swarm
↪ Optimization:\n")
#> Estimated parameters (A, B, C) using Particle Swarm
↪ Optimization:
print(estimated_theta)
#> [1] 4.06876562 1.42765613 0.01128767
```

```
# Plot data and fitted curve
plot(
    x,
    y,
    main = "Particle Swarm Optimization: Nonlinear Regression
    ↪ Optimization",
    pch = 19,
    cex = 0.5,
    xlab = "x",
    ylab = "y"
)
curve(
    nonlinear_model(estimated_theta, x),
    from = min(x),
    to = max(x),
    add = TRUE,
    col = "red",
    lwd = 2
)
legend(
    "topleft",
    legend = c("Data", "Fitted Curve"),
    pch = c(19, NA),
    lty = c(NA, 1),
    col = c("black", "red")
)
```

3.2.3.5 Evolutionary Strategies

Evolutionary Strategies (ES) are a class of evolutionary optimization algorithms that improve solutions by mutating and selecting individuals based on fitness. Unlike Genetic Algorithm, ES focuses on self-adaptive mutation rates and selection pressure rather than crossover. This makes ES particularly robust for continuous optimization problems like nonlinear regression.

ES is particularly useful when:

- The function is complex and noisy or lacks smooth gradients.
- Gradient-based methods fail due to non-differentiability.
- An adaptive approach to exploration and exploitation is needed.

The goal of ES is to find an optimal solution $\hat{\theta}$ that minimizes an objective function:

$$\hat{\theta} = \arg\min_{\theta} SSE(\theta),$$

where:

$$SSE(\theta) = \sum_{i=1}^{n} (y_i - f(x_i; \theta))^2.$$

1. **Population Representation**
   Each **individual** is a solution θ_i in the parameter space:

   $$\theta_i = (\theta_{i1}, \theta_{i2}, \theta_{i3}).$$

   The **population** consists of multiple individuals, each representing different candidate parameters.
2. **Mutation**
   New candidate solutions are generated by **adding random noise**:

   $$\theta_i' = \theta_i + \sigma \mathcal{N}(0, I),$$

   Where:

   - σ is the **mutation step size**, which adapts over time.
   - $\mathcal{N}(0, I)$ is a standard normal distribution.

3. **Selection**
   ES employs (μ, λ)-selection:

   - (μ, λ)-ES: Select the best μ solutions from λ offspring.
   - $(\mu + \lambda)$-ES: Combine parents and offspring, selecting the top μ.

4. **Step Size Adaptation**
   Mutation strength σ **self-adapts** using the **1/5 success rule**:

   $$\sigma_{t+1} = \begin{cases} \sigma_t / c, & \text{if success rate } > 1/5 \\ \sigma_t \cdot c, & \text{if success rate } < 1/5 \end{cases}$$

   where $c > 1$ is a scaling factor.

Evolutionary Strategies Algorithm
1. Initialize a population of λ solutions with random parameters.
2. Set mutation step size σ.
3. Repeat for max iterations:
   - Generate λ offspring by mutating parent solutions.
   - Evaluate fitness (SSE) of each offspring.
   - Select the best μ solutions for the next generation.
   - Adapt mutation step size based on success rate.
4. Return the best solution found (Fig. 3.18).

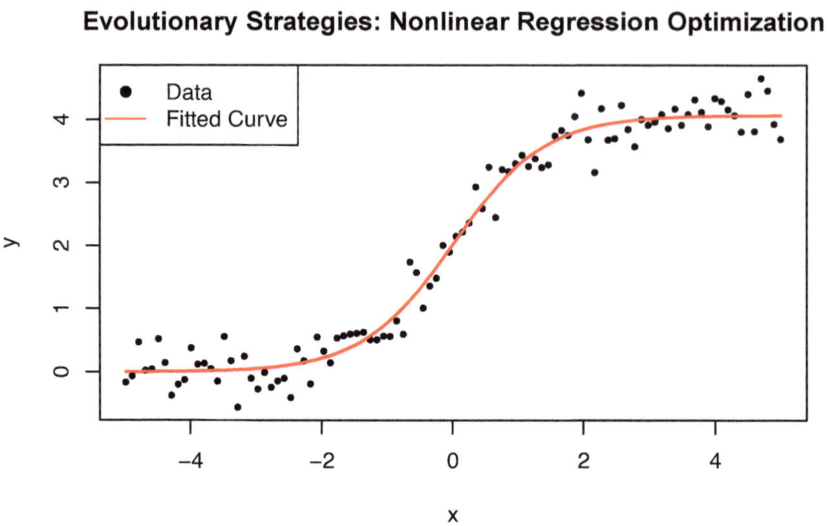

Fig. 3.18 Evolutionary Strategies: nonlinear regression optimization

```
# Load required library
library(DEoptim)

# Define a numerically stable logistic function
safe_exp <- function(x) {
    return(ifelse(x > 700, Inf, exp(pmin(x, 700))))  # Prevent
     ↪  overflow
}

# Define the logistic growth model
nonlinear_model <- function(theta, x) {
    return(theta[1] / (1 + safe_exp(-theta[2] * (x -
     ↪  theta[3]))))
}

# Define the Sum of Squared Errors (SSE) function for
 ↪  optimization
sse <- function(theta, x, y) {
    predictions <- nonlinear_model(theta, x)
    return(sum((y - predictions)^2))
}

# Evolutionary Strategies Optimization (Using Differential
 ↪  Evolution)
```

```
es_optimization <- function(x, y) {
    # Define fitness function (minimize SSE)
    fitness_function <- function(theta) {
        return(sse(theta, x, y))
    }

    # Set parameter bounds
    lower_bounds <- c(0.1, 0.01, min(x))  # Ensure positive
↪   scaling
    upper_bounds <- c(max(y) * 2, 10, max(x))

    # Run Differential Evolution (mimicking ES)
    es_result <- DEoptim(
        fn = fitness_function,
        lower = lower_bounds,
        upper = upper_bounds,
        # 50 individuals, 200 generations, suppress iteration
         ↪  output
        DEoptim.control(NP = 50, itermax = 200, trace = F)
    )

    return(es_result$optim$bestmem)  # Return optimized
     ↪  parameters
}

# Generate synthetic data
set.seed(123)
x <- seq(-5, 5, length.out = 100)
true_theta <- c(4, 1.5, 0)  # True parameters (A, B, C)
y <- nonlinear_model(true_theta, x) + rnorm(length(x), sd =
↪   0.3)

# Run Evolutionary Strategies Optimization
estimated_theta <- es_optimization(x, y)

# Display results
cat("Estimated parameters (A, B, C) using Evolutionary
↪   Strategies:\n")
#> Estimated parameters (A, B, C) using Evolutionary
↪   Strategies:
print(estimated_theta)
#>       par1       par2       par3
#> 4.06876561 1.42765613 0.01128767
```

```
# Plot data and fitted curve
plot(
    x, y,
    main = "Evolutionary Strategies: Nonlinear Regression
    ↪ Optimization",
    pch = 19,
    cex = 0.5,
    xlab = "x",
    ylab = "y"
)
curve(
    nonlinear_model(estimated_theta, x),
    from = min(x),
    to = max(x),
    add = TRUE,
    col = "red",
    lwd = 2
)
legend(
    "topleft",
    legend = c("Data", "Fitted Curve"),
    pch = c(19, NA),
    lty = c(NA, 1),
    col = c("black", "red")
)
```

3.2.4 Linearization

3.2.4.1 Taylor Series Approximation

Taylor Series Approximation is a fundamental tool in nonlinear optimization, enabling local approximation of complex functions using polynomial expansions. It is widely used to linearize nonlinear models, facilitate derivative-based optimization, and derive Newton-type methods.

Taylor Series Approximation is particularly useful when:

- A nonlinear function is difficult to compute directly.
- Optimization requires local gradient and curvature information.
- A simpler, polynomial-based approximation improves computational efficiency.

Given a differentiable function $f(\theta)$, its **Taylor series expansion** around a point θ_0 is:

$$f(\theta) = f(\theta_0) + f'(\theta_0)(\theta - \theta_0) + \frac{1}{2} f''(\theta_0)(\theta - \theta_0)^2 + \mathcal{O}((\theta - \theta_0)^3).$$

For optimization, we often use:

1. **First-order approximation (linear approximation):**

$$f(\theta) \approx f(\theta_0) + f'(\theta_0)(\theta - \theta_0).$$

2. **Second-order approximation (quadratic approximation):**

$$f(\theta) \approx f(\theta_0) + f'(\theta_0)(\theta - \theta_0) + \frac{1}{2} f''(\theta_0)(\theta - \theta_0)^2.$$

For **gradient-based optimization**, we use the **Newton-Raphson update**:

$$\theta^{(k+1)} = \theta^{(k)} - [H_f(\theta^{(k)})]^{-1} \nabla f(\theta^{(k)}),$$

Where:

- $\nabla f(\theta)$ is the **gradient** (first derivative).
- $H_f(\theta)$ is the **Hessian matrix** (second derivative).

For nonlinear regression, we approximate the **SSE**:

$$SSE(\theta) = \sum_{i=1}^{n} (y_i - f(x_i; \theta))^2.$$

1. **First-Order Approximation (Gradient Descent)**
   The **gradient** of SSE w.r.t. parameters θ is:

$$\nabla SSE(\theta) = -2 \sum_{i=1}^{n} (y_i - f(x_i; \theta)) \nabla f(x_i; \theta).$$

   Using **first-order Taylor approximation**, we update parameters via **Gradient Descent**:

$$\theta^{(k+1)} = \theta^{(k)} - \alpha \nabla SSE(\theta^{(k)}),$$

   where α is the **learning rate**.
2. **Second-Order Approximation (Newton's Method)**
   The **Hessian matrix** of SSE is:

$$H_{SSE}(\theta) = 2\sum_{i=1}^{n} \nabla f(x_i;\theta)\nabla f(x_i;\theta)^T - 2\sum_{i=1}^{n}(y_i - f(x_i;\theta))H_f(x_i;\theta).$$

The **Newton-Raphson update** becomes:

$$\theta^{(k+1)} = \theta^{(k)} - H_{SSE}(\theta)^{-1}\nabla SSE(\theta).$$

```
# Load required libraries
library(numDeriv)

# Define a numerically stable logistic function
safe_exp <- function(x) {
   return(ifelse(is.na(x) |
                       x > 700, Inf, exp(pmin(x, 700))))  #
↪  Prevent overflow
}

# Define the logistic growth model
nonlinear_model <- function(theta, x) {
   return(theta[1] / (1 + safe_exp(-theta[2] * (x -
    ↪  theta[3]))))
}

# Define the Sum of Squared Errors (SSE) function
sse <- function(theta, x, y) {
   predictions <- nonlinear_model(theta, x)
   return(sum((y - predictions) ^ 2, na.rm = TRUE))  # Avoid
    ↪  NA errors
}

# First-Order Approximation: Gradient Descent Optimization
gradient_descent <-
   function(x,
            y,
            alpha = 0.005,
            tol = 1e-6,
            max_iter = 5000) {
      theta <- c(2, 1, 0)  # Initial guess
      for (i in 1:max_iter) {
         grad_sse <-
            grad(function(t)
               sse(t, x, y), theta)  # Compute gradient
         theta_new <-
```

```
            theta - alpha * grad_sse  # Update parameters

        if (sum(abs(theta_new - theta)) < tol)
            break  # Check convergence
        theta <- theta_new
    }
    return(theta)
}

# Second-Order Approximation: Newton's Method with
↪  Regularization
newton_method <-
    function(x,
             y,
             tol = 1e-6,
             max_iter = 100,
             lambda = 1e-4) {
    theta <- c(2, 1, 0)  # Initial guess
    for (i in 1:max_iter) {
        grad_sse <-
            grad(function(t)
                sse(t, x, y), theta)  # Compute gradient
        hessian_sse <-
            hessian(function(t)
                sse(t, x, y), theta)  # Compute Hessian

        # Regularize Hessian to avoid singularity
        hessian_reg <-
            hessian_sse + lambda * diag(length(theta))

        # Ensure Hessian is invertible
        if (is.na(det(hessian_reg)) ||
            det(hessian_reg) < 1e-10) {
            message("Singular Hessian found; increasing
             ↪  regularization.")
            lambda <- lambda * 10  # Increase
↪  regularization
            next
        }

        # Newton update
        theta_new <- theta - solve(hessian_reg) %*%
↪  grad_sse
```

```
            if (sum(abs(theta_new - theta)) < tol)
                break  # Check convergence
            theta <- theta_new
        }
        return(theta)
    }

# Generate synthetic data
set.seed(123)
x <- seq(-5, 5, length.out = 100)
true_theta <- c(4, 1.5, 0)  # True parameters (A, B, C)
y <- nonlinear_model(true_theta, x) + rnorm(length(x), sd =
↪  0.3)

# Run Gradient Descent
estimated_theta_gd <- gradient_descent(x, y)

# Run Newton's Method with Regularization
estimated_theta_newton <- newton_method(x, y)
#> Singular Hessian found; increasing regularization.
#> Singular Hessian found; increasing regularization.
#> Singular Hessian found; increasing regularization.
#> Singular Hessian found; increasing regularization.
#> Singular Hessian found; increasing regularization.
#> Singular Hessian found; increasing regularization.

# Display results
cat("Estimated parameters (A, B, C) using Gradient
↪  Descent:\n")
#> Estimated parameters (A, B, C) using Gradient Descent:
print(estimated_theta_gd)
#> [1] 4.06876224 1.42766371 0.01128539

cat("Estimated parameters (A, B, C) using Newton's Method:\n")
#> Estimated parameters (A, B, C) using Newton's Method:
print(estimated_theta_newton)
#>            [,1]
#> [1,] 4.06876368
#> [2,] 1.42766047
#> [3,] 0.01128636
```

Figure 3.19 shows the fitted curve.

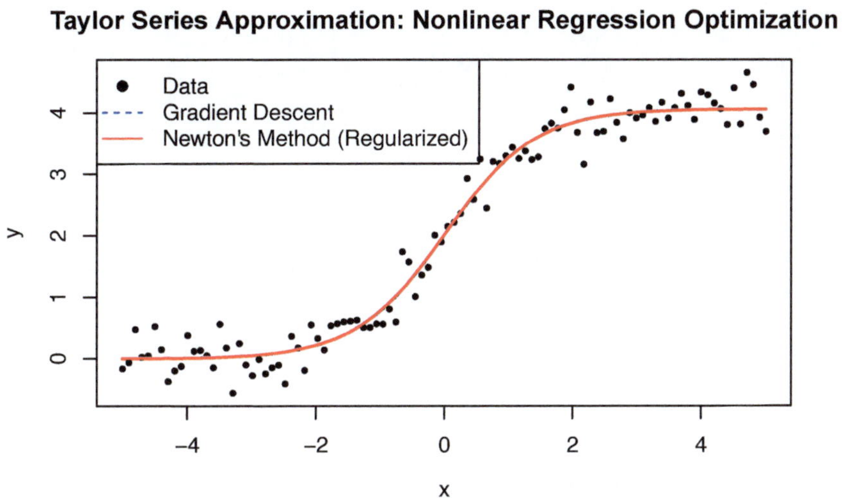

Fig. 3.19 Taylor Series Approximation: nonlinear regression optimization

```
# Plot data and fitted curve
plot(
    x,
    y,
    main = "Taylor Series Approximation: Nonlinear Regression
    ↪ Optimization",
    pch = 19,
    cex = 0.5,
    xlab = "x",
    ylab = "y"
)
curve(
    nonlinear_model(estimated_theta_gd, x),
    from = min(x),
    to = max(x),
    add = TRUE,
    col = "blue",
    lwd = 2,
    lty = 2  # Dashed line to differentiate Gradient Descent
)
curve(
    nonlinear_model(estimated_theta_newton, x),
    from = min(x),
    to = max(x),
    add = TRUE,
```

```
    col = "red",
    lwd = 2
)
legend(
    "topleft",
    legend = c("Data", "Gradient Descent", "Newton's Method
    ↪ (Regularized)"),
    pch = c(19, NA, NA),
    lty = c(NA, 2, 1),
    col = c("black", "blue", "red")
)
```

3.2.4.2 Log-Linearization

Log-Linearization is a mathematical technique used to transform nonlinear models into linear models by taking the logarithm of both sides. This transformation simplifies parameter estimation and enables the use of linear regression techniques on originally nonlinear functions.
Log-Linearization is particularly useful when:

- The model exhibits exponential, power-law, or logistic growth behavior.
- Linear regression methods are preferred over nonlinear optimization.
- A linearized version provides better interpretability and computational efficiency.

A **nonlinear model** can often be expressed in the form:

$$y = f(x; \theta).$$

Applying a **log transformation**, we obtain:

$$\log y = g(x; \theta),$$

where $g(x; \theta)$ is now **linear in parameters**. We then estimate θ using Ordinary Least Squares.

Example 1 Exponential Model
Consider an **exponential growth model**:

$$y = Ae^{Bx}.$$

Taking the natural logarithm:

$$\log y = \log A + Bx.$$

This is now **linear in** $\log y$, allowing estimation via **linear regression**.

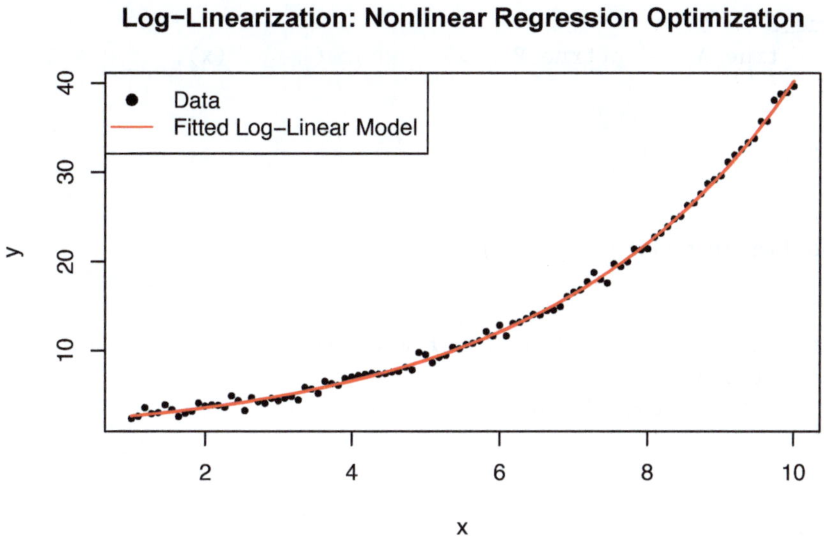

Fig. 3.20 Log-Linearization: nonlinear regression optimization

Example 2 Power Law Model
For a **power law function**:

$$y = Ax^B.$$

Taking logs:

$$\log y = \log A + B \log x.$$

Again, this is **linearized**, making it solvable via **OLS regression**.

Log-Linearization Algorithm
1. Apply the logarithm transformation to the dependent variable.
2. Transform the equation into a linear form.
3. Use linear regression (OLS) to estimate parameters.
4. Convert parameters back to original scale if necessary (Fig. 3.20).

```
# Load required library
library(stats)

# Generate synthetic data for an exponential model
set.seed(123)
x <- seq(1, 10, length.out = 100)
true_A <- 2
```

```r
true_B <- 0.3
y <- true_A * exp(true_B * x) + rnorm(length(x), sd = 0.5)

# Apply logarithmic transformation
log_y <- log(y)

# Fit linear regression model
log_linear_model <- lm(log_y ~ x)

# Extract estimated parameters
estimated_B <- coef(log_linear_model)[2]  # Slope in log-space
estimated_A <-
    exp(coef(log_linear_model)[1])  # Intercept
    ↪  (back-transformed)

# Display results
cat("Estimated parameters (A, B) using Log-Linearization:\n")
#> Estimated parameters (A, B) using Log-Linearization:
print(c(estimated_A, estimated_B))
#> (Intercept)           x
#>   2.0012577   0.3001223
```

```r
# Plot data and fitted curve
plot(
    x,
    y,
    main = "Log-Linearization: Nonlinear Regression
    ↪  Optimization",
    pch = 19,
    cex = 0.5,
    xlab = "x",
    ylab = "y"
)
curve(
    estimated_A * exp(estimated_B * x),
    from = min(x),
    to = max(x),
    add = TRUE,
    col = "red",
    lwd = 2
)
legend(
    "topleft",
```

```
    legend = c("Data", "Fitted Log-Linear Model"),
    pch = c(19, NA),
    lty = c(NA, 1),
    col = c("black", "red")
)
```

3.2.5 Hybrid

3.2.5.1 Adaptive Levenberg-Marquardt

The **Levenberg-Marquardt Algorithm (LMA)** is a powerful **nonlinear least squares optimization method** that adaptively combines:

- Gauss-Newton Algorithm for fast convergence near the solution.
- Steepest Descent (Gradient Descent) for stability when far from the solution.

The Adaptive Levenberg-Marquardt Algorithm further adjusts the damping parameter τ dynamically, making it more efficient in practice.
Given an objective function **SSE**:

$$SSE(\theta) = \sum_{i=1}^{n} (y_i - f(x_i; \theta))^2.$$

The **update rule for LMA** is:

$$\hat{\theta}^{(j+1)} = \hat{\theta}^{(j)} - \alpha_j [\mathbf{F}(\hat{\theta}^{(j)})' \mathbf{F}(\hat{\theta}^{(j)}) + \tau \mathbf{I}_{p \times p}] \frac{\partial \mathbf{Q}(\hat{\theta}^{(j)})}{\partial \theta}.$$

Where:

- τ is the **adaptive damping parameter**.
- $\mathbf{I}_{p \times p}$ is the **identity matrix**.
- $\mathbf{F}(\hat{\theta}^{(j)})$ is the **Jacobian matrix** of partial derivatives.
- $\frac{\partial \mathbf{Q}(\hat{\theta}^{(j)})}{\partial \theta}$ is the **gradient vector**.
- α_j is the **learning rate**.

The key **adaptation rule for** τ:

- **If the new step decreases SSE, reduce** τ:

$$\tau \leftarrow \tau/10.$$

- **Otherwise, increase** τ to ensure stability:

$$\tau \leftarrow 10\tau.$$

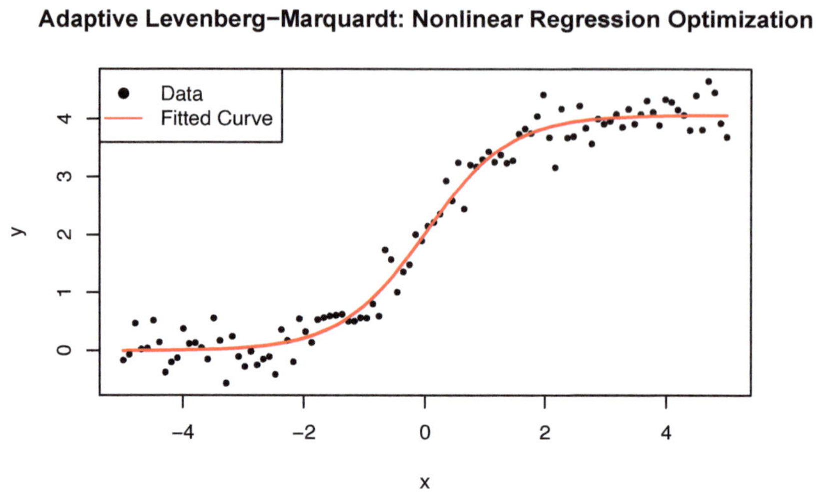

Fig. 3.21 Adaptive Levenberg-Marquardt: nonlinear regression optimization

This adjustment **ensures a balance between stability and efficiency**.

Adaptive Levenberg-Marquardt Algorithm

1. Initialize parameters θ_0, damping factor τ.
2. Compute Jacobian $\mathbf{F}(\hat{\theta}^{(j)})$.
3. Compute step direction using modified Gauss-Newton update.
4. Adjust τ dynamically:
   - Decrease τ if SSE improves.
   - Increase τ if SSE worsens.
5. Repeat until convergence (Fig. 3.21).

```
# Load required libraries
library(numDeriv)

# Define a numerically stable logistic function
safe_exp <- function(x) {
  return(ifelse(x > 700, Inf, exp(pmin(x, 700))))  # Prevent
  ↪ overflow
}

# Define the logistic growth model
nonlinear_model <- function(theta, x) {
  return(theta[1] / (1 + safe_exp(-theta[2] * (x -
  ↪ theta[3]))))
}
```

```
# Define the Sum of Squared Errors (SSE) function
sse <- function(theta, x, y) {
    predictions <- nonlinear_model(theta, x)
    return(sum((y - predictions) ^ 2))
}

# Adaptive Levenberg-Marquardt Optimization
adaptive_lm_optimization <-
    function(x, y, tol = 1e-6, max_iter = 100) {
        theta <- c(2, 1, 0)  # Initial parameter guess
        tau <- 1e-3  # Initial damping parameter
        alpha <- 1  # Step size scaling
        iter <- 0

        while (iter < max_iter) {
            iter <- iter + 1

            # Compute Jacobian numerically
            J <- jacobian(function(t)
                nonlinear_model(t, x), theta)

            # Compute gradient of SSE
            residuals <- y - nonlinear_model(theta, x)
            grad_sse <- -2 * t(J) %*% residuals

            # Compute Hessian approximation
            H <- 2 * t(J) %*% J + tau * diag(length(theta))

            # Compute parameter update step
            delta_theta <- solve(H, grad_sse)

            # Trial step
            theta_new <- theta - alpha * delta_theta

            # Compute SSE for new parameters
            if (sse(theta_new, x, y) < sse(theta, x, y)) {
                # Accept step, decrease tau
                theta <- theta_new
                tau <- tau / 10
            } else {
                # Reject step, increase tau
                tau <- tau * 10
            }
```

```
        # Check convergence
        if (sum(abs(delta_theta)) < tol)
            break
    }

    return(theta)
}

# Generate synthetic data
set.seed(123)
x <- seq(-5, 5, length.out = 100)
true_theta <- c(4, 1.5, 0)  # True parameters (A, B, C)
y <- nonlinear_model(true_theta, x) + rnorm(length(x), sd =
↪  0.3)

# Run Adaptive Levenberg-Marquardt Optimization
estimated_theta <- adaptive_lm_optimization(x, y)

# Display results
cat("Estimated parameters (A, B, C) using Adaptive
↪  Levenberg-Marquardt:\n")
#> Estimated parameters (A, B, C) using Adaptive
↪  Levenberg-Marquardt:
print(estimated_theta)
#>            [,1]
#> [1,] 4.06876562
#> [2,] 1.42765612
#> [3,] 0.01128767
```

```
# Plot data and fitted curve
plot(
    x,
    y,
    main = "Adaptive Levenberg-Marquardt: Nonlinear Regression
    ↪  Optimization",
    pch = 19,
    cex = 0.5,
    xlab = "x",
    ylab = "y"
)
curve(
    nonlinear_model(estimated_theta, x),
    from = min(x),
```

```
    to = max(x),
    add = TRUE,
    col = "red",
    lwd = 2
)
legend(
    "topleft",
    legend = c("Data", "Fitted Curve"),
    pch = c(19, NA),
    lty = c(NA, 1),
    col = c("black", "red")
)
```

3.2.6 Comparison of Nonlinear Optimizers

See Fig. 3.22.

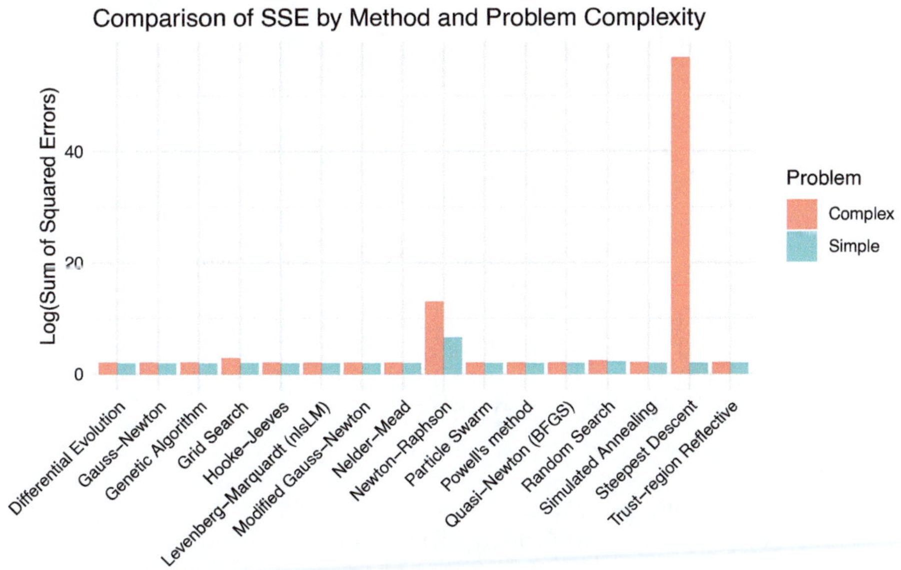

Fig. 3.22 Comparison of SSE by Method and Problem Complexity

```
# ALL-IN-ONE R SCRIPT COMPARING MULTIPLE NONLINEAR-REGRESSION
↪  OPTIMIZERS

library(minpack.lm) # nlsLM (Levenberg-Marquardt)
library(dfoptim)    # Powell (nmk), Hooke-Jeeves
library(nloptr)     # trust-region reflective
library(GA)         # genetic algorithm
library(DEoptim)    # differential evolution
library(GenSA)      # simulated annealing
library(pso)        # particle swarm
library(MASS)       # for ginv fallback
library(microbenchmark)
library(ggplot2)
library(dplyr)
#>
#> Attaching package: 'dplyr'
#> The following object is masked from 'package:MASS':
#>
#>     select
#> The following objects are masked from 'package:stats':
#>
#>     filter, lag
#> The following objects are masked from 'package:base':
#>
#>     intersect, setdiff, setequal, union

# -- 1) DEFINE MODELS (SIMPLE VS COMPLEX) ---

# 3-parameter logistic
f_logistic <- function(theta, x) {
  A <- theta[1]
  B <- theta[2]
  C <- theta[3]
  A / (1 + exp(-B * (x - C)))
}
sse_logistic <-
  function(theta, x, y)
    sum((y - f_logistic(theta, x)) ^ 2)

# 4-parameter "extended" model
f_complex <- function(theta, x) {
  A <- theta[1]
  B <- theta[2]
  C <- theta[3]
```

```
  D <- theta[4]
  A / (1 + exp(-B * (x - C))) + D * exp(-0.5 * x)
}
sse_complex <-
  function(theta, x, y)
    sum((y - f_complex(theta, x)) ^ 2)

# Generate synthetic data
set.seed(123)
n <- 100
x_data <- seq(-5, 5, length.out = n)
# "simple" scenario
true_theta_simple <- c(4, 1.5, 0)
y_data_simple <-
  f_logistic(true_theta_simple, x_data) + rnorm(n, sd = 0.3)
# "complex" scenario
true_theta_complex <- c(4, 1.2, -1, 0.5)
y_data_complex <-
  f_complex(true_theta_complex, x_data) + rnorm(n, sd = 0.3)

# -- 2) OPTIMIZERS (EXCEPT BISECTION) ----
#
# All methods share signature:
#   FUN(par, x, y, sse_fn, model_fn, lower=NULL, upper=NULL,
↪   ...)
# Some do not strictly use lower/upper if unconstrained.

# 2.1 Gauss-Newton
gauss_newton_fit <- function(par,
                             x,
                             y,
                             sse_fn,
                             model_fn,
                             lower = NULL,
                             upper = NULL,
                             max_iter = 100,
                             tol = 1e-6)
{
  theta <- par
  for (iter in seq_len(max_iter)) {
    eps <- 1e-6
    nP  <- length(theta)
    Fmat <- matrix(0, nrow = length(x), ncol = nP)
    for (p in seq_len(nP)) {
```

```
      pert <- theta
      pert[p] <- pert[p] + eps
      Fmat[, p] <-
        (model_fn(pert, x) - model_fn(theta, x)) / eps
    }
    r <- y - model_fn(theta, x)
    delta <- tryCatch(
      solve(t(Fmat) %*% Fmat, t(Fmat) %*% r),
      error = function(e) {
        # fallback to pseudoinverse
        MASS::ginv(t(Fmat) %*% Fmat) %*% (t(Fmat) %*% r)
      }
    )
    theta_new <- theta + delta
    if (sum(abs(theta_new - theta)) < tol)
      break
    theta <- theta_new
  }
  theta
}

# 2.2 Modified Gauss-Newton (with step halving)
modified_gauss_newton_fit <- function(par,
                                      x,
                                      y,
                                      sse_fn,
                                      model_fn,
                                      lower = NULL,
                                      upper = NULL,
                                      max_iter = 100,
                                      tol = 1e-6)
{
  theta <- par
  for (iter in seq_len(max_iter)) {
    eps <- 1e-6
    nP  <- length(theta)
    Fmat <- matrix(0, nrow = length(x), ncol = nP)
    for (p in seq_len(nP)) {
      pert <- theta
      pert[p] <- pert[p] + eps
      Fmat[, p] <-
        (model_fn(pert, x) - model_fn(theta, x)) / eps
    }
    r <- y - model_fn(theta, x)
```

```
    lhs <- t(Fmat) %*% Fmat
    rhs <- t(Fmat) %*% r
    delta <- tryCatch(
      solve(lhs, rhs),
      error = function(e)
        MASS::ginv(lhs) %*% rhs
    )
    sse_old <- sse_fn(theta, x, y)
    alpha <- 1
    for (k in 1:10) {
      new_sse <- sse_fn(theta + alpha * delta, x, y)
      if (new_sse < sse_old)
        break
      alpha <- alpha / 2
    }
    theta_new <- theta + alpha * delta
    if (sum(abs(theta_new - theta)) < tol)
      break
    theta <- theta_new
  }
  theta
}

# 2.3 Steepest Descent (Gradient Descent)
steepest_descent_fit <- function(par,
                                 x,
                                 y,
                                 sse_fn,
                                 model_fn,
                                 lower = NULL,
                                 upper = NULL,
                                 lr = 0.001,
                                 max_iter = 5000,
                                 tol = 1e-6)
{
  theta <- par
  for (iter in seq_len(max_iter)) {
    eps <- 1e-6
    f0  <- sse_fn(theta, x, y)
    grad <- numeric(length(theta))
    for (p in seq_along(theta)) {
      pert <- theta
      pert[p] <- pert[p] + eps
      grad[p] <- (sse_fn(pert, x, y) - f0) / eps
```

```
    }
    theta_new <- theta - lr * grad
    if (sum(abs(theta_new - theta)) < tol)
      break
    theta <- theta_new
  }
  theta
}

# 2.4 Levenberg-Marquardt (nlsLM)
lm_fit <- function(par,
                   x,
                   y,
                   sse_fn,
                   model_fn,
                   lower = NULL,
                   upper = NULL,
                   form = c("simple", "complex"))
{
  form <- match.arg(form)
  if (form == "simple") {
    fit <- nlsLM(y ~ A / (1 + exp(-B * (x - C))),
                 start = list(A = par[1],
                              B = par[2],
                              C = par[3]))
  } else {
    fit <- nlsLM(y ~ A / (1 + exp(-B * (x - C))) + D *
↪   exp(-0.5 * x),
                 start = list(
                   A = par[1],
                   B = par[2],
                   C = par[3],
                   D = par[4]
                 ))
  }
  coef(fit)
}

# 2.5 Newton-Raphson (with numeric Hessian, fallback if
↪  singular)
newton_raphson_fit <- function(par,
                               x,
                               y,
                               sse_fn,
```

```
                                            model_fn,
                                            lower = NULL,
                                            upper = NULL,
                                            max_iter = 50,
                                            tol = 1e-6)
{
  theta <- par
  for (i in seq_len(max_iter)) {
    eps <- 1e-6
    f0  <- sse_fn(theta, x, y)
    grad <- numeric(length(theta))
    for (p in seq_along(theta)) {
      pert <- theta
      pert[p] <- pert[p] + eps
      grad[p] <- (sse_fn(pert, x, y) - f0) / eps
    }
    Hess <- matrix(0, length(theta), length(theta))
    for (p in seq_along(theta)) {
      pert_p <- theta
      pert_p[p] <- pert_p[p] + eps
      f_p <- sse_fn(pert_p, x, y)
      for (q in seq_along(theta)) {
        pert_q <- pert_p
        pert_q[q] <- pert_q[q] + eps
        Hess[p, q] <- (sse_fn(pert_q, x, y) -
                                f_p - (f0 - sse_fn(theta, x, y))) /
↪   (eps ^
                                        2)
  }
  }
  delta <- tryCatch(
      solve(Hess, grad),
      error = function(e)
        MASS::ginv(Hess) %*% grad
    )
    theta_new <- theta - delta
    if (sum(abs(theta_new - theta)) < tol)
      break
    theta <- theta_new
  }
  theta
}

# 2.6 Quasi-Newton (BFGS via optim)
```

```
quasi_newton_fit <- function(par,
                             x,
                             y,
                             sse_fn,
                             model_fn,
                             lower = NULL,
                             upper = NULL)
{
  fn <- function(pp)
    sse_fn(pp, x, y)
  res <- optim(par, fn, method = "BFGS")
  res$par
}

# 2.7 Trust-region reflective (nloptr)
trust_region_fit <- function(par,
                             x,
                             y,
                             sse_fn,
                             model_fn,
                             lower = NULL,
                             upper = NULL)
{
  # numeric gradient
  grad_numeric <- function(pp, eps = 1e-6) {
    g  <- numeric(length(pp))
    f0 <- sse_fn(pp, x, y)
    for (i in seq_along(pp)) {
      p2 <- pp
      p2[i] <- p2[i] + eps
      g[i] <- (sse_fn(p2, x, y) - f0) / eps
    }
    g
  }
  eval_f <- function(pp) {
    val <- sse_fn(pp, x, y)
    gr  <- grad_numeric(pp)
    list(objective = val, gradient = gr)
  }
  lb <- if (is.null(lower))
    rep(-Inf, length(par))
  else
    lower
  ub <- if (is.null(upper))
```

```
    rep(Inf, length(par))
  else
    upper
  res <- nloptr(
    x0 = par,
    eval_f = eval_f,
    lb = lb,
    ub = ub,
    opts = list(
      algorithm = "NLOPT_LD_TNEWTON",
      maxeval = 500,
      xtol_rel = 1e-6
    )
  )
  res$solution
}

# 2.8 Grid search
grid_search_fit <- function(par,
                            x,
                            y,
                            sse_fn,
                            model_fn,
                            lower = NULL,
                            upper = NULL,
                            grid_defs = NULL)
{
  if (is.null(grid_defs))
    stop("Must provide grid_defs for multi-parameter grid
     ↪ search.")
  g <- expand.grid(grid_defs)
  g$SSE <-
    apply(g, 1, function(rowp)
      sse_fn(as.numeric(rowp), x, y))
  best_idx <- which.min(g$SSE)
  as.numeric(g[best_idx, seq_along(grid_defs)])
}

# 2.9 Nelder-Mead
nelder_mead_fit <- function(par,
                            x,
                            y,
                            sse_fn,
                            model_fn,
```

```
                                        lower = NULL,
                                        upper = NULL)
{
  fn <- function(pp)
    sse_fn(pp, x, y)
  res <- optim(par, fn, method = "Nelder-Mead")
  res$par
}

# 2.10 Powell-s method (dfoptim::nmk for unconstrained)
powell_fit <-
  function(par,
           x,
           y,
           sse_fn,
           model_fn,
           lower = NULL,
           upper = NULL) {
    fn <- function(pp)
      sse_fn(pp, x, y)
    dfoptim::nmk(par, fn)$par
  }

# 2.11 Hooke-Jeeves (dfoptim::hjkb)
hooke_jeeves_fit <-
  function(par,
           x,
           y,
           sse_fn,
           model_fn,
           lower = NULL,
           upper = NULL) {
    fn <- function(pp)
      sse_fn(pp, x, y)
    dfoptim::hjkb(par, fn)$par
  }

# 2.12 Random Search
random_search_fit <- function(par,
                              x,
                              y,
                              sse_fn,
                              model_fn,
                              lower,
```

```
                                  upper,
                                  max_iter = 2000,
                                  ...)
{
  best_par <- NULL
  best_sse <- Inf
  dimp <- length(lower)
  for (i in seq_len(max_iter)) {
    candidate <- runif(dimp, min = lower, max = upper)
    val <- sse_fn(candidate, x, y)
    if (val < best_sse) {
      best_sse <- val
      best_par <- candidate
    }
  }
  best_par
}

# 2.13 Differential Evolution (DEoptim)
diff_evo_fit <- function(par,
                         x,
                         y,
                         sse_fn,
                         model_fn,
                         lower,
                         upper,
                         max_iter = 100,
                         ...)
{
  fn <- function(v)
    sse_fn(v, x, y)
  out <- DEoptim(fn,
                 lower = lower,
                 upper = upper,
                 DEoptim.control(NP = 50, itermax = max_iter,
                 ↪  trace = F))
  out$optim$bestmem
}

# 2.14 Simulated Annealing (GenSA)
sim_anneal_fit <- function(par,
                           x,
                           y,
                           sse_fn,
```

```
                            model_fn,
                            lower = NULL,
                            upper = NULL,
                            ...)
{
  fn <- function(pp)
    sse_fn(pp, x, y)
  lb <- if (is.null(lower))
    rep(-Inf, length(par))
  else
    lower
  ub <- if (is.null(upper))
    rep(Inf, length(par))
  else
    upper
  # GenSA requires: GenSA(par, fn, lower, upper,
  ↪  control=list(...))
  out <-
    GenSA(
      par,
      fn,
      lower = lb,
      upper = ub,
      control = list(max.call = 10000)
    )
  out$par
}

# 2.15 Genetic Algorithm (GA)
genetic_fit <- function(par,
                        x,
                        y,
                        sse_fn,
                        model_fn,
                        lower,
                        upper,
                        max_iter = 100,
                        ...)
{
  fitness_fun <- function(pp)
    - sse_fn(pp, x, y)
  gares <- ga(
    type = "real-valued",
    fitness = fitness_fun,
```

```r
    lower = lower,
    upper = upper,
    popSize = 50,
    maxiter = max_iter,
    run = 50
  )
  gares@solution[1,]
}

# 2.16 Particle Swarm (pso)
particle_swarm_fit <- function(par,
                               x,
                               y,
                               sse_fn,
                               model_fn,
                               lower,
                               upper,
                               max_iter = 100,
                               ...)
{
  fn <- function(pp)
    sse_fn(pp, x, y)
  res <- psoptim(
    par = (lower + upper) / 2,
    fn = fn,
    lower = lower,
    upper = upper,
    control = list(maxit = max_iter)
  )
  res$par
}


# -- 3) RUN METHOD WRAPPER ---
run_method <- function(method_name,
                       FUN,
                       par_init,
                       x,
                       y,
                       sse_fn,
                       model_fn,
                       lower = NULL,
                       upper = NULL,
                       ...)
```

```
{
  mb <- microbenchmark(result = {
    out <- FUN(par_init, x, y, sse_fn, model_fn, lower, upper,
↪   ...)
    out
  }, times = 1)
  final_par <-
    FUN(par_init, x, y, sse_fn, model_fn, lower, upper, ...)
  if (is.null(final_par)) {
    # e.g. placeholders that return NULL
    return(data.frame(
      Method = method_name,
      Parameters = "N/A",
      SSE = NA,
      Time_ms = NA
    ))
  }
  data.frame(
    Method      = method_name,
    Parameters = paste(round(final_par, 4), collapse = ", "),
    SSE        = round(sse_fn(final_par, x, y), 6),
    Time_ms    = median(mb$time) / 1e6
  )
}

# -- 4) MASTER FUNCTION TO COMPARE ALL METHODS (SIMPLE /
↪  COMPLEX) ---

compare_all_methods <- function(is_complex = FALSE) {
  if (!is_complex) {
    # SIMPLE (3-param logistic)
    x <- x_data
    y <- y_data_simple
    sse_fn    <- sse_logistic
    model_fn <- f_logistic
    init_par <- c(3, 1, 0.5)
    grid_defs <- list(
      A = seq(2, 6, length.out = 10),
      B = seq(0.5, 2, length.out = 10),
      C = seq(-1, 1, length.out = 10)
    )
    lower <- c(1, 0.1,-3)
    upper <- c(6, 3,  3)
    lm_form <- "simple"
```

```
} else {
  # COMPLEX (4-param model)
  x <- x_data
  y <- y_data_complex
  sse_fn   <- sse_complex
  model_fn <- f_complex
  init_par <- c(3, 1,-0.5, 0.2)
  grid_defs <- list(
    A = seq(2, 6, length.out = 8),
    B = seq(0.5, 2, length.out = 8),
    C = seq(-2, 2, length.out = 8),
    D = seq(0, 2, length.out = 8)
  )
  lower <- c(1, 0.1,-3, 0)
  upper <- c(6, 3,  3, 2)
  lm_form <- "complex"
}

# RUN each method
out <- bind_rows(
  run_method(
    "Gauss-Newton",
    gauss_newton_fit,
    init_par,
    x,
    y,
    sse_fn,
    model_fn
  ),
  run_method(
    "Modified Gauss-Newton",
    modified_gauss_newton_fit,
    init_par,
    x,
    y,
    sse_fn,
    model_fn
  ),
  run_method(
    "Steepest Descent",
    steepest_descent_fit,
    init_par,
    x,
    y,
```

```
    sse_fn,
    model_fn
  ),
  run_method(
    "Levenberg-Marquardt (nlsLM)",
    lm_fit,
    init_par,
    x,
    y,
    sse_fn,
    model_fn,
    form = lm_form
  ),
  run_method(
    "Newton-Raphson",
    newton_raphson_fit,
    init_par,
    x,
    y,
    sse_fn,
    model_fn
  ),
  run_method(
    "Quasi-Newton (BFGS)",
    quasi_newton_fit,
    init_par,
    x,
    y,
    sse_fn,
    model_fn
  ),
  run_method(
    "Trust-region Reflective",
    trust_region_fit,
    init_par,
    x,
    y,
    sse_fn,
    model_fn,
    lower,
    upper
  ),
  run_method(
    "Grid Search",
```

```
    grid_search_fit,
    NULL,
    x,
    y,
    sse_fn,
    model_fn,
    grid_defs = grid_defs
  ),
  run_method(
    "Nelder-Mead",
    nelder_mead_fit,
    init_par,
    x,
    y,
    sse_fn,
    model_fn
  ),
  run_method("Powell's method",
             powell_fit,
             init_par,
             x,
             y,
             sse_fn,
             model_fn),
  run_method(
    "Hooke-Jeeves",
    hooke_jeeves_fit,
    init_par,
    x,
    y,
    sse_fn,
    model_fn
  ),
  run_method(
    "Random Search",
    random_search_fit,
    NULL,
    x,
    y,
    sse_fn,
    model_fn,
    lower,
    upper,
    max_iter = 1000
```

```
  ),
  run_method(
    "Differential Evolution",
    diff_evo_fit,
    NULL,
    x,
    y,
    sse_fn,
    model_fn,
    lower,
    upper,
    max_iter = 50
  ),
  run_method(
    "Simulated Annealing",
    sim_anneal_fit,
    init_par,
    x,
    y,
    sse_fn,
    model_fn,
    lower,
    upper
  ),
  run_method(
    "Genetic Algorithm",
    genetic_fit,
    NULL,
    x,
    y,
    sse_fn,
    model_fn,
    lower,
    upper,
    max_iter = 50
  ),
  run_method(
    "Particle Swarm",
    particle_swarm_fit,
    NULL,
    x,
    y,
    sse_fn,
    model_fn,
```

```
      lower,
      upper,
      max_iter = 50
    )
  )
  out
}

# -- 5) RUN & VISUALIZE ----

# Compare "simple" logistic (3 params)
results_simple  <- compare_all_methods(is_complex = FALSE)
results_simple$Problem <- "Simple"

# Compare "complex" (4 params)
results_complex <- compare_all_methods(is_complex = TRUE)
results_complex$Problem <- "Complex"

# Combine
all_results <- rbind(results_simple, results_complex)
# print(all_results)
# DT::datatable(all_results)
```

```
# Example: SSE by method & problem
ggplot(all_results, aes(x = Method, y = log(SSE), fill =
↪  Problem)) +
  geom_bar(stat = "identity", position = "dodge") +
  theme_minimal(base_size = 11) +
  labs(title = "Comparison of SSE by Method and Problem
  ↪  Complexity",
       x = "", y = "Log(Sum of Squared Errors)") +
  theme(axis.text.x = element_text(angle = 45, hjust = 1))
```

```
# Example: Time (ms) by method & problem
ggplot(all_results, aes(x = Method, y = Time_ms, fill =
↪  Problem)) +
  geom_bar(stat = "identity", position = "dodge") +
  theme_minimal(base_size = 11) +
```

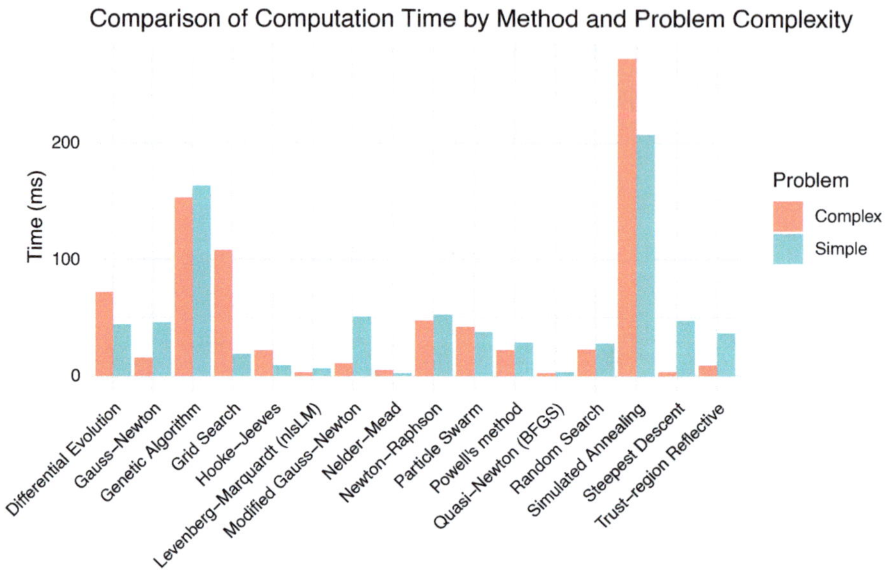

Fig. 3.23 Comparison of computation time by method and problem complexity

```
labs(title = "Comparison of Computation Time by Method and
↪  Problem Complexity",
     x = "", y = "Time (ms)") +
theme(axis.text.x = element_text(angle = 45, hjust = 1))
```

See Fig. 3.23.

3.3 Practical Considerations

For optimization algorithms to converge, they require good initial estimates of the parameters. The choice of starting values, constraints, and the complexity of the model all play a role in whether an optimization algorithm successfully finds a suitable solution.

3.3.1 Selecting Starting Values

Choosing good starting values can significantly impact the efficiency and success of optimization algorithms. Several approaches can be used:

- **Prior or theoretical information**: If prior knowledge about the parameters is available, it should be incorporated into the choice of initial values.
- **Grid search or graphical inspection of** $SSE(\theta)$: Evaluating the SSE across a grid of possible values can help identify promising starting points.
- Ordinary Least Squares **estimates**: If a linear approximation of the model exists, using OLS to obtain initial estimates can be effective.
- **Model interpretation**: Understanding the structure and behavior of the model can provide intuition for reasonable starting values.
- **Expected value parameterization**: Reformulating the model based on expected values may improve the interpretability and numerical stability of the estimation.

3.3.1.1 Grid Search for Optimal Starting Values

```
# Set seed for reproducibility
set.seed(123)

# Generate x as 100 integers using seq function
x <- seq(0, 100, 1)

# Generate coefficients for exponential function
a <- runif(1, 0, 20)  # Random coefficient a
b <- runif(1, 0.005, 0.075)  # Random coefficient b
c <- runif(101, 0, 5)  # Random noise

# Generate y as a * e^(b*x) + c
y <- a * exp(b * x) + c

# Print the generated parameters
cat("Generated coefficients:\n")
#> Generated coefficients:
cat("a =", a, "\n")
#> a = 5.75155
cat("b =", b, "\n")
#> b = 0.06018136

# Define our data frame
datf <- data.frame(x, y)

# Define our model function
mod <- function(a, b, x) {
  a * exp(b * x)
}
```

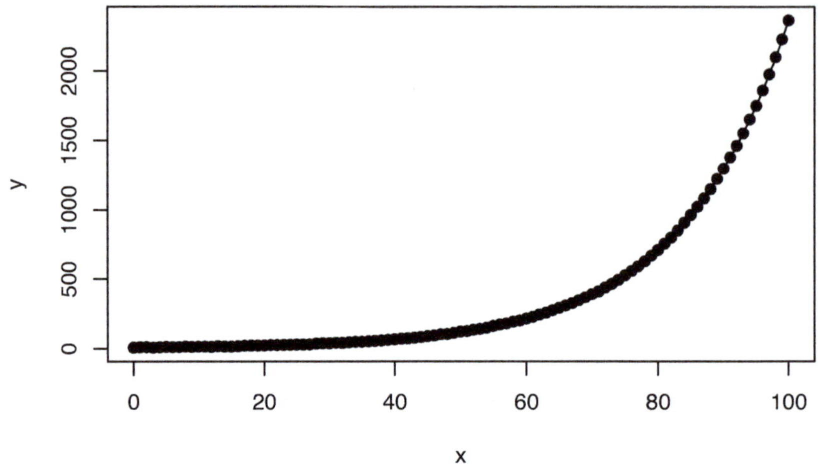

Fig. 3.24 Exponential growth fit

Figure 3.24. plots the original data.

```
# Ensure all y values are positive (avoid log issues)
y_adj <-
  ifelse(y > 0, y, min(y[y > 0]) + 1e-3)  # Shift small values
  ↪ slightly

# Create adjusted dataframe
datf_adj <- data.frame(x, y_adj)

# Linearize by taking log(y)
lin_mod <- lm(log(y_adj) ~ x, data = datf_adj)

# Extract starting values
astrt <-
  exp(coef(lin_mod)[1])  # Convert intercept back from log
  ↪ scale
bstrt <- coef(lin_mod)[2]  # Slope remains the same
cat("Starting values for non-linear fit:\n")
print(c(astrt, bstrt))

# Fit nonlinear model with these starting values
nlin_mod <- nls(y ~ mod(a, b, x),
                start = list(a = astrt, b = bstrt),
```

```
                 data = datf)

# Model summary
summary(nlin_mod)

# Plot original data
plot(
  x,
  y,
  main = "Exponential Growth Fit",
  col = "blue",
  pch = 16,
  xlab = "x",
  ylab = "y"
)

# Add fitted curve in red
lines(x, predict(nlin_mod), col = "red", lwd = 2)

# Add legend
legend(
  "topleft",
  legend = c("Original Data", "Fitted Model"),
  col = c("blue", "red"),
  pch = c(16, NA),
  lwd = c(NA, 2)
)
```

```
# Define grid of possible parameter values
aseq <- seq(10, 18, 0.2)
bseq <- seq(0.001, 0.075, 0.001)

na <- length(aseq)
nb <- length(bseq)
SSout <- matrix(0, na * nb, 3)  # Matrix to store SSE values
cnt <- 0

# Evaluate SSE across grid
for (k in 1:na) {
  for (j in 1:nb) {
    cnt <- cnt + 1
    ypred <-
      # Evaluate model at these parameter values
```

```
      mod(aseq[k], bseq[j], x)

    # Compute SSE
    ss <- sum((y - ypred) ^ 2)
    SSout[cnt, 1] <- aseq[k]
    SSout[cnt, 2] <- bseq[j]
    SSout[cnt, 3] <- ss
  }
}

# Identify optimal starting values
mn_indx <- which.min(SSout[, 3])
astrt <- SSout[mn_indx, 1]
bstrt <- SSout[mn_indx, 2]

# Fit nonlinear model using optimal starting values
nlin_modG <-
  nls(y ~ mod(a, b, x), start = list(a = astrt, b = bstrt))

# Display model results
summary(nlin_modG)
#>
#> Formula: y ~ mod(a, b, x)
#>
#> Parameters:
#>    Estimate Std. Error t value Pr(>|t|)
#> a 5.889e+00  1.986e-02   296.6   <2e-16 ***
#> b 5.995e-02  3.644e-05  1645.0   <2e-16 ***
#> ---
#> Signif. codes:
#> 0 '***' 0.001 '**' 0.01 '*' 0.05 '.' 0.1 ' ' 1
#>
#> Residual standard error: 2.135 on 99 degrees of freedom
#>
#> Number of iterations to convergence: 4
#> Achieved convergence tolerance: 7.213e-06
```

Note The `nls_multstart` package can perform a grid search more efficiently without requiring manual looping.

Visualizing Prediction Intervals

Once the model is fitted, it is useful to visualize prediction intervals to assess model uncertainty.

```
# Load necessary package
library(nlstools)
#>
#> 'nlstools' has been loaded.
#> IMPORTANT NOTICE: Most nonlinear regression models and data
↪   set examples
#> related to predictive microbiolgy have been moved to the
↪   package 'nlsMicrobio'

# Plot fitted model with confidence and prediction intervals
investr::plotFit(
  nlin_modG,
  interval = "both",
  pch = 19,
  shade = TRUE,
  col.conf = "skyblue4",
  col.pred = "lightskyblue2",
  data = datf
)
```

3.3.1.2 Using Programmed Starting Values in `nls`

Many nonlinear models have well-established functional forms, allowing for programmed starting values in the `nls` function. For example, models such as **logistic growth** and **asymptotic regression** have built-in self-starting functions.

To explore available self-starting models in R, use:

```
apropos("^SS")
```

This command lists functions with names starting with SS, which typically denote self-starting functions for nonlinear regression.

3.3.1.3 Custom Self-Starting Functions

If your model does not match any built-in `nls` functions, you can define your own **self-starting function**. Self-starting functions in R automate the process of estimating initial values, which helps in fitting nonlinear models more efficiently.

If needed, a self-starting function should:

1. Define the nonlinear equation.
2. Implement a method for computing starting values.
3. Return the function structure in an appropriate format.

3.3.2 *Handling Constrained Parameters*

In some cases, parameters must satisfy constraints (e.g., $\theta_i > a$ or $a < \theta_i < b$). The following strategies help address constrained parameter estimation:

1. **Fit the model without constraints first**: If the unconstrained parameter estimates satisfy the desired constraints, no further action is needed.
2. **Reparameterization**: If the estimated parameters violate constraints, consider reparameterizing the model to naturally enforce the required bounds.

3.3.3 *Failure to Converge*

Several factors can cause an algorithm to fail to converge:

- **A "flat" SSE function**: If the sum of squared errors $SSE(\theta)$ is relatively constant in the neighborhood of the minimum, the algorithm may struggle to locate an optimal solution.
- **Poor starting values**: Trying different or better initial values can help.
- **Overly complex models**: If the model is too complex relative to the data, consider simplifying it.

3.3.4 *Convergence to a Local Minimum*

- **Linear least squares models** have a well-defined, unique minimum because the SSE function is quadratic:

$$SSE(\theta) = (Y - X\beta)'(Y - X\beta)$$

- **Nonlinear least squares models** may have multiple local minima.
- **Testing different starting values** can help identify a global minimum.
- **Graphing** $SSE(\theta)$ as a function of individual parameters (if feasible) can provide insights.
- **Alternative optimization algorithms** such as Genetic Algorithm or Particle Swarm Optimization may be useful in non-convex problems.

3.3.5 *Model Adequacy and Estimation Considerations*

Assessing the adequacy of a **nonlinear model** involves checking its **nonlinearity**, **goodness of fit**, and **residual behavior**. Unlike linear models, nonlinear models do not always have a direct equivalent of R^2, and issues such as collinearity, leverage, and residual heteroscedasticity must be carefully evaluated.

3.3.5.1 Components of Nonlinearity

Bates and Watts (1980) define two key aspects of **nonlinearity** in statistical modeling:

1. **Intrinsic nonlinearity**

- Measures the **bending and twisting** in the function $f(\theta)$.
- Assumes that the function is relatively **flat (planar)** in the neighborhood of $\hat{\theta}$.
- If severe, the **distribution of residuals** will be **distorted**.
- Leads to:
  - **Slow convergence** of optimization algorithms.
  - **Difficulties in identifying** parameter estimates.
- Solution approaches:
  - Higher-order **Taylor expansions** for estimation.
  - **Bayesian methods** for parameter estimation.

```
library(MASS)

# Check intrinsic curvature
modD <- deriv3(~ a * exp(b * x), c("a", "b"), function(a, b,
↪  x) NULL)

nlin_modD <- nls(y ~ modD(a, b, x),
                 start = list(a = astrt, b = bstrt),
                 data = datf)

rms.curv(nlin_modD)  # Function from the MASS package to
↪  assess curvature
#> Parameter effects: c^theta x sqrt(F) = 0.0564
#>         Intrinsic: c^iota  x sqrt(F) = 9e-04
```

2. **Parameter-effects nonlinearity**

- Measures **how the curvature** (nonlinearity) depends on the parameterization.
- Strong parameter-effects nonlinearity can cause **problems with inference on** $\hat{\theta}$.
- Can be assessed using:
  - `rms.curv` function from `MASS`.

  - Bootstrap-based inference.
- **Solution:** Try **reparameterization** to stabilize the function.

3.3.5.2 Goodness of Fit in Nonlinear Models

In **linear regression**, we use the standard **coefficient of determination** (R^2):

$$R^2 = \frac{SSR}{SSTO} = 1 - \frac{SSE}{SSTO}$$

Where:

- SSR = regression sum of squares
- SSE = error sum of squares
- $SSTO$ = total sum of squares

However, in **nonlinear models**, the error and model sum of squares do not necessarily add up to the total corrected sum of squares:

$$SSR + SSE \neq SST$$

Thus, R^2 is not directly valid in the nonlinear case. Instead, we use a pseudo-R^2:

$$R^2_{pseudo} = 1 - \frac{\sum_{i=1}^n (Y_i - \hat{Y})^2}{\sum_{i=1}^n (Y_i - \bar{Y})^2}$$

- Unlike true R^2, this cannot be interpreted as the proportion of variability explained by the model.
- Should be used only for relative model comparison (e.g., comparing different nonlinear models).

3.3.5.3 Residual Analysis in Nonlinear Models

Residual plots help assess model adequacy, particularly when **intrinsic curvature is small** (Fig. 3.25).

In nonlinear models, the **studentized residuals** are:

$$r_i = \frac{e_i}{s\sqrt{1 - \hat{c}_i}}$$

Where:

- e_i = residual for observation i

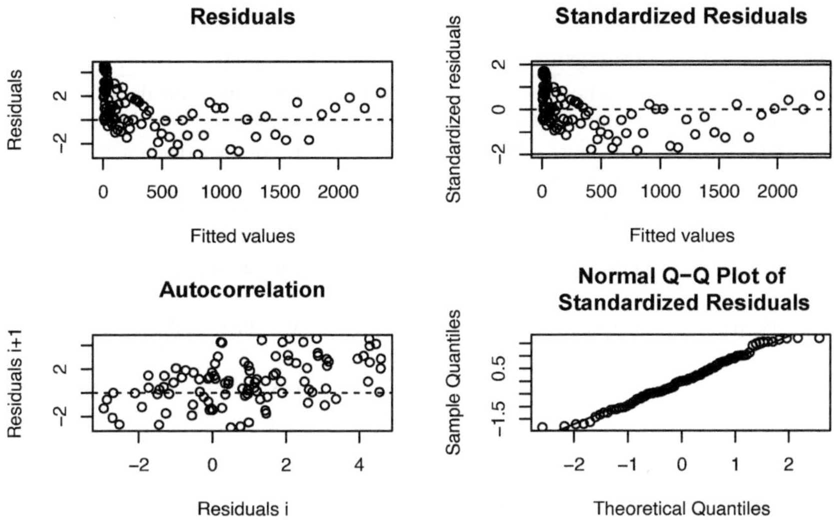

Fig. 3.25 Diagnostic plots for regression residuals

- $\hat{c}_i$ = ith diagonal element of the **tangent-plane hat matrix**:

$$\hat{\mathbf{H}} = \mathbf{F}(\hat{\boldsymbol{\theta}})[\mathbf{F}(\hat{\boldsymbol{\theta}})'\mathbf{F}(\hat{\boldsymbol{\theta}})]^{-1}\mathbf{F}(\hat{\boldsymbol{\theta}})'$$

```
# Residual diagnostics for nonlinear models
library(nlstools)
resid_nls <- nlsResiduals(nlin_modD)

# Generate residual plots
plot(resid_nls)
```

3.3.5.4 Potential Issues in Nonlinear Regression Models

Table 3.4 shows common issues in nonlinear regression and their solutions.

3.3.5.4.1 Collinearity

- Measures how correlated the model's predictors are.
- In nonlinear models, collinearity is assessed using the condition number of:

$$[\mathbf{F}(\hat{\boldsymbol{\theta}})'\mathbf{F}(\hat{\boldsymbol{\theta}})]^{-1}$$

Table 3.4 Common issues in nonlinear regression and their solutions

Issue	Description	Solution
Intrinsic nonlinearity	Function curvature independent of parameterization	Bayesian estimation, Taylor expansion
Parameter-effects nonlinearity	Curvature influenced by parameterization	Reparameterization, bootstrap
Collinearity	High correlation among predictors	Reparameterization, condition number check
Leverage	Influential points affecting model fit	Assess tangent-plane hat matrix
Heterogeneous errors	Unequal variance in residuals	Weighted nonlinear least squares
Correlated errors	Autocorrelated residuals	GNLS, nonlinear mixed-effects models, Bayesian methods

- If condition number > 30, collinearity is a concern.
- Solution: Consider reparameterization (Magel and Hertsgaard, 1987).

3.3.5.4.2 Leverage

- Similar to leverage in Ordinary Least Squares.
- In nonlinear models, leverage is assessed using the **tangent-plane hat matrix**:

$$\hat{\mathbf{H}} = \mathbf{F}(\hat{\boldsymbol{\theta}})[\mathbf{F}(\hat{\boldsymbol{\theta}})'\mathbf{F}(\hat{\boldsymbol{\theta}})]^{-1}\mathbf{F}(\hat{\boldsymbol{\theta}})'$$

- **Solution:** Identify influential points and assess their impact on parameter estimates (St Laurent and Cook, 1992).

3.3.5.4.3 Heterogeneous Errors

- Non-constant variance across observations.
- **Solution:** Use **Weighted nonlinear least squares (WNLS)**.

3.3.5.4.4 Correlated Errors

- Residuals may be autocorrelated.
- **Solution approaches:**
  - **Generalized nonlinear least squares (GNLS)**
  - **Nonlinear mixed-effects models (NLMEM)**
  - **Bayesian methods**

3.4 Application

3.4.1 Nonlinear Estimation Using Gauss-Newton Algorithm

This section demonstrates nonlinear parameter estimation using the Gauss-Newton algorithm and compares results with `nls()`. The model is given by:

$$y_i = \frac{\theta_0 + \theta_1 x_i}{1 + \theta_2 \exp(0.4 x_i)} + \epsilon_i$$

Where:

- $i = 1, \ldots, n$
- $\theta_0, \theta_1, \theta_2$ are the unknown parameters.
- ϵ_i represents errors.

1. Loading and Visualizing the Data (Fig. 3.26)

```
library(dplyr)
#>
#> Attaching package: 'dplyr'
#> The following objects are masked from 'package:stats':
#>
```

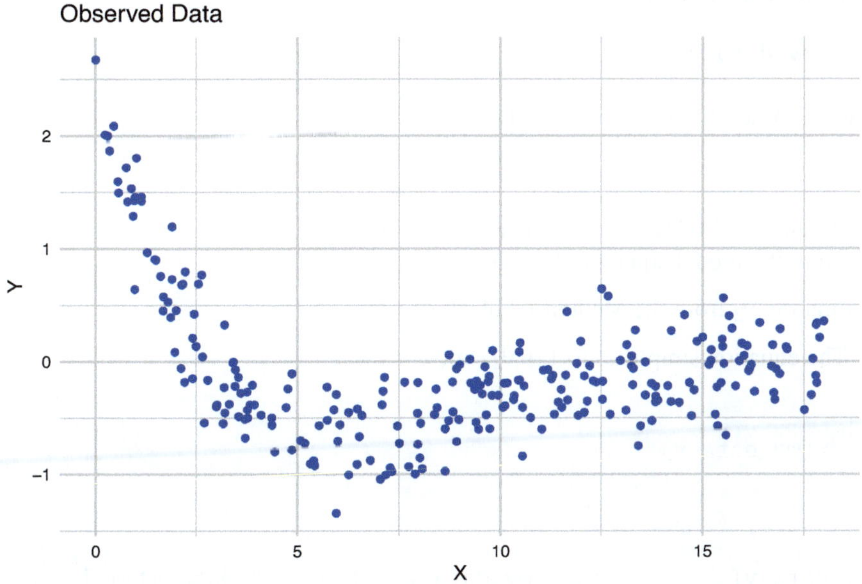

Fig. 3.26 Observed data

```
#>     filter, lag
#> The following objects are masked from 'package:base':
#>
#>     intersect, setdiff, setequal, union
library(ggplot2)
library(investr)

# Load the dataset
my_data <- read.delim("images/S21hw1pr4.txt", header = FALSE,
↪  sep = "") %>%
  dplyr::rename(x = V1, y = V2)
```

```
# Plot data
ggplot(my_data, aes(x = x, y = y)) +
  geom_point(color = "blue") +
  labs(title = "Observed Data", x = "X", y = "Y") +
  theme_minimal()
```

2. Deriving Starting Values for Parameters

Since nonlinear optimization is sensitive to starting values, we estimate reasonable initial values based on model interpretation.

Finding the Maximum Y Value

```
max(my_data$y)
#> [1] 2.6722
my_data$x[which.max(my_data$y)]
#> [1] 0.0094
```

- When $y = 2.6722$, the corresponding $x = 0.0094$.
- From the model equation: $\theta_0 + 0.0094\theta_1 = 2.6722$.

Estimating θ_2 from the Median y Value

- The equation simplifies to $1 + \theta_2 \exp(0.4x) = 2$.

```
# find mean y
mean(my_data$y)
#> [1] -0.0747864

# find y closest to its mean
my_data$y[which.min(abs(my_data$y - (mean(my_data$y))))]
#> [1] -0.0773
```

```r
# find x closest to the mean y
my_data$x[which.min(abs(my_data$y - (mean(my_data$y))))]
#> [1] 11.0648
```

- This yields the equation: $83.58967\theta_2 = 1$.

Finding the Value of θ_0 and θ_1

```r
# find value of x closet to 1
my_data$x[which.min(abs(my_data$x - 1))]
#> [1] 0.9895

# find index of x closest to 1
match(my_data$x[which.min(abs(my_data$x - 1))], my_data$x)
#> [1] 14

# find y value
my_data$y[match(my_data$x[which.min(abs(my_data$x - 1))],
↪  my_data$x)]
#> [1] 1.4577
```

- This provides another equation: $\theta_0 + \theta_1 \times 0.9895 - 2.164479\theta_2 = 1.457$.

Solving for $\theta_0, \theta_1, \theta_2$

```r
library(matlib)

# Define coefficient matrix
A = matrix(
  c(0, 0.0094, 0, 0, 0, 83.58967, 1, 0.9895, -2.164479),
  nrow = 3,
  ncol = 3,
  byrow = T
)

# Define constant vector
b <- c(2.6722, 1, 1.457)

# Display system of equations
showEqn(A, b)
#> 0*x1 + 0.0094*x2             + 0*x3  =  2.6722
#> 0*x1           + 0*x2 + 83.58967*x3  =       1
#> 1*x1 + 0.9895*x2 - 2.164479*x3  =   1.457
```

```
# Solve for parameters
theta_start <- Solve(A, b, fractions = FALSE)
#> x1       =  -279.80879739
#>    x2     =   284.27659574
#>       x3  =      0.0119632
theta_start
#> [1] "x1          =  -279.80879739" "   x2      =   284.27659574"
#> [3] "      x3  =      0.0119632"
```

3. Implementing the Gauss-Newton Algorithm

Using these estimates, we manually implement the **Gauss-Newton optimization**.

Defining the Model and Its Derivatives

```
# Starting values
theta_0_strt <- as.numeric(gsub(".*=\\s*", "",
↪   theta_start[1]))
theta_1_strt <- as.numeric(gsub(".*=\\s*", "",
↪   theta_start[2]))
theta_2_strt <- as.numeric(gsub(".*=\\s*", "",
↪   theta_start[3]))

# Model function
mod_4 <- function(theta_0, theta_1, theta_2, x) {
  (theta_0 + theta_1 * x) / (1 + theta_2 * exp(0.4 * x))
}

# Define function expression
f_4 = expression((theta_0 + theta_1 * x) / (1 + theta_2 *
↪   exp(0.4 * x)))

# First derivatives
df_4.d_theta_0 <- D(f_4, 'theta_0')
df_4.d_theta_1 <- D(f_4, 'theta_1')
df_4.d_theta_2 <- D(f_4, 'theta_2')
```

Iterative Gauss-Newton Optimization

```
# Initialize
theta_vec <- matrix(c(theta_0_strt, theta_1_strt,
↪  theta_2_strt))
delta <- matrix(NA, nrow = 3, ncol = 1)
i <- 1

# Evaluate function at initial estimates
f_theta <- as.matrix(eval(f_4, list(
  x = my_data$x,
  theta_0 = theta_vec[1, 1],
  theta_1 = theta_vec[2, 1],
  theta_2 = theta_vec[3, 1]
)))

repeat {
  # Compute Jacobian matrix
  F_theta_0 <- as.matrix(cbind(
    eval(df_4.d_theta_0, list(
      x = my_data$x,
      theta_0 = theta_vec[1, i],
      theta_1 = theta_vec[2, i],
      theta_2 = theta_vec[3, i]
    )),
    eval(df_4.d_theta_1, list(
      x = my_data$x,
      theta_0 = theta_vec[1, i],
      theta_1 = theta_vec[2, i],
      theta_2 = theta_vec[3, i]
    )),
    eval(df_4.d_theta_2, list(
      x = my_data$x,
      theta_0 = theta_vec[1, i],
      theta_1 = theta_vec[2, i],
      theta_2 = theta_vec[3, i]
    ))
  ))

  # Compute parameter updates
  delta[, i] <- (solve(t(F_theta_0)%*%F_theta_0))%*%t
↪  (F_theta_0)%*%(my_data$y-f_theta[,i])


  # Update parameter estimates
  theta_vec <- cbind(theta_vec, theta_vec[, i] + delta[, i])
```

```
  theta_vec[, i + 1] = theta_vec[, i] + delta[, i]

  # Increment iteration counter
  i <- i + 1

  # Compute new function values
  f_theta <- cbind(f_theta, as.matrix(eval(f_4, list(
    x = my_data$x,
    theta_0 = theta_vec[1, i],
    theta_1 = theta_vec[2, i],
    theta_2 = theta_vec[3, i]
  ))))

  delta = cbind(delta, matrix(NA, nrow = 3, ncol = 1))

  # Convergence criteria based on SSE
  if (abs(sum((my_data$y - f_theta[, i])^2) - sum((my_data$y
  ↪  - f_theta[, i - 1])^2)) /
      sum((my_data$y - f_theta[, i - 1])^2) < 0.001) {
    break
  }
}

# Final parameter estimates
theta_vec[, ncol(theta_vec)]
#> [1]  3.6335135 -1.3055166  0.5043502
```

4. Checking Convergence and Variance

```
# Final objective function value (SSE)
sum((my_data$y - f_theta[, i])^2)
#> [1] 19.80165

sigma2 <- 1 / (nrow(my_data) - 3) *
  (t(my_data$y - f_theta[, ncol(f_theta)]) %*%
   (my_data$y - f_theta[, ncol(f_theta)]))  # p = 3

# Asymptotic variance-covariance matrix
as.numeric(sigma2)*as.matrix(solve(crossprod(F_theta_0)))
#>             [,1]        [,2]        [,3]
#> [1,]  0.11552571 -0.04817428  0.02685848
#> [2,] -0.04817428  0.02100861 -0.01158212
#> [3,]  0.02685848 -0.01158212  0.00703916
```

5. Validating with `nls()`

```
nlin_4 <- nls(
  y ~ mod_4(theta_0, theta_1, theta_2, x),
  start = list(
    theta_0 = as.numeric(gsub(".*=\\s*", "", theta_start[1])),
    theta_1 = as.numeric(gsub(".*=\\s*", "", theta_start[2])),
    theta_2 = as.numeric(gsub(".*=\\s*", "", theta_start[3]))
  ),
  data = my_data
)
summary(nlin_4)
#>
#> Formula: y ~ mod_4(theta_0, theta_1, theta_2, x)
#>
#> Parameters:
#>          Estimate Std. Error t value Pr(>|t|)
#> theta_0   3.63591    0.36528   9.954  < 2e-16 ***
#> theta_1  -1.30639    0.15561  -8.395 3.65e-15 ***
#> theta_2   0.50528    0.09215   5.483 1.03e-07 ***
#> ---
#> Signif. codes:
#> 0 '***' 0.001 '**' 0.01 '*' 0.05 '.' 0.1 ' ' 1
#>
#> Residual standard error: 0.2831 on 247 degrees of freedom
#>
#> Number of iterations to convergence: 9
#> Achieved convergence tolerance: 2.307e-07
```

3.4.2 Logistic Growth Model

A classic logistic growth model follows the equation:

$$P = \frac{K}{1 + \exp(P_0 + rt)} + \epsilon$$

Where:

- P = population at time t
- K = carrying capacity (maximum population)
- r = population growth rate
- P_0 = initial population log-ratio

However, R's built-in `SSlogis` function uses a slightly different parameterization:

$$P = \frac{asym}{1 + \exp\left(\frac{xmid - t}{scal}\right)}$$

Where:

- $asym$ = carrying capacity (K)
- $xmid$ = the x-value at the inflection point of the curve
- $scal$ = scaling parameter

This gives the parameter relationships:

- $K = asym$
- $r = -1/scal$
- $P_0 = -r \cdot xmid$

```
# Simulated time-series data
time <- c(1, 2, 3, 5, 10, 15, 20, 25, 30, 35)
population <- c(2.8, 4.2, 3.5, 6.3, 15.7, 21.3, 23.7, 25.1,
↪  25.8, 25.9)
```

Figure 3.27 shows the raw data.

```
# Plot data points
plot(time, population, las = 1, pch = 16, main = "Logistic
↪  Growth Model")
```

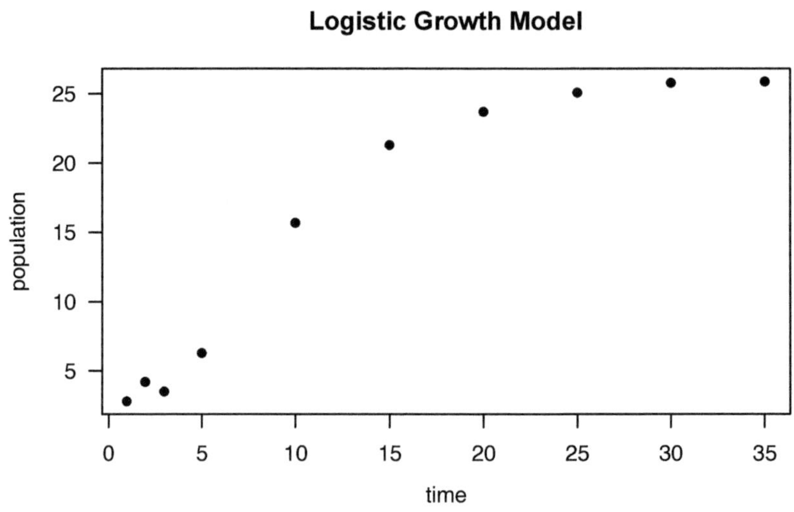

Fig. 3.27 Logistic growth model

```
# Fit the logistic growth model using programmed starting
↪  values
logisticModelSS <- nls(population ~ SSlogis(time, Asym, xmid,
↪  scal))

# Model summary
summary(logisticModelSS)
#>
#> Formula: population ~ SSlogis(time, Asym, xmid, scal)
#>
#> Parameters:
#>      Estimate Std. Error t value Pr(>|t|)
#> Asym  25.5029     0.3666   69.56 3.34e-11 ***
#> xmid   8.7347     0.3007   29.05 1.48e-08 ***
#> scal   3.6353     0.2186   16.63 6.96e-07 ***
#> ---
#> Signif. codes:
#> 0 '***' 0.001 '**' 0.01 '*' 0.05 '.' 0.1 ' ' 1
#>
#> Residual standard error: 0.6528 on 7 degrees of freedom
#>
#> Number of iterations to convergence: 1
#> Achieved convergence tolerance: 1.906e-06

# Extract parameter estimates
coef(logisticModelSS)
#>      Asym      xmid      scal
#> 25.502890  8.734698  3.635333
```

To fit the model using an alternative parameterization (K, r, P_0), we convert the estimated coefficients:

```
# Convert parameter estimates to alternative logistic model
↪  parameters
Ks <- as.numeric(coef(logisticModelSS)[1])  # Carrying
↪  capacity (K)
rs <- -1 / as.numeric(coef(logisticModelSS)[3])  # Growth rate
↪  (r)
Pos <- -rs * as.numeric(coef(logisticModelSS)[2])  # P_0

# Fit the logistic model with the alternative parameterization
logisticModel <- nls(
   population ~ K / (1 + exp(Po + r * time)),
```

```
    start = list(Po = Pos, r = rs, K = Ks)
)

# Model summary
summary(logisticModel)
#>
#> Formula: population ~ K/(1 + exp(Po + r * time))
#>
#> Parameters:
#>     Estimate Std. Error t value Pr(>|t|)
#> Po  2.40272    0.12702   18.92 2.87e-07 ***
#> r  -0.27508    0.01654  -16.63 6.96e-07 ***
#> K  25.50289    0.36665   69.56 3.34e-11 ***
#> ---
#> Signif. codes:
#> 0 '***' 0.001 '**' 0.01 '*' 0.05 '.' 0.1 ' ' 1
#>
#> Residual standard error: 0.6528 on 7 degrees of freedom
#>
#> Number of iterations to convergence: 0
#> Achieved convergence tolerance: 1.91e-06
```

Visualizing the Logistic Model Fit (Fig. 3.28)

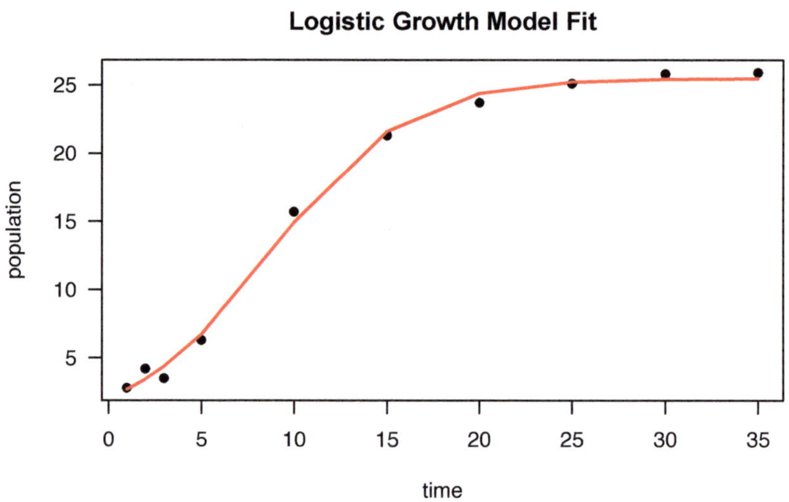

Fig. 3.28 Logistic growth model fit

```
# Plot original data
plot(time,
     population,
     las = 1,
     pch = 16,
     main = "Logistic Growth Model Fit")

# Overlay the fitted logistic curve
lines(time,
      predict(logisticModel),
      col = "red",
      lwd = 2)
```

3.4.3 Nonlinear Plateau Model

This example is based on Schabenberger and Pierce (2001) and demonstrates the use of a **plateau model** to estimate the relationship between soil nitrate (NO_3) concentration and relative yield percent (RYP) at two different depths (30 cm and 60 cm) (Fig. 3.29).

```
# Load data
dat <- read.table("images/dat.txt", header = TRUE)

# Plot NO3 concentration vs. relative yield percent, colored
↪ by depth
library(ggplot2)
dat.plot <- ggplot(dat) +
  geom_point(aes(x = no3, y = ryp, color = as.factor(depth)))
  ↪ +
  labs(color = 'Depth (cm)') +
  xlab('Soil NO3 Concentration') +
  ylab('Relative Yield Percent') +
  theme_minimal()

# Display plot
dat.plot
```

The **suggested nonlinear plateau model** is given by:

$$E(Y_{ij}) = (\beta_{0j} + \beta_{1j}N_{ij})I_{N_{ij}\leq\alpha_j} + (\beta_{0j} + \beta_{1j}\alpha_j)I_{N_{ij}>\alpha_j}$$

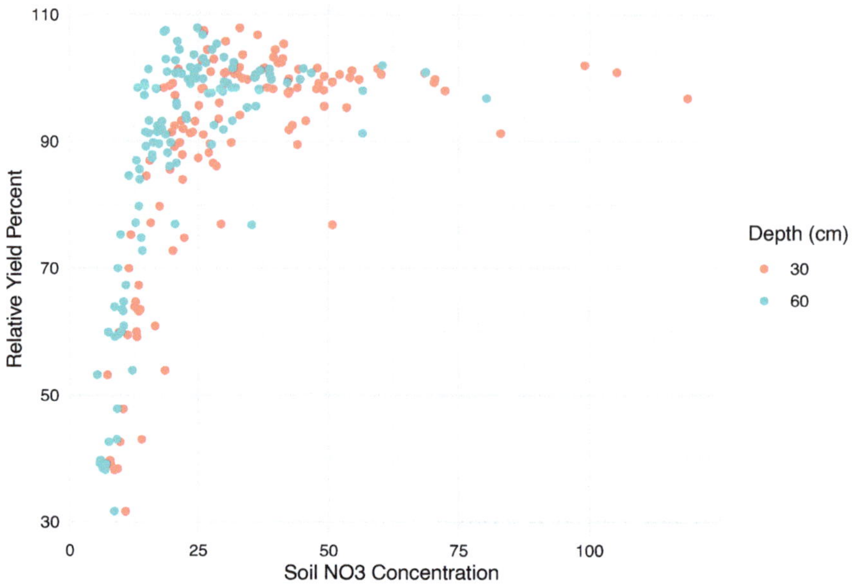

Fig. 3.29 Nonlinear plateau plot

Where:

- N_{ij} represents the soil nitrate (NO_3) concentration for observation i at depth j.
- i indexes individual observations.
- $j = 1, 2$ corresponds to depths **30 cm** and **60 cm**.

This model assumes a **linear increase** up to a threshold (α_j), beyond which the response **levels off (plateaus).**
Defining the Plateau Model as a Function

```
# Define the nonlinear plateau model function
nonlinModel <- function(predictor, b0, b1, alpha) {
  ifelse(predictor <= alpha,
         b0 + b1 * predictor,  # Linear growth below threshold
         b0 + b1 * alpha)      # Plateau beyond threshold
}
```

Creating a Self-Starting Function for `nls`
Since the model is **piecewise linear**, we can estimate starting values using:

1. A **linear regression** on the **first half** of sorted predictor values to estimate b_0 and b_1.
2. The **last predictor value** used in the regression as the plateau threshold (α)

```r
# Define initialization function for self-starting plateau
↪ model
nonlinModelInit <- function(mCall, LHS, data) {
  # Sort data by increasing predictor value
  xy <- sortedXyData(mCall[['predictor']], LHS, data)
  n <- nrow(xy)

  # Fit a simple linear model using the first half of the
  ↪ sorted data
  lmFit <- lm(xy[1:(n / 2), 'y'] ~ xy[1:(n / 2), 'x'])

  # Extract initial estimates
  b0 <- coef(lmFit)[1]  # Intercept
  b1 <- coef(lmFit)[2]  # Slope
  alpha <- xy[(n / 2), 'x']  # Last x-value in the fitted
↪ linear range

  # Return initial parameter estimates
  value <- c(b0, b1, alpha)
  names(value) <- mCall[c('b0', 'b1', 'alpha')]
  value
}
```

Combining Model and Self-Start Function

```r
# Define a self-starting nonlinear model for nls
SS_nonlinModel <- selfStart(nonlinModel,
                            nonlinModelInit,
                            c('b0', 'b1', 'alpha'))
```

The `nls` function is used to estimate parameters separately for **each soil depth (30 cm and 60 cm).**

```r
# Fit the model for depth = 30 cm
sep30_nls <- nls(ryp ~ SS_nonlinModel(predictor = no3, b0, b1,
↪ alpha),
                 data = dat[dat$depth == 30,])
#> Warning in getInitial.selfStart(func, data, mCall =
#> as.list(match.call(func, : selfStart initializing functions
#> should have a final '...' argument since R 4.1.0

# Fit the model for depth = 60 cm
sep60_nls <- nls(ryp ~ SS_nonlinModel(predictor = no3, b0, b1,
↪ alpha),
```

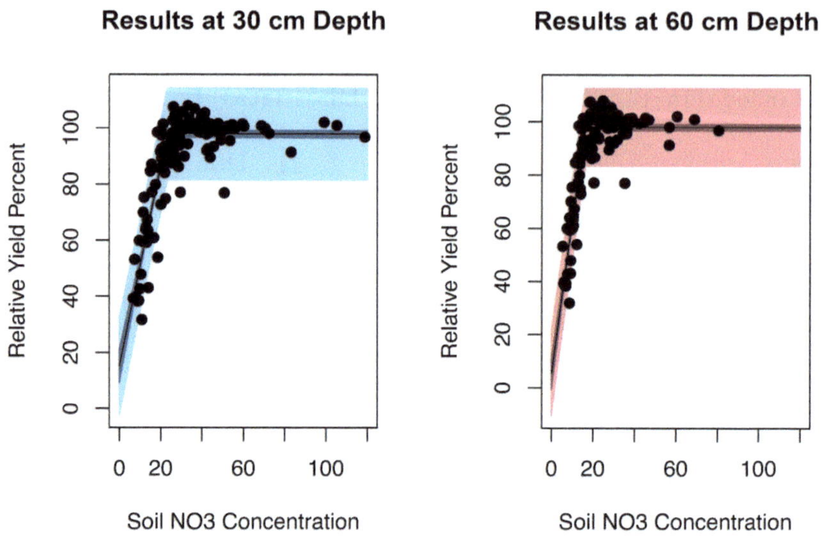

Fig. 3.30 Fitted curves for relative yield response by soil NO concentration at 30 cm and 60 cm depths

```
                    data = dat[dat$depth == 60,])
#> Warning in getInitial.selfStart(func, data, mCall =
#> as.list(match.call(func, : selfStart initializing functions
#> should have a final '...' argument since R 4.1.0
```

We generate separate plots for **30 cm** and **60 cm** depths, showing both **confidence** and **prediction intervals** (Fig. 3.30).

```
# Set plotting layout
par(mfrow = c(1, 2))

# Plot model fit for 30 cm depth
investr::plotFit(
  sep30_nls,
  interval = "both",
  pch = 19,
  shade = TRUE,
  col.conf = "skyblue4",
  col.pred = "lightskyblue2",
  data = dat[dat$depth == 30,],
  main = "Results at 30 cm Depth",
  ylab = "Relative Yield Percent",
  xlab = "Soil NO3 Concentration",
```

```
  xlim = c(0, 120)
)

# Plot model fit for 60 cm depth
investr::plotFit(
  sep60_nls,
  interval = "both",
  pch = 19,
  shade = TRUE,
  col.conf = "lightpink4",
  col.pred = "lightpink2",
  data = dat[dat$depth == 60,],
  main = "Results at 60 cm Depth",
  ylab = "Relative Yield Percent",
  xlab = "Soil NO3 Concentration",
  xlim = c(0, 120)
)
```

```
summary(sep30_nls)
#>
#> Formula: ryp ~ SS_nonlinModel(predictor = no3, b0, b1,
↪   alpha)
#>
#> Parameters:
#>       Estimate Std. Error t value Pr(>|t|)
#> b0     15.1943     2.9781   5.102 6.89e-07 ***
#> b1      3.5760     0.1853  19.297  < 2e-16 ***
#> alpha  23.1324     0.5098  45.373  < 2e-16 ***
#> ---
#> Signif. codes:
#> 0 '***' 0.001 '**' 0.01 '*' 0.05 '.' 0.1 ' ' 1
#>
#> Residual standard error: 8.258 on 237 degrees of freedom
#>
#> Number of iterations to convergence: 6
#> Achieved convergence tolerance: 1.166e-08
summary(sep60_nls)
#>
#> Formula: ryp ~ SS_nonlinModel(predictor = no3, b0, b1,
↪   alpha)
#>
#> Parameters:
```

```
#>        Estimate Std. Error t value Pr(>|t|)
#> b0       5.4519     2.9785    1.83   0.0684 .
#> b1       5.6820     0.2529   22.46   <2e-16 ***
#> alpha   16.2863     0.2818   57.80   <2e-16 ***
#> ---
#> Signif. codes:
#> 0 '***' 0.001 '**' 0.01 '*' 0.05 '.' 0.1 ' ' 1
#>
#> Residual standard error: 7.427 on 237 degrees of freedom
#>
#> Number of iterations to convergence: 5
#> Achieved convergence tolerance: 2.196e-08
```

Modeling Soil Depths Together and Comparing Models
Instead of fitting separate models for different soil depths, we first fit a **combined model** where all observations share a **common slope, intercept, and plateau**. We then test whether modeling the two depths separately provides a significantly better fit.

1. **Fitting a Reduced (Combined) Model**

The reduced model assumes that **all soil depths follow the same nonlinear relationship** (Fig. 3.31).

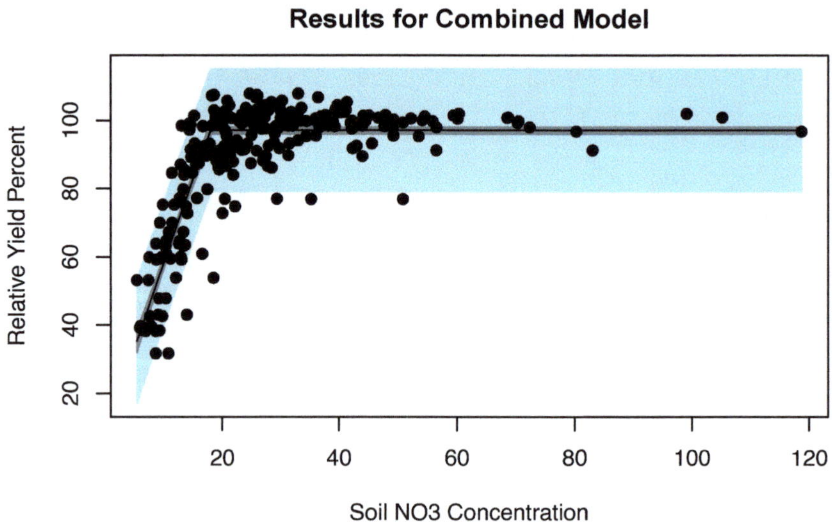

Fig. 3.31 Results for combined model

```
# Fit the combined model (common parameters across all depths)
red_nls <- nls(
  ryp ~ SS_nonlinModel(predictor = no3, b0, b1, alpha),
  data = dat
)
#> Warning in getInitial.selfStart(func, data, mCall =
#> as.list(match.call(func, : selfStart initializing functions
#> should have a final '...' argument since R 4.1.0

# Display model summary
summary(red_nls)
#>
#> Formula: ryp ~ SS_nonlinModel(predictor = no3, b0, b1,
↪  alpha)
#>
#> Parameters:
#>       Estimate Std. Error t value Pr(>|t|)
#> b0      8.7901     2.7688   3.175   0.0016 **
#> b1      4.8995     0.2207  22.203   <2e-16 ***
#> alpha  18.0333     0.3242  55.630   <2e-16 ***
#> ---
#> Signif. codes:
#> 0 '***' 0.001 '**' 0.01 '*' 0.05 '.' 0.1 ' ' 1
#>
#> Residual standard error: 9.13 on 477 degrees of freedom
#>
#> Number of iterations to convergence: 7
#> Achieved convergence tolerance: 8.568e-09
```

```
# Visualizing the combined model fit
par(mfrow = c(1, 1))
plotFit(
  red_nls,
  interval = "both",
  pch = 19,
  shade = TRUE,
  col.conf = "lightblue4",
  col.pred = "lightblue2",
  data = dat,
  main = 'Results for Combined Model',
  ylab = 'Relative Yield Percent',
```

```
  xlab = 'Soil NO3 Concentration'
)
```

2. **Examining Residuals for the Combined Model**

Checking residuals helps diagnose potential **lack of fit**.

```
library(nlstools)
#>
#> 'nlstools' has been loaded.
#> IMPORTANT NOTICE: Most nonlinear regression models and data
↪  set examples
#> related to predictive microbiolgy have been moved to the
↪  package 'nlsMicrobio'

# Residual diagnostics using nlstools
resid <- nlsResiduals(red_nls)
```

```
# Plot residuals
plot(resid)
```

If there is a **pattern** in the residuals (e.g., systematic deviations based on soil depth), this suggests that a **separate model for each depth** may be necessary (Fig. 3.32).

3. **Testing Whether Depths Require Separate Models**

To formally test whether soil depth significantly affects the model parameters, we introduce a **parameterization where depth-specific parameters are increments from a baseline model** (30 cm depth):

$$\beta_{02} = \beta_{01} + d_0$$

$$\beta_{12} = \beta_{11} + d_1$$

$$\alpha_2 = \alpha_1 + d_a$$

Where:

- $\beta_{01}, \beta_{11}, \alpha_1$ are parameters for **30 cm depth**.
- d_0, d_1, d_a represent **depth-specific differences** for **60 cm depth**.
- If d_0, d_1, d_a are significantly **different from 0**, the **two depths should be modeled separately**.

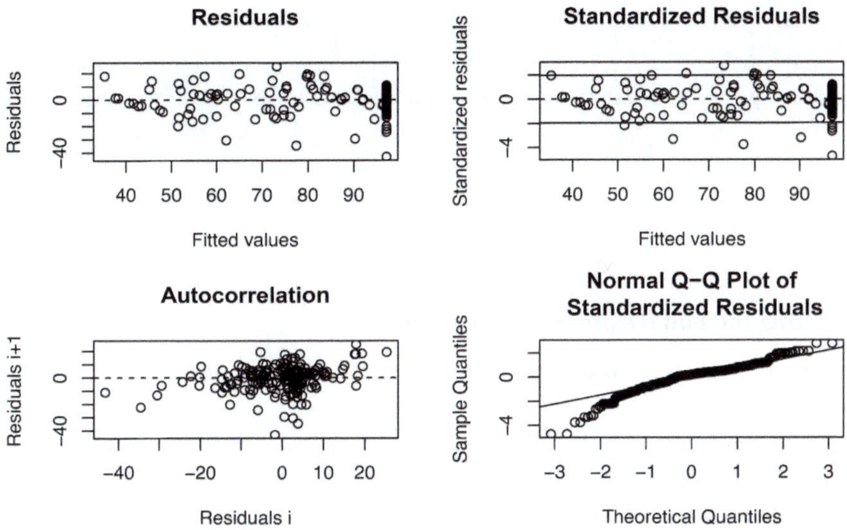

Fig. 3.32 Residual plot nonlinear example

4. **Defining the Full (Depth-Specific) Model**

```
nonlinModelF <- function(predictor, soildep, b01, b11, a1, d0,
↪  d1, da) {

  # Define parameters for 60 cm depth as increments from 30 cm
  ↪  parameters
  b02 <- b01 + d0
  b12 <- b11 + d1
  a2 <- a1 + da

  # Compute model output for 30 cm depth
  y1 <- ifelse(
    predictor <= a1,
    b01 + b11 * predictor,
    b01 + b11 * a1
  )

  # Compute model output for 60 cm depth
  y2 <- ifelse(
    predictor <= a2,
    b02 + b12 * predictor,
    b02 + b12 * a2
  )
```

```
  # Assign correct model output based on depth
  y <- y1 * (soildep == 30) + y2 * (soildep == 60)

  return(y)
}
```

5. **Fitting the Full (Depth-Specific) Model**

The starting values are taken from the **separately fitted models** for each depth.

```
Soil_full <- nls(
  ryp ~ nonlinModelF(
    predictor = no3,
    soildep = depth,
    b01,
    b11,
    a1,
    d0,
    d1,
    da
  ),
  data = dat,
  start = list(
    b01 = 15.2,     # Intercept for 30 cm depth
    b11 = 3.58,     # Slope for 30 cm depth
    a1 = 23.13,     # Plateau cutoff for 30 cm depth
    d0 = -9.74,     # Intercept difference (60 cm - 30 cm)
    d1 = 2.11,      # Slope difference (60 cm - 30 cm)
    da = -6.85      # Plateau cutoff difference (60 cm - 30 cm)
  )
)

# Display model summary
summary(Soil_full)
#>
#> Formula: ryp ~ nonlinModelF(predictor = no3, soildep =
↪  depth, b01, b11,
#>     a1, d0, d1, da)
#>
#> Parameters:
#>     Estimate Std. Error t value Pr(>|t|)
```

```
#> b01   15.1943     2.8322    5.365 1.27e-07 ***
#> b11    3.5760     0.1762   20.291  < 2e-16 ***
#> a1    23.1324     0.4848   47.711  < 2e-16 ***
#> d0    -9.7424     4.2357   -2.300   0.0219 *
#> d1     2.1060     0.3203    6.575 1.29e-10 ***
#> da    -6.8461     0.5691  -12.030  < 2e-16 ***
#> ---
#> Signif. codes:
#> 0 '***' 0.001 '**' 0.01 '*' 0.05 '.' 0.1 ' ' 1
#>
#> Residual standard error: 7.854 on 474 degrees of freedom
#>
#> Number of iterations to convergence: 1
#> Achieved convergence tolerance: 3.742e-06
```

6. **Model Comparison: Does Depth Matter?**

- If d_0, d_1, d_a **are significantly different from 0**, the depths should be modeled separately.
- The **p-values** for these parameters indicate whether depth-specific modeling is necessary.

Chapter 4
Generalized Linear Models

Generalized linear models (GLMs) extend the traditional linear regression framework to accommodate response variables that do not necessarily follow a normal distribution. They provide a flexible approach to modeling relationships between a set of predictors and various types of dependent variables.

While Ordinary Least Squares regression assumes that the response variable is continuous and normally distributed, GLMs allow for response variables that follow distributions from the exponential family, such as binomial, Poisson, and gamma distributions. This flexibility makes them particularly useful in a wide range of business and research applications.

A GLM consists of three key components:

1. **A random component**: The response variable Y_i follows a distribution from the exponential family (e.g., binomial, Poisson, gamma).
2. **A systematic component**: A linear predictor $\eta_i = \mathbf{x_i'}\beta$, where $\mathbf{x_i'}$ is a vector of observed covariates (predictor variables) and β is a vector of parameters to be estimated.
3. **A link function**: A function $g(\cdot)$ that relates the expected value of the response variable, $\mu_i = E(Y_i)$, to the linear predictor (i.e., $\eta_i = g(\mu_i)$).

Although the relationship between the predictors and the outcome *may appear nonlinear* on the original outcome scale (due to the link function), a GLM is still considered "linear" in the statistical sense because it remains linear in the parameters β. Consequently, GLMs are **not** generally classified as nonlinear regression models. They "generalize" the traditional linear model by allowing for a broader range of response variable distributions and link functions but retain linearity in their parameters.

The choice of **distribution** and **link function** depends on the nature of the response variable. In the following sections, we will explore several important GLM variants:

M. Nguyen, *Regression Techniques for Data Analysis*,
https://doi.org/10.1007/978-3-032-01835-9_4

- Logistic regression: Used for binary response variables (e.g., customer churn, loan defaults).
- Probit regression: Similar to logistic regression but assumes a normal distribution for the underlying probability.
- Poisson regression: Used for modeling count data (e.g., number of purchases, call center inquiries).
- Negative binomial regression: An extension of Poisson regression that accounts for overdispersion in count data.
- Quasi-Poisson regression: A variation of Poisson regression that adjusts for overdispersion by allowing the variance to be a linear function of the mean.
- Multinomial logistic regression: A generalization of logistic regression for categorical response variables with more than two outcomes.
- Generalization of generalized linear model: A flexible generalization of ordinary linear regression that allows for response variables with different distributions (e.g., normal, binomial, Poisson).

4.1 Logistic Regression

Logistic regression is a widely used generalized linear model designed for modeling binary response variables. It is particularly useful in applications such as credit scoring, medical diagnosis, and customer churn prediction.

4.1.1 Logistic Model

Given a set of predictor variables $\mathbf{x}_i$, the probability of a positive outcome (e.g., success, event occurring) is modeled as:

$$p_i = f(\mathbf{x}_i; \beta) = \frac{\exp(\mathbf{x_i}'\boldsymbol{\beta})}{1 + \exp(\mathbf{x_i}'\boldsymbol{\beta})}$$

where:

- $p_i = \mathbb{E}[Y_i]$ is the probability of success for observation i.
- $\mathbf{x_i}$ is the vector of predictor variables.
- β is the vector of model coefficients.

4.1.1.1 Logit Transformation

The logistic function can be rewritten in terms of the **log odds**, also known as the **logit function**:

$$\text{logit}(p_i) = \log\left(\frac{p_i}{1-p_i}\right) = \mathbf{x_i'}\boldsymbol{\beta}$$

where:

- $\frac{p_i}{1-p_i}$ represents the **odds** of success (the ratio of the probability of success to the probability of failure).
- The logit function **ensures linearity** in the parameters, which aligns with the GLM framework.

Thus, logistic regression belongs to the family of **generalized linear models** because **a function of the mean response (logit) is linear in the predictors**.

4.1.2 Likelihood Function

Since Y_i follows a **Bernoulli distribution** with probability p_i, the likelihood function for n independent observations is:

$$L(p_i) = \prod_{i=1}^{n} p_i^{Y_i}(1-p_i)^{1-Y_i}$$

By substituting the logistic function for p_i:

$$p_i = \frac{\exp(\mathbf{x_i'}\boldsymbol{\beta})}{1+\exp(\mathbf{x_i'}\boldsymbol{\beta})}, \quad 1-p_i = \frac{1}{1+\exp(\mathbf{x_i'}\boldsymbol{\beta})}$$

we obtain:

$$L(\beta) = \prod_{i=1}^{n}\left(\frac{\exp(\mathbf{x_i'}\boldsymbol{\beta})}{1+\exp(\mathbf{x_i'}\boldsymbol{\beta})}\right)^{Y_i}\left(\frac{1}{1+\exp(\mathbf{x_i'}\boldsymbol{\beta})}\right)^{1-Y_i}$$

Taking the natural logarithm of the likelihood function gives the **log-likelihood function**:

$$Q(\beta) = \log L(\beta) = \sum_{i=1}^{n} Y_i\mathbf{x_i'}\boldsymbol{\beta} - \sum_{i=1}^{n}\log(1+\exp(\mathbf{x_i'}\boldsymbol{\beta}))$$

Since this function is **concave**, we can maximize it numerically using **iterative optimization techniques**, such as:

- Newton-Raphson Algorithm
- **Fisher-Scoring Algorithm**

These methods allow us to obtain the Maximum Likelihood Estimates of the parameters, $\hat{\beta}$.

Under standard regularity conditions, the **MLEs of logistic regression parameters are asymptotically normal**:

$$\hat{\beta} \dot{\sim} AN(\beta, [\mathbf{I}(\beta)]^{-1})$$

where:

- $\mathbf{I}(\beta)$ is the Fisher information matrix, which determines the variance-covariance structure of $\hat{\beta}$.

4.1.3 Fisher Information Matrix

The **Fisher Information Matrix** quantifies the amount of information that an observable random variable carries about the **unknown parameter** β. It is crucial in estimating the **variance-covariance matrix** of the estimated coefficients in logistic regression.

Mathematically, the Fisher information matrix is defined as:

$$\mathbf{I}(\beta) = E\left[\frac{\partial \log L(\beta)}{\partial \beta}\frac{\partial \log L(\beta)}{\partial \beta'}\right]$$

which expands to:

$$\mathbf{I}(\beta) = E\left[\left(\frac{\partial \log L(\beta)}{\partial \beta_i}\frac{\partial \log L(\beta)}{\partial \beta_j}\right)_{ij}\right]$$

Under **regularity conditions**, the Fisher information matrix is equivalent to the **negative expected Hessian matrix**:

$$\mathbf{I}(\beta) = -E\left[\frac{\partial^2 \log L(\beta)}{\partial \beta \partial \beta'}\right]$$

which further expands to:

$$\mathbf{I}(\beta) = -E\left[\left(\frac{\partial^2 \log L(\beta)}{\partial \beta_i \partial \beta_j}\right)_{ij}\right]$$

This representation is particularly useful because it allows us to compute the Fisher information matrix directly from the Hessian of the log-likelihood function.

Example (Fisher Information Matrix in Logistic Regression) Consider a **simple logistic regression model** with one predictor:

$$x_i'\beta = \beta_0 + \beta_1 x_i$$

From the log-likelihood function, the second-order partial derivatives are:

$$-\frac{\partial^2 \ln(L(\beta))}{\partial \beta_0^2} = \sum_{i=1}^{n} \frac{\exp(x_i'\beta)}{1+\exp(x_i'\beta)} - \left[\frac{\exp(x_i'\beta)}{1+\exp(x_i'\beta)}\right]^2 \qquad \text{Intercept}$$

$$= \sum_{i=1}^{n} p_i(1-p_i)$$

$$-\frac{\partial^2 \ln(L(\beta))}{\partial \beta_1^2} = \sum_{i=1}^{n} \frac{x_i^2 \exp(x_i'\beta)}{1+\exp(x_i'\beta)} - \left[\frac{x_i \exp(x_i'\beta)}{1+\exp(x_i'\beta)}\right]^2 \qquad \text{Slope}$$

$$= \sum_{i=1}^{n} x_i^2 p_i(1-p_i)$$

$$-\frac{\partial^2 \ln(L(\beta))}{\partial \beta_0 \partial \beta_1} = \sum_{i=1}^{n} \frac{x_i \exp(x_i'\beta)}{1+\exp(x_i'\beta)} - x_i\left[\frac{\exp(x_i'\beta)}{1+\exp(x_i'\beta)}\right]^2 \qquad \text{Cross-derivative}$$

$$= \sum_{i=1}^{n} x_i p_i(1-p_i)$$

Combining these elements, the **Fisher information matrix** for the logistic regression model is:

$$\mathbf{I}(\beta) = \begin{bmatrix} \sum_{i=1}^{n} p_i(1-p_i) & \sum_{i=1}^{n} x_i p_i(1-p_i) \\ \sum_{i=1}^{n} x_i p_i(1-p_i) & \sum_{i=1}^{n} x_i^2 p_i(1-p_i) \end{bmatrix}$$

where:

- $p_i = \frac{\exp(x_i'\beta)}{1+\exp(x_i'\beta)}$ represents the predicted probability.
- $p_i(1-p_i)$ is the **variance of the Bernoulli response variable**.
- The diagonal elements represent the variances of the estimated coefficients.
- The off-diagonal elements represent the covariances between β_0 and β_1.

The inverse of the Fisher information matrix provides the **variance-covariance matrix** of the estimated coefficients:

$$\mathbf{Var}(\hat{\beta}) = \mathbf{I}(\hat{\beta})^{-1}$$

This matrix is essential for:

- **Estimating standard errors** of the logistic regression coefficients.
- **Constructing confidence intervals** for β.
- **Performing hypothesis tests** (e.g., Wald test).

```
# Load necessary library
library(stats)

# Simulated dataset
set.seed(123)
n <- 100
x <- rnorm(n)
y <- rbinom(n, 1, prob = plogis(0.5 + 1.2 * x))

# Fit logistic regression model
model <- glm(y ~ x, family = binomial)

# Extract the Fisher Information Matrix (Negative Hessian)
fisher_info <- summary(model)$cov.unscaled

# Display the Fisher Information Matrix
print(fisher_info)
#>             (Intercept)          x
#> (Intercept)  0.05718171 0.01564322
#> x            0.01564322 0.10302992
```

4.1.4 *Inference in Logistic Regression*

Once we estimate the model parameters $\hat{\beta}$ using Maximum Likelihood Estimation, we can conduct inference to assess the significance of predictors, construct confidence intervals, and perform hypothesis testing. The two most common inference approaches in logistic regression are:

1. Likelihood ratio test
2. Wald statistics

These tests rely on the **asymptotic normality** of MLEs and the properties of the Fisher information matrix.

4.1.4.1 Likelihood Ratio Test

The **likelihood ratio test** compares two models:

- **Restricted model**: A simpler model where some parameters are constrained to specific values.
- **Unrestricted model**: The full model without constraints.

To test a hypothesis about a subset of parameters β_1, we leave β_2 (nuisance parameters) unspecified.

Hypothesis setup:

$$H_0 : \beta_1 = \beta_{1,0}$$

where $\beta_{1,0}$ is a specified value (often zero). Let:

- $\hat{\beta}_{2,0}$ be the **MLE of** β_2 under the constraint $\beta_1 = \beta_{1,0}$.
- $\hat{\beta}_1, \hat{\beta}_2$ be the **MLEs under the full model**.

The **likelihood ratio test statistic** is:

$$-2 \log \Lambda = -2[\log L(\beta_{1,0}, \hat{\beta}_{2,0}) - \log L(\hat{\beta}_1, \hat{\beta}_2)]$$

where:

- The first term is the **log-likelihood of the restricted model**.
- The second term is the **log-likelihood of the unrestricted model**.

Under the null hypothesis:

$$-2 \log \Lambda \sim \chi^2_{\upsilon}$$

where υ is the number of restricted parameters. We **reject** H_0 if:

$$-2 \log \Lambda > \chi^2_{\upsilon,1-\alpha}$$

Interpretation If the likelihood ratio test statistic is large, this suggests that the restricted model (under H_0) fits significantly worse than the full model, leading us to reject the null hypothesis.

4.1.4.2 Wald Test

The **Wald test** is based on the asymptotic normality of MLEs:

$$\hat{\beta} \sim AN(\beta, [\mathbf{I}(\beta)]^{-1})$$

We test:

$$H_0 : \mathbf{L}\hat{\beta} = 0$$

Table 4.1 Comparing likelihood ratio and Wald tests

Test	Best used when...
Likelihood ratio test	More accurate in small samples, providing better control of error rates. Recommended when sample sizes are small
Wald test	Easier to compute but may be inaccurate in small samples. Recommended when computational efficiency is a priority

where $\mathbf{L}$ is a $q \times p$ matrix with q linearly independent rows (often used to test multiple coefficients simultaneously). The **Wald test statistic** is:

$$W = (\mathbf{L}\hat{\boldsymbol{\beta}})'(\mathbf{L}[\mathbf{I}(\hat{\boldsymbol{\beta}})]^{-1}\mathbf{L}')^{-1}(\mathbf{L}\hat{\boldsymbol{\beta}})$$

Under H_0:

$$W \sim \chi_q^2$$

Interpretation If W is large, the null hypothesis is rejected, suggesting that at least one of the tested coefficients is significantly different from zero (Table 4.1).

```
# Load necessary library
library(stats)

# Simulate some binary outcome data
set.seed(123)
n <- 100
x <- rnorm(n)
y <- rbinom(n, 1, prob = plogis(0.5 + 1.2 * x))

# Fit logistic regression model
model <- glm(y ~ x, family = binomial)

# Display model summary (includes Wald tests)
summary(model)
#>
#> Call:
#> glm(formula = y ~ x, family = binomial)
#>
#> Coefficients:
#>             Estimate Std. Error z value Pr(>|z|)
#> (Intercept)   0.7223     0.2391   3.020 0.002524 **
#> x             1.2271     0.3210   3.823 0.000132 ***
#> ---
#> Signif. codes:
#> 0 '***' 0.001 '**' 0.01 '*' 0.05 '.' 0.1 ' ' 1
```

```
#>
#> (Dispersion parameter for binomial family taken to be 1)
#>
#>     Null deviance: 128.21  on 99  degrees of freedom
#> Residual deviance: 108.29  on 98  degrees of freedom
#> AIC: 112.29
#>
#> Number of Fisher Scoring iterations: 4

# Perform likelihood ratio test using anova()
anova(model, test="Chisq")
#> Analysis of Deviance Table
#>
#> Model: binomial, link: logit
#>
#> Response: y
#>
#> Terms added sequentially (first to last)
#>
#>
#>      Df Deviance Resid. Df Resid. Dev  Pr(>Chi)
#> NULL                    99     128.21
#> x     1   19.913        98     108.29 8.105e-06 ***
#> ---
#> Signif. codes:
#> 0 '***' 0.001 '**' 0.01 '*' 0.05 '.' 0.1 ' ' 1
```

4.1.4.3 Confidence Intervals for Coefficients

A **95% confidence interval** for a logistic regression coefficient β_i is given by:

$$\hat{\beta}_i \pm 1.96\hat{s}_{ii}$$

where:

- $\hat{\beta}_i$ is the estimated coefficient.
- $\hat{s}_{ii}$ is the standard error (square root of the diagonal element of $[\mathbf{I}(\hat{\boldsymbol{\beta}})]^{-1}$).

This confidence interval provides a **range of plausible values** for β_i. If the interval does not include **zero**, we conclude that β_i is statistically significant.

- **For large sample sizes**, the likelihood ratio test and Wald test yield similar results.

- **For small sample sizes**, the likelihood ratio test is preferred because the Wald test can be less reliable.

4.1.4.4 Interpretation of Logistic Regression Coefficients

For a **single predictor variable**, the logistic regression model is:

$$\text{logit}(\hat{p}_i) = \log\left(\frac{\hat{p}_i}{1-\hat{p}_i}\right) = \hat{\beta}_0 + \hat{\beta}_1 x_i$$

where:

- $\hat{p}_i$ is the predicted probability of success at x_i.
- $\hat{\beta}_1$ represents the **log-odds change** for a one-unit increase in x.

Interpreting β_1 in Terms of Odds
When the predictor variable increases by **one unit**, the logit of the probability changes by $\hat{\beta}_1$:

$$\text{logit}(\hat{p}_{x_i+1}) = \hat{\beta}_0 + \hat{\beta}_1(x_i + 1) = \text{logit}(\hat{p}_{x_i}) + \hat{\beta}_1$$

Thus, the difference in log odds is:

$$\begin{aligned}\text{logit}(\hat{p}_{x_i+1}) - \text{logit}(\hat{p}_{x_i}) &= \log(\text{odds}(\hat{p}_{x_i+1})) - \log(\text{odds}(\hat{p}_{x_i}))\\ &= \log\left(\frac{\text{odds}(\hat{p}_{x_i+1})}{\text{odds}(\hat{p}_{x_i})}\right)\\ &= \hat{\beta}_1\end{aligned}$$

Exponentiating both sides:

$$\exp(\hat{\beta}_1) = \frac{\text{odds}(\hat{p}_{x_i+1})}{\text{odds}(\hat{p}_{x_i})}$$

This quantity, $\exp(\hat{\beta}_1)$, is the **odds ratio**, which quantifies the effect of a one-unit increase in x on the odds of success.

Generalization: Odds Ratio for Any Change in x
For a difference of c units in the predictor x, the estimated odds ratio is:

$$\exp(c\hat{\beta}_1)$$

For multiple predictors, $\exp(\hat{\beta}_k)$ represents the odds ratio for x_k, holding all other variables constant.

4.1.4.5 Inference on the Mean Response

For a given set of predictor values $x_h = (1, x_{h1}, ..., x_{h,p-1})'$, the estimated **mean response** (probability of success) is:

$$\hat{p}_h = \frac{\exp(\mathbf{x_h'}\hat{\boldsymbol{\beta}})}{1 + \exp(\mathbf{x_h'}\hat{\boldsymbol{\beta}})}$$

The **variance of the estimated probability** is:

$$s^2(\hat{p}_h) = \mathbf{x_h'}[\mathbf{I}(\hat{\boldsymbol{\beta}})]^{-1}\mathbf{x_h}$$

where:

- $\mathbf{I}(\hat{\beta})^{-1}$ is the **variance-covariance matrix** of $\hat{\beta}$.
- $s^2(\hat{p}_h)$ provides an estimate of **uncertainty** in $\hat{p}_h$.

In many applications, logistic regression is used for **classification**, where we predict whether an observation belongs to **category 0 or 1**. A commonly used decision rule is:

- Assign $y = 1$ if $\hat{p}_h \geq \tau$
- Assign $y = 0$ if $\hat{p}_h < \tau$

where τ is a chosen cutoff threshold (typically $\tau = 0.5$).

```
# Load necessary library
library(stats)

# Simulated dataset
set.seed(123)
n <- 100
x <- rnorm(n)
y <- rbinom(n, 1, prob = plogis(0.5 + 1.2 * x))

# Fit logistic regression model
model <- glm(y ~ x, family = binomial)

# Display model summary
summary(model)
#>
#> Call:
#> glm(formula = y ~ x, family = binomial)
#>
#> Coefficients:
#>             Estimate Std. Error z value Pr(>|z|)
```

```
#> (Intercept)   0.7223     0.2391   3.020 0.002524 **
#> x             1.2271     0.3210   3.823 0.000132 ***
#> ---
#> Signif. codes:
#> 0 '***' 0.001 '**' 0.01 '*' 0.05 '.' 0.1 ' ' 1
#>
#> (Dispersion parameter for binomial family taken to be 1)
#>
#>     Null deviance: 128.21  on 99  degrees of freedom
#> Residual deviance: 108.29  on 98  degrees of freedom
#> AIC: 112.29
#>
#> Number of Fisher Scoring iterations: 4

# Extract coefficients and standard errors
coef_estimates <- coef(summary(model))
beta_hat <- coef_estimates[, 1]  # Estimated coefficients
se_beta  <- coef_estimates[, 2]  # Standard errors

# Compute 95% confidence intervals for coefficients
conf_intervals <- cbind(
  beta_hat - 1.96 * se_beta,
  beta_hat + 1.96 * se_beta
)

# Compute Odds Ratios
odds_ratios <- exp(beta_hat)

# Display results
print("Confidence Intervals for Coefficients:")
#> [1] "Confidence Intervals for Coefficients:"
print(conf_intervals)
#>                  [,1]     [,2]
#> (Intercept) 0.2535704 1.190948
#> x           0.5979658 1.856218

print("Odds Ratios:")
#> [1] "Odds Ratios:"
print(odds_ratios)
#> (Intercept)           x
#>    2.059080    3.411295

# Predict probability for a new observation (e.g., x = 1)
new_x <- data.frame(x = 1)
```

```
predicted_prob <- predict(model, newdata = new_x, type =
  ↪  "response")

print("Predicted Probability for x = 1:")
#> [1] "Predicted Probability for x = 1:"
print(predicted_prob)
#>         1
#> 0.8753759
```

4.1.5 *Application: Logistic Regression*

In this section, we demonstrate the application of **logistic regression** using simulated data. We explore model fitting, inference, residual analysis, and goodness-of-fit testing.

1. Load Required Libraries

```
library(kableExtra)
library(dplyr)
#>
#> Attaching package: 'dplyr'
#> The following object is masked from 'package:kableExtra':
#>
#>     group_rows
#> The following objects are masked from 'package:stats':
#>
#>     filter, lag
#> The following objects are masked from 'package:base':
#>
#>     intersect, setdiff, setequal, union
library(pscl)
#> Classes and Methods for R originally developed in the
#> Political Science Computational Laboratory
#> Department of Political Science
#> Stanford University (2002-2015),
#> by and under the direction of Simon Jackman.
#> hurdle and zeroinfl functions by Achim Zeileis.
library(ggplot2)
library(faraway)
library(nnet)
library(agridat)
```

```
library(nlstools)
#>
#> 'nlstools' has been loaded.
#> IMPORTANT NOTICE: Most nonlinear regression models and data
↪  set examples
#> related to predictive microbiolgy have been moved to the
↪  package 'nlsMicrobio'
```

2. Data Generation
We generate a dataset where the predictor variable X follows a **uniform distribution**:

$$x \sim Unif(-0.5, 2.5)$$

The **linear predictor** is given by:

$$\eta = 0.5 + 0.75x$$

Passing η into the inverse-logit function, we obtain:

$$p = \frac{\exp(\eta)}{1 + \exp(\eta)}$$

which ensures that $p \in [0, 1]$. We then generate the binary response variable:

$$y \sim Bernoulli(p)$$

```
set.seed(23) # Set seed for reproducibility
x <- runif(1000, min = -0.5, max = 2.5)  # Generate X values
eta1 <- 0.5 + 0.75 * x                   # Compute linear
↪  predictor
p <- exp(eta1) / (1 + exp(eta1))          # Compute
↪  probabilities
y <- rbinom(1000, 1, p)                   # Generate binary
↪  response
BinData <- data.frame(X = x, Y = y)       # Create data frame
```

3. Model Fitting
We fit a **logistic regression model** to the simulated data:

$$\text{logit}(p) = \beta_0 + \beta_1 X$$

```
Logistic_Model <- glm(formula = Y ~ X,
                      # Specifies the response distribution
                      family = binomial,
                      data = BinData)

summary(Logistic_Model) # Model summary
#>
#> Call:
#> glm(formula = Y ~ X, family = binomial, data = BinData)
#>
#> Coefficients:
#>             Estimate Std. Error z value Pr(>|z|)
#> (Intercept)  0.46205    0.10201   4.530 5.91e-06 ***
#> X            0.78527    0.09296   8.447  < 2e-16 ***
#> ---
#> Signif. codes:
#> 0 '***' 0.001 '**' 0.01 '*' 0.05 '.' 0.1 ' ' 1
#>
#> (Dispersion parameter for binomial family taken to be 1)
#>
#>     Null deviance: 1106.7  on 999  degrees of freedom
#> Residual deviance: 1027.4  on 998  degrees of freedom
#> AIC: 1031.4
#>
#> Number of Fisher Scoring iterations: 4
nlstools::confint2(Logistic_Model) # Confidence intervals
#>                 2.5 %    97.5 %
#> (Intercept) 0.2618709 0.6622204
#> X           0.6028433 0.9676934

# Compute odds ratios
OddsRatio <- coef(Logistic_Model) %>% exp
OddsRatio
#> (Intercept)           X
#>    1.587318    2.192995
```

Interpretation of the Odds Ratio

- When $x = 0$, the **odds of success** are **1.59**.
- When $x = 1$, the **odds of success increase by a factor of 2.19**, indicating a **119.29% increase**.

4. Deviance Test

We assess the model's significance using the **deviance test**, which compares:

- H_0: No predictors are related to the response (intercept-only model).
- H_1: At least one predictor is related to the response.

The **test statistic** is:

$$D = \text{Null Deviance} - \text{Residual Deviance}$$

```
Test_Dev <- Logistic_Model$null.deviance -
↪  Logistic_Model$deviance
p_val_dev <- 1 - pchisq(q = Test_Dev, df = 1)
p_val_dev
#> [1] 0
```

Since the **p-value is approximately 0**, we **reject** H_0, confirming that X is significantly related to Y.

5. Residual Analysis

We compute **deviance residuals** and plot them against X.

```
Logistic_Resids <- residuals(Logistic_Model, type =
↪  "deviance")

plot(
    y = Logistic_Resids,
    x = BinData$X,
    xlab = 'X',
    ylab = 'Deviance Residuals'
)
```

This plot is not very informative. A more insightful approach is **binned residual plots** (Fig. 4.1).

6. Binned Residual Plot

We group residuals into **bins** based on predicted values (Fig. 4.2).

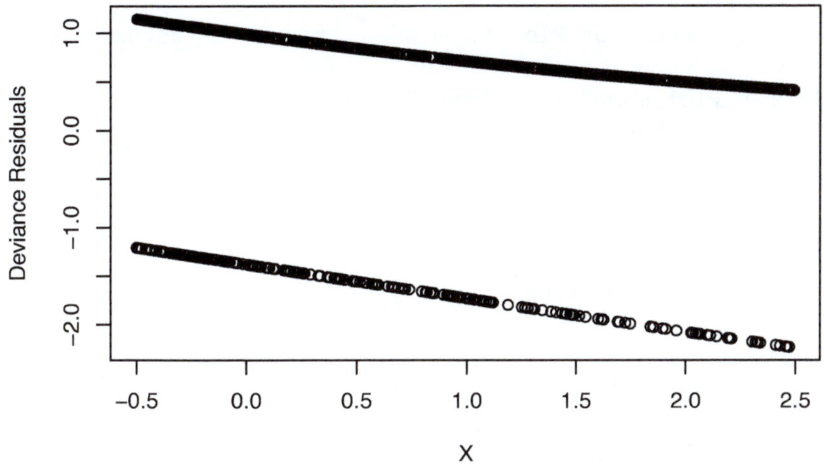

Fig. 4.1 Deviance residuals

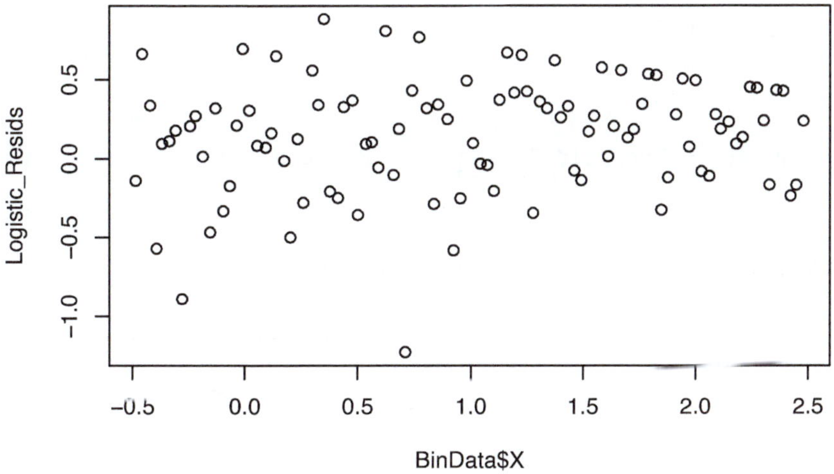

Fig. 4.2 Binned residual plot

```
plot_bin <- function(Y,
                     X,
                     bins = 100,
                     return.DF = FALSE) {
  Y_Name <- deparse(substitute(Y))
  X_Name <- deparse(substitute(X))

  Binned_Plot <- data.frame(Plot_Y = Y, Plot_X = X)
  Binned_Plot$bin <-
```

```
    cut(Binned_Plot$Plot_X, breaks = bins) %>% as.numeric

  Binned_Plot_summary <- Binned_Plot %>%
    group_by(bin) %>%
    summarise(
      Y_ave = mean(Plot_Y),
      X_ave = mean(Plot_X),
      Count = n()
    ) %>% as.data.frame

  plot(
    y = Binned_Plot_summary$Y_ave,
    x = Binned_Plot_summary$X_ave,
    ylab = Y_Name,
    xlab = X_Name
  )

  if (return.DF)
    return(Binned_Plot_summary)
}

plot_bin(Y = Logistic_Resids, X = BinData$X, bins = 100)
```

We also examine **predicted values vs residuals** (Fig. 4.3):

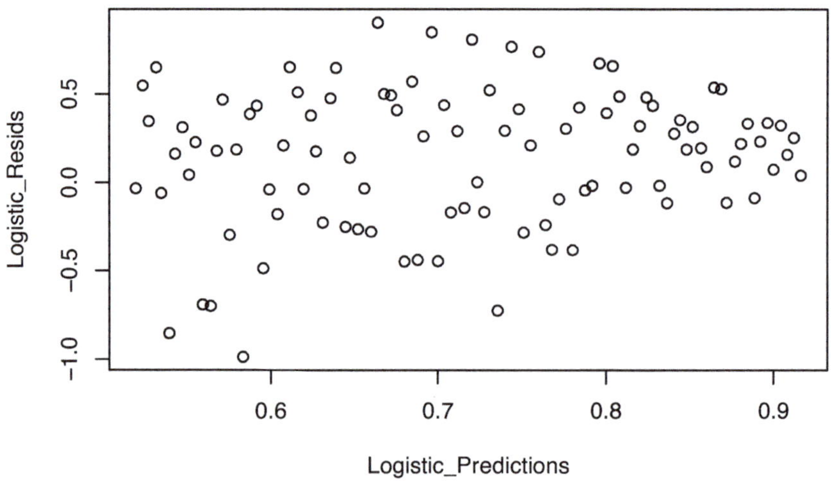

Fig. 4.3 Predicted values versus residuals

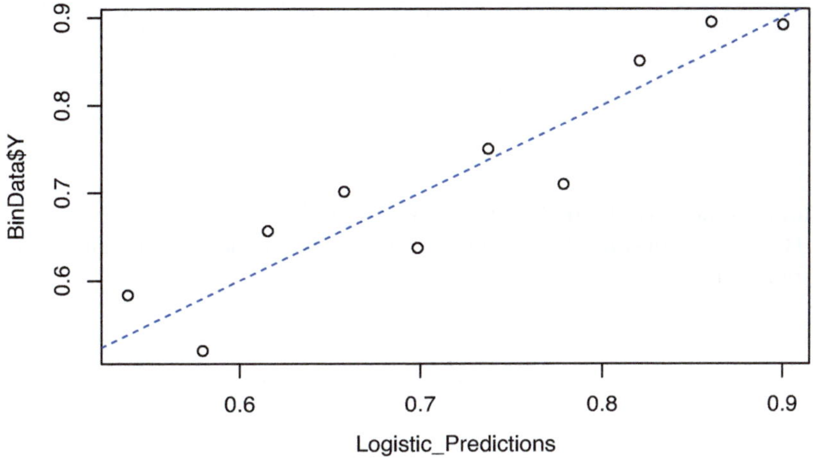

Fig. 4.4 Predicted versus actual values

```
Logistic_Predictions <- predict(Logistic_Model, type =
 ↪  "response")
plot_bin(Y = Logistic_Resids, X = Logistic_Predictions, bins
 ↪  = 100)
```

Finally, we compare **predicted probabilities** to actual outcomes (Fig. 4.4):

```
NumBins <- 10
Binned_Data <- plot_bin(
    Y = BinData$Y,
    X = Logistic_Predictions,
    bins = NumBins,
    return.DF = TRUE
)
Binned_Data
#>     bin      Y_ave     X_ave Count
#> 1     1 0.5833333 0.5382095    72
#> 2     2 0.5200000 0.5795887    75
#> 3     3 0.6567164 0.6156540    67
#> 4     4 0.7014925 0.6579674    67
#> 5     5 0.6373626 0.6984765    91
#> 6     6 0.7500000 0.7373341    72
#> 7     7 0.7096774 0.7786747    93
```

```
#> 8      8 0.8503937 0.8203819    127
#> 9      9 0.8947368 0.8601232    133
#> 10    10 0.8916256 0.9004734    203
abline(0, 1, lty = 2, col = 'blue')
```

7. Model Goodness-of-Fit: Hosmer-Lemeshow Test

The **Hosmer-Lemeshow test** evaluates whether the model fits the data well. The test statistic is:

$$X^2_{HL} = \sum_{j=1}^{J} \frac{(y_j - m_j \hat{p}_j)^2}{m_j \hat{p}_j (1 - \hat{p}_j)}$$

where:

- y_j is the observed number of successes in bin j.
- m_j is the number of observations in bin j.
- $\hat{p}_j$ is the predicted probability in bin j.

Under H_0, we assume:

$$X^2_{HL} \sim \chi^2_{J-1}$$

```
HL_BinVals <- (Binned_Data$Count * Binned_Data$Y_ave -
                 Binned_Data$Count * Binned_Data$X_ave) ^ 2 /
                 (Binned_Data$Count * Binned_Data$X_ave * (1 -
↪  Binned_Data$X_ave))

HLpval <- pchisq(q = sum(HL_BinVals),
                 df = NumBins - 1,
                 lower.tail = FALSE)
HLpval
#> [1] 0.4150004
```

Conclusion

- Since p-value = 0.99, we **fail to reject** H_0.
- This indicates that **the model fits the data well**.

4.2 Probit Regression

Probit regression is a type of generalized linear models used for binary outcome variables. Unlike logistic regression, which uses the logit function, probit regression assumes that the probability of success is determined by an **underlying normally distributed latent variable**.

4.2.1 Probit Model

Let Y_i be a binary response variable:

$$Y_i = \begin{cases} 1, & \text{if success occurs} \\ 0, & \text{otherwise} \end{cases}$$

We assume that Y_i follows a **Bernoulli distribution**:

$$Y_i \sim \text{Bernoulli}(p_i), \quad \text{where } p_i = P(Y_i = 1|\mathbf{x_i})$$

Instead of the logit function in logistic regression:

$$\text{logit}(p_i) = \log\left(\frac{p_i}{1-p_i}\right) = \mathbf{x_i'}\boldsymbol{\beta}$$

Probit regression uses the **inverse standard normal CDF**:

$$\Phi^{-1}(p_i) = \mathbf{x_i'}\boldsymbol{\theta}$$

where:

- $\Phi(\cdot)$ is the **CDF of the standard normal distribution**.
- $\mathbf{x_i}$ is the vector of predictors.
- θ is the vector of regression coefficients.

Thus, the probability of success is:

$$p_i = P(Y_i = 1|\mathbf{x_i}) = \Phi(\mathbf{x_i'}\boldsymbol{\theta})$$

where:

$$\Phi(z) = \int_{-\infty}^{z} \frac{1}{\sqrt{2\pi}} e^{-t^2/2} dt$$

4.2.2 Application: Probit Regression

```
# Load necessary library
library(ggplot2)

# Set seed for reproducibility
set.seed(123)

# Simulate data
n <- 1000
x1 <- rnorm(n)
x2 <- rnorm(n)
latent <- 0.5 * x1 + 0.7 * x2 + rnorm(n)  # Linear combination
y <- ifelse(latent > 0, 1, 0)  # Binary outcome

# Create dataframe
data <- data.frame(y, x1, x2)

# Fit Probit model
probit_model <-
    glm(y ~ x1 + x2, family = binomial(link = "probit"), data
     ↪ = data)
summary(probit_model)
#>
#> Call:
#> glm(formula = y ~ x1 + x2, family = binomial(link =
↪  "probit"),
#>     data = data)
#>
#> Coefficients:
#>             Estimate Std. Error z value Pr(>|z|)
#> (Intercept) -0.09781    0.04499  -2.174   0.0297 *
#> x1           0.43838    0.04891   8.963   <2e-16 ***
#> x2           0.75538    0.05306  14.235   <2e-16 ***
#> ---
#> Signif. codes:
#> 0 '***' 0.001 '**' 0.01 '*' 0.05 '.' 0.1 ' ' 1
#>
#> (Dispersion parameter for binomial family taken to be 1)
#>
#>     Null deviance: 1385.1  on 999  degrees of freedom
#> Residual deviance: 1045.3  on 997  degrees of freedom
#> AIC: 1051.3
```

```
#>
#> Number of Fisher Scoring iterations: 5

# Fit Logit model
logit_model <-
    glm(y ~ x1 + x2, family = binomial(link = "logit"), data
    ↪  = data)
summary(logit_model)
#>
#> Call:
#> glm(formula = y ~ x1 + x2, family = binomial(link =
↪  "logit"),
#>     data = data)
#>
#> Coefficients:
#>             Estimate Std. Error z value Pr(>|z|)
#> (Intercept) -0.16562    0.07600  -2.179   0.0293 *
#> x1           0.73234    0.08507   8.608   <2e-16 ***
#> x2           1.25220    0.09486  13.201   <2e-16 ***
#> ---
#> Signif. codes:
#> 0 '***' 0.001 '**' 0.01 '*' 0.05 '.' 0.1 ' ' 1
#>
#> (Dispersion parameter for binomial family taken to be 1)
#>
#>     Null deviance: 1385.1  on 999  degrees of freedom
#> Residual deviance: 1048.4  on 997  degrees of freedom
#> AIC: 1054.4
#>
#> Number of Fisher Scoring iterations: 4

# Compare Coefficients
coef_comparison <- data.frame(
    Variable = names(coef(probit_model)),
    Probit_Coef = coef(probit_model),
    Logit_Coef = coef(logit_model),
    Logit_Probit_Ratio = coef(logit_model) /
    ↪  coef(probit_model)
)

print(coef_comparison)
#>                Variable Probit_Coef Logit_Coef
#> (Intercept) (Intercept) -0.09780689 -0.1656216
#> x1                   x1  0.43837627  0.7323392
```

```
#> x2                      x2  0.75538259  1.2522008
#>                 Logit_Probit_Ratio
#> (Intercept)               1.693353
#> x1                        1.670572
#> x2                        1.657704

# Compute predicted probabilities
data$probit_pred <- predict(probit_model, type = "response")
data$logit_pred <- predict(logit_model, type = "response")
```

```
# Plot Probit vs Logit predictions
ggplot(data, aes(x = probit_pred, y = logit_pred)) +
  geom_point(alpha = 0.5) +
  geom_abline(slope = 1,
              intercept = 0,
              col = "red") +
  labs(title = "Comparison of Predicted Probabilities",
       x = "Probit Predictions", y = "Logit Predictions")
```

See Fig. 4.5.

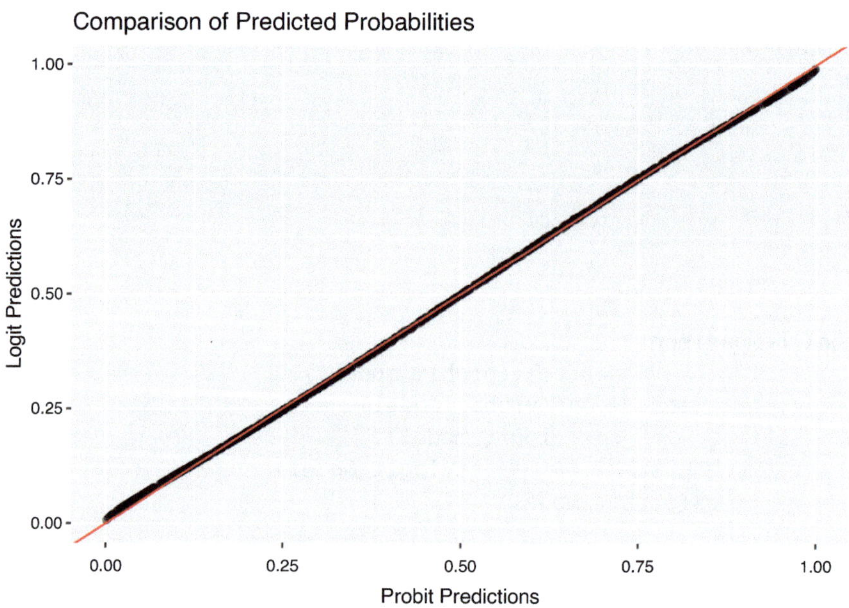

Fig. 4.5 Comparison of predicted probabilities

```
# Classification Accuracy
threshold <- 0.5
data$probit_class <- ifelse(data$probit_pred > threshold, 1,
↪  0)
data$logit_class <- ifelse(data$logit_pred > threshold, 1, 0)

probit_acc <- mean(data$probit_class == data$y)
logit_acc <- mean(data$logit_class == data$y)

print(paste("Probit Accuracy:", round(probit_acc, 4)))
#> [1] "Probit Accuracy: 0.71"
print(paste("Logit Accuracy:", round(logit_acc, 4)))
#> [1] "Logit Accuracy: 0.71"
```

4.3 Binomial Regression

In previous sections, we introduced **binomial regression models**, including both logistic regression and probit regression, and discussed their theoretical foundations. Now, we apply these methods to real-world data using the `esoph` dataset, which examines the relationship between esophageal cancer and potential risk factors such as **alcohol consumption** and **age group**.

4.3.1 Dataset Overview

The `esoph` dataset consists of (Fig. 4.6):

- **Successes (`ncases`)**: The number of individuals diagnosed with esophageal cancer
- **Failures (`ncontrols`)**: The number of individuals in the control group (without cancer)
- **Predictors**:
  - `agegp`: Age group of individuals
  - `alcgp`: Alcohol consumption category
  - `tobgp`: Tobacco consumption category

Before fitting our models, let's inspect the dataset and visualize some key relationships.

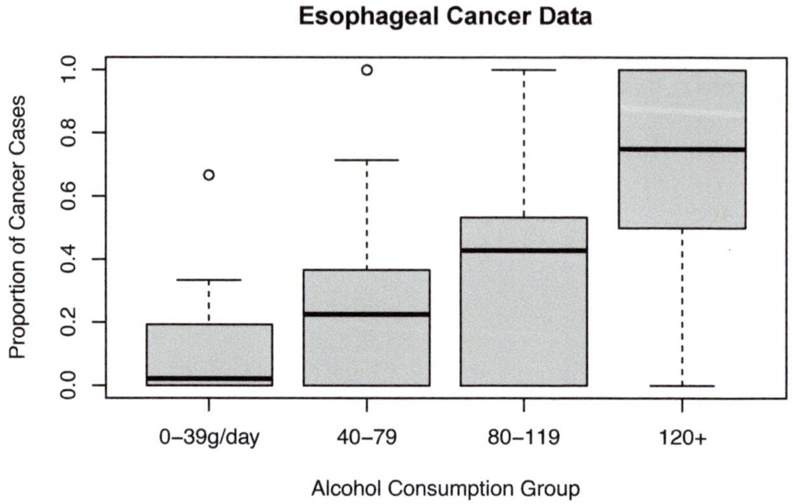

Fig. 4.6 Esophageal cancer data

```
# Load and inspect the dataset
data("esoph")
head(esoph, n = 3)
#>   agegp     alcgp    tobgp ncases ncontrols
#> 1 25-34 0-39g/day 0-9g/day      0        40
#> 2 25-34 0-39g/day    10-19      0        10
#> 3 25-34 0-39g/day    20-29      0         6
```

```
# Visualizing the proportion of cancer cases by alcohol
↪  consumption
plot(
  esoph$ncases / (esoph$ncases + esoph$ncontrols) ~
↪  esoph$alcgp,
  ylab = "Proportion of Cancer Cases",
  xlab = "Alcohol Consumption Group",
  main = "Esophageal Cancer Data"
)
```

```
# Ensure categorical variables are treated as factors
class(esoph$agegp) <- "factor"
class(esoph$alcgp) <- "factor"
class(esoph$tobgp) <- "factor"
```

4.3.2 Apply Logistic Model

We first fit a **logistic regression model**, where the response variable is the proportion of cancer cases (`ncases`) relative to total observations (`ncases + ncontrols`).

```
# Logistic regression using alcohol consumption as a predictor
model <-
    glm(cbind(ncases, ncontrols) ~ alcgp,
        data = esoph,
        family = binomial)

# Summary of the model
summary(model)
#>
#> Call:
#> glm(formula = cbind(ncases, ncontrols) ~ alcgp, family =
↪  binomial,
#>     data = esoph)
#>
#> Coefficients:
#>             Estimate Std. Error z value Pr(>|z|)
#> (Intercept)  -2.5885     0.1925 -13.444  < 2e-16 ***
#> alcgp40-79    1.2712     0.2323   5.472 4.46e-08 ***
#> alcgp80-119   2.0545     0.2611   7.868 3.59e-15 ***
#> alcgp120+     3.3042     0.3237  10.209  < 2e-16 ***
#> ---
#> Signif. codes:
#> 0 '***' 0.001 '**' 0.01 '*' 0.05 '.' 0.1 ' ' 1
#>
#> (Dispersion parameter for binomial family taken to be 1)
#>
#>     Null deviance: 367.95  on 87  degrees of freedom
#> Residual deviance: 221.46  on 84  degrees of freedom
#> AIC: 344.51
#>
#> Number of Fisher Scoring iterations: 5
```

Interpretation

- The coefficients represent the **log odds** of having esophageal cancer relative to the baseline alcohol consumption group.
- **P-values** indicate whether alcohol consumption levels significantly influence cancer risk.

Model Diagnostics

```
# Convert coefficients to odds ratios
exp(coefficients(model))
#> (Intercept)  alcgp40-79 alcgp80-119   alcgp120+
#>  0.07512953  3.56527094  7.80261593 27.22570533

# Model goodness-of-fit measures
deviance(model) / df.residual(model)  # Closer to 1 suggests a
↪  better fit
#> [1] 2.63638
model$aic  # Lower AIC is preferable for model comparison
#> [1] 344.5109
```

To improve our model, we include **age group (agegp)** as an additional predictor.

```
# Logistic regression with alcohol consumption and age
better_model <- glm(
    cbind(ncases, ncontrols) ~ agegp + alcgp,
    data = esoph,
    family = binomial
)

# Summary of the improved model
summary(better_model)
#>
#> Call:
#> glm(formula = cbind(ncases, ncontrols) ~ agegp + alcgp,
↪  family = binomial,
#>     data = esoph)
#>
#> Coefficients:
#>             Estimate Std. Error z value Pr(>|z|)
#> (Intercept)  -6.1472     1.0419  -5.900 3.63e-09 ***
#> agegp35-44    1.6311     1.0800   1.510 0.130973
#> agegp45-54    3.4258     1.0389   3.297 0.000976 ***
#> agegp55-64    3.9435     1.0346   3.811 0.000138 ***
#> agegp65-74    4.3568     1.0413   4.184 2.87e-05 ***
#> agegp75+      4.4242     1.0914   4.054 5.04e-05 ***
#> alcgp40-79    1.4343     0.2448   5.859 4.64e-09 ***
#> alcgp80-119   2.0071     0.2776   7.230 4.84e-13 ***
#> alcgp120+     3.6800     0.3763   9.778  < 2e-16 ***
#> ---
#> Signif. codes:
#> 0 '***' 0.001 '**' 0.01 '*' 0.05 '.' 0.1 ' ' 1
```

```
#>
#> (Dispersion parameter for binomial family taken to be 1)
#>
#>     Null deviance: 367.95  on 87  degrees of freedom
#> Residual deviance: 105.88  on 79  degrees of freedom
#> AIC: 238.94
#>
#> Number of Fisher Scoring iterations: 6

# Model evaluation
better_model$aic # Lower AIC is better
#> [1] 238.9361

# Convert coefficients to odds ratios
# exp(coefficients(better_model))
data.frame(`Odds Ratios` = exp(coefficients(better_model)))
#>              Odds.Ratios
#> (Intercept)  0.002139482
#> agegp35-44   5.109601844
#> agegp45-54  30.748594216
#> agegp55-64  51.596634690
#> agegp65-74  78.005283850
#> agegp75+    83.448437749
#> alcgp40-79   4.196747169
#> alcgp80-119  7.441782227
#> alcgp120+   39.646885126

# Compare models using likelihood ratio test (Chi-square test)
pchisq(
    q = model$deviance - better_model$deviance,
    df = model$df.residual - better_model$df.residual,
    lower.tail = FALSE
)
#> [1] 2.713923e-23
```

Key Takeaways

- **AIC reduction**: A lower AIC suggests that adding age as a predictor improves the model.
- **Likelihood ratio test**: This test compares the two models and determines whether the improvement is statistically significant.

4.3.3 Apply Probit Model

As discussed earlier, the **probit** model is an alternative to logistic regression, using a cumulative normal distribution instead of the logistic function.

```
# Probit regression model
Prob_better_model <- glm(
    cbind(ncases, ncontrols) ~ agegp + alcgp,
    data = esoph,
    family = binomial(link = probit)
)

# Summary of the probit model
summary(Prob_better_model)
#>
#> Call:
#> glm(formula = cbind(ncases, ncontrols) ~ agegp + alcgp,
↪   family = binomial(link = probit),
#>     data = esoph)
#>
#> Coefficients:
#>              Estimate Std. Error z value Pr(>|z|)
#> (Intercept)   -3.3741     0.4922  -6.855 7.13e-12 ***
#> agegp35-44     0.8562     0.5081   1.685 0.092003 .
#> agegp45-54     1.7829     0.4904   3.636 0.000277 ***
#> agegp55-64     2.1034     0.4876   4.314 1.61e-05 ***
#> agegp65-74     2.3374     0.4930   4.741 2.13e-06 ***
#> agegp75+       2.3694     0.5275   4.491 7.08e-06 ***
#> alcgp40-79     0.8080     0.1330   6.076 1.23e-09 ***
#> alcgp80-119    1.1399     0.1558   7.318 2.52e-13 ***
#> alcgp120+      2.1204     0.2060  10.295  < 2e-16 ***
#> ---
#> Signif. codes:
#> 0 '***' 0.001 '**' 0.01 '*' 0.05 '.' 0.1 ' ' 1
#>
#> (Dispersion parameter for binomial family taken to be 1)
#>
#>     Null deviance: 367.95  on 87  degrees of freedom
#> Residual deviance: 104.48  on 79  degrees of freedom
#> AIC: 237.53
#>
#> Number of Fisher Scoring iterations: 6
```

Why Consider a Probit Model?

- Like logistic regression, probit regression estimates probabilities, but it assumes a **normal distribution of the latent variable**.
- While the **interpretation of coefficients differs**, model comparisons can still be made using **AIC**.

4.4 Poisson Regression

4.4.1 *The Poisson Distribution*

Poisson regression is used for modeling **count data**, where the response variable represents the number of occurrences of an event within a fixed period, space, or other unit. The Poisson distribution is defined as:

$$f(Y_i) = \frac{\mu_i^{Y_i} \exp(-\mu_i)}{Y_i!}, \quad Y_i = 0, 1, 2, \ldots$$

$$E(Y_i) = \mu_i$$

$$\text{Var}(Y_i) = \mu_i$$

where:

- Y_i is the count variable.
- μ_i is the expected count for the ith observation.
- The **mean and variance are equal** $E(Y_i) = \text{Var}(Y_i)$, making Poisson regression suitable when variance follows this property.

However, real-world count data often exhibit **overdispersion**, where the variance exceeds the mean. We will discuss remedies such as quasi-Poisson and negative binomial regression later.

4.4.2 *Poisson Model*

We model the expected count μ_i as a function of predictors $\mathbf{x_i}$ and parameters $\boldsymbol{\theta}$:

$$\mu_i = f(\mathbf{x_i}; \theta)$$

4.4.3 *Link Function Choices*

Since μ_i must be positive, we often use a **log-link function**:

$$\log(\mu_i) = \mathbf{x_i'}\boldsymbol{\theta}$$

This ensures that the predicted counts are always non-negative. This is analogous to logistic regression, where we use the logit link for binary outcomes.

Rewriting:

$$\mu_i = \exp(\mathbf{x_i}'\boldsymbol{\theta})$$

which ensures $\mu_i > 0$ for all parameter values.

4.4.4 Application: Poisson Regression

We apply Poisson regression to the **bioChemists** dataset (from the `pscl` package), which contains information on academic productivity in terms of published articles.

4.4.4.1 Dataset Overview

```
library(tidyverse)
#> -- Attaching core tidyverse packages ---- tidyverse 2.0.0
↪ --
#> v forcats   1.0.0     v stringr   1.5.1
#> v lubridate 1.9.4     v tibble    3.2.1
#> v purrr     1.0.4     v tidyr     1.3.1
#> v readr     2.1.5
#> -- Conflicts ---------------------- tidyverse_conflicts()
↪ --
#> x dplyr::filter()     masks stats::filter()
#> x dplyr::group_rows() masks kableExtra::group_rows()
#> x dplyr::lag()        masks stats::lag()
#> i Use the conflicted package
↪ (<http://conflicted.r-lib.org/>) to force all conflicts to
↪ become errors
# Load dataset
data(bioChemists, package = "pscl")

# Rename columns for clarity
bioChemists <- bioChemists %>%
  rename(
    Num_Article = art,
    # Number of articles in last 3 years
    Sex = fem,
    # 1 if female, 0 if male
    Married = mar,
```

```
    # 1 if married, 0 otherwise
    Num_Kid5 = kid5,
    # Number of children under age 6
    PhD_Quality = phd,
    # Prestige of PhD program
    Num_MentArticle = ment   # Number of articles by mentor in
    ↪  last 3 years
  )
```

```
# Visualize response variable distribution
hist(bioChemists$Num_Article,
     breaks = 25,
     main = "Number of Articles Published")
```

The **distribution of the number of articles** is right-skewed, which suggests a Poisson model may be appropriate (Fig. 4.7).

4.4.4.2 Fitting a Poisson Regression Model

We model the number of articles published (`Num_Article`) as a function of various predictors.

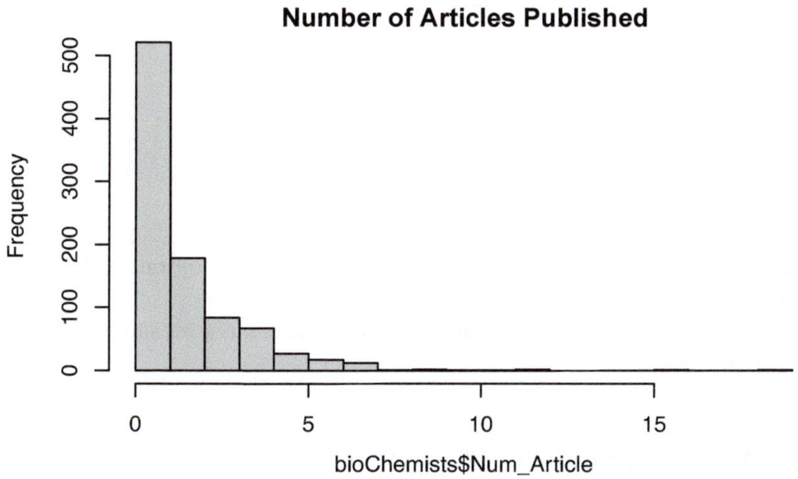

Fig. 4.7 Number of articles published

```
# Poisson regression model
Poisson_Mod <-
  glm(Num_Article ~ .,
      family = poisson,
      data = bioChemists)

# Summary of the model
summary(Poisson_Mod)
#>
#> Call:
#> glm(formula = Num_Article ~ ., family = poisson, data =
↪  bioChemists)
#>
#> Coefficients:
#>                    Estimate Std. Error z value Pr(>|z|)
#> (Intercept)        0.304617   0.102981   2.958   0.0031 **
#> SexWomen          -0.224594   0.054613  -4.112 3.92e-05 ***
#> MarriedMarried     0.155243   0.061374   2.529   0.0114 *
#> Num_Kid5          -0.184883   0.040127  -4.607 4.08e-06 ***
#> PhD_Quality        0.012823   0.026397   0.486   0.6271
#> Num_MentArticle    0.025543   0.002006  12.733  < 2e-16 ***
#> ---
#> Signif. codes:
#> 0 '***' 0.001 '**' 0.01 '*' 0.05 '.' 0.1 ' ' 1
#>
#> (Dispersion parameter for poisson family taken to be 1)
#>
#>     Null deviance: 1817.4  on 914  degrees of freedom
#> Residual deviance: 1634.4  on 909  degrees of freedom
#> AIC: 3314.1
#>
#> Number of Fisher Scoring iterations: 5
```

Interpretation:

- Coefficients are on the log scale, meaning they represent log-rate ratios.
- Exponentiating the coefficients gives the rate ratios.
- Statistical significance tells us whether each variable has a meaningful impact on publication count.

4.4.4.3 Model Diagnostics: Goodness of Fit

4.4.4.3.1 Pearson's Chi-Square Test for Overdispersion

We compute the **Pearson chi-square statistic** to check whether the variance significantly exceeds the mean:

$$X^2 = \sum \frac{(Y_i - \hat{\mu}_i)^2}{\hat{\mu}_i}$$

```
# Compute predicted means
Predicted_Means <- predict(Poisson_Mod, type = "response")

# Pearson chi-square test
X2 <-
  sum((bioChemists$Num_Article - Predicted_Means) ^ 2 /
  ↪ Predicted_Means)
X2
#> [1] 1662.547
pchisq(X2, Poisson_Mod$df.residual, lower.tail = FALSE)
#> [1] 7.849882e-47
```

- If p-value is small, overdispersion is present.
- Large X^2 statistic suggests the model may not adequately capture variability.

4.4.4.3.2 Overdispersion Check: Ratio of Deviance to Degrees of Freedom

We compute:

$$\hat{\phi} = \frac{\text{deviance}}{\text{degrees of freedom}}$$

```
# Overdispersion check
Poisson_Mod$deviance / Poisson_Mod$df.residual
#> [1] 1.797988
```

- If $\hat{\phi} > 1$, overdispersion is likely present.
- A value significantly above 1 suggests the need for an alternative model.

4.4.4.4 Addressing Overdispersion

4.4.4.4.1 Including Interaction Terms

One possible remedy is to incorporate **interaction terms**, capturing complex relationships between predictors.

```
# Adding two-way and three-way interaction terms
Poisson_Mod_All2way <-
  glm(Num_Article ~ . ^ 2, family = poisson, data =
   ↪ bioChemists)
Poisson_Mod_All3way <-
  glm(Num_Article ~ . ^ 3, family = poisson, data =
   ↪ bioChemists)
```

This may improve model fit but can lead to overfitting.

4.4.4.4.2 Quasi-Poisson Model (Adjusting for Overdispersion)

A quick fix is to allow the variance to scale by introducing $\hat{\phi}$:

$$\mathrm{Var}(Y_i) = \hat{\phi}\mu_i$$

```
# Estimate dispersion parameter
phi_hat = Poisson_Mod$deviance / Poisson_Mod$df.residual

# Adjusting Poisson model to account for overdispersion
summary(Poisson_Mod, dispersion = phi_hat)
#>
#> Call:
#> glm(formula = Num_Article ~ ., family = poisson, data =
↪  bioChemists)
#>
#> Coefficients:
#>                  Estimate Std. Error z value Pr(>|z|)
#> (Intercept)       0.30462    0.13809   2.206  0.02739 *
#> SexWomen         -0.22459    0.07323  -3.067  0.00216 **
#> MarriedMarried    0.15524    0.08230   1.886  0.05924 .
#> Num_Kid5         -0.18488    0.05381  -3.436  0.00059 ***
#> PhD_Quality       0.01282    0.03540   0.362  0.71715
#> Num_MentArticle   0.02554    0.00269   9.496  < 2e-16 ***
#> ---
#> Signif. codes:
#> 0 '***' 0.001 '**' 0.01 '*' 0.05 '.' 0.1 ' ' 1
#>
#> (Dispersion parameter for poisson family taken to be
↪  1.797988)
#>
```

```
#>     Null deviance: 1817.4  on 914  degrees of freedom
#> Residual deviance: 1634.4  on 909  degrees of freedom
#> AIC: 3314.1
#>
#> Number of Fisher Scoring iterations: 5
```

Alternatively, we refit using a quasi-Poisson model, which adjusts standard errors:

```
# Quasi-Poisson model
quasiPoisson_Mod <- glm(Num_Article ~ ., family =
↪  quasipoisson, data = bioChemists)
summary(quasiPoisson_Mod)
#>
#> Call:
#> glm(formula = Num_Article ~ ., family = quasipoisson, data
↪  = bioChemists)
#>
#> Coefficients:
#>                  Estimate Std. Error t value Pr(>|t|)
#> (Intercept)      0.304617   0.139273   2.187 0.028983 *
#> SexWomen        -0.224594   0.073860  -3.041 0.002427 **
#> MarriedMarried   0.155243   0.083003   1.870 0.061759 .
#> Num_Kid5        -0.184883   0.054268  -3.407 0.000686 ***
#> PhD_Quality      0.012823   0.035700   0.359 0.719544
#> Num_MentArticle  0.025543   0.002713   9.415  < 2e-16 ***
#> ---
#> Signif. codes:
#> 0 '***' 0.001 '**' 0.01 '*' 0.05 '.' 0.1 ' ' 1
#>
#> (Dispersion parameter for quasipoisson family taken to be
↪  1.829006)
#>
#>     Null deviance: 1817.4  on 914  degrees of freedom
#> Residual deviance: 1634.4  on 909  degrees of freedom
#> AIC: NA
#>
#> Number of Fisher Scoring iterations: 5
```

While quasi-Poisson corrects standard errors, it does not introduce an extra parameter for overdispersion.

4.4.4.4.3 Negative Binomial Regression (Preferred Approach)

A negative binomial regression explicitly models overdispersion by introducing a dispersion parameter θ:

$$\mathrm{Var}(Y_i) = \mu_i + \theta\mu_i^2$$

This extends Poisson regression by allowing the variance to grow quadratically rather than linearly.

```
# Load MASS package
library(MASS)
#>
#> Attaching package: 'MASS'
#> The following object is masked from 'package:dplyr':
#>
#>     select

# Fit Negative Binomial regression
NegBin_Mod <- glm.nb(Num_Article ~ ., data = bioChemists)

# Model summary
summary(NegBin_Mod)
#>
#> Call:
#> glm.nb(formula = Num_Article ~ ., data = bioChemists,
↪  init.theta = 2.264387695,
#>     link = log)
#>
#> Coefficients:
#>                  Estimate Std. Error z value Pr(>|z|)
#> (Intercept)      0.256144   0.137348   1.865 0.062191 .
#> SexWomen        -0.216418   0.072636  -2.979 0.002887 **
#> MarriedMarried   0.150489   0.082097   1.833 0.066791 .
#> Num_Kid5        -0.176415   0.052813  -3.340 0.000837 ***
#> PhD_Quality      0.015271   0.035873   0.426 0.670326
#> Num_MentArticle  0.029082   0.003214   9.048  < 2e-16 ***
#> ---
#> Signif. codes:
#> 0 '***' 0.001 '**' 0.01 '*' 0.05 '.' 0.1 ' ' 1
#>
#> (Dispersion parameter for Negative Binomial(2.2644) family
↪  taken to be 1)
#>
#>     Null deviance: 1109.0  on 914  degrees of freedom
```

```
#> Residual deviance: 1004.3  on 909  degrees of freedom
#> AIC: 3135.9
#>
#> Number of Fisher Scoring iterations: 1
#>
#>
#>               Theta:  2.264
#>           Std. Err.:  0.271
#>
#>  2 x log-likelihood:  -3121.917
```

This model is generally **preferred over quasi-Poisson**, as it explicitly accounts for **heterogeneity** in the data.

4.5 Negative Binomial Regression

When modeling **count data**, Poisson regression assumes that the **mean and variance are equal**:

$$\text{Var}(Y_i) = E(Y_i) = \mu_i$$

However, in many real-world datasets, the variance exceeds the mean—a phenomenon known as **overdispersion**. When overdispersion is present, the Poisson model underestimates the variance, leading to:

- Inflated test statistics (small p-values).
- Overconfident predictions.
- Poor model fit.

4.5.1 Negative Binomial Distribution

To address overdispersion, **negative binomial regression** introduces an extra **dispersion parameter** θ to allow variance to be greater than the mean:

$$\text{Var}(Y_i) = \mu_i + \theta\mu_i^2$$

where:

- $\mu_i = \exp(\mathbf{x_i'}\boldsymbol{\theta})$ is the expected count.
- θ is the dispersion parameter.
- When $\theta \to 0$, the NB model reduces to the Poisson model.

Thus, negative binomial regression is a **generalization of Poisson regression** that accounts for overdispersion.

4.5.2 *Application: Negative Binomial Regression*

We apply **negative binomial regression** to the `bioChemists` dataset to model the number of research articles (`Num_Article`) as a function of several predictors.

4.5.2.1 Fitting the Negative Binomial Model

```
# Load necessary package
library(MASS)

# Fit Negative Binomial model
NegBinom_Mod <- MASS::glm.nb(Num_Article ~ ., data =
↪  bioChemists)

# Model summary
summary(NegBinom_Mod)
#>
#> Call:
#> MASS::glm.nb(formula = Num_Article ~ ., data = bioChemists,
↪   init.theta = 2.264387695,
#>     link = log)
#>
#> Coefficients:
#>                  Estimate Std. Error z value Pr(>|z|)
#> (Intercept)       0.256144   0.137348   1.865 0.062191 .
#> SexWomen         -0.216418   0.072636  -2.979 0.002887 **
#> MarriedMarried    0.150489   0.082097   1.833 0.066791 .
#> Num_Kid5         -0.176415   0.052813  -3.340 0.000837 ***
#> PhD_Quality       0.015271   0.035873   0.426 0.670326
#> Num_MentArticle   0.029082   0.003214   9.048  < 2e-16 ***
#> ---
#> Signif. codes:
#> 0 '***' 0.001 '**' 0.01 '*' 0.05 '.' 0.1 ' ' 1
#>
#> (Dispersion parameter for Negative Binomial(2.2644) family
↪   taken to be 1)
#>
#>     Null deviance: 1109.0  on 914  degrees of freedom
```

```
#> Residual deviance: 1004.3  on 909  degrees of freedom
#> AIC: 3135.9
#>
#> Number of Fisher Scoring iterations: 1
#>
#>
#>               Theta:  2.264
#>           Std. Err.:  0.271
#>
#>  2 x log-likelihood:  -3121.917
```

Interpretation:

- The coefficients are on the log scale.
- The dispersion parameter θ (also called size parameter in some contexts) is estimated as 2.264 with a standard error of 0.271. Check overdispersion for more detail.
- Since θ is significantly different from 1, this confirms overdispersion, validating the choice of the negative binomial model over Poisson regression.

4.5.2.2 Model Comparison: Poisson vs. Negative Binomial

4.5.2.2.1 Checking Overdispersion in Poisson Model

Before using NB regression, we confirm **overdispersion** by computing:

$$\hat{\phi} = \frac{\text{deviance}}{\text{degrees of freedom}}$$

```
# Overdispersion check for Poisson model
Poisson_Mod$deviance / Poisson_Mod$df.residual
#> [1] 1.797988
```

- If $\hat{\phi} > 1$, overdispersion is present.
- A large value suggests that Poisson regression underestimates variance.

4.5.2.2.2 Likelihood Ratio Test: Poisson vs. Negative Binomial

We compare the Poisson and negative binomial models using a **likelihood ratio test**, where:

$$G^2 = 2 \times (\log L_{NB} - \log L_{Poisson})$$

with df = 1

```
# Likelihood ratio test between Poisson and Negative Binomial
pchisq(2 * (logLik(NegBinom_Mod) - logLik(Poisson_Mod)),
       df = 1,
       lower.tail = FALSE)
#> 'log Lik.' 4.391728e-41 (df=7)
```

- Small p-value (< 0.05) → negative binomial model is significantly better.
- Large p-value (> 0.05) → Poisson model is adequate.

  Since overdispersion is confirmed, the negative binomial model is preferred.

4.5.2.3 Model Diagnostics and Evaluation

4.5.2.3.1 Checking Dispersion Parameter θ

The negative binomial dispersion parameter θ can be retrieved:

```
# Extract dispersion parameter estimate
NegBinom_Mod$theta
#> [1] 2.264388
```

- A large θ suggests that overdispersion is **not extreme**, while a small θ (close to 0) would indicate the Poisson model is reasonable.

4.5.2.4 Predictions and Rate Ratios

In negative binomial regression, exponentiating the coefficients gives rate ratios:

```
# Convert coefficients to rate ratios
data.frame(`Odds Ratios` = exp(coef(NegBinom_Mod)))
#>                  Odds.Ratios
#> (Intercept)        1.2919388
#> SexWomen           0.8053982
#> MarriedMarried     1.1624030
#> Num_Kid5           0.8382698
#> PhD_Quality        1.0153884
#> Num_MentArticle    1.0295094
```

A rate ratio of:

- > 1 → increases expected article count.

- < 1 → decreases expected article count.
- = 1 → no effect.

For example:

- If `PhD_Quality` has an exponentiated coefficient of 1.5, individuals from higher-quality PhD programs are expected to publish 50% more articles.
- If `Sex` has an exponentiated coefficient of 0.8, females publish 20% fewer articles than males, all else equal.

4.5.2.5 Alternative Approach: Zero-Inflated Models

If a dataset has excess zeros (many individuals publish no articles), **zero-inflated negative binomial (ZINB) models** may be required.

$$\mathrm{P}(Y_i = 0) = p + (1 - p) f(Y_i = 0|\mu, \theta)$$

where:

- p is the probability of always being a zero (e.g., inactive researchers).
- $f(Y_i)$ follows the negative binomial distribution.

4.5.3 Fitting a Zero-Inflated Negative Binomial Model

```
# Load package for zero-inflated models
library(pscl)

# Fit ZINB model
ZINB_Mod <- zeroinfl(Num_Article ~ ., data = biochemists, dist
↪   = "negbin")

# Model summary
summary(ZINB_Mod)
#>
#> Call:
#> zeroinfl(formula = Num_Article ~ ., data = bioChemists,
#>     dist = "negbin")
#>
#> Pearson residuals:
#>     Min      1Q  Median      3Q     Max
#> -1.2942 -0.7601 -0.2909  0.4448  6.4155
#>
```

```
#> Count model coefficients (negbin with log link):
#>                   Estimate Std. Error z value Pr(>|z|)
#> (Intercept)      0.4167465  0.1435966   2.902  0.00371 **
#> SexWomen        -0.1955068  0.0755926  -2.586  0.00970 **
#> MarriedMarried   0.0975826  0.0844520   1.155  0.24789
#> Num_Kid5        -0.1517325  0.0542061  -2.799  0.00512 **
#> PhD_Quality     -0.0007001  0.0362697  -0.019  0.98460
#> Num_MentArticle  0.0247862  0.0034927   7.097 1.28e-12 ***
#> Log(theta)       0.9763565  0.1354695   7.207 5.71e-13 ***
#>
#> Zero-inflation model coefficients (binomial with logit
↪  link):
#>                 Estimate Std. Error z value Pr(>|z|)
#> (Intercept)     -0.19169    1.32282  -0.145  0.88478
#> SexWomen         0.63593    0.84892   0.749  0.45379
#> MarriedMarried  -1.49947    0.93867  -1.597  0.11017
#> Num_Kid5         0.62843    0.44278   1.419  0.15582
#> PhD_Quality     -0.03771    0.30801  -0.122  0.90254
#> Num_MentArticle -0.88229    0.31623  -2.790  0.00527 **
#> ---
#> Signif. codes:  0 '***' 0.001 '**' 0.01 '*' 0.05 '.' 0.1 ' ' 1
#>
#> Theta = 2.6548
#> Number of iterations in BFGS optimization: 43
#> Log-likelihood: -1550 on 13 Df
```

This model accounts for:

- Structural zero inflation.
- Overdispersion.

ZINB is often preferred when many observations are zero. However, since ZINB does not fall under the GLM framework, we will discuss it further in nonlinear and generalized linear mixed models.

Why ZINB Is Not a GLM?

- Unlike GLMs, which assume a single response distribution from the exponential family, ZINB is a mixture model with two components:
  - Count model—A negative binomial regression for the main count process.
  - Inflation model—A logistic regression for excess zeros.

Because ZINB combines two distinct processes rather than using a single exponential family distribution, it does not fit within the standard GLM framework.

ZINB is part of finite mixture models and is sometimes considered within generalized linear mixed models (GLMMs) or semi-parametric models.

4.6 Quasi-Poisson Regression

Poisson regression assumes that the **mean and variance are equal**:

$$\text{Var}(Y_i) = E(Y_i) = \mu_i$$

However, many real-world datasets exhibit **overdispersion**, where the variance exceeds the mean:

$$\text{Var}(Y_i) = \phi\mu_i$$

where ϕ (the **dispersion parameter**) allows the variance to scale beyond the Poisson assumption.

To correct for this, we use quasi-Poisson regression, which:

- Follows the generalized linear models structure but is not a strict GLM.
- Uses a variance function proportional to the mean: $\text{Var}(Y_i) = \phi\mu_i$.
- Does not assume a specific probability distribution, unlike Poisson or negative binomial models.

4.6.1 Is Quasi-Poisson Regression a Generalized Linear Model?

Yes, quasi-Poisson is GLM-like:

- **Linear predictor:** Like Poisson regression, it models the log of the expected count as a function of predictors:

$$\log(E(Y)) = X\beta$$

- **Canonical link function:** It typically uses a log-link function, just like standard Poisson regression.
- **Variance structure:** Unlike standard Poisson, which assumes $\text{Var}(Y) = E(Y)$, quasi-Poisson allows for **overdispersion**:

$$\text{Var}(Y) = \phi E(Y)$$

where ϕ is estimated rather than assumed to be 1.

No, quasi-Poisson is not a strict GLM because:

- GLMs require a full probability distribution from the exponential family:
  - Standard Poisson regression assumes a Poisson distribution (which belongs to the exponential family).

  - Quasi-Poisson does not assume a full probability distribution, only a mean-variance relationship.

- It does not use Maximum Likelihood Estimation:
  - Standard GLMs use MLE to estimate parameters.
  - Quasi-Poisson uses quasi-likelihood methods, which require specifying only the mean and variance, but not a full likelihood function.

- Likelihood-based inference is not valid:
  - AIC, BIC, and likelihood ratio tests cannot be used with quasi-Poisson regression.

When to Use Quasi-Poisson

- When **data exhibit overdispersion** (variance > mean), making standard Poisson regression inappropriate.
- When negative binomial regression is not preferred, but an alternative is needed to handle overdispersion.
- If overdispersion is present, negative binomial regression is often a better alternative because it is a true GLM with a full likelihood function, whereas quasi-Poisson is only a quasi-likelihood approach.

4.6.2 Application: Quasi-Poisson Regression

We analyze the `bioChemists` dataset, modeling the number of published articles (`Num_Article`) as a function of various predictors.

4.6.2.1 Checking Overdispersion in the Poisson Model

We first fit a Poisson regression model and check for overdispersion using the deviance-to-degrees-of-freedom ratio:

```
# Fit Poisson regression model
Poisson_Mod <-
    glm(Num_Article ~ ., family = poisson, data = bioChemists)

# Compute dispersion parameter
dispersion_estimate <-
    Poisson_Mod$deviance / Poisson_Mod$df.residual
dispersion_estimate
#> [1] 1.797988
```

- If $\hat{\phi} > 1$, the Poisson model underestimates variance.
- A large value (» 1) suggests that Poisson regression is not appropriate.

4.6.2.2 Fitting the Quasi-Poisson Model

Since overdispersion is present, we refit the model using quasi-Poisson regression, which scales standard errors by ϕ.

```
# Fit Quasi-Poisson regression model
quasiPoisson_Mod <-
    glm(Num_Article ~ ., family = quasipoisson, data =
    ↪ bioChemists)

# Summary of the model
summary(quasiPoisson_Mod)
#>
#> Call:
#> glm(formula = Num_Article ~ ., family = quasipoisson, data
↪   = bioChemists)
#>
#> Coefficients:
#>                   Estimate Std. Error t value Pr(>|t|)
#> (Intercept)       0.304617   0.139273   2.187 0.028983 *
#> SexWomen         -0.224594   0.073860  -3.041 0.002427 **
#> MarriedMarried    0.155243   0.083003   1.870 0.061759 .
#> Num_Kid5         -0.184883   0.054268  -3.407 0.000686 ***
#> PhD_Quality       0.012823   0.035700   0.359 0.719544
#> Num_MentArticle   0.025543   0.002713   9.415  < 2e-16 ***
#> ---
#> Signif. codes:
#> 0 '***' 0.001 '**' 0.01 '*' 0.05 '.' 0.1 ' ' 1
#>
#> (Dispersion parameter for quasipoisson family taken to be
↪   1.829006)
#>
#>     Null deviance: 1817.4  on 914  degrees of freedom
#> Residual deviance: 1634.4  on 909  degrees of freedom
#> AIC: NA
#>
#> Number of Fisher Scoring iterations: 5
```

Interpretation:

- The coefficients remain the same as in Poisson regression.
- Standard errors are inflated to account for overdispersion.
- P-values increase, leading to more conservative inference.

4.6.2.3 Comparing Poisson and Quasi-Poisson

To see the effect of using quasi-Poisson, we compare standard errors:

```
# Extract coefficients and standard errors
poisson_se <- summary(Poisson_Mod)$coefficients[, 2]
quasi_se <- summary(quasiPoisson_Mod)$coefficients[, 2]

# Compare standard errors
se_comparison <- data.frame(Poisson = poisson_se,
                            Quasi_Poisson = quasi_se)
se_comparison
#>                    Poisson Quasi_Poisson
#> (Intercept)    0.102981443   0.139272885
#> SexWomen       0.054613488   0.073859696
#> MarriedMarried 0.061374395   0.083003199
#> Num_Kid5       0.040126898   0.054267922
#> PhD_Quality    0.026397045   0.035699564
#> Num_MentArticle 0.002006073  0.002713028
```

- Quasi-Poisson has larger standard errors than Poisson.
- This leads to wider confidence intervals, reducing the likelihood of false positives.

4.6.2.4 Model Diagnostics: Checking Residuals

We examine residuals to assess model fit (Fig. 4.8):

```
# Residual plot
plot(
    quasiPoisson_Mod$fitted.values,
    residuals(quasiPoisson_Mod, type = "pearson"),
    xlab = "Fitted Values",
    ylab = "Pearson Residuals",
    main = "Residuals vs. Fitted Values (Quasi-Poisson)"
)
abline(h = 0, col = "red")
```

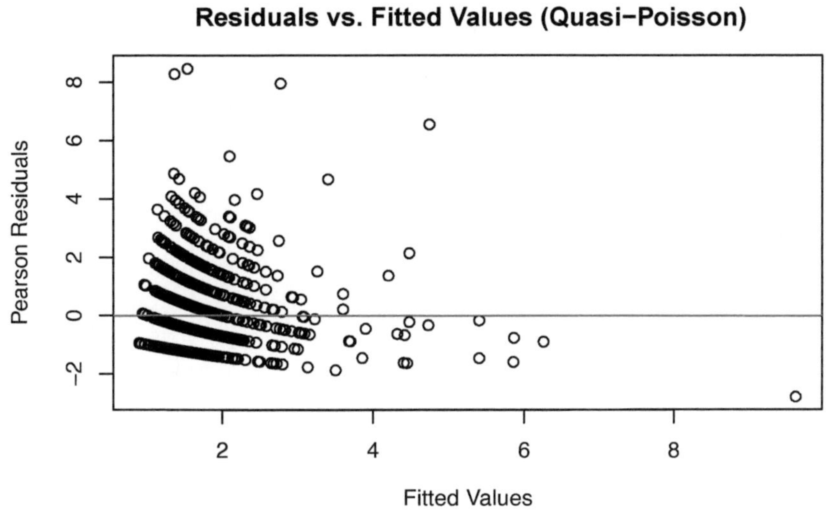

Fig. 4.8 Residuals vs. fitted values (quasi-Poisson)

- If residuals show a pattern, additional predictors or transformations may be needed.
- Random scatter around zero suggests a well-fitting model.

4.6.2.5 Alternative: Negative Binomial vs. Quasi-Poisson

If overdispersion is **severe**, negative binomial regression may be preferable because it explicitly models dispersion:

```
# Fit Negative Binomial model
library(MASS)
NegBinom_Mod <- glm.nb(Num_Article ~ ., data = bioChemists)

# Model summaries
summary(quasiPoisson_Mod)
#>
#> Call:
#> glm(formula = Num_Article ~ ., family = quasipoisson, data
↪   = bioChemists)
#>
#> Coefficients:
#>                  Estimate Std. Error t value Pr(>|t|)
#> (Intercept)      0.304617   0.139273   2.187 0.028983 *
```

```
#> SexWomen          -0.224594   0.073860  -3.041 0.002427 **
#> MarriedMarried     0.155243   0.083003   1.870 0.061759 .
#> Num_Kid5          -0.184883   0.054268  -3.407 0.000686 ***
#> PhD_Quality        0.012823   0.035700   0.359 0.719544
#> Num_MentArticle    0.025543   0.002713   9.415  < 2e-16 ***
#> ---
#> Signif. codes:
#> 0 '***' 0.001 '**' 0.01 '*' 0.05 '.' 0.1 ' ' 1
#>
#> (Dispersion parameter for quasipoisson family taken to be
↪  1.829006)
#>
#>     Null deviance: 1817.4  on 914  degrees of freedom
#> Residual deviance: 1634.4  on 909  degrees of freedom
#> AIC: NA
#>
#> Number of Fisher Scoring iterations: 5
summary(NegBinom_Mod)
#>
#> Call:
#> glm.nb(formula = Num_Article ~ ., data = bioChemists,
↪  init.theta = 2.264387695,
#>     link = log)
#>
#> Coefficients:
#>                  Estimate Std. Error z value Pr(>|z|)
#> (Intercept)       0.256144   0.137348   1.865 0.062191 .
#> SexWomen         -0.216418   0.072636  -2.979 0.002887 **
#> MarriedMarried    0.150489   0.082097   1.833 0.066791 .
#> Num_Kid5         -0.176415   0.052813  -3.340 0.000837 ***
#> PhD_Quality       0.015271   0.035873   0.426 0.670326
#> Num_MentArticle   0.029082   0.003214   9.048  < 2e-16 ***
#> ---
#> Signif. codes:
#> 0 '***' 0.001 '**' 0.01 '*' 0.05 '.' 0.1 ' ' 1
#>
#> (Dispersion parameter for Negative Binomial(2.2644) family
↪  taken to be 1)
#>
#>     Null deviance: 1109.0  on 914  degrees of freedom
#> Residual deviance: 1004.3  on 909  degrees of freedom
#> AIC: 3135.9
#>
#> Number of Fisher Scoring iterations: 1
```

```
#>
#>
#>               Theta:  2.264
#>           Std. Err.:  0.271
#>
#>  2 x log-likelihood:  -3121.917
```

4.6.2.6 Key Differences: Quasi-Poisson vs. Negative Binomial

When to Choose (Table 4.2)

- Use quasi-Poisson when you only need robust standard errors and do not require model selection via AIC/BIC.
- Use negative binomial when overdispersion is large and you want a true likelihood-based model.

While quasi-Poisson is a quick fix, negative binomial is generally the better choice for modeling count data with overdispersion.

4.7 Multinomial Logistic Regression

When dealing with categorical response variables with more than two possible outcomes, the **multinomial logistic regression** is a natural extension of the binary logistic model.

4.7.1 The Multinomial Distribution

Suppose we have a categorical response variable Y_i that can take values in $\{1, 2, \ldots, J\}$. For each observation i, the probability that it falls into category j is given by:

Table 4.2 Comparison of quasi-Poisson and negative binomial models for overdispersed count data

Feature	Quasi-Poisson	Negative binomial
Handles overdispersion?	Yes	Yes
Uses a full probability distribution?	No	Yes
MLE-based?	No (quasi-likelihood)	Yes
Can use AIC/BIC for model selection?	No	Yes
Better for model interpretation?	Yes	Yes
Best for severe overdispersion?	No	Yes

$$p_{ij} = P(Y_i = j), \quad \text{where} \quad \sum_{j=1}^{J} p_{ij} = 1.$$

The response follows a **multinomial distribution**:

$$Y_i \sim \text{Multinomial}(1; p_{i1}, p_{i2}, ..., p_{iJ}).$$

This means that each observation belongs to exactly one of the J categories.

4.7.2 Modeling Probabilities Using Log Odds

We cannot model the probabilities p_{ij} directly because they must sum to 1. Instead, we use a **logit transformation**, comparing each category j to a **baseline category** (typically the first category, $j = 1$):

$$\eta_{ij} = \log \frac{p_{ij}}{p_{i1}}, \quad j = 2, \ldots, J.$$

Using a **linear function of covariates** $\mathbf{x}_i$, we define:

$$\eta_{ij} = \mathbf{x}_i' \beta_j = \beta_{j0} + \sum_{p=1}^{P} \beta_{jp} x_{ip}.$$

Rearranging to express probabilities explicitly:

$$p_{ij} = p_{i1} \exp(\mathbf{x}_i' \beta_j).$$

Since all probabilities must sum to 1:

$$p_{i1} + \sum_{j=2}^{J} p_{ij} = 1.$$

Substituting for p_{ij}:

$$p_{i1} + \sum_{j=2}^{J} p_{i1} \exp(\mathbf{x}_i' \beta_j) = 1.$$

Solving for p_{i1}:

$$p_{i1} = \frac{1}{1 + \sum_{j=2}^{J} \exp(\mathbf{x}_i' \beta_j)}.$$

Thus, the probability for category j is:

$$p_{ij} = \frac{\exp(\mathbf{x}_i' \beta_j)}{1 + \sum_{l=2}^{J} \exp(\mathbf{x}_i' \beta_l)}, \quad j = 2, \ldots, J.$$

This formulation is known as the **multinomial logit model**.

4.7.3 *Softmax Representation*

An alternative formulation avoids choosing a baseline category and instead treats all J categories **symmetrically** using the **softmax function**:

$$P(Y_i = j | X_i = x) = \frac{\exp(\beta_{j0} + \sum_{p=1}^{P} \beta_{jp} x_p)}{\sum_{l=1}^{J} \exp(\beta_{l0} + \sum_{p=1}^{P} \beta_{lp} x_p)}.$$

This representation is often used in **neural networks** and general machine learning models.

4.7.4 *Log-Odds Ratio Between Two Categories*

The **log-odds ratio** between two categories k and k' is:

$$\log \frac{P(Y = k | X = x)}{P(Y = k' | X = x)} = (\beta_{k0} - \beta_{k'0}) + \sum_{p=1}^{P} (\beta_{kp} - \beta_{k'p}) x_p.$$

This equation tells us that:

- If $\beta_{kp} > \beta_{k'p}$, then increasing x_p increases the odds of choosing category k over k'.
- If $\beta_{kp} < \beta_{k'p}$, then increasing x_p decreases the odds of choosing k over k'.

4.7.5 *Estimation*

To estimate the parameters β_j, we use Maximum Likelihood Estimation.

Given n independent observations (Y_i, X_i), the likelihood function is:

$$L(\beta) = \prod_{i=1}^{n} \prod_{j=1}^{J} p_{ij}^{Y_{ij}}.$$

Taking the log-likelihood:

$$\log L(\beta) = \sum_{i=1}^{n} \sum_{j=1}^{J} Y_{ij} \log p_{ij}.$$

Since there is no closed-form solution, numerical methods (see nonlinear least squares estimation) are used for estimation.

4.7.6 Interpretation of Coefficients

- Each β_{jp} represents the effect of x_p on the log odds of category j relative to the baseline.
- Positive coefficients mean increasing x_p makes category j more likely relative to the baseline.
- Negative coefficients mean increasing x_p makes category j less likely relative to the baseline.

4.7.7 Application: Multinomial Logistic Regression

1. Load Necessary Libraries and Data

```
library(faraway)  # For the dataset
library(dplyr)    # For data manipulation
library(ggplot2)  # For visualization
library(nnet)     # For multinomial logistic regression

# Load and inspect data
data(nes96, package="faraway")
head(nes96, 3)
#>   popul TVnews selfLR ClinLR DoleLR     PID age  educ
#> 1     0      7 extCon extLib    Con  strRep  36    HS
#> 2   190      1 sliLib sliLib sliCon weakDem  20  Coll
#> 3    31      7    Lib    Lib    Con weakDem  24 BAdeg
#>      income    vote
```

```
#> 1 $3Kminus    Dole
#> 2 $3Kminus Clinton
#> 3 $3Kminus Clinton
```

The dataset `nes96` contains survey responses, including political party identification (`PID`), age (`age`), and education level (`educ`).

2. Define Political Strength Categories
We classify political strength into three categories:

1. **Strong**: Strong democrat or strong republican
2. **Weak**: Weak democrat or weak republican
3. **Neutral**: Independents and other affiliations

```
# Check distribution of political identity
table(nes96$PID)
#>
#>  strDem weakDem  indDem  indind  indRep weakRep  strRep
#>     200     180     108      37      94     150     175

# Define Political Strength variable
nes96 <- nes96 %>%
  mutate(Political_Strength = case_when(
    PID %in% c("strDem", "strRep") ~ "Strong",
    PID %in% c("weakDem", "weakRep") ~ "Weak",
    PID %in% c("indDem", "indind", "indRep") ~ "Neutral",
    TRUE ~ NA_character_
  ))

# Summarize
nes96 %>% group_by(Political_Strength) %>% summarise(Count =
↪  n())
#> # A tibble: 3 x 2
#>   Political_Strength Count
#>   <chr>              <int>
#> 1 Neutral              239
#> 2 Strong               375
#> 3 Weak                 330
```

3. Visualizing Political Strength by Age
We visualize the proportion of each political strength category across age groups (Fig. 4.9).

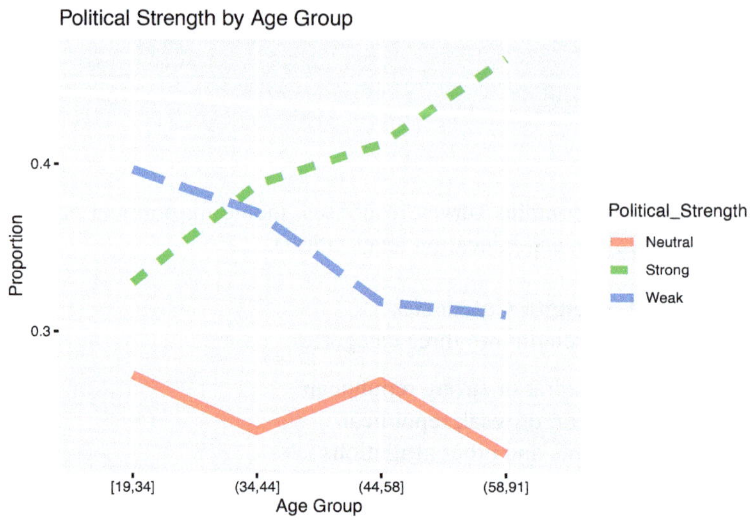

Fig. 4.9 Political strength by age group

```
# Prepare data for visualization
Plot_DF <- nes96 %>%
    mutate(Age_Grp = cut_number(age, 4)) %>%
    group_by(Age_Grp, Political_Strength) %>%
    summarise(count = n(), .groups = 'drop') %>%
    group_by(Age_Grp) %>%
    mutate(etotal = sum(count), proportion = count / etotal)

# Plot age vs political strength
Age_Plot <- ggplot(
    Plot_DF,
    aes(
        x        = Age_Grp,
        y        = proportion,
        group    = Political_Strength,
        linetype = Political_Strength,
        color    = Political_Strength
    )
) +
    geom_line(size = 2) +
    labs(title = "Political Strength by Age Group",
         x = "Age Group",
         y = "Proportion")
```

```
#> Warning: Using `size` aesthetic for lines was deprecated in
↪  ggplot2
#> 3.4.0.
#> i Please use `linewidth` instead.
#> This warning is displayed once every 8 hours.
#> Call `lifecycle::last_lifecycle_warnings()` to see where
#> this warning was generated.
```

```
# Display plot
Age_Plot
```

4. Fit a Multinomial Logistic Model
We model **political strength** as a function of **age** and **education** (Fig. 4.10).

```
# Fit multinomial logistic regression
Multinomial_Model <-
    multinom(Political_Strength ~ age + educ,
             data = nes96,
             trace = FALSE)
summary(Multinomial_Model)
#> Call:
#> multinom(formula = Political_Strength ~ age + educ, data =
↪  nes96,
```

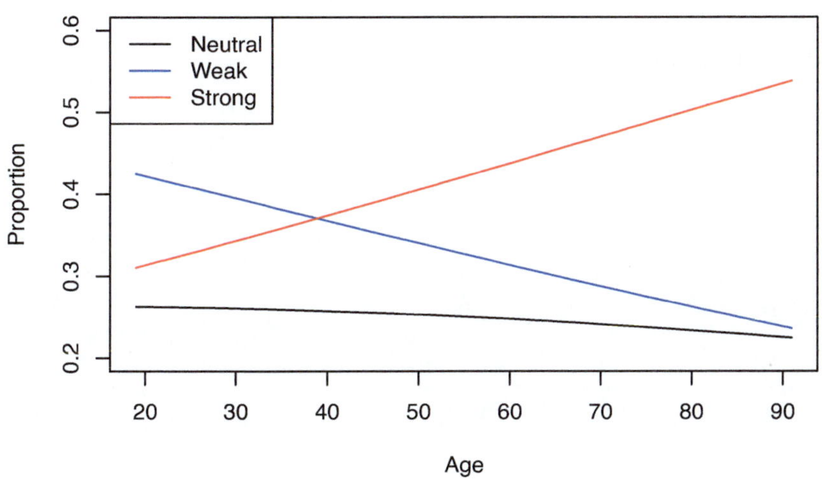

Fig. 4.10 Age group by predicted probabilities

```
#>     trace = FALSE)
#>
#> Coefficients:
#>        (Intercept)          age     educ.L     educ.Q
#> Strong -0.08788729  0.010700364 -0.1098951 -0.2016197
#> Weak    0.51976285 -0.004868771 -0.1431104 -0.2405395
#>            educ.C      educ^4     educ^5     educ^6
#> Strong -0.1757739 -0.02116307 -0.1664377 -0.1359449
#> Weak   -0.2411795  0.18353634 -0.1489030 -0.2173144
#>
#> Std. Errors:
#>        (Intercept)         age    educ.L    educ.Q
#> Strong   0.3017034 0.005280743 0.4586041 0.4318830
#> Weak     0.3097923 0.005537561 0.4920736 0.4616446
#>           educ.C    educ^4    educ^5    educ^6
#> Strong 0.3628837 0.2964776 0.2515012 0.2166774
#> Weak   0.3881003 0.3169149 0.2643747 0.2199186
#>
#> Residual Deviance: 2024.596
#> AIC: 2056.596
```

5. Stepwise Model Selection Based on AIC
We perform stepwise selection to find the best model.

```
Multinomial_Step <- step(Multinomial_Model, trace = 0)
#> trying - age
#> trying - educ
#> trying - age
Multinomial_Step
#> Call:
#> multinom(formula = Political_Strength ~ age, data = nes96,
↪  trace = FALSE)
#>
#> Coefficients:
#>        (Intercept)          age
#> Strong -0.01988977  0.009832916
#> Weak    0.59497046 -0.005954348
#>
#> Residual Deviance: 2030.756
#> AIC: 2038.756
```

Compare the best model to the full model based on **deviance**:

```
pchisq(
    q = deviance(Multinomial_Step) -
    ↪ deviance(Multinomial_Model),
    df = Multinomial_Model$edf - Multinomial_Step$edf,
    lower.tail = FALSE
)
#> [1] 0.9078172
```

A nonsignificant p-value suggests **no major difference** between the full and stepwise models.

6. Predictions and Visualization
Predicting Political Strength Probabilities by Age

```
# Create data for prediction
PlotData <- data.frame(age = seq(from = 19, to = 91))

# Get predicted probabilities
Preds <- PlotData %>%
    bind_cols(data.frame(predict(Multinomial_Step,
                                 PlotData,
                                 type = "probs")))
```

```
# Plot predicted probabilities across age
plot(
    x = Preds$age,
    y = Preds$Neutral,
    type = "l",
    ylim = c(0.2, 0.6),
    col = "black",
    ylab = "Proportion",
    xlab = "Age"
)

lines(x = Preds$age,
      y = Preds$Weak,
      col = "blue")
lines(x = Preds$age,
      y = Preds$Strong,
      col = "red")
```

```
legend(
    "topleft",
    legend = c("Neutral", "Weak", "Strong"),
    col = c("black", "blue", "red"),
    lty = 1
)
```

Predict for Specific Ages

```
# Predict class for a 34-year-old
predict(Multinomial_Step, data.frame(age = 34))
#> [1] Weak
#> Levels: Neutral Strong Weak

# Predict probabilities for 34 and 35-year-olds
predict(Multinomial_Step,
        data.frame(age = c(34, 35)),
        type = "probs")
#>     Neutral    Strong      Weak
#> 1 0.2597275 0.3556910 0.3845815
#> 2 0.2594080 0.3587639 0.3818281
```

4.7.8 Application: Gamma Regression

When response variables are **strictly positive**, we use **gamma regression**.

1. Load and Prepare Data

```
library(agridat)  # Agricultural dataset
library(ggplot2)

# Load and filter data
dat <- agridat::streibig.competition
gammaDat <- subset(dat, sseeds < 1)  # Keep only barley
gammaDat <-
    transform(gammaDat,
```

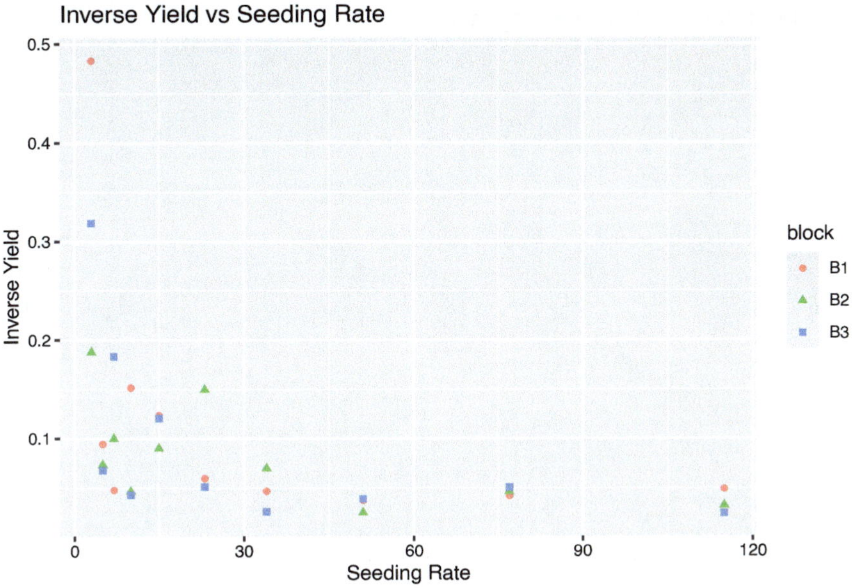

Fig. 4.11 Inverse yield vs. seeding rate

```
            x = bseeds,
            y = bdwt,
            block = factor(block))
```

2. Visualization of Inverse Yield (Fig. 4.11)

```
ggplot(gammaDat, aes(x = x, y = 1 / y)) +
    geom_point(aes(color = block, shape = block)) +
    labs(title = "Inverse Yield vs Seeding Rate",
         x = "Seeding Rate",
         y = "Inverse Yield")
```

3. Fit Gamma Regression Model
Gamma regression models **yield as a function of seeding rate** using an inverse link (Fig. 4.12):

$$\eta_{ij} = \beta_{0j} + \beta_{1j}x_{ij} + \beta_2 x_{ij}^2, \quad Y_{ij} = \eta_{ij}^{-1}$$

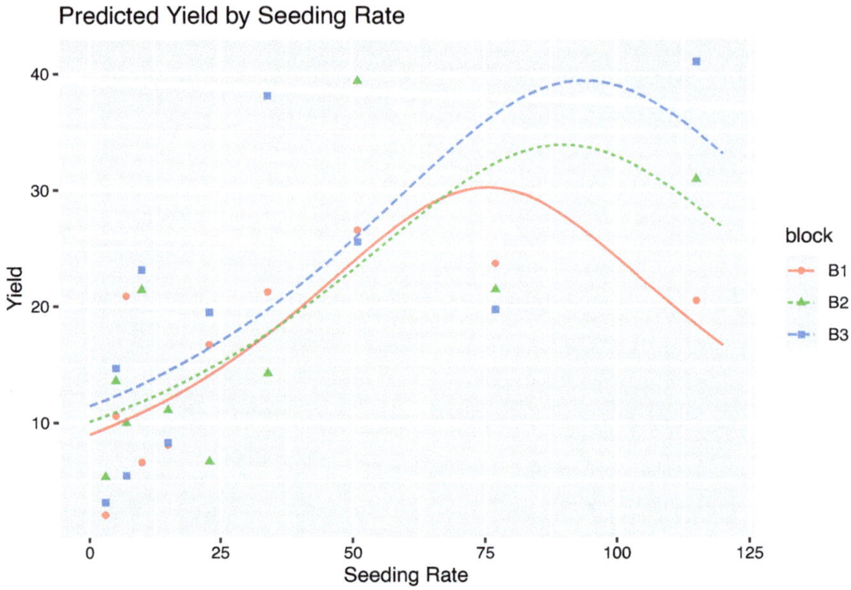

Fig. 4.12 Predicted yield by seeding rate

```
m1 <- glm(y ~ block + block * x + block * I(x^2),
          data = gammaDat, family = Gamma(link = "inverse"))

summary(m1)
#>
#> Call:
#> glm(formula = y ~ block + block * x + block * I(x^2),
↪  family = Gamma(link = "inverse"),
#>     data = gammaDat)
#>
#> Coefficients:
#>                  Estimate Std. Error t value Pr(>|t|)
#> (Intercept)     1.115e-01  2.870e-02   3.886 0.000854 ***
#> blockB2        -1.208e-02  3.880e-02  -0.311 0.758630
#> blockB3        -2.386e-02  3.683e-02  -0.648 0.524029
#> x              -2.075e-03  1.099e-03  -1.888 0.072884 .
#> I(x^2)          1.372e-05  9.109e-06   1.506 0.146849
#> blockB2:x       5.198e-04  1.468e-03   0.354 0.726814
#> blockB3:x       7.475e-04  1.393e-03   0.537 0.597103
```

```
#> blockB2:I(x^2) -5.076e-06  1.184e-05  -0.429 0.672475
#> blockB3:I(x^2) -6.651e-06  1.123e-05  -0.592 0.560012
#> ---
#> Signif. codes:
#> 0 '***' 0.001 '**' 0.01 '*' 0.05 '.' 0.1 ' ' 1
#>
#> (Dispersion parameter for Gamma family taken to be
↪  0.3232083)
#>
#>     Null deviance: 13.1677  on 29  degrees of freedom
#> Residual deviance:  7.8605  on 21  degrees of freedom
#> AIC: 225.32
#>
#> Number of Fisher Scoring iterations: 5
```

4. Predictions and Visualization

```
# Generate new data for prediction
newdf <-
    expand.grid(x = seq(0, 120, length = 50),
                block = factor(c("B1", "B2", "B3")))

# Predict responses
newdf$pred <- predict(m1, newdata = newdf, type = "response")
```

```
# Plot predictions
ggplot(gammaDat, aes(x = x, y = y)) +
    geom_point(aes(color = block, shape = block)) +
    geom_line(data = newdf, aes(
        x = x,
        y = pred,
        color = block,
        linetype = block
    )) +
    labs(title = "Predicted Yield by Seeding Rate",
         x = "Seeding Rate",
         y = "Yield")
```

4.8 Generalization of Generalized Linear Models

We have seen that Poisson regression bears similarities to logistic regression. This insight leads us to a broader class of models known as generalized linear models, introduced by Nelder and Wedderburn (1972). These models provide a unified framework for handling different types of response variables while maintaining the fundamental principles of linear modeling.

4.8.1 Exponential Family

The foundation of GLMs is built on the **exponential family of distributions**, which provides a flexible class of probability distributions that share a common form:

$$f(y;\theta,\phi)=\exp\left(\frac{\theta y-b(\theta)}{a(\phi)}+c(y,\phi)\right)$$

where:

- θ is the **natural parameter** (canonical parameter).
- ϕ is the **dispersion parameter**.
- $a(\phi)$, $b(\theta)$, and $c(y,\phi)$ are functions ensuring the proper distributional form.

Distributions in the Exponential Family That Can Be Used in GLMs
1. Normal distribution
2. Binomial distribution
3. Poisson distribution
4. Gamma distribution
5. Inverse Gaussian distribution
6. Negative binomial distribution (used in GLMs but requires overdispersion adjustments)
7. Multinomial distribution (for categorical response)

Exponential Family Distributions Not Commonly Used in GLMs
1. Beta distribution
2. Dirichlet distribution
3. Wishart distribution
4. Geometric distribution
5. Exponential distribution (can be used indirectly through survival models)

Example (Normal Distribution) Consider a normally distributed response variable $Y\sim N(\mu,\sigma^2)$. The probability density function (PDF) is:

$$f(y; \mu, \sigma^2) = \frac{1}{\sqrt{2\pi\sigma^2}} \exp\left(-\frac{(y-\mu)^2}{2\sigma^2}\right)$$
$$= \exp\left(-\frac{y^2 - 2y\mu + \mu^2}{2\sigma^2} - \frac{1}{2}\log(2\pi\sigma^2)\right)$$

Rewriting in exponential family form:

$$f(y; \mu, \sigma^2) = \exp\left(\frac{y\mu - \frac{\mu^2}{2}}{\sigma^2} - \frac{y^2}{2\sigma^2} - \frac{1}{2}\log(2\pi\sigma^2)\right)$$
$$= \exp\left(\frac{\theta y - b(\theta)}{a(\phi)} + c(y, \phi)\right)$$

where:

- **Natural parameter**: $\theta = \mu$
- **Function** $b(\theta)$: $b(\theta) = \frac{\mu^2}{2}$
- **Dispersion function**: $a(\phi) = \sigma^2 = \phi$
- **Function** $c(y, \phi)$: $c(y, \phi) = -\frac{1}{2}\left(\frac{y^2}{\phi} + \log(2\pi\sigma^2)\right)$

4.8.2 Properties of GLM Exponential Families

1. **Expected Value (Mean)**

$$E(Y) = b'(\theta)$$

   where $b'(\theta) = \frac{\partial b(\theta)}{\partial \theta}$. (Note: ' is "prime," not transpose).
2. **Variance**

$$\text{Var}(Y) = a(\phi)b''(\theta) = a(\phi)V(\mu)$$

   where

   - $V(\mu) = b''(\theta)$ is the **variance function**, though it only represents the variance when $a(\phi) = 1$.

3. If $a(\phi)$, $b(\theta)$, and $c(y, \phi)$ are identifiable, we can derive the expected value and variance of Y.

Examples of Exponential Family Distributions

1. Normal Distribution
For a normal distribution $Y \sim N(\mu, \sigma^2)$, the exponential family representation is:

$$f(y; \mu, \sigma^2) = \frac{1}{\sqrt{2\pi\sigma^2}} \exp\left(-\frac{(y-\mu)^2}{2\sigma^2}\right)$$

which can be rewritten in exponential form:

$$\exp\left(\frac{y\mu - \frac{1}{2}\mu^2}{\sigma^2} - \frac{y^2}{2\sigma^2} - \frac{1}{2}\log(2\pi\sigma^2)\right)$$

From this, we identify:

- $\theta = \mu$
- $b(\theta) = \frac{\theta^2}{2}$
- $a(\phi) = \sigma^2$

Computing derivatives:

$$b'(\theta) = \frac{\partial b(\theta)}{\partial \theta} = \mu, \quad V(\mu) = b''(\theta) = 1$$

Thus,

$$E(Y) = \mu, \quad \text{Var}(Y) = a(\phi)V(\mu) = \sigma^2$$

2. Poisson Distribution
For a Poisson-distributed response $Y \sim \text{Poisson}(\mu)$, the probability mass function is:

$$f(y; \mu) = \frac{\mu^y e^{-\mu}}{y!}$$

Rewriting in exponential form:

$$\exp(y \log \mu - \mu - \log y!)$$

Thus, we identify:

- $\theta = \log \mu$
- $b(\theta) = e^\theta$
- $a(\phi) = 1$
- $c(y, \phi) = \log y!$

Computing derivatives:

$$E(Y) = b'(\theta) = e^{\theta} = \mu, \quad \text{Var}(Y) = b''(\theta) = \mu$$

Since $\mu = E(Y)$, we confirm the variance function:

$$\text{Var}(Y) = V(\mu) = \mu$$

4.8.3 Structure of a Generalized Linear Model

In a GLM, we model the mean μ through a **link function** that connects it to a linear predictor:

$$g(\mu) = g(b'(\theta)) = \mathbf{x}'\boldsymbol{\beta}$$

Equivalently,

$$\mu = g^{-1}(\mathbf{x}'\boldsymbol{\beta})$$

where:

- $g(\cdot)$ is the **link function**, which ensures a transformation between the expected response $E(Y) = \mu$ and the linear predictor.
- $\eta = \mathbf{x}'\boldsymbol{\beta}$ is called the **linear predictor**.

4.8.4 Components of a GLM

A GLM consists of two main components:

4.8.4.1 Random Component

This describes the **distribution of the response variable** $Y_1, \ldots, Y_n$. The response variables are assumed to follow a distribution from the **exponential family**, which can be written as:

$$f(y_i; \theta_i, \phi) = \exp\left(\frac{\theta_i y_i - b(\theta_i)}{a(\phi)} + c(y_i, \phi)\right)$$

where:

- Y_i are **independent** random variables.
- The **canonical parameter** θ_i may differ for each observation.

- The **dispersion parameter** ϕ is assumed to be constant across all i.
- The mean response is given by:

$$\mu_i = E(Y_i)$$

4.8.4.2 Systematic Component

This specifies how the mean response μ is related to the explanatory variables $\mathbf{x}$ through a **linear predictor** η:

- The systematic component consists of:
  - A **link function** $g(\mu)$.
  - A **linear predictor** $\eta = \mathbf{x}'\boldsymbol{\beta}$.
- **Notation**:
  - We assume:

$$g(\mu_i) = \mathbf{x}'\boldsymbol{\beta} = \eta_i$$

  - The parameter vector $\boldsymbol{\beta} = (\beta_1, \ldots, \beta_p)'$ needs to be estimated.

4.8.5 Canonical Link

In a GLM, a link function $g(\cdot)$ relates the mean μ_i of the response Y_i to the linear predictor η_i via

$$\eta_i = g(\mu_i).$$

A **canonical link** is a special case of $g(\cdot)$ where

$$g(\mu_i) = \eta_i = \theta_i,$$

and θ_i is the **natural parameter** of the exponential family. In other words, the link function directly equates the linear predictor η_i with the distribution's natural parameter θ_i. Hence, $g(\mu)$ is **canonical** if $g(\mu) = \theta$ (Fig. 4.13, Table 4.3).

Exponential Family Components

- $b(\theta)$: the **cumulant (moment generating) function**, which defines the variance function.
- $g(\mu)$: the **link function**, which must be:

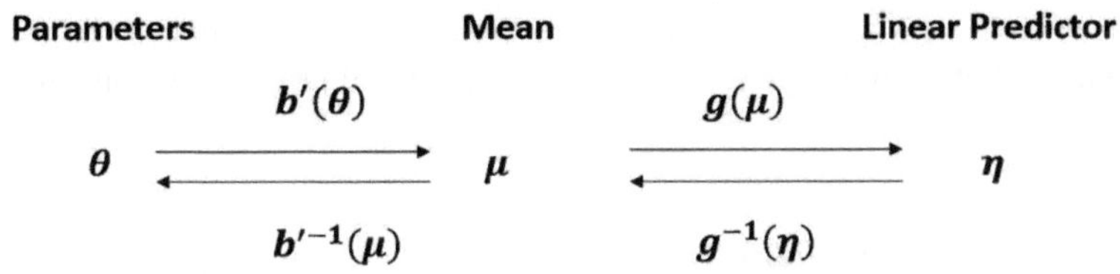

Fig. 4.13 GLM structure

Table 4.3 Common link functions in generalized linear models

Link	$\eta_i = g(\mu_i)$	$\mu_i = g^{-1}(\eta_i)$
Identity	μ_i	η_i
Log	$\log_e \mu_i$	e^{η_i}
Inverse	μ_i^{-1}	η_i^{-1}
Inverse square	μ_i^{-2}	$\eta_i^{-1/2}$
Square root	$\sqrt{\mu_i}$	η_i^2
Logit	$\log_e \left(\frac{\mu_i}{1-\mu_i}\right)$	$\frac{1}{1+e^{-\eta_i}}$
Probit	$\Phi^{-1}(\mu_i)$	$\Phi(\eta_i)$

– Monotonically increasing
– Continuously differentiable
– Invertible

For an exponential family distribution, the function $b(\theta)$ is called the *cumulant moment generating function*, and it relates θ to the mean via its derivative:

$$\mu = b'(\theta) \quad \Longleftrightarrow \quad \theta = b'^{-1}(\mu).$$

By defining the link so that $g(\mu) = \theta$, we impose $\eta_i = \theta_i$, which is why $g(\cdot)$ is termed **canonical** in this setting.

When the link is canonical, an equivalent way to express this is

$$\gamma^{-1} \circ g^{-1} = I,$$

indicating that the inverse link $g^{-1}(\cdot)$ directly maps the linear predictor (now the natural parameter θ) back to μ in a way that respects the structure of the exponential family.

Choosing $g(\cdot)$ to be canonical often simplifies mathematical derivations and computations—especially for parameter estimation—because the linear predictor η and the natural parameter θ coincide. Common examples of canonical links include:

- **Identity link** for the normal (Gaussian) distribution
- **Log link** for the Poisson distribution
- **Logit link** for the Bernoulli (binomial) distribution

In each case, setting $\eta = \theta$ streamlines the relationship between the mean and the linear predictor, making the model both elegant and practically convenient.

4.8.6 *Inverse Link Functions*

The **inverse link function** $g^{-1}(\eta)$ (also called the **mean function**) transforms the linear predictor η (which can take any real value) into a **valid mean response** μ.

Example 1 (Normal Distribution (Identity Link))

- Random component: $Y_i \sim N(\mu_i, \sigma^2)$.
- Mean response: $\mu_i = \theta_i$.
- Canonical link function:

$$g(\mu_i) = \mu_i$$

(i.e., the identity function).

Example 2 (Binomial Distribution (Logit Link))

- Random component: $Y_i \sim \text{Binomial}(n_i, p_i)$.
- Mean response:

$$\mu_i = \frac{n_i e^{\theta_i}}{1 + e^{\theta_i}}$$

- Canonical link function:

$$g(\mu_i) = \log\left(\frac{\mu_i}{n_i - \mu_i}\right)$$

(Logit link function).

Example 3 (Poisson Distribution (Log Link))

- Random component: $Y_i \sim \text{Poisson}(\mu_i)$.
- Mean response:

$$\mu_i = e^{\theta_i}$$

- Canonical link function:

$$g(\mu_i) = \log(\mu_i)$$

(Log link function).

Example 4 (Gamma Distribution (Inverse Link))

- Random component: $Y_i \sim \text{Gamma}(\alpha, \mu_i)$.
- Mean response:

$$\mu_i = -\frac{1}{\theta_i}$$

- Canonical link function:

$$g(\mu_i) = -\frac{1}{\mu_i}$$

(Inverse link function).

The following table presents common **GLM link functions** and their corresponding **inverse functions**.

where:

- μ_i is the expected value of the response.
- η_i is the linear predictor.
- $\Phi(\cdot)$ represents the CDF of the standard normal distribution.

4.8.7 Estimation of Parameters in GLMs

The GLM framework extends linear regression by allowing for response variables that follow **exponential family distributions**:

- Maximum Likelihood Estimation is used to estimate the parameters of the systematic component (β), providing a consistent and efficient approach. The derivation and computation processes are unified, thanks to the exponential form of the model, which simplifies mathematical treatment and implementation.
- However, this unification does not extend to the estimation of the dispersion parameter (ϕ), which requires separate treatment, often involving alternative estimation methods such as moment-based approaches or quasi-likelihood estimation.

In GLMs, the response variable Y_i follows an exponential family distribution characterized by the density function:

$$f(y_i; \theta_i, \phi) = \exp\left(\frac{\theta_i y_i - b(\theta_i)}{a(\phi)} + c(y_i, \phi)\right)$$

where:

- θ_i is the **canonical parameter**.
- ϕ is the **dispersion parameter** (which may be known or estimated separately).
- $b(\theta_i)$ determines the mean and variance of Y_i.

- $a(\phi)$ scales the variance.
- $c(y_i, \phi)$ ensures proper normalization.

  For this family, we obtain:

- **Mean of Y_i:**

$$E(Y_i) = \mu_i = b'(\theta_i)$$

- **Variance of Y_i:**

$$\text{Var}(Y_i) = b''(\theta_i)a(\phi) = V(\mu_i)a(\phi)$$

  where $V(\mu_i)$ is the variance function.
- **Systematic component (link function):**

$$g(\mu_i) = \eta_i = \mathbf{x}_i'\beta$$

The function $g(\cdot)$ is the link function, which connects the expected response μ_i to the linear predictor $\mathbf{x}_i'\beta$.

For a single observation Y_i, the log-likelihood function is:

$$l_i(\beta, \phi) = \frac{\theta_i y_i - b(\theta_i)}{a(\phi)} + c(y_i, \phi)$$

For n independent observations, the total log-likelihood is:

$$l(\beta, \phi) = \sum_{i=1}^{n} l_i(\beta, \phi)$$

Expanding this,

$$l(\beta, \phi) = \sum_{i=1}^{n} \left(\frac{\theta_i y_i - b(\theta_i)}{a(\phi)} + c(y_i, \phi) \right).$$

To estimate β, we maximize this log-likelihood function.

4.8.7.1 Estimation of Systematic Component (β)

To differentiate $l(\beta, \phi)$ with respect to β_j, we apply the chain rule:

$$\frac{\partial l_i(\beta, \phi)}{\partial \beta_j} = \underbrace{\frac{\partial l_i(\beta, \phi)}{\partial \theta_i}}_{\text{depends on } (y_i - \mu_i)} \times \underbrace{\frac{\partial \theta_i}{\partial \mu_i}}_{=1/V(\mu_i) \text{ if canonical link}} \times \underbrace{\frac{\partial \mu_i}{\partial \eta_i}}_{\text{depends on the link}} \times \underbrace{\frac{\partial \eta_i}{\partial \beta_j}}_{=x_{ij}}.$$

Let us see why these four pieces appear:

1. $l_i(\beta, \phi)$ depends on θ_i. So we start by computing $\frac{\partial l_i}{\partial \theta_i}$.
2. θ_i (the "natural parameter" in the exponential family) may in turn be a function of μ_i:
   - In **canonical** link GLMs, we often have $\theta_i = \eta_i$.
   - In more **general** link GLMs, θ_i is still some function of μ_i.
     Hence we need $\frac{\partial \theta_i}{\partial \mu_i}$.
3. μ_i (the mean) is a function of the linear predictor η_i. Typically, $\eta_i = g(\mu_i)$ implies $\mu_i = g^{-1}(\eta_i)$. So we need $\frac{\partial \mu_i}{\partial \eta_i}$.
4. Finally, $\eta_i = \mathbf{x}_i'\beta$. So the derivative $\frac{\partial \eta_i}{\partial \beta_j}$ is simply x_{ij}, the jth component of the covariate vector $\mathbf{x}_i$.

Let us look at each factor in turn.

First Term

$$\frac{\partial l_i(\beta, \phi)}{\partial \theta_i}$$

Recall

$$l_i(\theta_i, \phi) = \frac{\theta_i y_i - b(\theta_i)}{a(\phi)} + c(y_i, \phi).$$

1. Differentiate $\theta_i y_i - b(\theta_i)$ with respect to θ_i:

$$\frac{\partial}{\partial \theta_i} [\theta_i y_i - b(\theta_i)] = y_i - b'(\theta_i).$$

   But by exponential family definitions,$b'(\theta_i) = \mu_i$.
   So that is $y_i - \mu_i$.
2. Since everything is divided by $a(\phi)$, we get

$$\frac{\partial l_i}{\partial \theta_i} = \frac{1}{a(\phi)}[y_i - \mu_i].$$

Hence,

$$\boxed{\frac{\partial l_i(\beta, \phi)}{\partial \theta_i} = \frac{y_i - \mu_i}{a(\phi)}.}$$

Second Term

$$\frac{\partial \theta_i}{\partial \mu_i}$$

1. In an exponential family with **canonical link**, we have

$$\theta_i = \eta_i = \mathbf{x}_i' \beta.$$

Then θ_i is actually the same as η_i, which is the same as $g(\mu_i)$. Recall also that if $\mu_i = b'(\theta_i)$, then $d\mu_i/d\theta_i = b''(\theta_i)$. But $b''(\theta_i) = V(\mu_i)$. Hence

$$\frac{d\mu_i}{d\theta_i} = V(\mu_i) \quad \Longrightarrow \quad \frac{d\theta_i}{d\mu_i} = \frac{1}{V(\mu_i)}.$$

This identity is a well-known property of **canonical** links in the exponential family.

2. In more general (non-canonical) links, θ_i may be some other smooth function of μ_i. The key idea is that if $\mu_i = b'(\theta_i)$ but $\eta_i \neq \theta_i$, you would have to carefully derive $\partial\theta_i/\partial\mu_i$. Often, a canonical link is assumed to keep expressions simpler.

   If we **assume** a canonical link throughout, then

$$\boxed{\frac{\partial \theta_i}{\partial \mu_i} = \frac{1}{V(\mu_i)}.}$$

Third Term

$$\frac{\partial \mu_i}{\partial \eta_i}$$

Here we consider the link function $g(\cdot)$, defined by

$$\eta_i = g(\mu_i) \quad \Longrightarrow \quad \mu_i = g^{-1}(\eta_i).$$

For example,

- In a Bernoulli (logistic regression) model, $\mu g(\mu) = \log \frac{\mu}{1-\mu}$. So $\mu = g^{-1}(\eta) = \frac{1}{1+e^{-\eta}}$. Then $\frac{d\mu}{d\eta} = \mu(1-\mu)$.
- For a Poisson (log) link, $g(\mu) = \log(\mu)$. So $\mu = e^{\eta}$. Then $\frac{d\mu}{d\eta} = e^{\eta} = \mu$.
- For an identity link, $g(\mu) = \mu$. Then $\eta = \mu$ and $\frac{d\mu}{d\eta} = 1$.

In general,

$$\boxed{\frac{\partial \mu_i}{\partial \eta_i} = \frac{d}{d\eta}\left[g^{-1}(\eta)\right]\Big|_{\eta=\eta_i} = \left(g^{-1}\right)'(\eta_i).}$$

If the link is *also* canonical, then typically $\frac{\partial \mu_i}{\partial \eta_i} = V(\mu_i)$. Indeed, that is consistent with the second term result.

Fourth Term

$$\frac{\partial \eta_i}{\partial \beta_j}$$

Finally, the linear predictor is

$$\eta_i = \mathbf{x}_i'\beta = \sum_{k=1}^{p} x_{ik}\beta_k.$$

Hence, the derivative of η_i with respect to β_j is simply

$$\boxed{\frac{\partial \eta_i}{\partial \beta_j} = x_{ij}.}$$

Therefore, for the **entire** log-likelihood $l(\beta, \phi) = \sum_{i=1}^{n} l_i(\beta, \phi)$, we sum over i:

$$\boxed{\frac{\partial l(\beta, \phi)}{\partial \beta_j} = \sum_{i=1}^{n}\left[\frac{y_i - \mu_i}{a(\phi)} \times \frac{1}{V(\mu_i)} \times \frac{\partial \mu_i}{\partial \eta_i} \times x_{ij}\right]}$$

To simplify expressions, we define the weight:

$$w_i = \left(\left(\frac{\partial \eta_i}{\partial \mu_i}\right)^2 V(\mu_i)\right)^{-1}.$$

For canonical links, this often simplifies to $w_i = V(\mu_i)$, such as:

- **Bernoulli (logit link)**: $w_i = p_i(1 - p_i)$.
- **Poisson (log link)**: $w_i = \mu_i$.

Rewriting the score function in terms of w_i:

$$\frac{\partial l(\beta, \phi)}{\partial \beta_j} = \sum_{i=1}^{n}\left[\frac{y_i - \mu_i}{a(\phi)} \times w_i \times \frac{\partial \eta_i}{\partial \mu_i} \times x_{ij}\right].$$

To use the Newton-Raphson Algorithm for estimating β, we require the expected second derivative:

$$- E\left(\frac{\partial^2 l(\beta, \phi)}{\partial \beta_j \partial \beta_k}\right),$$

which corresponds to the (j, k)th element of the Fisher information matrix $\mathbf{I}(\beta)$:

$$\mathbf{I}_{jk}(\beta) = -E\left(\frac{\partial^2 l(\beta, \phi)}{\partial \beta_j \partial \beta_k}\right) = \sum_{i=1}^{n} \frac{w_i}{a(\phi)} x_{ij} x_{ik}.$$

Example (Bernoulli Model with Logit Link) For a Bernoulli response with a logit link function (canonical link), we have:

$$\begin{aligned}
b(\theta) &= \log(1 + \exp(\theta)) = \log(1 + \exp(\mathbf{x}'\beta)), \\
a(\phi) &= 1, \\
c(y_i, \phi) &= 0.
\end{aligned}$$

From the mean and link function:

$$\begin{aligned}
E(Y) = b'(\theta) &= \frac{\exp(\theta)}{1 + \exp(\theta)} = \mu = p, \\
\eta = g(\mu) &= \log\left(\frac{\mu}{1 - \mu}\right) = \theta = \log\left(\frac{p}{1 - p}\right) = \mathbf{x}'\beta.
\end{aligned}$$

The log-likelihood for Y_i is:

$$l_i(\beta, \phi) = \frac{y_i \theta_i - b(\theta_i)}{a(\phi)} + c(y_i, \phi) = y_i \mathbf{x}'_i \beta - \log(1 + \exp(\mathbf{x}'\beta)).$$

We also obtain:

$$\begin{aligned}
V(\mu_i) &= \mu_i(1 - \mu_i) = p_i(1 - p_i), \\
\frac{\partial \mu_i}{\partial \eta_i} &= p_i(1 - p_i).
\end{aligned}$$

Thus, the first derivative simplifies as:

$$\frac{\partial l(\beta,\phi)}{\partial \beta_j} = \sum_{i=1}^{n}\left[\frac{y_i-\mu_i}{a(\phi)}\times\frac{1}{V(\mu_i)}\times\frac{\partial \mu_i}{\partial \eta_i}\times x_{ij}\right]$$
$$= \sum_{i=1}^{n}(y_i-p_i)\times\frac{1}{p_i(1-p_i)}\times p_i(1-p_i)\times x_{ij}$$
$$= \sum_{i=1}^{n}(y_i-p_i)x_{ij}$$
$$= \sum_{i=1}^{n}\left(y_i-\frac{\exp(\mathbf{x}'_i\beta)}{1+\exp(\mathbf{x}'_i\beta)}\right)x_{ij}.$$

The weight term in this case is:

$$w_i = \left(\left(\frac{\partial \eta_i}{\partial \mu_i}\right)^2 V(\mu_i)\right)^{-1} = p_i(1-p_i).$$

Thus, the Fisher information matrix has elements:

$$\mathbf{I}_{jk}(\beta) = \sum_{i=1}^{n}\frac{w_i}{a(\phi)}x_{ij}x_{ik} = \sum_{i=1}^{n}p_i(1-p_i)x_{ij}x_{ik}.$$

The Fisher-Scoring algorithm for the Maximum Likelihood Estimate of $\boldsymbol{\beta}$ is given by:

$$\begin{pmatrix}\beta_1\\ \beta_2\\ \vdots\\ \beta_p\end{pmatrix}^{(m+1)} = \begin{pmatrix}\beta_1\\ \beta_2\\ \vdots\\ \beta_p\end{pmatrix}^{(m)} + \mathbf{I}^{-1}(\boldsymbol{\beta})\begin{pmatrix}\frac{\partial l(\beta,\phi)}{\partial \beta_1}\\ \frac{\partial l(\beta,\phi)}{\partial \beta_2}\\ \vdots\\ \frac{\partial l(\beta,\phi)}{\partial \beta_p}\end{pmatrix}\Bigg|_{\beta=\beta^{(m)}}$$

This is similar to the Newton-Raphson Algorithm, except that we replace the observed matrix of the second derivatives (Hessian) with its expected value.

In matrix representation, the score function (gradient of the log-likelihood) is:

$$\frac{\partial l}{\partial \beta} = \frac{1}{a(\phi)}\mathbf{X}'\mathbf{W}\boldsymbol{\Delta}(\mathbf{y}-\boldsymbol{\mu})$$
$$= \frac{1}{a(\phi)}\mathbf{F}'\mathbf{V}^{-1}(\mathbf{y}-\boldsymbol{\mu})$$

The expected Fisher information matrix is:

$$\mathbf{I}(\beta) = \frac{1}{a(\phi)}\mathbf{X}'\mathbf{W}\mathbf{X} = \frac{1}{a(\phi)}\mathbf{F}'\mathbf{V}^{-1}\mathbf{F}$$

where:

- $\mathbf{X}$ is an $n \times p$ matrix of covariates.
- $\mathbf{W}$ is an $n \times n$ diagonal matrix with (i, i)th element given by w_i.
- $\boldsymbol{\Delta}$ is an $n \times n$ diagonal matrix with (i, i)th element given by $\frac{\partial \eta_i}{\partial \mu_i}$.
- $\mathbf{F} = \frac{\partial \mu}{\partial \beta}$ is an $n \times p$ matrix, where the ith row is given by $\frac{\partial \mu_i}{\partial \beta} = (\frac{\partial \mu_i}{\partial \eta_i})\mathbf{x}_i'$.
- $\mathbf{V}$ is an $n \times n$ diagonal matrix with (i, i)th element given by $V(\mu_i)$.

Setting the derivative of the log-likelihood equal to zero gives the MLE equations:

$$\mathbf{F}'\mathbf{V}^{-1}\mathbf{y} = \mathbf{F}'\mathbf{V}^{-1}\boldsymbol{\mu}$$

Since all components of this equation (except for $\mathbf{y}$) depend on β, we solve iteratively.

Special Cases

1. Canonical link function
   If a canonical link is used, the estimating equations simplify to:

$$\mathbf{X}'\mathbf{y} = \mathbf{X}'\boldsymbol{\mu}$$

2. Identity link function

   If the identity link is used, the estimating equation reduces to:

$$\mathbf{X}'\mathbf{V}^{-1}\mathbf{y} = \mathbf{X}'\mathbf{V}^{-1}\mathbf{X}\hat{\boldsymbol{\beta}}$$

   which leads to the generalized least squares estimator:

$$\hat{\beta} = (\mathbf{X}'\mathbf{V}^{-1}\mathbf{X})^{-1}\mathbf{X}'\mathbf{V}^{-1}\mathbf{y}$$

Fisher-Scoring Algorithm in a General Form
The iterative update formula for Fisher-Scoring can be rewritten as:

$$\beta^{(m+1)} = \beta^{(m)} + (\hat{\mathbf{F}}'\hat{\mathbf{V}}^{-1}\hat{\mathbf{F}})^{-1}\hat{\mathbf{F}}'\hat{\mathbf{V}}^{-1}(\mathbf{y} - \hat{\boldsymbol{\mu}})$$

Since $\hat{F}, \hat{V}, \hat{\mu}$ depend on β, we evaluate them at $\beta^{(m)}$.
From an initial guess $\beta^{(0)}$, we iterate until convergence.

Table 4.4 Comparison of dispersion parameter estimation methods in GLMs

Approach	Description	Pros	Cons
MLE	Estimates ϕ by maximizing the likelihood function	Theoretically optimal	Computationally complex, lacks a general closed-form solution
Moment estimation	Uses a bias-corrected χ^2 method based on residual variance	Simpler, widely used in GLMs	Not as efficient as MLE

Notes

- If $a(\phi)$ is a constant or takes the form $m_i\phi$ with known m_i, then ϕ cancels from the equations, simplifying the estimation.

4.8.7.2 Estimation of Dispersion Parameter (ϕ)

There are two common approaches to estimating ϕ (Table 4.4):

1. Maximum Likelihood Estimation

   The derivative of the log-likelihood with respect to ϕ is:

$$\frac{\partial l_i}{\partial \phi} = \frac{(\theta_i y_i - b(\theta_i)a'(\phi))}{a^2(\phi)} + \frac{\partial c(y_i, \phi)}{\partial \phi}$$

   The MLE of ϕ satisfies the equation:

$$\frac{a^2(\phi)}{a'(\phi)} \sum_{i=1}^{n} \frac{\partial c(y_i, \phi)}{\partial \phi} = \sum_{i=1}^{n} (\theta_i y_i - b(\theta_i))$$

Challenges

- For distributions other than the normal case, the expression for $\frac{\partial c(y,\phi)}{\partial \phi}$ is often complicated.
- Even with a **canonical link function** and constant $a(\phi)$, there is no simple general expression for the expected Fisher information:

$$-E\left(\frac{\partial^2 l}{\partial \phi^2}\right)$$

This means that the unification GLMs provide for estimating β does not extend as neatly to ϕ.

2. Moment Estimation (Bias-Corrected χ^2 Method)

- The MLE is **not** the conventional approach for estimating ϕ in generalized linear models.

- For an exponential family distribution, the variance function is:

$$\text{Var}(Y) = V(\mu)a(\phi)$$

This implies the following moment-based estimator:

$$a(\phi) = \frac{\text{Var}(Y)}{V(\mu)} = \frac{E(Y-\mu)^2}{V(\mu)}$$
$$a(\hat{\phi}) = \frac{1}{n-p}\sum_{i=1}^{n}\frac{(y_i - \hat{\mu}_i)^2}{V(\hat{\mu}_i)}$$

where p is the number of parameters in β.
- For a **GLM with a canonical link function** $g(.) = (b'(.))^{-1}$:

$$g(\mu) = \theta = \eta = \mathbf{x}'\boldsymbol{\beta}$$
$$\mu = g^{-1}(\eta) = b'(\eta)$$

Using this, the moment estimator for $a(\phi) = \phi$ becomes:

$$\hat{\phi} = \frac{1}{n-p}\sum_{i=1}^{n}\frac{(y_i - g^{-1}(\hat{\eta}_i))^2}{V(g^{-1}(\hat{\eta}_i))}$$

4.8.8 Inference

The estimated variance of $\hat{\beta}$ is given by:

$$\hat{\text{var}}(\beta) = a(\phi)(\hat{\mathbf{F}}'\hat{\mathbf{V}}^{-1}\hat{\mathbf{F}})^{-1}$$

where:

- $\mathbf{V}$ is an $n \times n$ diagonal matrix with diagonal elements given by $V(\mu_i)$.
- $\mathbf{F}$ is an $n \times p$ matrix given by:

$$\mathbf{F} = \frac{\partial \mu}{\partial \beta}$$

- Since $\mathbf{V}$ and $\mathbf{F}$ depend on the mean μ (and thus on β), their estimates $\hat{\mathbf{V}}$ and $\hat{\mathbf{F}}$ depend on $\hat{\beta}$.

To test a general hypothesis:

$$H_0 : \mathbf{L}\boldsymbol{\beta} = \mathbf{d}$$

where $\mathbf{L}$ is a $q \times p$ matrix, we use the **Wald test**:

$$W = (\mathbf{L}\hat{\boldsymbol{\beta}} - \mathbf{d})'(\mathbf{a}(\boldsymbol{\phi})\mathbf{L}(\hat{\mathbf{F}}'\hat{\mathbf{V}}^{-1}\hat{\mathbf{F}})\mathbf{L}')^{-1}(\mathbf{L}\hat{\boldsymbol{\beta}} - \mathbf{d})$$

Under H_0, the Wald statistic follows a **chi-square** distribution:

$$W \sim \chi_q^2$$

where q is the rank of $\mathbf{L}$.

Special Case: Testing a Single Coefficient
For a hypothesis of the form:

$$H_0 : \beta_j = 0$$

the Wald test simplifies to:

$$W = \frac{\hat{\beta}_j^2}{\hat{\text{var}}(\hat{\beta}_j)} \sim \chi_1^2$$

asymptotically.

Another common test is the likelihood ratio test, which compares the likelihoods of a **full model** and a **reduced model**:

$$\Lambda = 2\big(l(\hat{\beta}_f) - l(\hat{\beta}_r)\big) \sim \chi_q^2$$

where:

- $l(\hat{\beta}_f)$ is the log-likelihood of the **full** model.
- $l(\hat{\beta}_r)$ is the log-likelihood of the **reduced** model.
- q is the number of constraints used in fitting the reduced model.

While the Wald test is more convenient, the likelihood ratio test is often preferred when sample sizes are small, as it tends to have better statistical properties (Table 4.5).

Table 4.5 Comparison of Wald test and likelihood ratio test in model evaluation

Test	Pros	Cons
Wald test	Easy to compute, does not require fitting two models	May perform poorly in small samples
Likelihood ratio test	More accurate, especially for small samples	Requires fitting both full and reduced models

4.8.9 Deviance

Deviance plays a crucial role in:

- **Goodness-of-fit assessment**: Checking how well a model explains the observed data.
- **Statistical inference**: Used in hypothesis testing, particularly likelihood ratio tests.
- **Estimating dispersion parameters**: Helps in refining variance estimates.
- **Model comparison**: Facilitates selection between competing models.

Assuming the dispersion parameter ϕ is known, let:

- $\tilde{\theta}$ be the **MLE** under the **full model**.
- $\hat{\theta}$ be the **MLE** under the **reduced model**.

The **likelihood ratio statistic** (twice the difference in log-likelihoods) is:

$$2\sum_{i=1}^{n} \frac{y_i(\tilde{\theta}_i - \hat{\theta}_i) - b(\tilde{\theta}_i) + b(\hat{\theta}_i)}{a_i(\phi)}$$

For exponential family distributions, the mean parameter is:

$$\mu = E(y) = b'(\theta)$$

Thus, the **natural parameter** is a function of μ:

$$\theta = \theta(\mu) = b'^{-1}(\mu)$$

Rewriting the likelihood ratio statistic in terms of μ:

$$2\sum_{i=1}^{n} \frac{y_i\{\theta(\tilde{\mu}_i) - \theta(\hat{\mu}_i)\} - b(\theta(\tilde{\mu}_i)) + b(\theta(\hat{\mu}_i))}{a_i(\phi)}.$$

A **saturated model** provides the fullest possible fit, where each observation is perfectly predicted:

$$\tilde{\mu}_i = y_i, \quad i = 1, \ldots, n.$$

Setting $\tilde{\theta}_i^* = \theta(y_i)$ and $\hat{\theta}_i^* = \theta(\hat{\mu}_i)$, the likelihood ratio simplifies to:

$$2\sum_{i=1}^{n} \frac{y_i(\tilde{\theta}_i^* - \hat{\theta}_i^*) - b(\tilde{\theta}_i^*) + b(\hat{\theta}_i^*)}{a_i(\phi)}.$$

Following McCullagh (2019), for $a_i(\phi) = \phi$, we define the **scaled deviance** as:

$$D^*(\mathbf{y}, \hat{\boldsymbol{\mu}}) = \frac{2}{\phi} \sum_{i=1}^{n} \{y_i(\tilde{\theta}_i^* - \hat{\theta}_i^*) - b(\tilde{\theta}_i^*) + b(\hat{\theta}_i^*)\}$$

and the **deviance** as:

$$D(\mathbf{y}, \hat{\boldsymbol{\mu}}) = \phi D^*(\mathbf{y}, \hat{\boldsymbol{\mu}})$$

where:

- $D^*(\mathbf{y}, \hat{\boldsymbol{\mu}})$ is **scaled deviance.**
- $D(\mathbf{y}, \hat{\boldsymbol{\mu}})$ is **deviance.**

In some models, $a_i(\phi) = \phi m_i$, where m_i is a known scalar that varies across observations. Then, the **scaled deviance** is:

$$D^*(\mathbf{y}, \hat{\boldsymbol{\mu}}) = \sum_{i=1}^{n} \frac{2\{y_i(\tilde{\theta}_i^* - \hat{\theta}_i^*) - b(\tilde{\theta}_i^*) + b(\hat{\theta}_i^*)\}}{\phi m_i}.$$

The deviance is often decomposed into **deviance contributions**:

$$D^*(\mathbf{y}, \hat{\boldsymbol{\mu}}) = \sum_{i=1}^{n} d_i,$$

where d_i is the **deviance contribution** from the ith observation.

Uses of Deviance

- D is used for model selection.
- D^* is used for goodness-of-fit tests, as it is a likelihood ratio statistic:

$$D^*(\mathbf{y}, \hat{\boldsymbol{\mu}}) = 2\{l(\mathbf{y}, \tilde{\boldsymbol{\mu}}) - l(\mathbf{y}, \hat{\boldsymbol{\mu}})\}$$

- The individual deviance contributions d_i are used to form deviance residuals.

Deviance for Normal Distribution

For the normal distribution:

$$\theta = \mu,$$

$$\phi = \sigma^2,$$

$$b(\theta) = \frac{1}{2}\theta^2,$$

$$a(\phi) = \phi.$$

The MLEs are:

$$\tilde{\theta}_i = y_i,$$
$$\hat{\theta}_i = \hat{\mu}_i = g^{-1}(\hat{\eta}_i).$$

The **deviance** simplifies to:

$$\begin{aligned} D &= 2\sum_{i=1}^{n}\left(y_i^2 - y_i\hat{\mu}_i - \frac{1}{2}y_i^2 + \frac{1}{2}\hat{\mu}_i^2\right) \\ &= \sum_{i=1}^{n}(y_i^2 - 2y_i\hat{\mu}_i + \hat{\mu}_i^2) \\ &= \sum_{i=1}^{n}(y_i - \hat{\mu}_i)^2. \end{aligned}$$

Thus, for the normal model, the deviance corresponds to the residual sum of squares.

Deviance for Poisson Distribution
For the Poisson distribution:

$$\begin{aligned} f(y) &= \exp\{y\log(\mu) - \mu - \log(y!)\}, \\ \theta &= \log(\mu), \\ b(\theta) &= \exp(\theta), \\ a(\phi) &= 1. \end{aligned}$$

MLEs:

$$\begin{aligned} \tilde{\theta}_i &= \log(y_i), \\ \hat{\theta}_i &= \log(\hat{\mu}_i), \\ \hat{\mu}_i &= g^{-1}(\hat{\eta}_i). \end{aligned}$$

The **deviance** simplifies to:

$$\begin{aligned} D &= 2\sum_{i=1}^{n}\left(y_i\log(y_i) - y_i\log(\hat{\mu}_i) - y_i + \hat{\mu}_i\right) \\ &= 2\sum_{i=1}^{n}\left[y_i\log\left(\frac{y_i}{\hat{\mu}_i}\right) - (y_i - \hat{\mu}_i)\right]. \end{aligned}$$

The **deviance contribution** for each observation:

$$d_i = 2\left[y_i \log\left(\frac{y_i}{\hat{\mu}_i}\right) - (y_i - \hat{\mu}_i)\right].$$

4.8.9.1 Analysis of Deviance

The **analysis of deviance** is a likelihood-based approach for comparing nested models in GLM. It is analogous to the [analysis of variance (ANOVA)] in linear models.

When comparing a **reduced model** (denoted by $\hat{\mu}_r$) and a **full model** (denoted by $\hat{\mu}_f$), the difference in **scaled deviance** follows an asymptotic chi-square distribution:

$$D^*(\mathbf{y}; \hat{\boldsymbol{\mu}}_\mathbf{r}) - D^*(\mathbf{y}; \hat{\boldsymbol{\mu}}_\mathbf{f}) = 2\{l(\mathbf{y}; \hat{\boldsymbol{\mu}}_\mathbf{f}) - l(\mathbf{y}; \hat{\boldsymbol{\mu}}_\mathbf{r})\} \sim \chi^2_q,$$

where q is the difference in the number of free parameters between the two models.

This test provides a means to assess whether the additional parameters in the full model significantly improve model fit.

The dispersion parameter ϕ is estimated as:

$$\hat{\phi} = \frac{D(\mathbf{y}, \hat{\boldsymbol{\mu}})}{n - p},$$

where:

- $D(\mathbf{y}, \hat{\boldsymbol{\mu}})$ is the deviance of the fitted model.
- n is the total number of observations.
- p is the number of estimated parameters.

Caution The frequent use of χ^2 tests can be problematic due to their reliance on asymptotic approximations, especially for small samples or overdispersed data (McCullagh, 2019).

4.8.9.2 Deviance Residuals

Since deviance plays a role in model evaluation, we define **deviance residuals** to examine individual data points. Given that the total deviance is:

$$D = \sum_{i=1}^{n} d_i,$$

the **deviance residual** for observation i is:

$$r_{D_i} = \text{sign}(y_i - \hat{\mu}_i)\sqrt{d_i},$$

where:

- d_i is the deviance contribution from observation i.
- The **sign function** preserves the direction of the residual.

To account for varying leverage, we define the **standardized deviance residual**:

$$r_{s,i} = \frac{y_i - \hat{\mu}_i}{\hat{\sigma}(1 - h_{ii})^{1/2}},$$

where:

- h_{ii} is the **leverage** of observation i.
- $\hat{\sigma}$ is an estimate of the standard deviation.

Alternatively, using the **GLM hat matrix**:

$$\mathbf{H}^{GLM} = \mathbf{W}^{1/2} X (X' W X)^{-1} X' \mathbf{W}^{-1/2},$$

where $\mathbf{W}$ is an $n \times n$ diagonal matrix with (i, i)th element given by w_i (see Eestimation of systematic component (β)), we express standardized deviance residuals as:

$$r_{s,D_i} = \frac{r_{D_i}}{\{\hat{\phi}(1 - h_{ii}^{glm})\}^{1/2}},$$

where h_{ii}^{glm} is the ith diagonal element of $\mathbf{H}^{GLM}$.

4.8.9.3 Pearson Chi-Square Residuals

Another goodness-of-fit statistic is the **Pearson Chi-Square** statistic, defined as:

$$X^2 = \sum_{i=1}^{n} \frac{(y_i - \hat{\mu}_i)^2}{V(\hat{\mu}_i)},$$

where:

- $\hat{\mu}_i$ is the fitted mean response.
- $V(\hat{\mu}_i)$ is the variance function of the response.

The **scaled Pearson** χ^2 statistic is:

$$\frac{X^2}{\phi} \sim \chi^2_{n-p},$$

where p is the number of estimated parameters.

The **Pearson residuals** are:

$$X_i^2 = \frac{(y_i - \hat{\mu}_i)^2}{V(\hat{\mu}_i)}.$$

These residuals assess the difference between observed and fitted values, standardized by the variance.

If the assumptions hold:

- **Independent samples**
- **No overdispersion**: If $\phi = 1$, we expect:

$$\frac{D(\mathbf{y}; \hat{\boldsymbol{\mu}})}{n-p} \approx 1, \quad \frac{X^2}{n-p} \approx 1.$$

  A value **substantially greater than 1** suggests **overdispersion** or **model misspecification.**
- **Multiple groups**: If the model includes categorical predictors, separate calculations may be needed for each group.

Under these assumptions, both:

$$\frac{X^2}{\phi} \quad \text{and} \quad D^*(\mathbf{y}; \hat{\boldsymbol{\mu}})$$

follow a χ^2_{n-p} distribution.

```
set.seed(123)
n <- 100
x <- rnorm(n)
y <- rpois(n, lambda = exp(0.5 + 0.3 * x))

# Poisson response
# Fit Poisson GLM
fit <- glm(y ~ x, family = poisson(link = "log"))

# Extract deviance and Pearson residuals
deviance_residuals <- residuals(fit, type = "deviance")
pearson_residuals  <- residuals(fit, type = "pearson")
# Display residual summaries
summary(deviance_residuals)
#>    Min. 1st Qu.  Median    Mean 3rd Qu.    Max.
#> -2.1286 -1.0189 -0.1441 -0.1806  0.6171  1.8942
summary(pearson_residuals)
```

```
#>       Min.   1st Qu.     Median       Mean   3rd Qu.      Max.
↪
#> -1.505136 -0.837798 -0.140873  0.002003  0.658076  2.366618
```

- The residuals suggest a reasonably **well-fitted model** with a **slight underprediction** tendency.
- No extreme deviations or severe skewness.
- The largest residuals indicate **potential outliers**, but they are not extreme enough to immediately suggest model inadequacy.

4.8.10 Diagnostic Plots

Standardized residual plots help diagnose potential issues with model specification, such as an incorrect link function or variance structure. Common residual plots include:

- **Plot of Standardized Deviance Residuals vs. Fitted Values**

  $$\text{plot}(r_{s,D_i}, \hat{\mu}_i)$$

  or

  $$\text{plot}(r_{s,D_i}, T(\hat{\mu}_i)),$$

  where $T(\hat{\mu}_i)$ is a **constant information scale transformation**, which adjusts $\hat{\mu}_i$ to maintain a stable variance across observations.
- **Plot of Standardized Deviance Residuals vs. Linear Predictor**

  $$\text{plot}(r_{s,D_i}, \hat{\eta}_i),$$

  where $\hat{\eta}_i$ represents the linear predictor before applying the link function.

The following table summarizes the commonly used transformations $T(\hat{\mu}_i)$ for different random components (Table 4.6):
Interpretation of Residual Plots

Table 4.6 Variance-stabilizing transformations for common distributions in GLMs

Random component	$T(\hat{\mu}_i)$
Normal	$\hat{\mu}$
Poisson	$2\sqrt{\hat{\mu}}$
Binomial	$2\sin^{-1}(\sqrt{\hat{\mu}})$
Gamma	$2\log(\hat{\mu})$
Inverse Gaussian	$-2\hat{\mu}^{-1/2}$

- Trend in residuals: Indicates an incorrect link function or an inappropriate scale transformation.
- Systematic change in residual range with $T(\hat{\mu})$: Suggests an incorrect choice of random component—i.e., the assumed variance function does not match the data.
- Plot of absolute residuals vs. fitted values:

$$\text{plot}(|r_{D_i}|, \hat{\mu}_i).$$

This checks whether the variance function is correctly specified.

4.8.11 Goodness of Fit

To assess how well the model fits the data, we use:

- Deviance
- Pearson Chi-square residuals

For **nested models**, likelihood-based information criteria provide a comparison metric:

$$AIC = -2l(\hat{\boldsymbol{\mu}}) + 2p$$
$$AICC = -2l(\hat{\boldsymbol{\mu}}) + 2p\left(\frac{n}{n-p-1}\right)$$
$$BIC = -2l(\hat{\boldsymbol{\mu}}) + p\log(n),$$

where:

- $l(\hat{\mu})$ is the log-likelihood at the estimated parameters.
- p is the number of model parameters.
- n is the number of observations.

Important Considerations

- The **same data and model structure** (i.e., same link function and assumed random component distribution) must be used for comparisons.
- Models can differ in the **number of parameters** but must remain consistent in other respects.

While traditional R^2 is not directly applicable to GLMs, an analogous measure is:

$$R_p^2 = 1 - \frac{l(\hat{\mu})}{l(\hat{\mu}_0)},$$

Table 4.7 Modeling overdispersion in binomial and Poisson distributions

Random component	Standard assumption (var(Y))	Alternative model allowing overdispersion ($V(\mu)$)
Random component	Standard assumption (var(Y))	Alternative model allowing overdispersion ($V(\mu)$)
Binomial	$n\mu(1-\mu)$	$\phi n\mu(1-\mu)$, where $m_i = n$
Poisson	μ	$\phi\mu$

where $l(\hat{\mu}_0)$ is the log-likelihood of a model with only an intercept.

For binary response models, a **rescaled generalized** R^2 is often used:

$$\bar{R}^2 = \frac{R_*^2}{\max(R_*^2)} = \frac{1 - \exp\left(-\frac{2}{n}(l(\hat{\mu}) - l(\hat{\mu}_0))\right)}{1 - \exp\left(\frac{2}{n}l(\hat{\mu}_0)\right)},$$

where the denominator ensures the maximum possible R^2 scaling.

4.8.12 Overdispersion

Overdispersion occurs when the observed variance exceeds what the assumed model allows. This is common in Poisson and binomial models (Table 4.7).

Variance Assumptions for Common Random Components

- **By default,** $\phi = 1$, meaning the variance follows the assumed model.
- If $\phi \neq 1$, the variance differs from the expectation:
  - $\phi > 1$: **Overdispersion** (greater variance than expected).
  - $\phi < 1$: **Underdispersion** (less variance than expected).

This discrepancy suggests the assumed **random component distribution** may be inappropriate.

To account for dispersion issues, we can:

1. **Choose a more flexible random component distribution**
   - Negative binomial for overdispersed Poisson data.
   - Conway-Maxwell Poisson for both over- and underdispersed count data.

2. **Use mixed models to account for random effects**

   - Nonlinear and generalized linear mixed models introduce random effects, which can help capture additional variance.

```
# Load necessary libraries
library(ggplot2)
library(MASS)   # For negative binomial models
library(MuMIn)  # For R^2 in GLMs
library(glmmTMB) # For handling overdispersion

# Generate Example Data
set.seed(123)
n <- 100
x <- rnorm(n)
mu <- exp(0.5 + 0.3 * x)  # True mean function
y <- rpois(n, lambda = mu)  # Poisson outcome

# Fit a Poisson GLM
model_pois <- glm(y ~ x, family = poisson(link = "log"))

# Compute residuals
resid_dev <- residuals(model_pois, type = "deviance")
fitted_vals <- fitted(model_pois)
```

```
# Standardized Residual Plot: Residuals vs Fitted Values
ggplot(data = data.frame(fitted_vals, resid_dev),
       aes(x = fitted_vals, y = resid_dev)) +
    geom_point(alpha = 0.6) +
    geom_smooth(method = "loess",
                color = "red",
                se = FALSE) +
    theme_minimal() +
    labs(title = "Standardized Deviance Residuals vs Fitted
     ↪  Values",
         x = "Fitted Values", y = "Standardized Deviance
          ↪  Residuals")
#> `geom_smooth()` using formula = 'y ~ x'
```

```
# Absolute Residuals vs Fitted Values (Variance Function
↪  Check)
ggplot(data = data.frame(fitted_vals, abs_resid =
↪  abs(resid_dev)),
       aes(x = fitted_vals, y = abs_resid)) +
    geom_point(alpha = 0.6) +
    geom_smooth(method = "loess",
                color = "blue",
                se = FALSE) +
    theme_minimal() +
    labs(title = "Absolute Deviance Residuals vs Fitted
     ↪  Values",
         x = "Fitted Values", y = "|Residuals|")
#> `geom_smooth()` using formula = 'y ~ x'
```

```
# Goodness-of-Fit Metrics
AIC(model_pois)    # Akaike Information Criterion
#> [1] 322.9552
BIC(model_pois)    # Bayesian Information Criterion
#> [1] 328.1655
logLik(model_pois) # Log-likelihood
#> 'log Lik.' -159.4776 (df=2)

# R-squared for GLM
r_squared <-
    1 - (logLik(model_pois) / logLik(glm(y ~ 1, family =
     ↪  poisson)))
r_squared
#> 'log Lik.' 0.05192856 (df=2)

# Overdispersion Check
resid_pearson <- residuals(model_pois, type = "pearson")
# Estimated dispersion parameter
phi_hat <-
    sum(resid_pearson ^ 2) / (n - length(coef(model_pois)))
phi_hat
#> [1] 1.055881

# If phi > 1,
# fit a Negative Binomial Model to correct for overdispersion
if (phi_hat > 1) {
    model_nb <- glm.nb(y ~ x)
```

```
    summary(model_nb)
}
#>
#> Call:
#> glm.nb(formula = y ~ x, init.theta = 50.70707605, link =
↪   log)
#>
#> Coefficients:
#>             Estimate Std. Error z value Pr(>|z|)
#> (Intercept)  0.42913    0.08514   5.040 4.65e-07 ***
#> x            0.35262    0.08654   4.075 4.61e-05 ***
#> ---
#> Signif. codes:
#> 0 '***' 0.001 '**' 0.01 '*' 0.05 '.' 0.1 ' ' 1
#>
#> (Dispersion parameter for Negative Binomial(50.7071) family
↪   taken to be 1)
#>
#>     Null deviance: 135.71  on 99  degrees of freedom
#> Residual deviance: 118.92  on 98  degrees of freedom
#> AIC: 324.91
#>
#> Number of Fisher Scoring iterations: 1
#>
#>
#>               Theta:  51
#>           Std. Err.:  233
#>
#>  2 x log-likelihood:  -318.906
```

1. Standardized Residuals vs. Fitted Values (Fig. 4.14)

- If there is a clear trend in the residual plot (e.g., a curve), this suggests an incorrect link function.
- If the residual spread increases with fitted values, the variance function may be misspecified.

2. Absolute Residuals vs. Fitted Values (Fig. 4.15)

- A systematic pattern (e.g., funnel shape) suggests heteroscedasticity (i.e., changing variance).

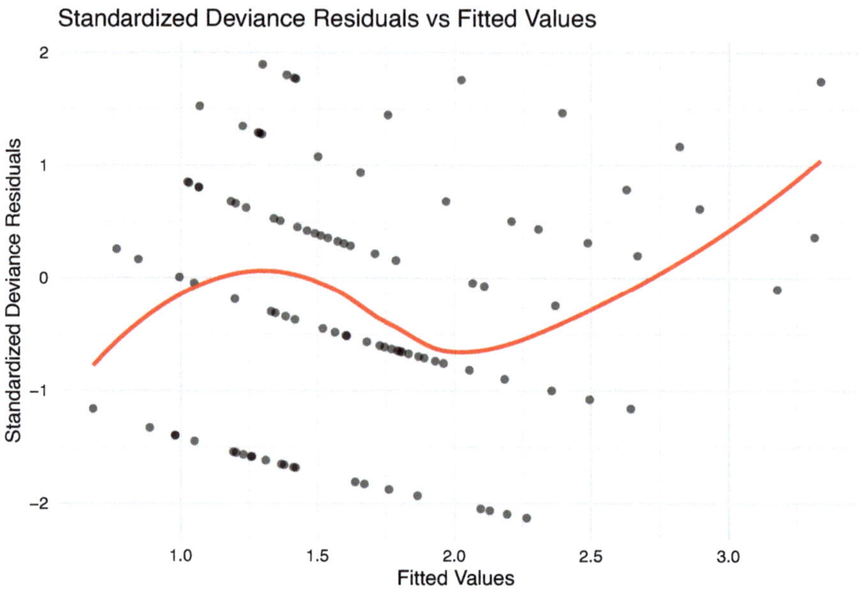

Fig. 4.14 Standardized Deviance Residuals vs Fitted Values

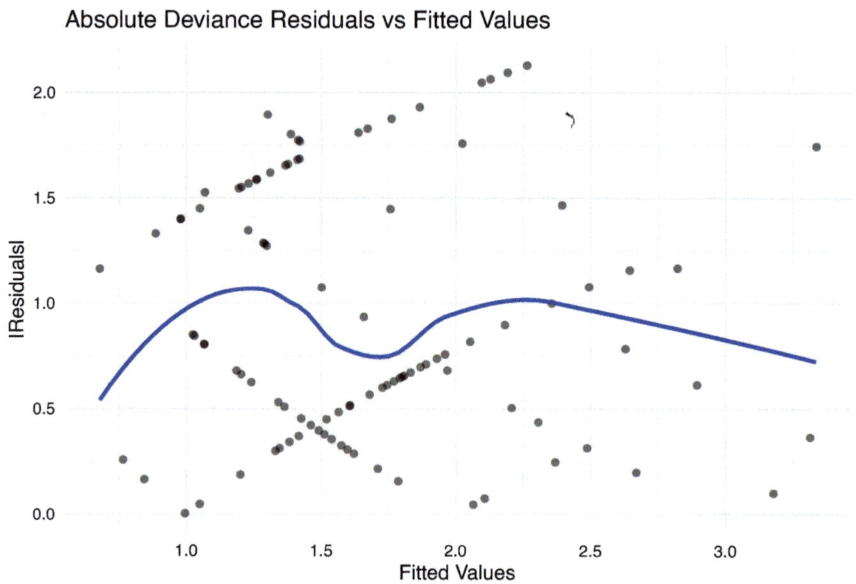

Fig. 4.15 Absolute deviance residuals vs. fitted values

- If the residuals increase with fitted values, a different variance function (e.g., negative binomial instead of Poisson) may be needed.

3. Goodness-of-Fit Metrics

- AIC/BIC: Lower values indicate a better model but must be compared across nested models.
- Log-likelihood: Higher values suggest a better-fitting model.
- R^2 for GLMs: Since traditional R^2 is not available, the likelihood-based R^2 is used to measure improvement over a null model.

4. Overdispersion Check (ϕ)

- If $\phi \approx 1$, the Poisson assumption is valid.
- If $\phi > 1$, there is overdispersion, meaning a negative binomial model may be more appropriate.
- If $\phi < 1$, underdispersion is present, requiring alternative distributions like the Conway-Maxwell Poisson.

Chapter 5
Linear Mixed Models

Recognizing clustered and longitudinal data structures, this chapter introduces linear mixed models (LMMs). We review random effect specification, restricted maximum likelihood (REML) estimation, and inference under unbalanced designs. Special attention is paid to split-plot experiments, repeated-measure designs, and cross-classified data common in education and panel studies. We outline model selection heuristics and present information-criteria benchmarks. Practical guidance on software and diagnostics (BLUPs, conditional residuals) rounds out the chapter.

5.1 Dependent Data

In many real-world applications, observations are not independent but exhibit **correlations** due to shared characteristics. Below are common forms of dependent data:

- **Multivariate measurements on individuals**: Multiple attributes measured on the same person may be correlated (e.g., blood pressure, cholesterol, and BMI).
- **Clustered measurements**: Individuals within a shared environment (e.g., families, schools, hospitals) often exhibit correlated responses.
- **Repeated measurements**: When the same individual is measured multiple times, correlations arise naturally.
  - *Example*: Tracking cholesterol levels of a person over time.
  - If these repeated measurements follow an experimental design where treatments were applied initially, they are referred to as repeated measures (Schabenberger & Pierce, 2001).
- **Longitudinal data**: Repeated measurements taken over time in an observational study are known as longitudinal data (Schabenberger and Pierce, 2001).

M. Nguyen, *Regression Techniques for Data Analysis*,
https://doi.org/10.1007/978-3-032-01835-9_5

- **Spatial data**: Measurements taken from individuals in nearby locations (e.g., residents of the same neighborhood) often exhibit spatial correlation.

Since standard linear regression assumes independent observations, these correlations violate its assumptions. Thus, linear mixed models (LMMs) provide a framework to account for these dependencies.

A linear mixed model (also called a **mixed linear model**) consists of two components:

- **Fixed effects**: Parameters associated with variables that have the same effect across all observations (e.g., gender, age, diet, time).
- **Random effects**: Individual-specific variations or correlation structures (e.g., subject-specific effects, spatial correlations), leading to dependent (correlated) errors.

A key advantage of LMMs is that they model random subject-specific effects, rather than including individual dummy variables. This provides:

- A reduction in the number of parameters to estimate, avoiding overfitting.
- A framework for inference on a population level, rather than restricting conclusions to observed individuals.

5.1.1 Motivation: A Repeated Measurement Example

Consider a scenario where we analyze **repeated measurements** of a response variable Y_{ij} for the i-th subject at time j:

- $i = 1, \ldots, N$ (subjects)
- $j = 1, \ldots, n_i$ (measurements per subject)

We define the response vector for subject i as

$$\mathbf{Y}_i = \begin{bmatrix} Y_{i1} \\ Y_{i2} \\ \vdots \\ Y_{in_i} \end{bmatrix}$$

To model this data, we use a two-stage hierarchical approach.

5.1.1.1 Stage 1: Regression Model (Within-Subject Variation)

We first model how the response changes over time for each subject:

$$\mathbf{Y}_i = Z_i \beta_i + \epsilon_i$$

where:

- Z_i is an $n_i \times q$ matrix of known covariates (e.g., time, treatment).
- β_i is a $q \times 1$ vector of subject-specific regression coefficients.
- ϵ_i represents random errors, often assumed to follow $\epsilon_i \sim N(0, \sigma^2 I)$.

At this stage, each individual has their own unique regression coefficients β_i. However, estimating a separate β_i for each subject is impractical when N is large. Thus, we introduce a second stage to impose structure on β_i.

5.1.1.2 Stage 2: Parameter Model (Between-Subject Variation)

We assume that the subject-specific coefficients β_i arise from a common population distribution:

$$\beta_i = K_i \beta + b_i$$

where:

- K_i is a $q \times p$ matrix of known covariates.
- β is a $p \times 1$ vector of global parameters (fixed effects).
- $\mathbf{b}_i$ are random effects, assumed to follow $\mathbf{b}_i \sim N(0, D)$.

Thus, each individual's regression coefficients β_i are modeled as deviations from a population-level mean β, with subject-specific variations b_i. This hierarchical structure enables:

- Borrowing of strength: Individual estimates β_i are informed by the overall population distribution.
- Improved generalization: The model captures both fixed and random variability efficiently.

The full linear mixed model can then be written as:

$$\mathbf{Y}_i = Z_i K_i \beta + Z_i b_i + \epsilon_i$$

where:

- $Z_i K_i \beta$ represents the **fixed effects** component.
- $Z_i b_i$ represents the **random effects** component.
- ϵ_i represents the residual errors.

This formulation accounts for both **within-subject correlations** (via random effects) and **between-subject variability** (via fixed effects), making it a powerful tool for analyzing dependent data.

5.1.2 *Example: Linear Mixed Model for Repeated Measurements*

5.1.2.1 Stage 1: Subject-Specific Model

The first stage models the response variable Y_{ij} for subject i at time t_{ij}:

$$Y_{ij} = \beta_{1i} + \beta_{2i} t_{ij} + \epsilon_{ij}$$

where:

- $j = 1, \ldots, n_i$ represents different time points for subject i.
- β_{1i} is the subject-specific intercept (baseline response for subject i).
- β_{2i} is the subject-specific slope (rate of change over time).
- $\epsilon_{ij} \sim N(0, \sigma^2)$ are independent errors.

In matrix notation, the model is written as:

$$\mathbf{Y}_i = \begin{bmatrix} Y_{i1} \\ Y_{i2} \\ \vdots \\ Y_{in_i} \end{bmatrix}, \quad \mathbf{Z}_i = \begin{bmatrix} 1 & t_{i1} \\ 1 & t_{i2} \\ \vdots & \vdots \\ 1 & t_{in_i} \end{bmatrix},$$

$$\beta_i = \begin{bmatrix} \beta_{1i} \\ \beta_{2i} \end{bmatrix}, \quad \epsilon_i = \begin{bmatrix} \epsilon_{i1} \\ \epsilon_{i2} \\ \vdots \\ \epsilon_{in_i} \end{bmatrix}.$$

Thus, the model can be rewritten as:

$$\mathbf{Y_i} = \mathbf{Z_i} \boldsymbol{\beta}_\mathbf{i} + \boldsymbol{\epsilon}_\mathbf{i}.$$

5.1.2.2 Stage 2: Population-Level Model

Since estimating a separate β_{1i} and β_{2i} for each subject is impractical, we assume they follow a population distribution:

$$\begin{aligned} \beta_{1i} &= \beta_0 + b_{1i}, \\ \beta_{2i} &= \beta_1 L_i + \beta_2 H_i + \beta_3 C_i + b_{2i}. \end{aligned}$$

where:

- L_i, H_i, C_i are indicator variables for treatment groups:

- $L_i = 1$ if the subject belongs to treatment group 1, else 0.
- $H_i = 1$ if the subject belongs to treatment group 2, else 0.
- $C_i = 1$ if the subject belongs to treatment group 3, else 0.

- β_0 represents the average baseline response across all subjects.
- $\beta_1, \beta_2, \beta_3$ are the average time effects for the respective treatment groups.
- b_{1i}, b_{2i} are random effects representing subject-specific deviations.

This structure implies that while the intercept β_{1i} varies randomly across subjects, the slope β_{2i} depends both on treatment and random subject-specific deviations.

In matrix form, this is:

$$\mathbf{K}_i = \begin{bmatrix} 1 & 0 & 0 & 0 \\ 0 & L_i & H_i & C_i \end{bmatrix},$$

$$\beta = \begin{bmatrix} \beta_0 \\ \beta_1 \\ \beta_2 \\ \beta_3 \end{bmatrix},$$

$$\mathbf{b}_i = \begin{bmatrix} b_{1i} \\ b_{2i} \end{bmatrix},$$

$$\beta_i = \mathbf{K_i}\,\boldsymbol{\beta} + \mathbf{b_i}.$$

5.1.2.3 Final Mixed Model Formulation

Substituting Stage 2 into Stage 1, we obtain the full mixed model:

$$\mathbf{Y}_i = \mathbf{Z}_i(\mathbf{K}_i\beta + \mathbf{b}_i) + \epsilon_i.$$

Expanding:

$$\mathbf{Y}_i = \mathbf{Z}_i\mathbf{K}_i\beta + \mathbf{Z}_i\mathbf{b}_i + \epsilon_i.$$

Interpretation:

- $\mathbf{Z}_i\mathbf{K}_i\beta$ represents the **fixed effects** (population-level trends).
- $\mathbf{Z}_i\mathbf{b}_i$ represents the **random effects** (subject-specific variations).
- ϵ_i represents the **residual errors**.

Assumptions

- Random effects: $\mathbf{b}_i \sim N(0, D)$, where D is the variance-covariance matrix.
- Residual errors: $\epsilon_i \sim N(0, \sigma^2 I)$.
- Independence: $\mathbf{b}_i$ and ϵ_i are independent.

To estimate $\hat{\beta}$, one might consider a sequential approach:

1. Estimate $\hat{\beta}_i$ for each subject in the first stage.
2. Estimate $\hat{\beta}$ in the second stage by replacing β_i with $\hat{\beta}_i$.

However, this method introduces several problems:

- Loss of information: Summarizing $\mathbf{Y}_i$ solely by $\hat{\beta}_i$ discards valuable within-subject variability.
- Ignoring uncertainty: Treating estimated values $\hat{\beta}_i$ as known values leads to underestimated variability.
- Unequal sample sizes: Subjects may have different numbers of observations (n_i), which affects variance estimation.

To address these issues, we adopt the linear mixed model (LMM) framework (Laird and Ware, 1982).

5.1.2.4 Reformulating the Linear Mixed Model

Substituting Stage 2 into Stage 1, we obtain:

$$\mathbf{Y}_i = \mathbf{Z}_i\mathbf{K}_i\beta + \mathbf{Z}_i\mathbf{b}_i + \epsilon_i.$$

Defining $\mathbf{X}_i = \mathbf{Z}_i\mathbf{K}_i$ as an $n_i \times p$ matrix of covariates, we rewrite the model as:

$$\mathbf{Y}_i = \mathbf{X}_i\beta + \mathbf{Z}_i\mathbf{b}_i + \epsilon_i.$$

where:

- $i = 1, \ldots, N$ (subjects).
- β are **fixed effects**, common to all subjects.
- $\mathbf{b}_i$ are **subject-specific random effects**, assumed to follow:

$$\mathbf{b}_i \sim N_q(\mathbf{0}, \mathbf{D}).$$

- ϵ_i represents residual errors:

$$\epsilon_i \sim N_{n_i}(\mathbf{0}, \mathbf{\Sigma_i}).$$

- Independence assumption: $\mathbf{b}_i$ and ϵ_i are independent.
- Dimension notation:
  - $\mathbf{Z}_i$ is an $n_i \times q$ matrix of known covariates for random effects.
  - $\mathbf{X}_i$ is an $n_i \times p$ matrix of known covariates for fixed effects.

5.1.2.5 Hierarchical (Conditional) Formulation

Rewriting the LMM in **hierarchical form**:

$$\mathbf{Y}_i|\mathbf{b}_i \sim N(\mathbf{X}_i\beta + \mathbf{Z}_i\mathbf{b}_i, \mathbf{\Sigma}_i),$$
$$\mathbf{b}_i \sim N(\mathbf{0}, \mathbf{D}).$$

where:

- The first equation states that, given the subject-specific random effects $\mathbf{b}_i$, the response follows a normal distribution with mean $\mathbf{X}_i\beta + \mathbf{Z}_i\mathbf{b}_i$ and covariance $\mathbf{\Sigma}_i$.
- The second equation states that the random effects $\mathbf{b}_i$ follow a **multivariate normal** distribution with mean zero and covariance D.

  We denote the respective probability density functions as:

$$f(\mathbf{Y}_i|\mathbf{b}_i) \quad \text{and} \quad f(\mathbf{b}_i).$$

  Using the general **marginalization formula**:

$$f(A, B) = f(A|B)f(B)$$
$$f(A) = \int f(A, B)dB = \int f(A|B)f(B)dB$$

we obtain the **marginal density** of $\mathbf{Y}_i$:

$$f(\mathbf{Y}_i) = \int f(\mathbf{Y}_i|\mathbf{b}_i)f(\mathbf{b}_i)d\mathbf{b}_i.$$

  Solving this integral, we obtain:

$$\mathbf{Y}_i \sim N(\mathbf{X_i}\boldsymbol{\beta}, \mathbf{Z_i}\mathbf{D}\mathbf{Z_i'} + \mathbf{\Sigma_i}).$$

Interpretation:

- **Mean structure:** The expectation remains $\mathbf{X}_i\beta$, the fixed effects.
- **Variance structure:** The covariance now incorporates random effect variability:

$$\mathbf{Z_i}\mathbf{D}\mathbf{Z_i'} + \mathbf{\Sigma}_i.$$

  This shows that the random effects contribute additional correlation between observations.

Key Takeaway:
The marginal formulation of LMM no longer includes $Z_i b_i$ in the mean, but instead incorporates it into the covariance structure, introducing marginal dependence in $\mathbf{Y}_i$.

Technically, rather than "averaging out" the random effect b_i, we add its contribution to the variance-covariance matrix.

Continue with the example:

$$Y_{ij} = (\beta_0 + b_{1i}) + (\beta_1 L_i + \beta_2 H_i + \beta_3 C_i + b_{2i})t_{ij} + \epsilon_{ij}.$$

For each treatment group:

$$Y_{ij} = \begin{cases} \beta_0 + b_{1i} + (\beta_1 + b_{2i})t_{ij} + \epsilon_{ij}, & L_i = 1 \\ \beta_0 + b_{1i} + (\beta_2 + b_{2i})t_{ij} + \epsilon_{ij}, & H_i = 1 \\ \beta_0 + b_{1i} + (\beta_3 + b_{2i})t_{ij} + \epsilon_{ij}, & C_i = 1. \end{cases}$$

Interpretation:

- **Intercepts and slopes are subject-specific**: Each subject has their own baseline response and rate of change.
- **Treatment groups affect slopes, but not intercepts**:
  - β_0 is the common intercept across groups.
  - Slopes differ by treatment: β_1 for L, β_2 for H, β_3 for C.
- **Random effects** introduce within-subject correlation:
  - b_{1i} allows individual variation in baseline response.
  - b_{2i} allows subject-specific deviations in slopes.

In the **hierarchical model form**, we again express the LMM as:

$$\mathbf{Y}_i | \mathbf{b}_i \sim N(\mathbf{X}_i \beta + \mathbf{Z}_i \mathbf{b}_i, \mathbf{\Sigma}_i)$$
$$\mathbf{b}_i \sim N(\mathbf{0}, \mathbf{D})$$

where:

- $\mathbf{X}_i$ and $\mathbf{Z}_i$ are design matrices for **fixed** and **random** effects, respectively.
- $\mathbf{b}_i$ represents **subject-specific random effects**.
- $\mathbf{\Sigma}_i$ is the **residual error covariance matrix**.
- $\mathbf{D}$ is the **random effects covariance matrix**.

The fixed effect parameter vector is:

$$\beta = (\beta_0, \beta_1, \beta_2, \beta_3)'.$$

From the model structure, we define:

$$\mathbf{X}_i = \mathbf{Z}_i \mathbf{K}_i.$$

Expanding:

$$\mathbf{Z}_i = \begin{bmatrix} 1 & t_{i1} \\ 1 & t_{i2} \\ \vdots & \vdots \\ 1 & t_{in_i} \end{bmatrix}, \quad \mathbf{K}_i = \begin{bmatrix} 1 & 0 & 0 & 0 \\ 0 & L_i & H_i & C_i \end{bmatrix}.$$

Multiplying:

$$\mathbf{X}_i = \begin{bmatrix} 1 & t_{i1}L_i & t_{i1}H_i & t_{i1}C_i \\ 1 & t_{i2}L_i & t_{i2}H_i & t_{i2}C_i \\ \vdots & \vdots & \vdots & \vdots \\ 1 & t_{in_i}L_i & t_{in_i}H_i & t_{in_i}C_i \end{bmatrix}.$$

The random effects vector is:

$$\mathbf{b}_i = \begin{bmatrix} b_{1i} \\ b_{2i} \end{bmatrix}.$$

The covariance matrix **D** for random effects is:

$$\mathbf{D} = \begin{bmatrix} d_{11} & d_{12} \\ d_{12} & d_{22} \end{bmatrix}.$$

We assume **conditional independence**:

$$\mathbf{\Sigma}_i = \sigma^2 \mathbf{I}_{n_i}.$$

This means that, **given the random effects** $\mathbf{b}_i$ and β, the responses for subject i are independent.

To derive the **marginal model**, we integrate out the random effects $\mathbf{b}_i$. This leads to:

$$Y_{ij} = \beta_0 + \beta_1 L_i t_{ij} + \beta_2 H_i t_{ij} + \beta_3 C_i t_{ij} + \eta_{ij},$$

where:

$$\eta_i \sim N(\mathbf{0}, \mathbf{Z}_i \mathbf{D} \mathbf{Z}_i' + \mathbf{\Sigma}_i).$$

Thus, the full marginal model is:

$$\mathbf{Y_i} \sim N(\mathbf{X}_i \beta, \mathbf{Z}_i \mathbf{D} \mathbf{Z}_i' + \mathbf{\Sigma}_i).$$

Example (Case Where $n_i = 2$) For a subject with **two observations**, we compute:

$$\mathbf{Z}_i = \begin{bmatrix} 1 & t_{i1} \\ 1 & t_{i2} \end{bmatrix}.$$

Then:

$$\mathbf{Z}_i \mathbf{D} \mathbf{Z}_i' = \begin{bmatrix} 1 & t_{i1} \\ 1 & t_{i2} \end{bmatrix} \begin{bmatrix} d_{11} & d_{12} \\ d_{12} & d_{22} \end{bmatrix} \begin{bmatrix} 1 & 1 \\ t_{i1} & t_{i2} \end{bmatrix}.$$

Expanding the multiplication:

$$\mathbf{Z}_i \mathbf{D} \mathbf{Z}_i' = \begin{bmatrix} d_{11} + 2d_{12}t_{i1} + d_{22}t_{i1}^2 & d_{11} + d_{12}(t_{i1} + t_{i2}) + d_{22}t_{i1}t_{i2} \\ d_{11} + d_{12}(t_{i1} + t_{i2}) + d_{22}t_{i1}t_{i2} & d_{11} + 2d_{12}t_{i2} + d_{22}t_{i2}^2 \end{bmatrix}.$$

Finally, incorporating the residual variance:

$$\text{Var}(Y_{i1}) = d_{11} + 2d_{12}t_{i1} + d_{22}t_{i1}^2 + \sigma^2.$$

Interpretation of the marginal model:

1. **Correlation in the Errors**:
   - The off-diagonal elements of $\mathbf{Z}_i \mathbf{D} \mathbf{Z}_i'$ introduce **correlation** between observations within the same subject.
   - This accounts for the fact that repeated measurements on the same individual are not independent.
2. **Variance Structure**:
   - The variance function of Y_{ij} is quadratic in time:

   $$\text{Var}(Y_{ij}) = d_{11} + 2d_{12}t_{ij} + d_{22}t_{ij}^2 + \sigma^2.$$

   - The curvature of this variance function is determined by d_{22}.
   - If $d_{22} > 0$, variance increases over time.

Key Takeaway:
In the hierarchical model, random effects contribute to the mean structure.
In the marginal model, they affect the covariance structure, leading to heteroskedasticity (changing variance over time) and correlation between repeated measures.

5.1.3 Random-Intercepts Model

The random-intercepts model is obtained by **removing random slopes**, meaning:

- All subject-specific variability in slopes is attributed only to treatment differences.
- The model allows each subject to have their own intercept, but within each treatment group, all subjects share the same slope.

The hierarchical (conditional) model is:

$$\mathbf{Y}_i | b_i \sim N(\mathbf{X}_i \beta + 1 b_i, \mathbf{\Sigma}_i),$$
$$b_i \sim N(0, d_{11}).$$

where:

- b_i is the **random intercept** for subject i, assumed to follow $N(0, d_{11})$.
- $\mathbf{X}_i$ contains the **fixed effects**, which include treatment and time.
- $\mathbf{\Sigma}_i$ represents **residual variance**, typically assumed to be $\sigma^2\mathbf{I}$ (independent errors).

Since there are no **random slopes**, the only source of **subject-specific variability** is the intercept (Table 5.1).

Integrating out b_i and assuming conditional independence $\mathbf{\Sigma}_i = \sigma^2\mathbf{I}_{n_i}$, the **marginal distribution** of $\mathbf{Y}_i$ is:

$$\mathbf{Y}_i \sim N(\mathbf{X}_i \beta, 11' d_{11} + \sigma^2 \mathbf{I}).$$

The **marginal covariance matrix** is:

$$\text{Cov}(\mathbf{Y}_i) = 11' d_{11} + \sigma^2 I$$
$$= \begin{bmatrix} d_{11}+\sigma^2 & d_{11} & d_{11} & \dots & d_{11} \\ d_{11} & d_{11}+\sigma^2 & d_{11} & \dots & d_{11} \\ d_{11} & d_{11} & d_{11}+\sigma^2 & \dots & d_{11} \\ \vdots & \vdots & \vdots & \ddots & \vdots \\ d_{11} & d_{11} & d_{11} & \dots & d_{11}+\sigma^2 \end{bmatrix}.$$

The **correlation matrix** is obtained by standardizing the covariance matrix:

Table 5.1 Summary of the random-intercepts model

Feature	Explanation
Intercepts	Random, subject-specific (b_i)
Slopes	Fixed within each treatment group
Covariance Structure	Compound symmetry (constant variance, equal correlation)
Intraclass Correlation (ρ)	Measures between-subject variability relative to total variability
Interpretation	Large $\rho \rightarrow$ Strong subject-level differences, Small $\rho \rightarrow$ Mostly residual noise

$$\mathrm{Corr}(\mathbf{Y}_i) = \begin{bmatrix} 1 & \rho & \rho & \dots & \rho \\ \rho & 1 & \rho & \dots & \rho \\ \rho & \rho & 1 & \dots & \rho \\ \vdots & \vdots & \vdots & \ddots & \vdots \\ \rho & \rho & \rho & \dots & 1 \end{bmatrix},$$

where:

$$\rho = \frac{d_{11}}{d_{11} + \sigma^2}.$$

This **correlation structure** is known as **compound symmetry**, meaning:

- **Constant variance**: $\mathrm{Var}(Y_{ij}) = d_{11} + \sigma^2$ for all j.
- **Equal correlation**: $\mathrm{Corr}(Y_{ij}, Y_{ik}) = \rho$ for all $j \neq k$.
- **Positive correlation**: Any two observations from the same subject are equally correlated.

Interpretation of ρ (intraclass correlation)

- ρ is called the intraclass correlation coefficient (ICC).
- It measures the proportion of total variability that is due to between-subject variability:

$$\rho = \frac{\text{Between-Subject Variance}}{\text{Total Variance}} = \frac{d_{11}}{d_{11} + \sigma^2}.$$

- If ρ is large:
  - The inter-subject variability (d_{11}) is large relative to the intra-subject variability (σ^2).
  - This means subjects differ substantially in their intercepts.
  - Responses from the same subject are highly correlated.
- If ρ is small:
  - The residual error variance dominates, meaning individual differences in intercepts are small.
  - Measurements from the same subject are weakly correlated.

5.1.4 Covariance Models in Linear Mixed Models

Previously, we assumed that the **within-subject errors** are **conditionally independent**, meaning:

$$\mathbf{\Sigma}_i = \sigma^2 \mathbf{I}_{n_i}.$$

However, real-world data often exhibit **correlated errors**, particularly in **longitudinal studies** where observations over time are influenced by underlying stochastic processes.

To model this dependence, we decompose the error term into two components:

$$\epsilon_i = \epsilon_{(1)i} + \epsilon_{(2)i},$$

where:

- $\epsilon_{(1)i}$ (Serial Correlation Component):
  - Represents subject-specific response to time-varying stochastic processes.
  - Captures dependency across observations for the same subject.
- $\epsilon_{(2)i}$ (Measurement Error Component):
  - Represents pure measurement error, assumed independent of $\epsilon_{(1)i}$.

Thus, the full LMM formulation becomes:

$$\mathbf{Y_i} = \mathbf{X_i}\boldsymbol{\beta} + \mathbf{Z_i}\mathbf{b_i} + \boldsymbol{\epsilon}_{\mathbf{(1)i}} + \boldsymbol{\epsilon}_{\mathbf{(2)i}}.$$

where:

- Random effects: $\mathbf{b_i} \sim N(\mathbf{0}, \mathbf{D})$.
- Measurement errors: $\epsilon_{(2)i} \sim N(\mathbf{0}, \boldsymbol{\sigma}^2\mathbf{I_{n_i}})$.
- Serial correlation errors: $\epsilon_{(1)i} \sim N(\mathbf{0}, \boldsymbol{\tau}^2\mathbf{H_i})$.
- Independence assumption: $\mathbf{b}_i$ and ϵ_i are mutually independent.

To model the correlation structure of the **serial correlation component** $\epsilon_{(1)i}$, we define the $n_i \times n_i$ correlation (or covariance) matrix $\mathbf{H}_i$.

The (j, k)th element of $\mathbf{H}_i$ is denoted as:

$$h_{ijk} = g(t_{ij}, t_{ik}),$$

where:

- h_{ijk} represents the covariance (or correlation) between time points t_{ij} and t_{ik}.
- $g(t_{ij}, t_{ik})$ is a decreasing function that defines the correlation between time points t_{ij} and t_{ik}.
- Typically, we assume that this function depends only on the time difference (stationarity assumption):

  $$h_{ijk} = g(|t_{ij} - t_{ik}|)$$

  meaning that the correlation only depends on the absolute time lag.
- To be a valid correlation matrix, we require:

  $$g(0) = 1.$$

This ensures that each observation has a perfect correlation with itself.

Common Choices for $g(|t_{ij} - t_{ik}|)$

Several functions can be used to define the **decay in correlation** as time differences increase.

1. Exponential Correlation (Continuous-Time AR(1))

$$g(|t_{ij} - t_{ik}|) = \exp(-\phi|t_{ij} - t_{ik}|)$$

- **Decay rate:** Controlled by $\phi > 0$.
- **Interpretation:**
  - Observations closer in time are more correlated.
  - The correlation decreases exponentially as time separation increases.
- **Use case:** Often used in biological and economic time-series models.

2. Gaussian Correlation (Squared Exponential Kernel)

$$g(|t_{ij} - t_{ik}|) = \exp(-\phi(t_{ij} - t_{ik})^2)$$

- **Decay rate:** Faster than exponential.
- **Interpretation:**
  - If ϕ is large, correlation drops sharply as time separation increases.
  - Produces smooth correlation structures.
- **Use case:** Used in spatial statistics and machine learning (Gaussian processes).

3. Autoregressive (AR(1)) Correlation

A **First-Order Autoregressive Model (AR(1))** assumes that the value at time t depends on its previous value:

$$\alpha_t = \phi\alpha_{t-1} + \eta_t,$$

where

- $\eta_t \sim$ i.i.d. $N(0, \sigma_\eta^2)$ (white noise process).
- ϕ is the **autocorrelation coefficient**.

Then, the covariance between two observations at different times is:

$$\mathrm{Cov}(\alpha_t, \alpha_{t+h}) = \frac{\sigma_\eta^2 \phi^{|h|}}{1 - \phi^2}.$$

Thus, the **correlation** between time points t and $t + h$ is:

$$\text{Corr}(\alpha_t, \alpha_{t+h}) = \phi^{|h|}.$$

For a sequence of T time points, the resulting **Toeplitz correlation matrix** is:

$$\text{Corr}(\alpha_T) = \begin{bmatrix} 1 & \phi^1 & \phi^2 & \dots & \phi^{T-1} \\ \phi^1 & 1 & \phi^1 & \dots & \phi^{T-2} \\ \phi^2 & \phi^1 & 1 & \dots & \phi^{T-3} \\ \vdots & \vdots & \vdots & \ddots & \vdots \\ \phi^{T-1} & \phi^{T-2} & \phi^{T-3} & \dots & 1 \end{bmatrix}.$$

- **Decay rate:** ϕ controls how fast correlation decreases.
- **Use case:**
  - Common in **time series models**.
  - Appropriate when **observations are equally spaced in time**.

Properties of the AR(1) Structure:

- Correlation decreases with time lag:
  - Observations closer in time are more correlated.
  - Decay rate is controlled by ϕ.
- Toeplitz structure:
  - The covariance matrix exhibits a banded diagonal pattern.
  - Useful in longitudinal and time-series data.
- Applicability:
  - Small ϕ (≈ 0) $\rightarrow$ Weak autocorrelation (errors are mostly independent).
  - Large ϕ (≈ 1) $\rightarrow$ Strong autocorrelation (highly dependent observations).

5.1.5 Covariance Structures in Mixed Models

Table 5.2 shows Covariance structures in mixed models. Table 5.3 shows how to choose the right covariance structure.

5.2 Estimation in Linear Mixed Models

The general linear mixed model is:

$$\mathbf{Y}_i = \mathbf{X}_i\beta + \mathbf{Z}_i\mathbf{b}_i + \epsilon_i,$$

Table 5.2 Covariance structures in mixed models

Structure	Covariance Function $g(\lvert t_{ij} - t_{ik}\rvert)$	Decay Behavior	Correlation Matrix Pattern	Key Properties	Common Use Cases
Compound Symmetry	Constant: $g(\lvert t_{ij} - t_{ik}\rvert) = \rho$	No decay	Constant off-diagonal	Equal correlation across all time points (homogeneous)	Random-intercepts model, repeated measures
AR(1) (Autoregressive)	$\phi^{\lvert t_{ij}-t_{ik}\rvert}$	Exponential decay	Toeplitz (banded diagonal)	Strong correlation for nearby observations, weak for distant ones	Time series, longitudinal data
Exponential	$\exp(-\phi\lvert t_{ij} - t_{ik}\rvert)$	Smooth decay	Spatially continuous	Models continuous correlation decline over time/space	Growth curves, ecological models
Gaussian (Squared Exponential)	$\exp(-\phi(t_{ij} - t_{ik})^2)$	Rapid decay	Spatially continuous	Smooth decay but stronger locality effect	Spatial processes, geostatistics
Toeplitz	Varies by lag, $g(\lvert t_{ij} - t_{ik}\rvert) = c_h$	General pattern	General symmetric matrix	Arbitrary structure, allows irregular spacing	Irregular time points, flexible spatial models
Unstructured	Fully parameterized, no constraints	No fixed pattern	General symmetric matrix	Allows any correlation pattern, but many parameters	Small datasets with unknown correlation
Banded	Zero correlation beyond a certain lag	Abrupt cutoff	Banded diagonal	Assumes only close observations are correlated	Large datasets, high-dimensional time series
Spatial Power	$\phi^{\lvert s_{ij}-s_{ik}\rvert}$	Exponential decay	Distance-based structure	Used when correlation depends on spatial distance	Spatial statistics, environmental data
Matérn Covariance	Function of $\lvert t_{ij} - t_{ik}\rvert^{\nu}$	Flexible decay	Spatially continuous	Generalization of Gaussian and exponential structures	Geostatistics, spatiotemporal models

Table 5.3 Choosing the right covariance structure

Scenario	Recommended Structure
Repeated measures with equal correlation	Compound Symmetry
Longitudinal data with time dependence	AR(1), Toeplitz
Continuous-time process	Exponential, Gaussian
Spatial correlation	Spatial Power, Matérn
Irregularly spaced time points	Toeplitz, Unstructured
High-dimensional data	Banded, AR(1)

where:

- β: Fixed effects (parameters shared across subjects).
- $\mathbf{b}_i$: Random effects (subject-specific deviations).
- $\mathbf{X}_i$: Design matrix for fixed effects.
- $\mathbf{Z}_i$: Design matrix for random effects.
- $\mathbf{D}$: Covariance matrix of the random effects.
- $\mathbf{\Sigma}_i$: Covariance matrix of the residual errors.

Since β, $\mathbf{b}_i$, $\mathbf{D}$, and $\mathbf{\Sigma}_i$ are **unknown**, they must be **estimated** from data.

- β, $\mathbf{D}$, $\mathbf{\Sigma}_i$ are **fixed parameters** $\rightarrow$ must be **estimated**.
- $\mathbf{b}_i$ is a **random variable** $\rightarrow$ must be **predicted** (not estimated). In other words, random thing/variable can't be estimated.

Thus, we define:

- **Estimator of** β: $\hat{\beta}$ (fixed effect estimation).
- **Predictor of** $\mathbf{b}_i$: $\hat{\mathbf{b}}_i$ (random effect prediction).

Then:

- The **population-level estimate** of $\mathbf{Y}_i$ is:

$$\hat{\mathbf{Y}}_i = \mathbf{X}_i\hat{\beta}.$$

- The **subject-specific prediction** is:

$$\hat{\mathbf{Y}}_i = \mathbf{X}_i\hat{\beta} + \mathbf{Z}_i\hat{\mathbf{b}}_i.$$

For all N subjects, we stack the equations into the mixed model equations (Henderson, 1975):

$$\mathbf{Y} = \mathbf{X}\beta + \mathbf{Z}\mathbf{b} + \epsilon.$$

and

$$\mathbf{Y} \sim N(\mathbf{X}\boldsymbol{\beta}, \mathbf{Z}\mathbf{B}\mathbf{Z}' + \mathbf{\Sigma})$$

where:

$$\mathbf{Y} = \begin{bmatrix} \mathbf{y}_1 \\ \vdots \\ \mathbf{y}_N \end{bmatrix}, \quad \mathbf{X} = \begin{bmatrix} \mathbf{X}_1 \\ \vdots \\ \mathbf{X}_N \end{bmatrix}, \quad \mathbf{b} = \begin{bmatrix} \mathbf{b}_1 \\ \vdots \\ \mathbf{b}_N \end{bmatrix}, \quad \boldsymbol{\epsilon} = \begin{bmatrix} \boldsymbol{\epsilon}_1 \\ \vdots \\ \boldsymbol{\epsilon}_N \end{bmatrix}.$$

The covariance structure is:

$$\mathrm{Cov}(\mathbf{b}) = \mathbf{B}, \quad \mathrm{Cov}(\epsilon) = \boldsymbol{\Sigma}, \quad \mathrm{Cov}(\mathbf{b}, \epsilon) = 0.$$

Expanding **Z** and **B** as block diagonal matrices:

$$\mathbf{Z} = \begin{bmatrix} \mathbf{Z}_1 & 0 & \dots & 0 \\ 0 & \mathbf{Z}_2 & \dots & 0 \\ \vdots & \vdots & \ddots & \vdots \\ 0 & 0 & \dots & \mathbf{Z}_N \end{bmatrix}, \quad \mathbf{B} = \begin{bmatrix} \mathbf{D} & 0 & \dots & 0 \\ 0 & \mathbf{D} & \dots & 0 \\ \vdots & \vdots & \ddots & \vdots \\ 0 & 0 & \dots & \mathbf{D} \end{bmatrix}.$$

The best linear unbiased estimator (BLUE) for β and the best linear unbiased predictor (BLUP) for **b** are obtained by solving (Henderson, 1975):

$$\begin{bmatrix} \hat{\beta} \\ \hat{\mathbf{b}} \end{bmatrix} = \begin{bmatrix} \mathbf{X}'\boldsymbol{\Sigma}^{-1}\mathbf{X} & \mathbf{X}'\boldsymbol{\Sigma}^{-1}\mathbf{Z} \\ \mathbf{Z}'\boldsymbol{\Sigma}^{-1}\mathbf{X} & \mathbf{Z}'\boldsymbol{\Sigma}^{-1}\mathbf{Z} + \mathbf{B}^{-1} \end{bmatrix}^{-1} \begin{bmatrix} \mathbf{X}'\boldsymbol{\Sigma}^{-1}\mathbf{Y} \\ \mathbf{Z}'\boldsymbol{\Sigma}^{-1}\mathbf{Y} \end{bmatrix}.$$

where:

- $\hat{\beta}$ is the Generalized Least Squares estimator for β.
- $\hat{\mathbf{b}}$ is the BLUP for **b**.

If we define:

$$\mathbf{V} = \mathbf{Z}\mathbf{B}\mathbf{Z}' + \boldsymbol{\Sigma}.$$

Then, the solutions to the **mixed model equations** are:

$$\hat{\beta} = (\mathbf{X}'\mathbf{V}^{-1}\mathbf{X})^{-1}\mathbf{X}'\mathbf{V}^{-1}\mathbf{Y},$$

$$\hat{\mathbf{b}} = \mathbf{B}\mathbf{Z}'\mathbf{V}^{-1}(\mathbf{Y} - \mathbf{X}\hat{\beta}).$$

where:

- $\hat{\beta}$ is obtained using generalized least squares.
- $\hat{\mathbf{b}}$ is a weighted least squares predictor, where weights come from **B** and **V** (Table 5.4).

Table 5.4 Key components and equations in linear mixed models

Component	Equation	Interpretation
Fixed effects ($\hat{\beta}$)	$(\mathbf{X}'\mathbf{V}^{-1}\mathbf{X})^{-1}\mathbf{X}'\mathbf{V}^{-1}\mathbf{Y}$	Generalized Least Squares estimator
Random effects ($\hat{\mathbf{b}}$)	$\mathbf{B}\mathbf{Z}'\mathbf{V}^{-1}(\mathbf{Y} - \mathbf{X}\hat{\boldsymbol{\beta}})$	Best Linear Unbiased Predictor (BLUP)
Variance of $\hat{\beta}$	$(\mathbf{X}'\mathbf{V}^{-1}\mathbf{X})^{-1}$	Accounts for uncertainty in fixed effect estimates
Variance of prediction error	$\mathbf{B} - \mathbf{B}\mathbf{Z}'\mathbf{V}^{-1}\mathbf{Z}\mathbf{B} + \mathbf{B}\mathbf{Z}'\mathbf{V}^{-1}\mathbf{X}(\mathbf{X}'\mathbf{V}^{-1}\mathbf{X})^{-1}\mathbf{X}'\mathbf{V}^{-1}\mathbf{B}$	Includes both variance and bias in prediction

Properties of the Estimators
For $\hat{\beta}$:

$$E(\hat{\beta}) = \beta, \quad \text{Var}(\hat{\beta}) = (\mathbf{X}'\mathbf{V}^{-1}\mathbf{X})^{-1}.$$

For $\hat{\mathbf{b}}$:

$$E(\hat{\mathbf{b}}) = 0.$$

The **variance of the prediction error** (mean squared prediction error, MSPE) is:

$$\text{Var}(\hat{\mathbf{b}} - \mathbf{b}) = \mathbf{B} - \mathbf{B}\mathbf{Z}'\mathbf{V}^{-1}\mathbf{Z}\mathbf{B} + \mathbf{B}\mathbf{Z}'\mathbf{V}^{-1}\mathbf{X}(\mathbf{X}'\mathbf{V}^{-1}\mathbf{X})^{-1}\mathbf{X}'\mathbf{V}^{-1}\mathbf{B}.$$

Key Insight:
Mean squared prediction error is more meaningful than $\text{Var}(\hat{\mathbf{b}})$ alone, since it accounts for both variance and bias in prediction.

5.2.1 Interpretation of the Mixed Model Equations

The system:

$$\begin{bmatrix} \mathbf{X}'\boldsymbol{\Sigma}^{-1}\mathbf{X} & \mathbf{X}'\boldsymbol{\Sigma}^{-1}\mathbf{Z} \\ \mathbf{Z}'\boldsymbol{\Sigma}^{-1}\mathbf{X} & \mathbf{Z}'\boldsymbol{\Sigma}^{-1}\mathbf{Z} + \mathbf{B}^{-1} \end{bmatrix} \begin{bmatrix} \beta \\ \mathbf{b} \end{bmatrix} = \begin{bmatrix} \mathbf{X}'\boldsymbol{\Sigma}^{-1}\mathbf{Y} \\ \mathbf{Z}'\boldsymbol{\Sigma}^{-1}\mathbf{Y} \end{bmatrix}.$$

can be understood as:

1. Fixed Effect Estimation ($\hat{\beta}$)
    - Uses Generalized Least Squares.
    - Adjusted for both random effects and correlated errors.

2. Random Effects Prediction ($\hat{\mathbf{b}}$)
   - Computed using the BLUP formula.
   - Shrinks subject-specific estimates toward the population mean.

5.2.2 Derivation of the Mixed Model Equations

To derive the **Mixed Model Equations**, consider:

$$\boldsymbol{\epsilon} = \mathbf{Y} - \mathbf{X}\boldsymbol{\beta} - \mathbf{Z}\mathbf{b}.$$

Define:

- $T = \sum_{i=1}^{N} n_i$ as **Total number of observations.**
- Nq as **Total number of random effects.**

The **joint distribution** of ($\mathbf{b}$, $\boldsymbol{\epsilon}$) is:

$$f(\mathbf{b}, \epsilon) = \frac{1}{(2\pi)^{(T+Nq)/2}} \begin{vmatrix} \mathbf{B} & 0 \\ 0 & \boldsymbol{\Sigma} \end{vmatrix}^{-1/2}$$

$$\exp\left(-\frac{1}{2}\begin{bmatrix} \mathbf{b} \\ \mathbf{Y} - \mathbf{X}\boldsymbol{\beta} - \mathbf{Z}\mathbf{b} \end{bmatrix}' \begin{bmatrix} \mathbf{B} & 0 \\ 0 & \boldsymbol{\Sigma} \end{bmatrix}^{-1} \begin{bmatrix} \mathbf{b} \\ \mathbf{Y} - \mathbf{X}\boldsymbol{\beta} - \mathbf{Z}\mathbf{b} \end{bmatrix}\right).$$

Maximizing $f(\mathbf{b}, \epsilon)$ with respect to $\mathbf{b}$ and β requires **minimization** of:

$$\begin{aligned} Q &= \begin{bmatrix} \mathbf{b} \\ \mathbf{Y} - \mathbf{X}\boldsymbol{\beta} - \mathbf{Z}\mathbf{b} \end{bmatrix}' \begin{bmatrix} \mathbf{B} & 0 \\ 0 & \boldsymbol{\Sigma} \end{bmatrix}^{-1} \begin{bmatrix} \mathbf{b} \\ \mathbf{Y} - \mathbf{X}\boldsymbol{\beta} - \mathbf{Z}\mathbf{b} \end{bmatrix} \\ &= \mathbf{b}'\mathbf{B}^{-1}\mathbf{b} + (\mathbf{Y} - \mathbf{X}\boldsymbol{\beta} - \mathbf{Z}\mathbf{b})'\boldsymbol{\Sigma}^{-1}(\mathbf{Y} - \mathbf{X}\boldsymbol{\beta} - \mathbf{Z}\mathbf{b}). \end{aligned}$$

Setting the derivatives of Q with respect to $\mathbf{b}$ and $\boldsymbol{\beta}$ to zero leads to the system of equations:

$$\mathbf{X}'\boldsymbol{\Sigma}^{-1}\mathbf{X}\boldsymbol{\beta} + \mathbf{X}'\boldsymbol{\Sigma}^{-1}\mathbf{Z}\mathbf{b} = \mathbf{X}'\boldsymbol{\Sigma}^{-1}\mathbf{Y}$$

$$(\mathbf{Z}'\boldsymbol{\Sigma}^{-1}\mathbf{Z} + \mathbf{B}^{-1})\mathbf{b} + \mathbf{Z}'\boldsymbol{\Sigma}^{-1}\mathbf{X}\boldsymbol{\beta} = \mathbf{Z}'\boldsymbol{\Sigma}^{-1}\mathbf{Y}$$

Rearranging

$$\begin{bmatrix} \mathbf{X}'\boldsymbol{\Sigma}^{-1}\mathbf{X} & \mathbf{X}'\boldsymbol{\Sigma}^{-1}\mathbf{Z} \\ \mathbf{Z}'\boldsymbol{\Sigma}^{-1}\mathbf{X} & \mathbf{Z}'\boldsymbol{\Sigma}^{-1}\mathbf{Z} + \mathbf{B}^{-1} \end{bmatrix} \begin{bmatrix} \beta \\ \mathbf{b} \end{bmatrix} = \begin{bmatrix} \mathbf{X}'\boldsymbol{\Sigma}^{-1}\mathbf{Y} \\ \mathbf{Z}'\boldsymbol{\Sigma}^{-1}\mathbf{Y} \end{bmatrix}$$

Thus, the solution to the mixed model equations give:

$$\begin{bmatrix} \hat{\beta} \\ \hat{\mathbf{b}} \end{bmatrix} = \begin{bmatrix} \mathbf{X}'\boldsymbol{\Sigma}^{-1}\mathbf{X} & \mathbf{X}'\boldsymbol{\Sigma}^{-1}\mathbf{Z} \\ \mathbf{Z}'\boldsymbol{\Sigma}^{-1}\mathbf{X} & \mathbf{Z}'\boldsymbol{\Sigma}^{-1}\mathbf{Z}+\mathbf{B}^{-1} \end{bmatrix}^{-1} \begin{bmatrix} \mathbf{X}'\boldsymbol{\Sigma}^{-1}\mathbf{Y} \\ \mathbf{Z}'\boldsymbol{\Sigma}^{-1}\mathbf{Y} \end{bmatrix}$$

5.2.3 Bayesian Interpretation of Linear Mixed Models

In a **Bayesian framework**, the posterior distribution of the **random effects b** given the observed data **Y** is derived using **Bayes' theorem**:

$$f(\mathbf{b}|\mathbf{Y}) = \frac{f(\mathbf{Y}|\mathbf{b})f(\mathbf{b})}{\int f(\mathbf{Y}|\mathbf{b})f(\mathbf{b})d\mathbf{b}}.$$

where:

- $f(\mathbf{Y}|\mathbf{b})$ is the likelihood function, describing how the data are generated given the random effects.
- $f(\mathbf{b})$ is the prior distribution of the random effects.
- The denominator $\int f(\mathbf{Y}|\mathbf{b})f(\mathbf{b})d\mathbf{b}$ is the normalizing constant that ensures the posterior integrates to 1.
- $f(\mathbf{b}|\mathbf{Y})$ is the posterior distribution, which updates our belief about **b** given the observed data **Y**.

In the linear mixed model, we assume:

$$\mathbf{Y}|\mathbf{b} \sim N(\mathbf{X}\beta + \mathbf{Z}\mathbf{b}, \Sigma),$$
$$\mathbf{b} \sim N(\mathbf{0}, \mathbf{B}).$$

This means:

- Likelihood: Given **b**, the data **Y** follows a multivariate normal distribution with mean $\mathbf{X}\beta + \mathbf{Z}\mathbf{b}$ and covariance $\boldsymbol{\Sigma}$.
- Prior for **b**: The random effects are assumed to follow a multivariate normal distribution with mean 0 and covariance **B**.

By applying **Bayes' theorem**, the posterior distribution of **b** given **Y** is:

$$\mathbf{b}|\mathbf{Y} \sim N(\mathbf{B}\mathbf{Z}'\mathbf{V}^{-1}(\mathbf{Y}-\mathbf{X}\boldsymbol{\beta}), (\mathbf{Z}'\boldsymbol{\Sigma}^{-1}\mathbf{Z}+\mathbf{B}^{-1})^{-1}).$$

where:

- Mean: $\mathbf{B}\mathbf{Z}'\mathbf{V}^{-1}(\mathbf{Y}-\mathbf{X}\boldsymbol{\beta})$
  - This is the BLUP.
  - It represents the expected value of **b** given **Y** under squared-error loss.

Table 5.5 Bayesian interpretation of linear mixed models

Step	Equation	Interpretation
Likelihood	$\mathbf{Y}\vert\mathbf{b} \sim N(\mathbf{X}\boldsymbol{\beta}+\mathbf{Zb}, \boldsymbol{\Sigma})$	Data given random effects
Prior	$\mathbf{b} \sim N(\mathbf{0}, \mathbf{B})$	Random effects distribution
Posterior	$\mathbf{b}\vert\mathbf{Y} \sim N(\mathbf{BZ}'\mathbf{V}^{-1}(\mathbf{Y}-\mathbf{X}\boldsymbol{\beta}), (\mathbf{Z}'\boldsymbol{\Sigma}^{-1}\mathbf{Z}+\mathbf{B}^{-1})^{-1})$	Updated belief about **b**
Posterior Mean (BLUP)	$E(\mathbf{b}\vert\mathbf{Y}) = \mathbf{BZ}'\mathbf{V}^{-1}(\mathbf{Y}-\mathbf{X}\boldsymbol{\beta})$	Best predictor (squared error loss)
Posterior Variance	$(\mathbf{Z}'\boldsymbol{\Sigma}^{-1}\mathbf{Z}+\mathbf{B}^{-1})^{-1}$	Uncertainty in predictions

- Covariance: $(\mathbf{Z}'\boldsymbol{\Sigma}^{-1}\mathbf{Z}+\mathbf{B}^{-1})^{-1}$
  - This posterior variance accounts for both prior uncertainty (**B**) and data uncertainty ($\boldsymbol{\Sigma}$).

Thus, the Bayesian posterior mean of **b** coincides with the BLUP predictor:

$$E(\mathbf{b}|\mathbf{Y}) = \mathbf{BZ}'\mathbf{V}^{-1}(\mathbf{Y}-\mathbf{X}\boldsymbol{\beta}).$$

Interpretation of the Posterior Distribution (Table 5.5)

1. Posterior Mean as a Shrinkage Estimator (BLUP)
   - The expectation $E(\mathbf{b}|\mathbf{Y})$ shrinks individual estimates toward the population mean.
   - Subjects with less data or more variability will have estimates closer to zero.
   - This is similar to ridge regression in penalized estimation.
2. Posterior Variance Quantifies Uncertainty
   - The matrix $(\mathbf{Z}'\boldsymbol{\Sigma}^{-1}\mathbf{Z}+\mathbf{B}^{-1})^{-1}$ captures the remaining uncertainty in **b** after seeing **Y**.
   - If $\mathbf{Z}'\boldsymbol{\Sigma}^{-1}\mathbf{Z}$ is large, the data provide strong information about **b**, reducing posterior variance.
   - If $\mathbf{B}^{-1}$ dominates, prior information heavily influences estimates.
3. Connection to Bayesian Inference
   - The random effects **b** follow a Gaussian posterior due to conjugacy.
   - This is analogous to Bayesian hierarchical models, where random effects are latent variables estimated from data.

5.2.4 Estimating the Variance-Covariance Matrix

If we have an estimate $\tilde{\mathbf{V}}$ for $\mathbf{V}$, we can estimate the fixed and random effects as:

$$\hat{\beta} = (\mathbf{X}'\tilde{\mathbf{V}}^{-1}\mathbf{X})^{-1}\mathbf{X}'\tilde{\mathbf{V}}^{-1}\mathbf{Y},$$

$$\hat{\mathbf{b}} = \mathbf{B}\mathbf{Z}'\tilde{\mathbf{V}}^{-1}(\mathbf{Y} - \mathbf{X}\hat{\boldsymbol{\beta}}).$$

where:

- $\hat{\beta}$ is the estimated fixed effects.
- $\hat{\mathbf{b}}$ is the empirical best linear unbiased predictor (EBLUP), also called the empirical Bayes estimate of **b**.

Properties of $\hat{\beta}$ and variance estimation

- **Consistency**: $\hat{\text{Var}}(\hat{\beta})$ is a consistent estimator of $\text{Var}(\hat{\beta})$ if $\tilde{\mathbf{V}}$ is a consistent estimator of **V**.
- **Bias Issue**: $\hat{\text{Var}}(\hat{\beta})$ is biased because it does not account for the uncertainty in estimating **V**.
- **Implication**: This means that $\hat{\text{Var}}(\hat{\beta})$ underestimates the true variability.

To estimate **V**, several approaches can be used:

- Maximum likelihood estimation (MLE)
- Restricted maximum likelihood (REML)
- Estimated generalized least squares (EGLS)
- Bayesian hierarchical models (BHM)

5.2.4.1 Maximum Likelihood Estimation

MLE finds parameter estimates by maximizing the likelihood function.

Define a parameter vector θ that includes all unknown variance components in $\boldsymbol{\Sigma}$ and **B**. Then, we assume:

$$\mathbf{Y} \sim N(\mathbf{X}\boldsymbol{\beta}, \mathbf{V}(\theta)).$$

The log-likelihood function (ignoring constant terms) is:

$$-2\log L(\mathbf{y}; \theta, \beta) = \log|\mathbf{V}(\theta)| + (\mathbf{Y} - \mathbf{X}\boldsymbol{\beta})'\mathbf{V}(\theta)^{-1}(\mathbf{Y} - \mathbf{X}\boldsymbol{\beta}).$$

Steps for MLE Estimation

1. Estimate $\hat{\beta}$, assuming θ is known:

$$\hat{\beta}_{MLE} = (\mathbf{X}'\mathbf{V}(\boldsymbol{\theta})^{-1}\mathbf{X})^{-1}\mathbf{X}'\mathbf{V}(\boldsymbol{\theta})^{-1}\mathbf{Y}.$$

2. Obtain $\hat{\theta}_{MLE}$ by maximizing the log-likelihood:

$$\hat{\theta}_{MLE} = \arg\max_{\theta} -2\log L(\mathbf{y}; \theta, \beta).$$

3. Substitute $\hat{\theta}_{MLE}$ to get updated estimates:

$$\hat{\beta}_{MLE} = (\mathbf{X}'\mathbf{V}(\hat{\boldsymbol{\theta}}_{\mathbf{MLE}})^{-1}\mathbf{X})^{-1}\mathbf{X}'\mathbf{V}(\hat{\boldsymbol{\theta}}_{\mathbf{MLE}})^{-1}\mathbf{Y}.$$

4. Predict random effects:

$$\hat{\mathbf{b}}_{MLE} = \mathbf{B}(\hat{\theta}_{MLE})\mathbf{Z}'\mathbf{V}(\hat{\theta}_{MLE})^{-1}(\mathbf{Y} - \mathbf{X}\hat{\boldsymbol{\beta}}_{\mathbf{MLE}}).$$

Key observations about MLE

- MLE tends to underestimate θ because it does not account for the estimation of fixed effects.
- Bias in variance estimates can be corrected using REML.

5.2.4.2 Restricted Maximum Likelihood

Restricted maximum likelihood (REML) is an estimation method that improves upon maximum likelihood estimation by accounting for the loss of degrees of freedom due to the estimation of fixed effects.

Unlike MLE, which estimates both fixed effects (β) and variance components (θ) simultaneously, REML focuses on estimating variance components by considering linear combinations of the data that are independent of the fixed effects.

Consider the linear mixed model:

$$\mathbf{y} = \mathbf{X}\beta + \mathbf{Z}\mathbf{b} + \epsilon,$$

where:

- $\mathbf{y}$: Response vector of length N
- $\mathbf{X}$: Design matrix for fixed effects ($N \times p$)
- β: Fixed effect parameter vector ($p \times 1$)
- $\mathbf{Z}$: Design matrix for random effects
- $\mathbf{b} \sim N(\mathbf{0}, \mathbf{D})$: Random effects
- $\epsilon \sim N(\mathbf{0}, \boldsymbol{\Sigma})$: Residual errors

The marginal distribution of $\mathbf{y}$ is:

$$\mathbf{y} \sim N(\mathbf{X}\beta, \mathbf{V}(\theta)),$$

where:

$$\mathbf{V}(\theta) = \mathbf{Z}\mathbf{D}\mathbf{Z}' + \boldsymbol{\Sigma}.$$

To eliminate dependence on β, consider linear transformations of $\mathbf{y}$ that are orthogonal to the fixed effects.

Let $\mathbf{K}$ be a full-rank contrast matrix of size $N \times (N - p)$ such that:

$$\mathbf{K}'\mathbf{X} = 0.$$

Then, we consider the transformed data:

$$\mathbf{K}'\mathbf{y} \sim N(\mathbf{0}, \mathbf{K}'\mathbf{V}(\theta)\mathbf{K}).$$

- This transformation removes β from the likelihood, focusing solely on the variance components θ.
- Importantly, the choice of $\mathbf{K}$ does not affect the final REML estimates.

The REML log-likelihood is:

$$-2\log L_{REML}(\theta) = \log|\mathbf{K}'\mathbf{V}(\theta)\mathbf{K}| + \mathbf{y}'\mathbf{K}(\mathbf{K}'\mathbf{V}(\theta)\mathbf{K})^{-1}\mathbf{K}'\mathbf{y}.$$

An equivalent form of the REML log-likelihood, avoiding explicit use of $\mathbf{K}$, is:

$$-2\log L_{REML}(\theta) = \log|\mathbf{V}(\theta)| + \log|\mathbf{X}'\mathbf{V}(\theta)^{-1}\mathbf{X}| + (\mathbf{y} - \mathbf{X}\hat{\beta})'\mathbf{V}(\theta)^{-1}(\mathbf{y} - \mathbf{X}\hat{\beta}),$$

where:

$$\hat{\beta} = (\mathbf{X}'\mathbf{V}(\theta)^{-1}\mathbf{X})^{-1}\mathbf{X}'\mathbf{V}(\theta)^{-1}\mathbf{y}.$$

This form highlights how REML adjusts for the estimation of fixed effects via the second term $\log|\mathbf{X}'\mathbf{V}^{-1}\mathbf{X}|$.

Steps for REML Estimation

1. Transform the data using $\mathbf{K}'\mathbf{y}$ to remove β from the likelihood.
2. Maximize the restricted likelihood to estimate $\hat{\theta}_{REML}$.
3. Estimate fixed effects using:

$$\hat{\beta}_{REML} = (\mathbf{X}'\mathbf{V}(\hat{\theta}_{REML})^{-1}\mathbf{X})^{-1}\mathbf{X}'\mathbf{V}(\hat{\theta}_{REML})^{-1}\mathbf{y}.$$

Properties of REML

1. **Unbiased Variance Component Estimates**: REML produces unbiased estimates of variance components by accounting for the degrees of freedom used to estimate fixed effects.
2. **Invariance to Fixed Effects**: The restricted likelihood is constructed to be independent of the fixed effects β.
3. **Asymptotic Normality**: REML estimates are consistent and asymptotically normal under standard regularity conditions.

Table 5.6 Comparison of REML and MLE

Criterion	MLE	REML
Approach	Maximizes full likelihood	Maximizes likelihood of contrasts (removes β)
Estimates Fixed Effects?	Yes	No (focuses on variance components)
Bias in Variance Estimates	Biased (underestimates variance components)	Unbiased (corrects for loss of degrees of freedom)
Effect of Changing **X**	Affects variance estimates	No effect on variance estimates
Consistency	Yes	Yes
Asymptotic Normality	Yes	Yes
Efficiency	Efficient under normality	More efficient for variance components
Model Comparison (AIC/BIC)	Suitable for comparing different fixed effect models	Not suitable (penalizes fixed effects differently)
Performance in Small Samples	Sensitive to small sample bias	More robust to small sample bias
Handling Outliers	More sensitive	Less sensitive
Equivalent to ANOVA?	No	Yes, in balanced designs

4. **Efficiency**: While REML estimates variance components efficiently, it does not maximize the joint likelihood of all parameters, so β estimates are slightly less efficient compared to MLE (Table 5.6).

5.2.4.3 Estimated Generalized Least Squares

MLE and REML rely on the Gaussian assumption, which may not always hold. EGLS provides an alternative by relying only on the first two moments (mean and variance).

The LMM framework is:

$$\mathbf{Y}_i = \mathbf{X}_i\beta + \mathbf{Z}_i\mathbf{b}_i + \epsilon_i.$$

where:

- Random effects: $\mathbf{b}_i \sim N(\mathbf{0}, \mathbf{D})$.
- Residual errors: $\epsilon_i \sim N(\mathbf{0}, \mathbf{\Sigma_i})$.
- Independence assumption: $\text{Cov}(\epsilon_i, \mathbf{b}_i) = 0$.

Thus, the first two moments are:

$$E(\mathbf{Y}_i) = \mathbf{X}_i\beta, \quad \text{Var}(\mathbf{Y}_i) = \mathbf{V}_i.$$

The EGLS estimator is:

$$\hat{\beta}_{GLS} = \left\{ \sum_{i=1}^{n} \mathbf{X_i'V_i}(\boldsymbol{\theta})^{-1}\mathbf{X_i} \right\}^{-1} \sum_{i=1}^{n} \mathbf{X_i'V_i}(\boldsymbol{\theta})^{-1}\mathbf{Y_i}.$$

Writing in matrix form:

$$\hat{\beta}_{GLS} = \left\{ \mathbf{X'V}(\boldsymbol{\theta})^{-1}\mathbf{X} \right\}^{-1} \mathbf{X'V}(\boldsymbol{\theta})^{-1}\mathbf{Y}.$$

Since $\mathbf{V}(\theta)$ is unknown, we estimate it as $\hat{\mathbf{V}}$, leading to the EGLS estimator:

$$\hat{\beta}_{EGLS} = \left\{ \mathbf{X'\hat{V}^{-1}X} \right\}^{-1} \mathbf{X'\hat{V}^{-1}Y}.$$

Key insights about EGLS

- Computational Simplicity:
  - EGLS does not require iterative maximization of a likelihood function, making it computationally attractive.
- Same Form as MLE/REML:
  - The fixed effect estimators for MLE, REML, and EGLS have the same form, differing only in how $\mathbf{V}$ is estimated.
- Robust to Non-Gaussian Data:
  - Since it only depends on first and second moments, it can handle cases where MLE and REML struggle with non-normality.

When to Use EGLS?

- When the normality assumption for MLE/REML is questionable.
- When $\mathbf{V}$ can be estimated efficiently without requiring complex optimization.
- In non-iterative approaches, where computational simplicity is a priority.

5.2.4.4 Bayesian Hierarchical Models

Bayesian methods offer a fully probabilistic framework to estimate $\mathbf{V}$ by incorporating prior distributions.

The joint distribution can be decomposed hierarchically:

$$f(A, B, C) = f(A|B, C) f(B|C) f(C).$$

Applying this to LMMs:

$$\begin{aligned} f(\mathbf{Y}, \boldsymbol{\beta}, \mathbf{b}, \boldsymbol{\theta}) &= f(\mathbf{Y}|\boldsymbol{\beta}, \mathbf{b}, \boldsymbol{\theta}) f(\mathbf{b}|\boldsymbol{\theta}, \boldsymbol{\beta}) f(\boldsymbol{\beta}|\boldsymbol{\theta}) f(\boldsymbol{\theta}) \\ &= f(\mathbf{Y}|\boldsymbol{\beta}, \mathbf{b}, \boldsymbol{\theta}) f(\mathbf{b}|\boldsymbol{\theta}) f(\boldsymbol{\beta}) f(\boldsymbol{\theta}). \end{aligned}$$

where:

- The first equality follows from probability decomposition.
- The second equality assumes conditional independence, meaning:
  - Given θ, no additional information about **b** is obtained from knowing β.

Using Bayes' theorem, the posterior distribution is:

$$f(\boldsymbol{\beta}, \mathbf{b}, \boldsymbol{\theta}|\boldsymbol{Y}) \propto f(\mathbf{Y}|\boldsymbol{\beta}, \mathbf{b}, \boldsymbol{\theta}) f(\mathbf{b}|\boldsymbol{\theta}) f(\boldsymbol{\beta}) f(\boldsymbol{\theta}).$$

where:

$$\mathbf{Y}|\boldsymbol{\beta}, \mathbf{b}, \boldsymbol{\theta} \sim N(\mathbf{X}\,\boldsymbol{\beta} + \boldsymbol{Z}\boldsymbol{b}, \boldsymbol{\Sigma}(\boldsymbol{\theta})),$$
$$\mathbf{b}|\boldsymbol{\theta} \sim N(\mathbf{0}, \mathbf{B}(\boldsymbol{\theta})).$$

To complete the Bayesian model, we specify prior distributions:

- $f(\beta)$: Prior on fixed effects.
- $f(\theta)$: Prior on variance components.

Since analytical solutions are generally unavailable, we use Markov chain Monte Carlo (MCMC) to sample from the posterior:

- Gibbs sampling (if conjugate priors are used).
- Hamiltonian Monte Carlo (HMC) (for complex models).

Advantages

1. Accounts for Parameter Uncertainty
   - Unlike MLE/REML, Bayesian methods propagate uncertainty in variance component estimation.
2. Flexible Model Specification
   - Can incorporate prior knowledge via informative priors.
   - Extends naturally beyond Gaussian assumptions (e.g., Student t distributions for heavy-tailed errors).
3. Robustness in Small Samples
   - Bayesian methods can stabilize variance estimation in small datasets where MLE/REML are unreliable.

Challenges

1. Computational Complexity
   - Requires MCMC algorithms, which can be computationally expensive.

Table 5.7 Comparison of estimation methods for **V**

Method	Assumptions	Computational Cost	Handles Non-Normality?	Best Use Case
MLE	Gaussian errors	High (iterative)	No	Model selection (AIC/BIC)
REML	Gaussian errors	High (iterative)	No	Variance estimation
EGLS	First two moments	Low (non-iterative)	Yes	Large-scale models with correlated errors
Bayesian (BHM)	Probabilistic	Very High (MCMC)	Yes	Small samples, prior information available

2. Convergence Issues
   - MCMC chains must be checked for convergence (e.g., using R-hat diagnostic).
3. Choice of Priors
   - Poorly chosen priors can bias estimates or slow down convergence.

Table 5.7 shows comparison of estimation methods for V.

5.3 Inference in Linear Mixed Models

5.3.1 *Inference for Fixed Effects (β)*

The goal is to test hypotheses about the fixed effect parameters β using various statistical tests:

- Wald Test
- F-Test
- Likelihood Ratio Test

5.3.1.1 Wald Test

The Wald test assesses whether certain linear combinations of fixed effects are equal to specified values.

Given:

$$\hat{\beta}(\theta) = \left(\mathbf{X}'\mathbf{V}^{-1}(\theta)\mathbf{X}\right)^{-1}\mathbf{X}'\mathbf{V}^{-1}(\theta)\mathbf{Y},$$

and its variance:

$$\text{Var}(\hat{\beta}(\theta)) = \left(\mathbf{X}'\mathbf{V}^{-1}(\theta)\mathbf{X}\right)^{-1}.$$

In practice, we substitute $\hat{\theta}$ (the estimate of θ) to obtain:

- Hypotheses:

$$H_0 : \mathbf{A}\boldsymbol{\beta} = \mathbf{d}$$

  where:

  - $\mathbf{A}$ is a contrast matrix specifying linear combinations of β.
  - $\mathbf{d}$ is a constant vector representing the null hypothesis values.

- Wald Test Statistic:

$$W = (\mathbf{A}\hat{\beta} - \mathbf{d})' \left[\mathbf{A}\left(\mathbf{X}'\hat{\mathbf{V}}^{-1}\mathbf{X}\right)^{-1}\mathbf{A}'\right]^{-1} (\mathbf{A}\hat{\beta} - \mathbf{d}).$$

- Distribution under H_0:

$$W \sim \chi^2_{\text{rank}(\mathbf{A})}.$$

Caution with Wald Test

- Underestimation of Variance:
  The Wald test ignores the variability from estimating $\hat{\theta}$, leading to underestimated standard errors and potentially inflated Type I error rates.
- Small Sample Issues:
  Less reliable in small samples or when variance components are near boundary values (e.g., variances close to zero).

5.3.1.2 F-Test

An alternative to the Wald test, the F-test adjusts for the estimation of σ^2 and provides better performance in small samples (Table 5.8).

Assume:

$$\text{Var}(\mathbf{Y}) = \sigma^2\mathbf{V}(\theta).$$

The F-statistic is:

$$F^* = \frac{(\mathbf{A}\hat{\beta} - \mathbf{d})' \left[\mathbf{A}\left(\mathbf{X}'\hat{\mathbf{V}}^{-1}\mathbf{X}\right)^{-1}\mathbf{A}'\right]^{-1} (\mathbf{A}\hat{\beta} - \mathbf{d})}{\hat{\sigma}^2\text{rank}(\mathbf{A})}.$$

Table 5.8 Comparison of Wald test and F-test under common evaluation criteria

Criterion	Wald Test	F-Test
Small Sample Performance	Poor (can inflate Type I error)	Better control of Type I error
Variance Estimation	Ignores variability in $\hat{\theta}$	Adjusts using $\hat{\sigma}^2$
Reduction to t-test	Yes (for single β)	Yes (when rank($\mathbf{A}$) = 1)

- Distribution under H_0:

$$F^* \sim F_{\text{rank}(\mathbf{A}), \text{df}_{\text{denominator}}}.$$

- Approximating Denominator Degrees of Freedom:
  - Satterthwaite approximation
  - Kenward-Roger approximation (provides bias-corrected standard errors)

F-Test Advantages:

- More accurate in small samples compared to the Wald test.
- Adjusts for variance estimation, reducing bias in hypothesis testing.

Wald Test vs. F-Test:

5.3.1.3 Likelihood Ratio Test

The likelihood ratio test (LRT) compares the fit of nested models:

- Null Hypothesis:

$$H_0 : \beta \in \Theta_{\beta,0}$$

  where $\Theta_{\beta,0}$ is a subset of the full parameter space Θ_β.
- Test Statistic:

$$-2 \log \lambda = -2 \log \left(\frac{\hat{L}_{ML,0}}{\hat{L}_{ML}} \right),$$

  where:
  - $\hat{L}_{ML,0}$ = Maximized likelihood under H_0 (restricted model)
  - $\hat{L}_{ML}$ = Maximized likelihood under the alternative (full model)
- Distribution under H_0:

$$-2 \log \lambda \sim \chi^2_{df}$$

  where $df = \dim(\Theta_\beta) - \dim(\Theta_{\beta,0})$ (the difference in the number of parameters).

Important Notes

- LRT is applicable only for ML estimates (not REML) when comparing models with different fixed effects.
- REML-based LRT can be used for comparing models that differ in random effects (variance components), but not fixed effects.

5.3.2 *Inference for Variance Components (θ)*

For ML and REML estimators:

$$\hat{\theta} \sim N(\theta, I(\theta)^{-1}),$$

where $I(\theta)$ is the Fisher information matrix.

This normal approximation holds well for large samples, enabling Wald-type tests and confidence intervals.

5.3.2.1 Wald Test for Variance Components

The Wald test for variance components follows the same structure as for fixed effects:

- Test Statistic:

$$W = \frac{(\hat{\theta} - \theta_0)^2}{\widehat{\mathrm{Var}}(\hat{\theta})}.$$

- Distribution under H_0:

$$W \sim \chi_1^2.$$

Limitations of Wald Test for Variance Components:

- Boundary Issues: The normal approximation fails when the true variance component is near zero (boundary of the parameter space).
- Less reliable for variance parameters than for covariance parameters.

5.3.2.2 Likelihood Ratio Test for Variance Components

LRT can also be applied to variance components:

- Test Statistic:

Table 5.9 Comparison of statistical tests for fixed and random effects in linear mixed models

Test	Best For	Strengths	Limitations
Wald Test	Fixed effects (β)	Simple, widely used	Underestimates variance, biased in small samples
F-Test	Fixed effects (β)	Better in small samples, adjusts df	Requires approximation for degrees of freedom
LRT (ML)	Fixed effects, nested models	Powerful, widely used	Not valid for REML with fixed effects
LRT (REML)	Variance components	Robust for random effects	Boundary issues when variances are near zero
Wald (Variance)	Variance components (θ)	Simple extension of Wald test	Fails near parameter space boundaries

$$-2\log\lambda = -2\log\left(\frac{\hat{L}_{REML,0}}{\hat{L}_{REML}}\right).$$

- Distribution under H_0:
  - Not always χ^2-distributed when variance components are on the boundary (e.g., testing if $\sigma^2 = 0$).
  - May require mixture distributions or adjusted critical values.

Table 5.9 shows Comparison of statistical tests for fixed and random effects in linear mixed models.

5.4 Information Criteria for Model Selection

Information criteria are statistical tools used to compare competing models by balancing model fit (likelihood) with model complexity (number of parameters). They help in identifying the most parsimonious model that adequately explains the data without overfitting.

The three most commonly used criteria are:

- Akaike information criterion (AIC)
- Corrected Akaike information criterion (AICc)
- Bayesian information criterion (BIC)

5.4.1 Akaike Information Criterion

The Akaike information criterion is derived from the Kullback-Leibler divergence, which measures the difference between the true data-generating process and the fitted model.

AIC Formula

$$AIC = -2l(\hat{\theta}, \hat{\beta}) + 2q$$

where:

- $l(\hat{\theta}, \hat{\beta})$: The maximized log-likelihood of the model, evaluated at the estimates $\hat{\theta}$ (variance components) and $\hat{\beta}$ (fixed effects).
- q: The effective number of parameters, including:
  - The number of fixed effects.
  - The number of variance-covariance parameters (random effects).
  - Excludes parameters constrained to boundary values (e.g., variances estimated as zero).

Key Points About AIC

- Model Selection Rule:
  - Lower AIC indicates a better model.
  - Occasionally, software may report AIC as $l - q$, in which case higher AIC is better (rare).
- Comparing Random Effects Models:
  - Not recommended when comparing models with different random effects because it's difficult to accurately count the effective number of parameters.
- Sample Size Considerations:
  - Requires large sample sizes for reliable comparisons.
  - In small samples, AIC tends to favor more complex models due to insufficient penalty for model complexity.
- Potential Bias:
  - Can be negatively biased (i.e., favoring overly complex models) when the sample size is small relative to the number of parameters.

  When to Use AIC:

- Comparing models with the same random effects structure but different fixed effects.
- Selecting covariance structures in mixed models when the sample size is large.

5.4.2 Corrected AIC

The Corrected AIC (AICc) addresses the bias in AIC for small sample sizes. It was developed by (Hurvich and Tsai, 1989).

AICc Formula

$$AICc = AIC + \frac{2q(q+1)}{n-q-1}$$

where:

- n: The sample size.
- q: The number of estimated parameters.

Key Points About AICc

- Small Sample Correction:
  - Provides a stronger penalty for model complexity when the sample size is small.
- Applicability:
  - Valid when comparing models with fixed covariance structures.
  - Not recommended for models with general covariance structures due to difficulties in bias correction.
- Model Selection Rule:
  - Lower AICc indicates a better model.

  When to Use AICc:

- Small sample sizes (n/q ratio is low).
- Models with fixed random effects or simple covariance structures.

5.4.3 Bayesian Information Criterion

The Bayesian information criterion is derived from a Bayesian framework and incorporates a stronger penalty for model complexity compared to AIC.

BIC Formula

$$BIC = -2l(\hat{\theta}, \hat{\beta}) + q \log(n)$$

where:

- n: The number of observations.
- q: The number of effective parameters.
- $l(\hat{\theta}, \hat{\beta})$: The maximized log-likelihood.

Key Points About BIC

- Model Selection Rule:
  - Lower BIC indicates a better model.
- Stronger Penalty:
  - The penalty term $q \log(n)$ grows with sample size, leading BIC to favor simpler models more than AIC.
- Applicability to MLE and REML:
  - BIC can be used with both MLE and REML, but:

    Use MLE when comparing models with different fixed effects.
    Use REML when comparing models with different random effects (same fixed effects).
- Consistency:
  - BIC is consistent, meaning that as the sample size increases, it will select the true model with probability 1 (if the true model is among the candidates).

  When to Use BIC:

- Large sample sizes where model simplicity is prioritized.
- Model selection for hypothesis testing (due to its connection to Bayesian inference).

Comparison of AIC, AICc, and BIC (Table 5.10)

Key Takeaways

1. AIC is suitable for large datasets and general model comparisons but may favor overly complex models in small samples.
2. AICc corrects AIC's bias in small sample sizes.

Table 5.10 Comparison of AIC, AICc, and BIC

Criterion	Formula	Penalty term	Best for	Model selection rule
AIC	$-2l + 2q$	$2q$	General model comparison (large n)	Lower is better
AICc	$AIC + \frac{2q(q+1)}{n-q-1}$	Adjusted for small samples	Small sample sizes, simple random effects	Lower is better
BIC	$-2l + q\log(n)$	$q\log(n)$	Large samples, model selection in hypothesis testing	Lower is better

3. BIC favors simpler models, especially as the sample size increases, making it suitable for hypothesis testing and situations where parsimony is essential.
4. Use MLE for comparing models with different fixed effects, and REML when comparing models with different random effects (same fixed effects).
5. When comparing random effects structures, AIC and BIC may not be reliable due to difficulty in counting effective parameters accurately.

5.4.4 Practical Example with Linear Mixed Models

Consider the linear mixed model:

$$Y_{ik} = \begin{cases} \beta_0 + b_{1i} + (\beta_1 + b_{2i})t_{ij} + \epsilon_{ij} & L \\ \beta_0 + b_{1i} + (\beta_2 + b_{2i})t_{ij} + \epsilon_{ij} & H \\ \beta_0 + b_{1i} + (\beta_3 + b_{2i})t_{ij} + \epsilon_{ij} & C \end{cases}$$

where:

- $i = 1, \ldots, N$ (subjects)
- $j = 1, \ldots, n_i$ (repeated measures at time t_{ij})

$$\mathbf{Y}_i | b_i \sim N(\mathbf{X}_i\beta + \mathbf{1}b_i, \sigma^2\mathbf{I})$$
$$b_i \sim N(0, d_{11})$$

We aim to **estimate**:

- **Fixed effects**: β
- **Variance components**: σ^2, d_{11}
- **Random effects**: Predict b_i

When comparing models (e.g., different random slopes or covariance structures), we can compute:

- **AIC**: Penalizes model complexity with $2q$.
- **BIC**: Stronger penalty via $q \log(n)$, favoring simpler models.
- **AICc**: Adjusted AIC for small sample sizes.

```
# Practical Example with Linear Mixed Models in R


# Load required libraries
library(lme4)      # For fitting linear mixed-effects models
#> Loading required package: Matrix
library(MuMIn)     # For calculating AICc
```

```
library(dplyr)    # For data manipulation
#>
#> Attaching package: 'dplyr'
#> The following objects are masked from 'package:stats':
#>
#>     filter, lag
#> The following objects are masked from 'package:base':
#>
#>     intersect, setdiff, setequal, union

# Set seed for reproducibility
set.seed(123)

# Simulate Data
N <- 50              # Number of subjects
n_i <- 5             # Number of repeated measures per subject
t_ij <- rep(1:n_i, N)  # Time points

# Treatment groups (L, H, C)
treatment <- rep(c("L", "H", "C"), length.out = N)
group <- factor(rep(treatment, each = n_i))

# Simulate random effects
b1_i <- rnorm(N, mean = 0, sd = 2)    # Random intercepts
b2_i <- rnorm(N, mean = 0, sd = 1)    # Random slopes

# Fixed effects
beta_0 <- 5
beta_1 <- 0.5
beta_2 <- 1
beta_3 <- 1.5

# Generate response variable Y based on the specified model
Y <- numeric(N * n_i)
subject_id <- rep(1:N, each = n_i)

for (i in 1:N) {
    for (j in 1:n_i) {
        idx <- (i - 1) * n_i + j
        time <- t_ij[idx]

        # Treatment-specific model
        if (group[idx] == "L") {
            Y[idx] <-
```

```
                beta_0 + b1_i[i] + (beta_1 + b2_i[i]) * time
 ↪  + rnorm(1, 0, 1)
        } else if (group[idx] == "H") {
            Y[idx] <-
                beta_0 + b1_i[i] + (beta_2 + b2_i[i]) * time
 ↪  + rnorm(1, 0, 1)
        } else {
            Y[idx] <-
                beta_0 + b1_i[i] + (beta_3 + b2_i[i]) * time
 ↪  + rnorm(1, 0, 1)
        }
    }
}

# Combine into a data frame
data <- data.frame(
    Y = Y,
    time = t_ij,
    group = group,
    subject = factor(subject_id)
)

# Fit Linear Mixed Models
# Model 1: Random Intercepts Only
model1 <-
    lmer(Y ~ time * group + (1 | subject), data = data, REML
     ↪  = FALSE)

# Model 2: Random Intercepts and Random Slopes
model2 <-
    lmer(Y ~ time * group + (1 + time |
                                 subject),
         data = data,
         REML = FALSE)

# Model 3: Simpler Model (No Interaction)
model3 <-
    lmer(Y ~ time + group + (1 | subject), data = data, REML
     ↪  = FALSE)

# Extract Information Criteria
results <- data.frame(
    Model = c(
        "Random Intercepts",
```

```
        "Random Intercepts + Slopes",
        "No Interaction"
    ),
    AIC = c(AIC(model1), AIC(model2), AIC(model3)),
    BIC = c(BIC(model1), BIC(model2), BIC(model3)),
    AICc = c(AICc(model1), AICc(model2), AICc(model3))
)

# Display the results
print(results)
#>                          Model       AIC      BIC      AICc
#> 1            Random Intercepts 1129.2064 1157.378 1129.8039
#> 2 Random Intercepts + Slopes  974.5514 1009.766  975.4719
#> 3               No Interaction 1164.2797 1185.408 1164.6253
```

Interpretation of Results

- Model 2 (with random intercepts and slopes) has the lowest AIC, BIC, and AICc, indicating the best fit among the models.
- Model 1 (random intercepts only) performs worse, suggesting that allowing random slopes improves model fit.
- Model 3 (simpler fixed effects without interaction) has the highest AIC/BIC/AICc, indicating poor fit compared to Models 1 and 2.

Model Selection Criteria (Table 5.11)

5.5 Split-Plot Designs

Split-plot designs are commonly used in experimental settings where there are **two or more factors**, and at least one of them requires **larger experimental units** compared to the other(s). This situation often arises in agricultural, industrial, and business experiments where certain treatments are harder or more expensive to apply.

Table 5.11 Model selection results based on information criteria

Criterion	Best model	Reason
AIC	Model 2	Best trade-off between fit and complexity
BIC	Model 2	Stronger penalty for complexity, still favored
AICc	Model 2	Adjusted for small samples, Model 2 still best

Key Characteristics

- Two factors with different experimental unit requirements:
  - Factor A (Whole-plot factor): Requires large experimental units (e.g., different fields, production batches).
  - Factor B (Sub-plot factor): Can be applied to smaller units within the larger experimental units (e.g., plots within fields, products within batches).
- Blocking: The experiment is typically divided into blocks (or replicates) to account for variability. However, unlike [randomized block designs], the randomization process in split-plot designs occurs at two levels:
  - Whole-plot randomization: Factor A is randomized across large units within each block.
  - Sub-plot randomization: Factor B is randomized within each whole plot.

5.5.1 *Example Setup*

- **Factor A:** three levels (applied to large units).
- **Factor B:** two levels (applied within the large units).
- **4 Blocks (replicates):** Each containing all combinations of A and B.

Unlike in [Randomized Block Designs], the randomization of Factor A is **restricted** due to the larger unit size. Factor A is applied **once per block**, while Factor B can be applied **multiple times within each block**.

5.5.2 *Statistical Model for Split-Plot Designs*

1. When Factor A is the primary focus (whole-plot analysis)

$$Y_{ij} = \mu + \rho_i + \alpha_j + e_{ij}$$

where:

- Y_{ij} = Response for the j-th level of factor A in the i-th block.
- μ = Overall mean.
- ρ_i = Random effect of the i-th block ($\rho_i \sim N(0, \sigma_\rho^2)$).
- α_j = Fixed effect of factor A (main effect).
- $e_{ij} \sim N(0, \sigma_e^2)$ = Whole-plot error (random), representing the variability within blocks due to factor A.

2. When Factor B is the primary focus (sub-plot analysis):

$$Y_{ijk} = \mu + \phi_{ij} + \beta_k + \epsilon_{ijk}$$

where:

- Y_{ijk} = Response for the k-th level of factor B within the j-th level of factor A and i-th block.
- ϕ_{ij} = Combined effect of block and factor A:

$$\phi_{ij} = \rho_i + \alpha_j + e_{ij}$$

- β_k = Fixed effect of factor B (main effect).
- $\epsilon_{ijk} \sim N(0, \sigma_\epsilon^2)$ = Sub-plot error (random), capturing variability within whole plots.

3. Full Split-Plot Model (Including Interaction)

$$Y_{ijk} = \mu + \rho_i + \alpha_j + e_{ij} + \beta_k + (\alpha\beta)_{jk} + \epsilon_{ijk}$$

where:

- i = Block (replication). - j = Level of factor A (whole-plot factor).
- k = Level of factor B (sub-plot factor).
- μ = Overall mean.
- ρ_i = Random effect of the i-th block.
- α_j = Fixed main effect of factor A.
- $e_{ij} \sim N(0, \sigma_e^2)$ = Whole-plot error (random).
- β_k = Fixed main effect of factor B.
- $(\alpha\beta)_{jk}$ = Fixed interaction effect between factors A and B.
- $\epsilon_{ijk} \sim N(0, \sigma_\epsilon^2)$ = Sub-plot error (random).

5.5.3 Approaches to Analyzing Split-Plot Designs

1. ANOVA Perspective

   Whole-Plot Comparisons:

- Factor A vs. Whole-Plot Error:
  Compare the variation due to factor A (α_j) against the whole-plot error (e_{ij}).
- Blocks vs. Whole-Plot Error:
  Compare the variation due to blocks (ρ_i) against the whole-plot error (e_{ij}).

Sub-Plot Comparisons:

- Factor B vs. Sub-Plot Error:
  Compare the variation due to factor B (β_k) against the sub-plot error (ϵ_{ijk}).
- Interaction (A $\times$ B) vs. Sub-Plot Error:
  Compare the interaction effect ($(\alpha\beta)_{jk}$) against the sub-plot error (ϵ_{ijk}).

2. **Mixed Model Perspective**

A more flexible approach is to treat split-plot designs using mixed-effect models, which can handle both fixed and random effects explicitly:

$$\mathbf{Y} = \mathbf{X}\boldsymbol{\beta} + \mathbf{Z}\mathbf{b} + \boldsymbol{\epsilon}$$

where:

- $\mathbf{Y}$ = Vector of observed responses.
- $\mathbf{X}$ = Design matrix for fixed effects (e.g., factors A, B, and their interaction).
- $\boldsymbol{\beta}$ = Vector of fixed effect coefficients (e.g., μ, α_j, β_k, $(\alpha\beta)_{jk}$).
- $\mathbf{Z}$ = Design matrix for random effects (e.g., blocks and whole-plot errors).
- $\mathbf{b}$ = Vector of random effect coefficients (e.g., ρ_i, e_{ij}).
- $\boldsymbol{\epsilon}$ = Vector of residuals (sub-plot error).

Mixed models are particularly useful when:

- There are **unbalanced designs** (missing data).
- You need to account for **complex correlation structures** within the data.

5.5.4 Application: Split-Plot Design

Consider an agricultural experiment designed to study the effects of irrigation and crop variety on yield. This scenario is well-suited for a split-plot design because irrigation treatments are applied to large plots (fields), while different crop varieties are planted within these plots.

5.5.4.1 Model Specification

The linear mixed-effect model is defined as:

$$y_{ijk} = \mu + i_i + v_j + (iv)_{ij} + f_k + \epsilon_{ijk}$$

where:

- y_{ijk} = Observed yield for the i-th irrigation, j-th variety, in the k-th field.
- μ = Overall mean yield.

- i_i = Fixed effect of the i-th irrigation level.
- v_j = Fixed effect of the j-th crop variety.
- $(iv)_{ij}$ = Interaction effect between irrigation and variety (fixed).
- $f_k \sim N(0, \sigma_f^2)$ = Random effect of field (captures variability between fields).
- $\epsilon_{ijk} \sim N(0, \sigma_\epsilon^2)$ = Residual error.

Note

Since each variety-field combination is observed only once, we cannot model a random interaction between variety and field.

5.5.4.2 Data Exploration

```
library(ggplot2)
data(irrigation, package = "faraway")  # Load the dataset

# Summary statistics and preview
summary(irrigation)
#>      field    irrigation variety      yield
#>  f1     :2   i1:4        v1:8    Min.   :34.80
#>  f2     :2   i2:4        v2:8    1st Qu.:37.60
#>  f3     :2   i3:4                Median :40.15
#>  f4     :2   i4:4                Mean   :40.23
#>  f5     :2                       3rd Qu.:42.73
#>  f6     :2                       Max.   :47.60
#>  (Other):4
head(irrigation, 4)
#>   field irrigation variety yield
#> 1    f1         i1      v1  35.4
#> 2    f1         i1      v2  37.9
#> 3    f2         i2      v1  36.7
#> 4    f2         i2      v2  38.2
```

```
# Exploratory plot
ggplot(irrigation,
       aes(
           x     = field,
           y     = yield,
           shape = irrigation,
           color = variety
       )) +
    geom_point(size = 3) +
    labs(title = "Yield by Field, Irrigation, and Variety",
```

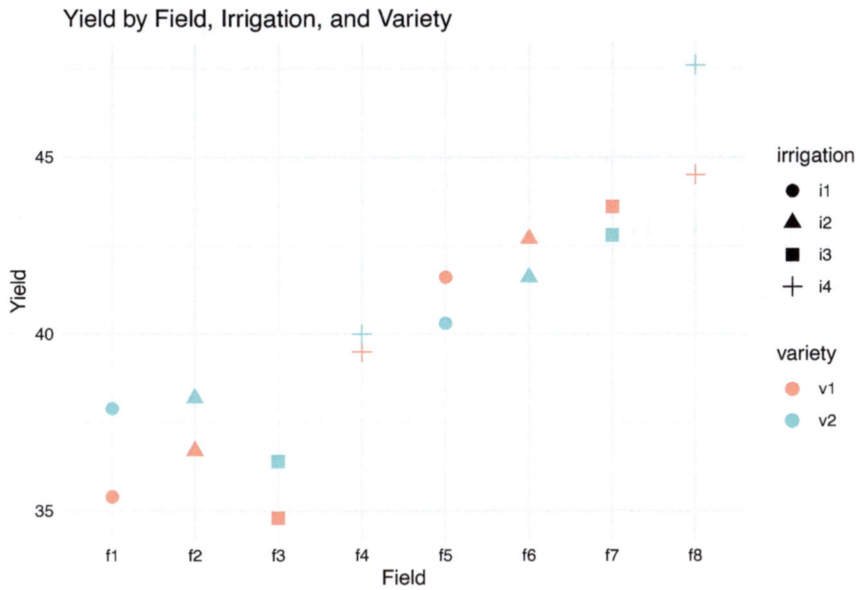

Fig. 5.1 Yield by field, irrigation, and variety

```
        x = "Field",
        y = "Yield") +
    theme_minimal()
```

This plot helps visualize how yield varies across fields, under different irrigation treatments, and for different varieties (Fig. 5.1).

5.5.4.3 Fitting the Initial Mixed-Effect Model

We fit a mixed-effect model where:

- Irrigation and variety (and their interaction) are fixed effects.
- Field is modeled as a random effect to account for variability between fields.

```
library(lmerTest)  # Provides p-values for lmer models
#>
#> Attaching package: 'lmerTest'
#> The following object is masked from 'package:lme4':
#>
#>     lmer
#> The following object is masked from 'package:stats':
#>
```

```
#>     step

# Full model with interaction term
sp_model <-
  lmer(yield ~ irrigation * variety
       + (1 | field), data = irrigation)
summary(sp_model)
#> Linear mixed model fit by REML. t-tests use
#>   Satterthwaite's method [lmerModLmerTest]
#> Formula: yield ~ irrigation * variety + (1 | field)
#>    Data: irrigation
#>
#> REML criterion at convergence: 45.4
#>
#> Scaled residuals:
#>     Min      1Q  Median      3Q     Max
#> -0.7448 -0.5509  0.0000  0.5509  0.7448
#>
#> Random effects:
#>  Groups   Name        Variance Std.Dev.
#>  field    (Intercept) 16.200   4.025
#>  Residual              2.107   1.452
#> Number of obs: 16, groups:  field, 8
#>
#> Fixed effects:
#>                          Estimate Std. Error     df t value
#> (Intercept)                38.500      3.026  4.487  12.725
#> irrigationi2                1.200      4.279  4.487   0.280
#> irrigationi3                0.700      4.279  4.487   0.164
#> irrigationi4                3.500      4.279  4.487   0.818
#> varietyv2                   0.600      1.452  4.000   0.413
#> irrigationi2:varietyv2     -0.400      2.053  4.000  -0.195
#> irrigationi3:varietyv2     -0.200      2.053  4.000  -0.097
#> irrigationi4:varietyv2      1.200      2.053  4.000   0.584
#>                          Pr(>|t|)
#> (Intercept)              0.000109 ***
#> irrigationi2             0.791591
#> irrigationi3             0.877156
#> irrigationi4             0.454584
#> varietyv2                0.700582
#> irrigationi2:varietyv2   0.855020
#> irrigationi3:varietyv2   0.927082
#> irrigationi4:varietyv2   0.590265
#> ---
```

```
#> Signif. codes:
#> 0 '***' 0.001 '**' 0.01 '*' 0.05 '.' 0.1 ' ' 1
#>
#> Correlation of Fixed Effects:
#>             (Intr) irrgt2 irrgt3 irrgt4 vrtyv2 irr2:2
#> irrigation2 -0.707
#> irrigation3 -0.707  0.500
#> irrigation4 -0.707  0.500  0.500
#> varietyv2   -0.240  0.170  0.170  0.170
#> irrgtn2:vr2  0.170 -0.240 -0.120 -0.120 -0.707
#> irrgtn3:vr2  0.170 -0.120 -0.240 -0.120 -0.707  0.500
#> irrgtn4:vr2  0.170 -0.120 -0.120 -0.240 -0.707  0.500
#>             irr3:2
#> irrigation2
#> irrigation3
#> irrigation4
#> varietyv2
#> irrgtn2:vr2
#> irrgtn3:vr2
#> irrgtn4:vr2  0.500

# ANOVA table using Kenward-Roger approximation for accurate
↪  p-values
anova(sp_model, ddf = "Kenward-Roger")
#> Registered S3 methods overwritten by 'broom':
#>   method          from
#>   nobs.fitdistr MuMIn
#>   nobs.multinom MuMIn
#> Type III Analysis of Variance Table with Kenward-Roger's
↪  method
#>                    Sum Sq Mean Sq NumDF DenDF F value
#> irrigation         2.4545 0.81818     3     4  0.3882
#> variety            2.2500 2.25000     1     4  1.0676
#> irrigation:variety 1.5500 0.51667     3     4  0.2452
#>                    Pr(>F)
#> irrigation         0.7685
#> variety            0.3599
#> irrigation:variety 0.8612
```

- Check the p-value of the interaction term (`irrigation:variety`).
  - If insignificant, this suggests no strong evidence of an interaction effect, and we may simplify the model by removing it.

5.5.4.4 Model Simplification: Testing for Additivity

We compare the **full model** (with interaction) to an **additive model** (without interaction):

```
library(lme4)

# Additive model (no interaction)
sp_model_additive <- lmer(yield ~ irrigation + variety + (1 |
↪   field), data = irrigation)

# Likelihood ratio test comparing the two models
anova(sp_model_additive, sp_model, ddf = "Kenward-Roger")
#> refitting model(s) with ML (instead of REML)
#> Data: irrigation
#> Models:
#> sp_model_additive: yield ~ irrigation + variety + (1 |
↪   field)
#> sp_model: yield ~ irrigation * variety + (1 | field)
#>                   npar    AIC    BIC  logLik -2*log(L)
#> sp_model_additive    7 83.959 89.368 -34.980    69.959
#> sp_model            10 88.609 96.335 -34.305    68.609
#>                    Chisq Df Pr(>Chisq)
#> sp_model_additive
#> sp_model          1.3503  3     0.7172
```

- **Hypotheses:**
  - H_0: The **additive model** (without interaction) fits the data adequately.
  - H_a: The **interaction model** provides a significantly better fit.
- **Interpretation:**
  - If the p-value is insignificant, we fail to reject H_0, meaning the simpler additive model is sufficient.
  - Check AIC and BIC: Lower values indicate a better-fitting model, supporting the use of the additive model if consistent with the hypothesis test.

5.5.4.5 Assessing the Random Effect: Exact Restricted Likelihood Ratio Test

To verify whether the random field effect is necessary, we conduct an exact RLRT:

- Hypotheses:

- H_0: $\sigma_f^2 = 0$ (no variability between fields; random effect is unnecessary).
- H_a: $\sigma_f^2 > 0$ (random field effect is significant).

```
library(RLRsim)

# RLRT for the random effect of field
exactRLRT(sp_model)
#>
#>  simulated finite sample distribution of RLRT.
#>
#>  (p-value based on 10000 simulated values)
#>
#> data:
#> RLRT = 6.1118, p-value = 0.01
```

Interpretation

- If the p-value is significant, we reject H_0, confirming that the random field effect is essential.
- A significant random effect implies substantial variability between fields that must be accounted for in the model.

5.6 Repeated Measures in Mixed Models

Repeated measures data arise when **multiple observations** are collected from the **same subjects** over time or under different conditions. This introduces **correlation** between observations from the same subject, which must be accounted for in the statistical model.

Mixed-effect models are particularly effective for repeated measures because they allow us to model both **fixed effects** (e.g., treatment, time) and **random effects** (e.g., subject-specific variability).

The general form of a mixed-effect model for repeated measures is:

$$Y_{ijk} = \mu + \alpha_i + \beta_j + (\alpha\beta)_{ij} + \delta_{i(k)} + \epsilon_{ijk}$$

where:

- Y_{ijk} = Response for the i-th group, j-th time point, and k-th subject.
- μ = Overall mean.
- α_i = Fixed effect of the i-th group (e.g., treatment group).
- β_j = Fixed effect of the j-th time point (repeated measure effect).
- $(\alpha\beta)_{ij}$ = Interaction effect between group and time (fixed).

- $\delta_{i(k)} \sim N(0, \sigma_\delta^2)$ = Random effect of the k-th subject within the i-th group (captures subject-specific deviations).
- $\epsilon_{ijk} \sim N(0, \sigma^2)$ = Residual error (independent across observations).

Here, $i = 1, \ldots, n_A$ (number of groups), $j = 1, \ldots, n_B$ (number of repeated measures), and $k = 1, \ldots, n_i$ (number of subjects in group i).

The **variance-covariance matrix** of the repeated observations for the k-th subject in the i-th group is given by:

$$\mathbf{Y}_{ik} = \begin{pmatrix} Y_{i1k} \\ Y_{i2k} \\ \vdots \\ Y_{in_Bk} \end{pmatrix}$$

1. Compound Symmetry (CS) Structure

Under the **compound symmetry** assumption (common in random-intercepts models), the covariance matrix is:

$$\mathbf{\Sigma}_{\text{subject}} = \begin{pmatrix} \sigma_\delta^2 + \sigma^2 & \sigma_\delta^2 & \cdots & \sigma_\delta^2 \\ \sigma_\delta^2 & \sigma_\delta^2 + \sigma^2 & \cdots & \sigma_\delta^2 \\ \vdots & \vdots & \ddots & \vdots \\ \sigma_\delta^2 & \sigma_\delta^2 & \cdots & \sigma_\delta^2 + \sigma^2 \end{pmatrix}$$

This matrix can be rewritten as:

$$\mathbf{\Sigma}_{\text{subject}} = (\sigma_\delta^2 + \sigma^2) \begin{pmatrix} 1 & \rho & \cdots & \rho \\ \rho & 1 & \cdots & \rho \\ \vdots & \vdots & \ddots & \vdots \\ \rho & \rho & \cdots & 1 \end{pmatrix}$$

Where:

- σ_δ^2 = Variance due to subject-specific random effects.
- σ^2 = Residual variance.
- $\rho = \frac{\sigma_\delta^2}{\sigma_\delta^2 + \sigma^2}$ = Intraclass correlation coefficient (ICC).

Key Points:

- Compound Symmetry Structure is the product of a scalar and a correlation matrix.
- The correlation between any two repeated measures from the same subject is constant (ρ).

- This structure assumes equal correlation across time points, which may not hold if measurements are collected over time.

Refer to random-intercepts model for a detailed discussion of compound symmetry.

2. Autoregressive (AR(1)) Structure

If repeated measures are collected over time, it may be more appropriate to assume an autoregressive correlation structure, where correlations decay as the time gap increases.

The AR(1) variance-covariance matrix is:

$$\boldsymbol{\Sigma}_{\text{subject}} = \sigma^2 \begin{pmatrix} 1 & \rho & \rho^2 & \cdots & \rho^{n_B-1} \\ \rho & 1 & \rho & \cdots & \rho^{n_B-2} \\ \rho^2 & \rho & 1 & \cdots & \rho^{n_B-3} \\ \vdots & \vdots & \vdots & \ddots & \vdots \\ \rho^{n_B-1} & \rho^{n_B-2} & \rho^{n_B-3} & \cdots & 1 \end{pmatrix}$$

where:

- σ^2 = Residual variance.
- ρ = Autoregressive parameter ($|\rho| < 1$), representing the correlation between consecutive time points.

Key Characteristics

- Correlations decrease exponentially as the time lag increases.
- Appropriate for longitudinal data where temporal proximity influences correlation.

  In matrix notation, the mixed model can be written as:

$$\mathbf{Y} = \mathbf{X}\boldsymbol{\beta} + \boldsymbol{\epsilon}$$

where:

- $\mathbf{Y}$ = Vector of observed responses.
- $\mathbf{X}$ = Design matrix for fixed effects (e.g., group, time, interaction).
- $\boldsymbol{\beta}$ = Vector of fixed effect coefficients.
- $\boldsymbol{\epsilon} \sim N(0, \sigma^2\boldsymbol{\Sigma})$ = Random error vector.
- $\boldsymbol{\Sigma}$ = Variance-covariance matrix of residuals:
  - Block diagonal structure if the covariance structure is identical for each subject.
  - Within each block (subject), the structure can be compound symmetry, AR(1), or another suitable structure depending on the data.

Choosing the Right Covariance Structure

- Compound Symmetry:
  - Suitable when correlations are constant across repeated measures (e.g., in randomized controlled trials).
  - Simple and interpretable but may be too restrictive for longitudinal data.
- Autoregressive (AR(1)):
  - Best when measurements are taken over equally spaced time intervals and correlations decay over time.
  - Assumes stronger correlation for adjacent time points.
- Unstructured (UN):
  - Allows different variances and covariances for each time point.
  - Provides maximum flexibility but requires more parameters and larger sample sizes.

Model selection criteria (AIC, BIC, likelihood ratio tests) can help determine the most appropriate covariance structure.

5.7 Unbalanced or Unequally Spaced Data

In many real-world applications, data are **unbalanced** (i.e., different numbers of observations per subject) or **unequally spaced** over time. This is common in longitudinal studies, clinical trials, and business analytics where subjects may be observed at irregular intervals or miss certain time points.

Mixed-effect models are flexible enough to handle such data structures, especially when we carefully model the **variance-covariance structure** of repeated measurements.

Consider the following mixed-effect model:

$$Y_{ikt} = \beta_0 + \beta_{0i} + \beta_1 t + \beta_{1i} t + \beta_2 t^2 + \beta_{2i} t^2 + \epsilon_{ikt}$$

where:

- Y_{ikt} = Response for the k-th subject in the i-th group at time t.
- $i = 1, 2$ = Groups (e.g., treatment vs. control).
- $k = 1, \ldots, n_i$ = Individuals within group i.
- $t = (t_1, t_2, t_3, t_4)$ = Time points (which may be unequally spaced).

Model components:

- **Fixed Effects:**
  - β_0 = Overall intercept (baseline).
  - β_1 = Common linear time trend.
  - β_2 = Common quadratic time trend.

- **Random Effects:**
  - β_{0i} = Random intercept for group i (captures group-specific baseline variation).
  - β_{1i} = Random slope for time in group i (captures group-specific linear trends).
  - β_{2i} = Random quadratic effect for group i (captures group-specific curvature over time).
- **Residual Error:**
  - $\epsilon_{ikt} \sim N(0, \sigma^2)$ = Measurement error, assumed independent of the random effects.

5.7.1 Variance-Covariance Structure: Power Model

Since observations are taken at unequally spaced time points, we cannot rely on simple structures like compound symmetry or AR(1). Instead, we use a power covariance model, which allows the correlation to depend on the distance between time points.

The variance-covariance matrix of the repeated measurements for subject k in group i is:

$$\mathbf{\Sigma}_{ik} = \sigma^2 \begin{pmatrix} 1 & \rho^{|t_2-t_1|} & \rho^{|t_3-t_1|} & \rho^{|t_4-t_1|} \\ \rho^{|t_2-t_1|} & 1 & \rho^{|t_3-t_2|} & \rho^{|t_4-t_2|} \\ \rho^{|t_3-t_1|} & \rho^{|t_3-t_2|} & 1 & \rho^{|t_4-t_3|} \\ \rho^{|t_4-t_1|} & \rho^{|t_4-t_2|} & \rho^{|t_4-t_3|} & 1 \end{pmatrix}$$

where:

- σ^2 = Residual variance.
- ρ = Correlation parameter ($0 < |\rho| < 1$), controlling how correlation decays with increasing time gaps.
- $|t_j - t_i|$ = Absolute time difference between measurements at times t_i and t_j.

Key Characteristics

- The correlation between observations decreases as the time difference increases, similar to AR(1), but flexible enough to handle unequal time intervals.
- This structure is sometimes referred to as a continuous-time autoregressive model or power covariance model.

After fitting the full model, we can evaluate whether all terms are necessary, focusing on the random effects:

- β_{0i} (Random Intercepts):
  Is there significant baseline variability between groups?
- β_{1i} (Random Slopes for Time):
  Do groups exhibit different linear trends over time?

- β_{2i} (Random Quadratic Terms):
  Is there group-specific curvature in the response over time?

  Model Comparison Approach:

1. **Fit the Full Model:**
   Includes all random effects.
2. **Fit Reduced Models:**
   Systematically remove random effects (e.g., quadratic terms) to create simpler models.
3. **Compare Models Using:**
   - **Likelihood Ratio Tests (LRT):**
     Test whether the more complex model significantly improves fit.
   - **Information Criteria (AIC, BIC):**
     Lower values indicate a better trade-off between fit and complexity.
4. **Assess Random Effects:**
   - Use the **exactRLRT** test to determine if random effects are significant.
   - Check variance estimates: if the variance of a random effect is near zero, it may not be necessary.

   In matrix notation, the model can be written as:

$$\mathbf{Y} = \mathbf{X}\boldsymbol{\beta} + \mathbf{Z}\mathbf{b} + \boldsymbol{\epsilon}$$

where:

- $\mathbf{Y}$ = Vector of observed responses.
- $\mathbf{X}$ = Design matrix for fixed effects (intercept, time, time^2, and group interactions).
- $\boldsymbol{\beta}$ = Vector of fixed effect coefficients.
- $\mathbf{Z}$ = Design matrix for random effects (random intercepts, slopes, etc.).
- $\mathbf{b} \sim N(0, \mathbf{G})$ = Vector of random effects with covariance matrix $\mathbf{G}$.
- $\boldsymbol{\epsilon} \sim N(0, \mathbf{R})$ = Vector of residual errors with covariance matrix $\mathbf{R}$, where $\mathbf{R}$ follows the power covariance structure.

5.8 Application: Mixed Models in Practice

Several R packages are available for fitting mixed-effect models, each with unique strengths:

- `nlme`
  - Supports nested and crossed random effects.
  - Flexible for complex covariance structures.
  - Less intuitive syntax compared to `lme4`.

- `lme4`
  - Computationally efficient and widely used.
  - User-friendly formula syntax.
  - Can handle non-normal responses (e.g., GLMMs).
  - For detailed documentation, refer to Bates et al. (2015).
- Others:
  - Bayesian mixed models: `MCMCglmm`, `brms`.
  - Genetics/plant breeding: `ASReml`.

5.8.1 Example 1: Pulp Brightness Analysis

5.8.1.1 Model Specification

We start with a random-intercepts model for pulp brightness:

$$y_{ij} = \mu + \alpha_i + \epsilon_{ij}$$

where:

- $i = 1, \ldots, a$ = Groups for random effect α_i.
- $j = 1, \ldots, n$ = Observations per group.
- μ = Overall mean brightness (fixed effect).
- $\alpha_i \sim N(0, \sigma_\alpha^2)$ = Group-specific random effect.
- $\epsilon_{ij} \sim N(0, \sigma_\epsilon^2)$ = Residual error.

This implies a compound symmetry structure, where the intraclass correlation coefficient is:

$$\rho = \frac{\sigma_\alpha^2}{\sigma_\alpha^2 + \sigma_\epsilon^2}$$

- If $\sigma_\alpha^2 \to 0$: Low correlation within groups ($\rho \to 0$) (i.e., when factor a doesn't explain much variation).
- If $\sigma_\alpha^2 \to \infty$: High correlation within groups ($\rho \to 1$).

5.8.1.2 Data Exploration

```
data(pulp, package = "faraway")
library(ggplot2)
library(dplyr)
```

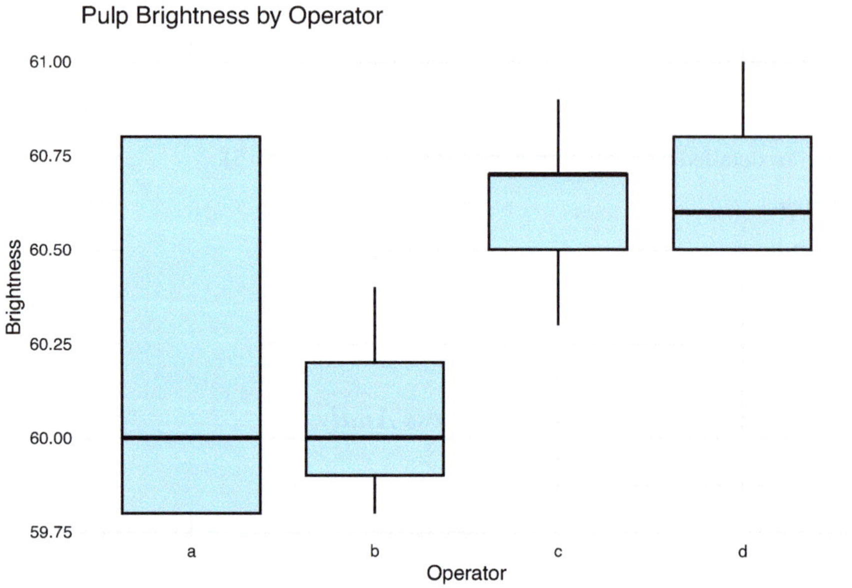

Fig. 5.2 Pulp brightness by operator

```
# Visualize brightness by operator
ggplot(pulp, aes(x = operator, y = bright)) +
    geom_boxplot(fill = "lightblue") +
    labs(title = "Pulp Brightness by Operator",
         x = "Operator",
         y = "Brightness") +
    theme_minimal()
```

See Fig. 5.2.

```
# Group-wise summary
pulp %>%
    group_by(operator) %>%
    summarise(average_brightness = mean(bright), .groups =
     ↪ 'drop')
#> # A tibble: 4 x 2
#>   operator average_brightness
#>   <fct>                 <dbl>
#> 1 a                      60.2
#> 2 b                      60.1
#> 3 c                      60.6
#> 4 d                      60.7
```

5.8.1.3 Fitting the Mixed Model with lme4

```
library(lme4)

# Random intercepts model: operator as a random effect
mixed_model <- lmer(bright ~ 1 + (1 | operator), data = pulp)

# Model summary
summary(mixed_model)
#> Linear mixed model fit by REML. t-tests use
#>   Satterthwaite's method [lmerModLmerTest]
#> Formula: bright ~ 1 + (1 | operator)
#>    Data: pulp
#>
#> REML criterion at convergence: 18.6
#>
#> Scaled residuals:
#>     Min      1Q  Median      3Q     Max
#> -1.4666 -0.7595 -0.1244  0.6281  1.6012
#>
#> Random effects:
#>  Groups   Name        Variance Std.Dev.
#>  operator (Intercept) 0.06808  0.2609
#>  Residual             0.10625  0.3260
#> Number of obs: 20, groups:  operator, 4
#>
#> Fixed effects:
#>             Estimate Std. Error      df t value Pr(>|t|)
#> (Intercept)  60.4000     0.1494  3.0000   404.2 3.34e-08
#>
#> (Intercept) ***
#> ---
#> Signif. codes:
#> 0 '***' 0.001 '**' 0.01 '*' 0.05 '.' 0.1 ' ' 1

# Fixed effects (overall mean)
fixef(mixed_model)
#> (Intercept)
#>        60.4

# Random effects (BLUPs)
ranef(mixed_model)
#> $operator
#>   (Intercept)
```

```
#> a   -0.1219403
#> b   -0.2591231
#> c    0.1676679
#> d    0.2133955
#>
#> with conditional variances for "operator"

# Variance components
VarCorr(mixed_model)
#>  Groups   Name        Std.Dev.
#>  operator (Intercept) 0.26093
#>  Residual             0.32596
re_dat <- as.data.frame(VarCorr(mixed_model))

# Intraclass Correlation Coefficient
rho <- re_dat[1, 'vcov'] / (re_dat[1, 'vcov'] + re_dat[2,
↪  'vcov'])
rho
#> [1] 0.3905354
```

5.8.1.4 Inference with `lmerTest`

To obtain p-values for fixed effects using Satterthwaite's approximation:

```
library(lmerTest)

# Model with p-values
summary(lmer(bright ~ 1 + (1 | operator), data =
↪  pulp))$coefficients
#>             Estimate Std. Error df  t value     Pr(>|t|)
#> (Intercept)     60.4  0.1494434  3 404.1664 3.340265e-08

# Confidence interval for the fixed effect
confint(mixed_model)[3,]
#> Computing profile confidence intervals ...
#>   2.5 %  97.5 %
#> 60.0713 60.7287
```

In this example, we can see that the confidence interval computed by `confint` from the `lme4` package is very close to the one computed by `confint` from the `lmerTest` package.

5.8.1.5 Bayesian Mixed Model with MCMCglmm

```
library(MCMCglmm)
#> Loading required package: coda
#> Loading required package: ape
#>
#> Attaching package: 'ape'
#> The following object is masked from 'package:dplyr':
#>
#>     where

# Bayesian mixed model
mixed_model_bayes <- MCMCglmm(
  bright ~ 1,
  random = ~ operator,
  data = pulp,
  verbose = FALSE
)

# Posterior summaries
summary(mixed_model_bayes)$solutions
#>             post.mean l-95% CI u-95% CI eff.samp pMCMC
#> (Intercept)  60.40789  60.1488 60.69595     1000 0.001
```

Bayesian credible intervals may differ slightly from frequentist confidence intervals due to prior assumptions.

5.8.1.6 Predictions

```
# Random effects predictions (BLUPs)
ranef(mixed_model)$operator
#>   (Intercept)
#> a  -0.1219403
#> b  -0.2591231
#> c   0.1676679
#> d   0.2133955

# Predictions per operator
fixef(mixed_model) + ranef(mixed_model)$operator
#>   (Intercept)
#> a    60.27806
#> b    60.14088
```

```
#> c      60.56767
#> d      60.61340

# Equivalent using predict()
predict(mixed_model,
        newdata = data.frame(operator = c('a', 'b', 'c',
        ↪  'd')))
#>        1        2        3        4
#> 60.27806 60.14088 60.56767 60.61340
```

For **bootstrap confidence intervals**, use:

```
bootMer(mixed_model, FUN = fixef, nsim = 100)
#>
#> PARAMETRIC BOOTSTRAP
#>
#>
#> Call:
#> bootMer(x = mixed_model, FUN = fixef, nsim = 100)
#>
#>
#> Bootstrap Statistics :
#>     original          bias    std. error
#> t1*     60.4 -0.0005452538    0.156374
#>
#> 10 message(s): boundary (singular) fit: see
↪  help('isSingular')
```

5.8.2 Example 2: Penicillin Yield (GLMM with Blocking)

```
data(penicillin, package = "faraway")
library(ggplot2)
```

```
# Visualize yield by treatment and blend
ggplot(penicillin, aes(
  y = yield,
  x = treat,
  shape = blend,
  color = blend
```

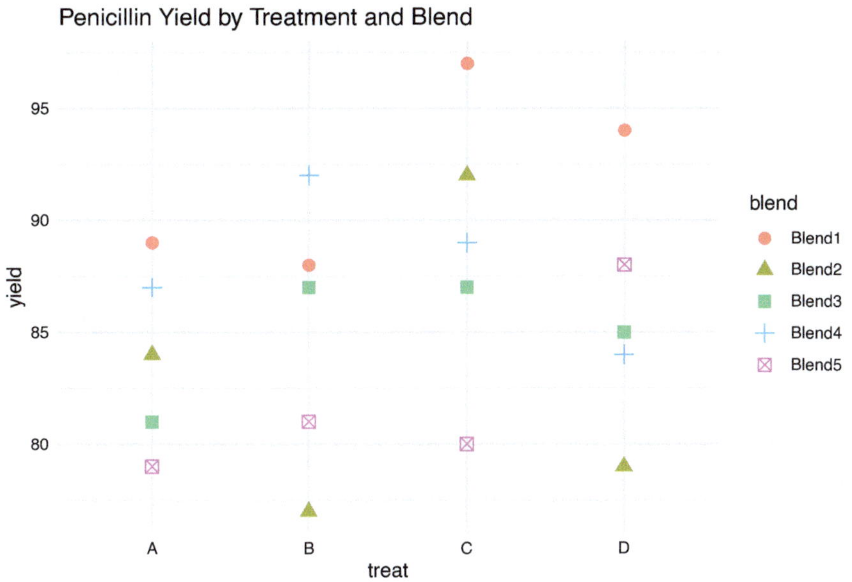

Fig. 5.3 Penicillin Yield by Treatment and Blend

```
)) +
  geom_point(size = 3) +
  labs(title = "Penicillin Yield by Treatment and Blend") +
  theme_minimal()
```

See Fig. 5.3.

```
# Mixed model: blend as random effect, treatment as fixed
mixed_model <- lmer(yield ~ treat + (1 | blend), data =
↪  penicillin)
summary(mixed_model)
#> Linear mixed model fit by REML. t-tests use
#>   Satterthwaite's method [lmerModLmerTest]
#> Formula: yield ~ treat + (1 | blend)
#>    Data: penicillin
#>
#> REML criterion at convergence: 103.8
#>
#> Scaled residuals:
#>     Min      1Q  Median      3Q     Max
#> -1.4152 -0.5017 -0.1644  0.6830  1.2836
#>
```

```
#> Random effects:
#>  Groups   Name        Variance Std.Dev.
#>  blend    (Intercept) 11.79    3.434
#>  Residual             18.83    4.340
#> Number of obs: 20, groups:  blend, 5
#>
#> Fixed effects:
#>             Estimate Std. Error     df t value Pr(>|t|)
#> (Intercept)   84.000      2.475 11.075  33.941 1.51e-12 ***
#> treatB         1.000      2.745 12.000   0.364   0.7219
#> treatC         5.000      2.745 12.000   1.822   0.0935 .
#> treatD         2.000      2.745 12.000   0.729   0.4802
#> ---
#> Signif. codes:
#> 0 '***' 0.001 '**' 0.01 '*' 0.05 '.' 0.1 ' ' 1
#>
#> Correlation of Fixed Effects:
#>        (Intr) treatB treatC
#> treatB -0.555
#> treatC -0.555  0.500
#> treatD -0.555  0.500  0.500

# BLUPs for each blend
ranef(mixed_model)$blend
#>        (Intercept)
#> Blend1   4.2878788
#> Blend2  -2.1439394
#> Blend3  -0.7146465
#> Blend4   1.4292929
#> Blend5  -2.8585859
```

Testing for Treatment Effect

```
# ANOVA for fixed effects
anova(mixed_model)
#> Type III Analysis of Variance Table with Satterthwaite's
↪  method
#>       Sum Sq Mean Sq NumDF DenDF F value Pr(>F)
#> treat     70  23.333     3    12  1.2389 0.3387
```

Since the p-value $> .05$, we fail to reject the null hypothesis (no treatment effect).

Model Comparison with Kenward-Roger Approximation

```r
library(pbkrtest)

# Full model vs. null model
full_model <-
    lmer(yield ~ treat + (1 | blend), penicillin, REML =
    ↪  FALSE)
null_model <-
    lmer(yield ~ 1 + (1 | blend), penicillin, REML = FALSE)

# Kenward-Roger approximation
KRmodcomp(full_model, null_model)
#> large : yield ~ treat + (1 | blend)
#> small : yield ~ 1 + (1 | blend)
#>           stat     ndf     ddf F.scaling p.value
#> Ftest  1.2389  3.0000 12.0000         1  0.3387
```

The results are consistent with the earlier ANOVA: **no significant treatment effect**.

5.8.3 *Example 3: Growth in Rats Over Time*

```r
rats <- read.csv(
    "images/rats.dat",
    header = FALSE,
    sep = ' ',
    col.names = c('Treatment', 'rat', 'age', 'y')
)

# Log-transformed time variable
rats$t <- log(1 + (rats$age - 45) / 10)
```

Model Fitting

```r
# Treatment as fixed effect, random intercepts for rats
rat_model <- lmer(y ~ t:Treatment + (1 | rat), data = rats)
summary(rat_model)
#> Linear mixed model fit by REML. t-tests use
#>   Satterthwaite's method [lmerModLmerTest]
#> Formula: y ~ t:Treatment + (1 | rat)
#>    Data: rats
#>
```

```
#> REML criterion at convergence: 932.4
#>
#> Scaled residuals:
#>      Min       1Q   Median       3Q      Max
#> -2.25574 -0.65898 -0.01163  0.58356  2.88309
#>
#> Random effects:
#>  Groups   Name        Variance Std.Dev.
#>  rat      (Intercept) 3.565    1.888
#>  Residual             1.445    1.202
#> Number of obs: 252, groups:  rat, 50
#>
#> Fixed effects:
#>                  Estimate Std. Error       df t value
#> (Intercept)       68.6074     0.3312  89.0275  207.13
#> t:Treatmentcon     7.3138     0.2808 247.2762   26.05
#> t:Treatmenthig     6.8711     0.2276 247.7097   30.19
#> t:Treatmentlow     7.5069     0.2252 247.5196   33.34
#>                  Pr(>|t|)
#> (Intercept)        <2e-16 ***
#> t:Treatmentcon     <2e-16 ***
#> t:Treatmenthig     <2e-16 ***
#> t:Treatmentlow     <2e-16 ***
#> ---
#> Signif. codes:
#> 0 '***' 0.001 '**' 0.01 '*' 0.05 '.' 0.1 ' ' 1
#>
#> Correlation of Fixed Effects:
#>             (Intr) t:Trtmntc t:Trtmnth
#> t:Tretmntcn -0.327
#> t:Tretmnthg -0.340  0.111
#> t:Tretmntlw -0.351  0.115     0.119
anova(rat_model)
#> Type III Analysis of Variance Table with Satterthwaite's
↪  method
#>             Sum Sq Mean Sq NumDF  DenDF F value    Pr(>F)
#> t:Treatment 3181.9  1060.6     3 223.21  734.11 < 2.2e-16
#>
#> t:Treatment ***
#> ---
#> Signif. codes:
#> 0 '***' 0.001 '**' 0.01 '*' 0.05 '.' 0.1 ' ' 1
```

Since the **p-value is significant**, we conclude that the **treatment effect varies over time**.

5.8.4 Example 4: Tree Water Use (Agridat)

```
library(agridat)
library(latticeExtra)
#> Loading required package: lattice
#>
#> Attaching package: 'latticeExtra'
#> The following object is masked from 'package:ggplot2':
#>
#>     layer
dat <- harris.wateruse
```

```
# Visualizing water use by species and age
useOuterStrips(
  xyplot(water ~ day | species * age,
         dat, group = tree,
         type = c('p', 'smooth'),
         main = "harris.wateruse 2 species, 2 ages (10 trees
         ↪  each)",
         as.table = TRUE)
)
```

See Fig. 5.4.
Remove outlier

```
dat <- subset(dat, day!=268)
```

Plot between water use and day for one age and species group (Fig. 5.5).

```
xyplot(
    water ~ day | tree,
    dat,
    subset    = age == "A2" & species == "S2",
    as.table = TRUE,
    type      = c('p', 'smooth'),
    ylab      = "Water use profiles of individual trees",
    main      = "harris.wateruse (Age 2, Species 2)"
)
```

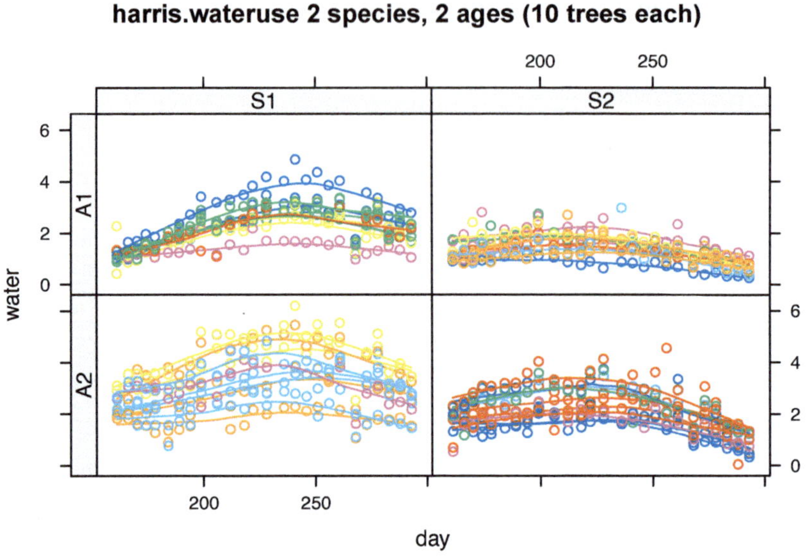

Fig. 5.4 Species by age

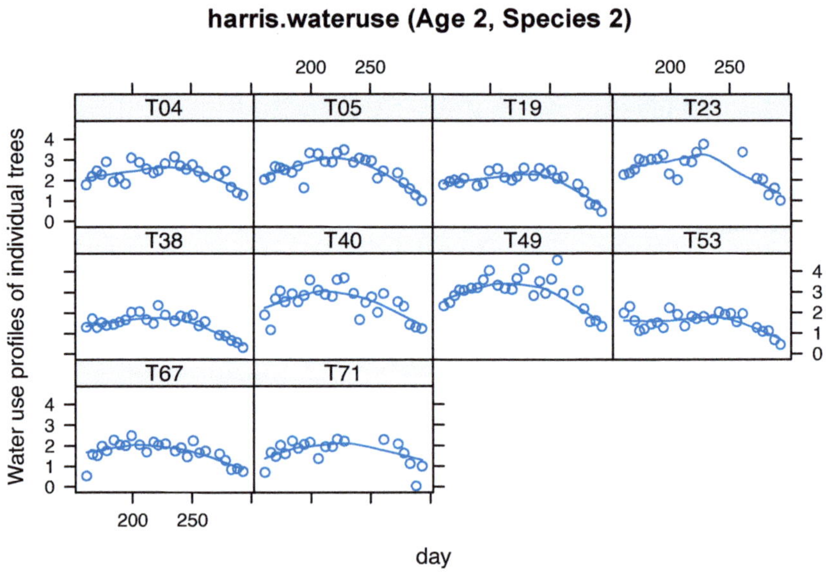

Fig. 5.5 harris.wateruse (Age 2, Species 2)

```
# Rescale day for nicer output, and convergence issues
dat <- transform(dat, ti = day / 100)
# add quadratic term
dat <- transform(dat, ti2 = ti * ti)

# Start with a subgroup: age 2, species 2
d22 <- droplevels(subset(dat, age == "A2" & species == "S2"))
```

Fitting with `nlme` using `lme`

```
library(nlme)
#>
#> Attaching package: 'nlme'
#> The following object is masked from 'package:dplyr':
#>
#>     collapse
#> The following object is masked from 'package:lme4':
#>
#>     lmList

## We use pdDiag() to get uncorrelated random effects
m1n <- lme(
    water ~ 1 + ti + ti2,
    #intercept, time and time-squared = fixed effects
    data = d22,
    na.action = na.omit,
    random = list(tree = pdDiag(~ 1 + ti + ti2))
    # random intercept, time
    # and time squared per tree = random effects
)

# for all trees
# m1n <- lme(
#   water ~ 1 + ti + ti2,
#   random = list(tree = pdDiag(~ 1 + ti + ti2)),
#   data = dat,
#   na.action = na.omit
# )

summary(m1n)
#> Linear mixed-effects model fit by REML
#>   Data: d22
#>        AIC      BIC    logLik
#>   276.5142 300.761 -131.2571
```

```
#>
#> Random effects:
#>  Formula: ~1 + ti + ti2 | tree
#>  Structure: Diagonal
#>         (Intercept)           ti          ti2  Residual
#> StdDev:   0.5187869 1.631223e-05 4.374982e-06 0.3836614
#>
#> Fixed effects:  water ~ 1 + ti + ti2
#>                  Value Std.Error  DF   t-value p-value
#> (Intercept) -10.798799 0.8814666 227 -12.25094       0
#> ti           12.346704 0.7827112 227  15.77428       0
#> ti2          -2.838503 0.1720614 227 -16.49704       0
#>  Correlation:
#>     (Intr) ti
#> ti  -0.979
#> ti2  0.970 -0.997
#>
#> Standardized Within-Group Residuals:
#>         Min          Q1         Med          Q3         Max
↪
#> -3.07588246 -0.58531056  0.01210209  0.65402695  3.88777402
↪
#>
#> Number of Observations: 239
#> Number of Groups: 10

ranef(m1n)
#>     (Intercept)           ti           ti2
#> T04   0.1985796  2.070606e-09  6.397103e-10
#> T05   0.3492827  3.199664e-10 -6.211457e-11
#> T19  -0.1978989 -9.879555e-10 -2.514502e-10
#> T23   0.4519003 -4.206418e-10 -3.094113e-10
#> T38  -0.6457494 -2.069198e-09 -4.227912e-10
#> T40   0.3739432  4.199061e-10 -3.260161e-11
#> T49   0.8620648  1.160387e-09 -6.925457e-12
#> T53  -0.5655049 -1.064849e-09 -5.870462e-11
#> T67  -0.4394623 -4.482549e-10  2.752922e-11
#> T71  -0.3871552  1.020034e-09  4.767595e-10

fixef(m1n)
#> (Intercept)          ti         ti2
#>  -10.798799   12.346704   -2.838503
```

Fitting with `lme4` using `lmer`

```
library(lme4)

m1lmer <-
    lmer(water ~ 1 + ti + ti2 + (ti + ti2 ||
                                        tree),
         data = d22,
         na.action = na.omit)
#> boundary (singular) fit: see help('isSingular')

# for all trees
# m1lmer <- lmer(water ~ 1 + ti + ti2 + (ti + ti2 || tree),
#                  data = dat, na.action = na.omit)

# summary(m1lmer)
ranef(m1lmer)
#> $tree
#>     (Intercept) ti ti2
#> T04   0.1985796  0   0
#> T05   0.3492827  0   0
#> T19  -0.1978989  0   0
#> T23   0.4519003  0   0
#> T38  -0.6457494  0   0
#> T40   0.3739432  0   0
#> T49   0.8620648  0   0
#> T53  -0.5655049  0   0
#> T67  -0.4394623  0   0
#> T71  -0.3871552  0   0
#>
#> with conditional variances for "tree"

fixef(m1lmer)
#> (Intercept)          ti         ti2
#>  -10.798799   12.346704   -2.838503
```

Notes:

- || double pipes = uncorrelated random effects
- To remove the intercept term:
  - `(0+ti|tree)`
  - `(ti-1|tree)`

```
m1l <-
    lmer(water ~ 1 + ti + ti2
         + (1 | tree) + (0 + ti | tree)
```

```
        + (0 + ti2 | tree), data = d22)
#> boundary (singular) fit: see help('isSingular')
ranef(m1l)
#> $tree
#>     (Intercept) ti ti2
#> T04   0.1985796  0   0
#> T05   0.3492827  0   0
#> T19  -0.1978989  0   0
#> T23   0.4519003  0   0
#> T38  -0.6457494  0   0
#> T40   0.3739432  0   0
#> T49   0.8620648  0   0
#> T53  -0.5655049  0   0
#> T67  -0.4394623  0   0
#> T71  -0.3871552  0   0
#>
#> with conditional variances for "tree"
fixef(m1l)
#> (Intercept)          ti         ti2
#>  -10.798799   12.346704   -2.838503
```

Adding correlation structure

```
m2n <- lme(
    water ~ 1 + ti + ti2,
    data = d22,
    random = ~ 1 | tree,
    cor = corExp(form =  ~ day | tree),
    na.action = na.omit
)

# for all trees
# m2n <- lme(
#   water ~ 1 + ti + ti2,
#   random = ~ 1 | tree,
#   cor = corExp(form = ~ day | tree),
#   data = dat,
#   na.action = na.omit
# )

summary(m2n)
#> Linear mixed-effects model fit by REML
#>   Data: d22
#>          AIC       BIC   logLik
```

```
#>   263.3081 284.0911 -125.654
#>
#> Random effects:
#>  Formula: ~1 | tree
#>         (Intercept)  Residual
#> StdDev:   0.5154042 0.3925777
#>
#> Correlation Structure: Exponential spatial correlation
#>  Formula: ~day | tree
#>  Parameter estimate(s):
#>    range
#> 3.794624
#> Fixed effects:  water ~ 1 + ti + ti2
#>                  Value Std.Error  DF   t-value p-value
#> (Intercept) -11.223310 1.0988725 227 -10.21348       0
#> ti           12.712094 0.9794235 227  12.97916       0
#> ti2          -2.913682 0.2148551 227 -13.56115       0
#>  Correlation:
#>     (Intr) ti
#> ti  -0.985
#> ti2  0.976 -0.997
#>
#> Standardized Within-Group Residuals:
#>         Min          Q1         Med          Q3         Max
↪
#> -3.04861039 -0.55703950  0.00278101  0.62558762  3.80676991
↪
#>
#> Number of Observations: 239
#> Number of Groups: 10
ranef(m2n)
#>     (Intercept)
#> T04   0.1929971
#> T05   0.3424631
#> T19  -0.1988495
#> T23   0.4538660
#> T38  -0.6413664
#> T40   0.3769378
#> T49   0.8410043
#> T53  -0.5528236
#> T67  -0.4452930
#> T71  -0.3689358
fixef(m2n)
```

```
#> (Intercept)          ti         ti2
#>  -11.223310   12.712094   -2.913682
```

Key Takeaways

- `lme4` is preferred for general mixed models due to efficiency and simplicity.
- `nlme` is powerful for complex correlation structures and nested designs.
- Bayesian models (e.g., `MCMCglmm`) offer flexible inference under uncertainty.
- Always consider model diagnostics and random effects structure carefully.

Chapter 6
Nonlinear and Generalized Linear Mixed Models

This chapter fuses the flexibility of nonlinear and generalized models with random effect structures, yielding NLMMs and GLMMs. We articulate their marginal and conditional interpretations, describe approximation strategies, and address identifiability in high-level random coefficients. The chapter systematically compares NLMMs and GLMMs, illustrating when each is preferred, and provides a full estimation workflow—model building, convergence assessment, and inferential reporting.

Nonlinear mixed models (NLMMs) and generalized linear mixed models (GLMMs) extend traditional models by incorporating both fixed effects and random effects, allowing for greater flexibility in modeling complex data structures.

- NLMMs extend nonlinear models to include both fixed and random effects, accommodating nonlinear relationships in the data.
- GLMMs extend generalized linear models to include random effects, allowing for correlated data and non-constant variance structures.

6.1 Nonlinear Mixed Models

A general form of a nonlinear mixed model is:

$$Y_{ij} = f(\mathbf{x}_{ij}, \boldsymbol{\theta}, \boldsymbol{\alpha}_i) + \epsilon_{ij}$$

for the j-th response from the i-th cluster (or subject), where:

- $i = 1, \ldots, n$ (number of clusters/subjects),
- $j = 1, \ldots, n_i$ (number of observations per cluster),
- $\boldsymbol{\theta}$ represents the fixed effects,
- $\boldsymbol{\alpha}_i$ are the random effects for cluster i,

M. Nguyen, *Regression Techniques for Data Analysis*,
https://doi.org/10.1007/978-3-032-01835-9_6

- $\mathbf{x}_{ij}$ are the regressors or design variables,
- $f(\cdot)$ is a nonlinear mean response function,
- ϵ_{ij} represents the residual error, often assumed to be normally distributed with mean 0.

NLMMs are particularly useful when the relationship between predictors and the response cannot be adequately captured by a linear model.

6.2 Generalized Linear Mixed Models

GLMMs extend GLMs by incorporating random effects, which allows for modeling data with hierarchical or clustered structures.

The conditional distribution of y_i given the random effects $\boldsymbol{\alpha}_i$ is:

$$y_i \mid \boldsymbol{\alpha}_i \sim \text{independent } f(y_i \mid \boldsymbol{\alpha})$$

where $f(y_i \mid \boldsymbol{\alpha})$ belongs to the exponential family of distributions:

$$f(y_i \mid \boldsymbol{\alpha}) = \exp\left(\frac{y_i\theta_i - b(\theta_i)}{a(\phi)} - c(y_i, \phi)\right)$$

- θ_i is the canonical parameter,
- $a(\phi)$ is a dispersion parameter,
- $b(\theta_i)$ and $c(y_i, \phi)$ are specific functions defining the exponential family.

The conditional mean of y_i is related to θ_i by:

$$\mu_i = \frac{\partial b(\theta_i)}{\partial \theta_i}$$

Applying a link function $g(\cdot)$, we relate the mean response to both fixed and random effects:

$$E(y_i \mid \boldsymbol{\alpha}) = \mu_i$$
$$g(\mu_i) = \mathbf{x}_i'\boldsymbol{\beta} + \mathbf{z}_i'\boldsymbol{\alpha}$$

- $g(\cdot)$ is a known link function,
- $\mathbf{x}_i$ and $\mathbf{z}_i$ are design matrices for fixed and random effects, respectively,
- $\boldsymbol{\beta}$ represents fixed effects, and $\boldsymbol{\alpha}$ represents random effects.

We also specify the distribution of the random effects:

$$\boldsymbol{\alpha} \sim f(\boldsymbol{\alpha})$$

This distribution is often assumed to be multivariate normal (law of large number applies to fixed effects) but can be chosen (subjectively) based on the context.

6.3 Relationship Between NLMMs and GLMMs

NLMMs can be viewed as a special case of GLMMs when the inverse link function corresponds to a nonlinear transformation of the linear predictor:

$$\mathbf{Y}_i = \mathbf{f}(\mathbf{x}_i, \boldsymbol{\theta}, \boldsymbol{\alpha}_i) + \boldsymbol{\epsilon}_i$$

$$\mathbf{Y}_i = g^{-1}(\mathbf{x}_i'\boldsymbol{\beta} + \mathbf{z}_i'\boldsymbol{\alpha}_i) + \boldsymbol{\epsilon}_i$$

Here, $g^{-1}(\cdot)$ represents the inverse link function, corresponding to a nonlinear transformation of the fixed and random effects.
Note:
We can't derive the analytical formulation of the marginal distribution because nonlinear combination of normal variables is not normally distributed, even in the case of additive error (ϵ_i) and random effects (α_i) are both normal.

6.4 Marginal Properties of GLMMs

6.4.1 Marginal Mean of y_i

The marginal mean is obtained by integrating over the distribution of the random effects:

$$E(y_i) = E_{\boldsymbol{\alpha}}(E(y_i \mid \boldsymbol{\alpha})) = E_{\boldsymbol{\alpha}}(\mu_i) = E\left(g^{-1}(\mathbf{x}_i'\boldsymbol{\beta} + \mathbf{z}_i'\boldsymbol{\alpha})\right)$$

Since $g^{-1}(\cdot)$ is nonlinear, this expectation cannot be simplified further without specific distributional assumptions.

6.4.1.1 Special Case: Log Link Function

For a log-link function, $g(\mu) = \log(\mu)$, the inverse link is the exponential function:

$$E(y_i) = E\left(\exp(\mathbf{x}_i'\boldsymbol{\beta} + \mathbf{z}_i'\boldsymbol{\alpha})\right)$$

Using properties of the moment-generating function (MGF):

$$E(y_i) = \exp(\mathbf{x}_i'\boldsymbol{\beta}) \cdot E\left(\exp(\mathbf{z}_i'\boldsymbol{\alpha})\right)$$

Here, $E(\exp(\mathbf{z}_i'\boldsymbol{\alpha}))$ is the MGF of $\boldsymbol{\alpha}$ evaluated at $\mathbf{z}_i$.

6.4.2 *Marginal Variance of* y_i

The variance decomposition formula applies:

$$\begin{aligned}\text{Var}(y_i) &= \text{Var}_{\boldsymbol{\alpha}}\left(E(y_i \mid \boldsymbol{\alpha})\right) + E_{\boldsymbol{\alpha}}\left(\text{Var}(y_i \mid \boldsymbol{\alpha})\right)\\ &= \text{Var}(\mu_i) + E\left(a(\phi)V(\mu_i)\right)\end{aligned}$$

Expressed explicitly:

$$\text{Var}(y_i) = \text{Var}\left(g^{-1}(\mathbf{x}_i'\boldsymbol{\beta} + \mathbf{z}_i'\boldsymbol{\alpha})\right) + E\left(a(\phi)V\left(g^{-1}(\mathbf{x}_i'\boldsymbol{\beta} + \mathbf{z}_i'\boldsymbol{\alpha})\right)\right)$$

Without specific assumptions about $g(\cdot)$ and the distribution of $\boldsymbol{\alpha}$, this is the most general form.

6.4.3 *Marginal Covariance of* $\mathbf{y}$

Random effects induce correlation between observations within the same cluster. The covariance between y_i and y_j is:

$$\begin{aligned}\text{Cov}(y_i, y_j) &= \text{Cov}_{\boldsymbol{\alpha}}\left(E(y_i \mid \boldsymbol{\alpha}), E(y_j \mid \boldsymbol{\alpha})\right) + E_{\boldsymbol{\alpha}}\left(\text{Cov}(y_i, y_j \mid \boldsymbol{\alpha})\right)\\ &= \text{Cov}(\mu_i, \mu_j) + E(0)\\ &= \text{Cov}\left(g^{-1}(\mathbf{x}_i'\boldsymbol{\beta} + \mathbf{z}_i'\boldsymbol{\alpha}), g^{-1}(\mathbf{x}_j'\boldsymbol{\beta} + \mathbf{z}_j'\boldsymbol{\alpha})\right)\end{aligned}$$

The second term vanishes when y_i and y_j are conditionally independent given $\boldsymbol{\alpha}$. This dependency structure is a hallmark of mixed models.

Example (Repeated Measurements with a Poisson GLMM) Consider repeated count measurements for subjects:

- Let y_{ij} be the j-th count for subject i.
- Assume $y_{ij} \mid \alpha_i \sim$ independent Poisson(μ_{ij}).

The model is specified as:

$$\log(\mu_{ij}) = \mathbf{x}_{ij}'\boldsymbol{\beta} + \alpha_i$$

where:

- $\alpha_i \sim$ i.i.d. $N(0, \sigma_\alpha^2)$ represents subject-specific random effects,
- This is a log-link GLMM with random intercepts for subjects.

The inclusion of α_i accounts for subject-level heterogeneity, capturing unobserved variability across individuals.

6.5 Estimation in Nonlinear and Generalized Linear Mixed Models

In linear mixed models, the marginal likelihood of the observed data $\mathbf{y}$ is derived by integrating out the random effects from the hierarchical formulation:

$$f(\mathbf{y}) = \int f(\mathbf{y} \mid \boldsymbol{\alpha}) f(\boldsymbol{\alpha}) d\boldsymbol{\alpha}$$

For LMMs, both component distributions—

- the conditional distribution $f(\mathbf{y} \mid \boldsymbol{\alpha})$, and
- the random effects distribution $f(\boldsymbol{\alpha})$—

are typically assumed to be Gaussian with linear relationships. These assumptions imply that the marginal distribution of $\mathbf{y}$ is also Gaussian, allowing the integral to be solved analytically using properties of the multivariate normal distribution.

In contrast:

- For GLMMs, the conditional distribution $f(\mathbf{y} \mid \boldsymbol{\alpha})$ belongs to the exponential family but is not Gaussian in general.
- For NLMMs, the relationship between the mean response and the random (and fixed) effects is nonlinear, complicating the integral.

In both cases, the marginal likelihood integral:

$$L(\boldsymbol{\beta}; \mathbf{y}) = \int f(\mathbf{y} \mid \boldsymbol{\alpha}) f(\boldsymbol{\alpha}) d\boldsymbol{\alpha}$$

cannot be solved analytically. Consequently, estimation requires:

- Numerical Integration
- Linearization of the Model
- Bayesian

6.5.1 *Estimation by Numerical Integration*

The marginal likelihood for parameter estimation is given by:

$$L(\boldsymbol{\beta}; \mathbf{y}) = \int f(\mathbf{y} \mid \boldsymbol{\alpha}) f(\boldsymbol{\alpha}) d\boldsymbol{\alpha}$$

To estimate the fixed effects $\boldsymbol{\beta}$, we often maximize the log-likelihood:

$$\ell(\boldsymbol{\beta}) = \log L(\boldsymbol{\beta}; \mathbf{y})$$

Optimization requires the score function (gradient):

$$\frac{\partial \ell(\boldsymbol{\beta})}{\partial \boldsymbol{\beta}}$$

Since the integral in $L(\boldsymbol{\beta}; \mathbf{y})$ is generally intractable, we rely on numerical techniques to approximate it.

6.5.1.1 Methods for Numerical Integration

1. Gaussian Quadrature

   - Suitable for low-dimensional random effects ($\dim(\boldsymbol{\alpha})$ is small).
   - Approximates the integral using weighted sums of function evaluations at specific points (nodes).
   - Gauss-Hermite quadrature is commonly used when random effects are normally distributed.

   Limitation: Computational cost grows exponentially with the dimension of $\boldsymbol{\alpha}$ (curse of dimensionality).
2. Laplace Approximation

   - Approximates the integral by expanding the log-likelihood around the mode of the integrand (i.e., the most likely value of $\boldsymbol{\alpha}$).
   - Provides accurate results for moderate-sized random effects and large sample sizes.
   - First-order Laplace approximation is commonly used; higher-order versions improve accuracy but increase complexity.

   Key Idea: Approximate the integral as:

$$\int e^{h(\boldsymbol{\alpha})} d\boldsymbol{\alpha} \approx e^{h(\hat{\boldsymbol{\alpha}})} \sqrt{\frac{(2\pi)^q}{|\mathbf{H}|}}$$

   where:

   - $\hat{\boldsymbol{\alpha}}$ is the mode of $h(\boldsymbol{\alpha})$,
   - $\mathbf{H}$ is the Hessian matrix of second derivatives at $\hat{\boldsymbol{\alpha}}$,
   - q is the dimension of $\boldsymbol{\alpha}$.

3. Monte Carlo Integration

   - Uses random sampling to approximate the integral.

Table 6.1 Comparison of numerical integration methods in Bayesian and likelihood-based estimation

Method	Dimensionality	Accuracy	Computational cost
Gaussian quadrature	Low-dimensional ($q \leq 3$)	High (with sufficient nodes)	High (exponential growth with q)
Laplace approximation	Moderate-dimensional	Moderate to high	Moderate
Monte Carlo methods	High-dimensional	Variable (depends on sample size)	High (but scalable)

- Importance sampling improves efficiency by sampling from a distribution that better matches the integrand.
- Markov chain Monte Carlo methods, such as Gibbs sampling or Metropolis-Hastings, are used when the posterior distribution is complex.

Advantage: Scales better with high-dimensional random effects compared to quadrature methods.
Limitation: Computationally intensive and variance of estimates can be high without careful tuning.

6.5.1.2 Choosing an Integration Method

Table 6.1 shows comparison between numerical integration methods.

- For small random effect dimensions, quadrature methods are effective.
- For moderate dimensions, Laplace approximation offers a good balance.
- For high dimensions or complex models, Monte Carlo techniques are often the method of choice.

6.5.2 Estimation by Linearization

When estimating parameters in NLMMs and GLMMs, one common and effective approach is linearization. This technique approximates the nonlinear or non-Gaussian components with linear counterparts, enabling the use of standard LMM estimation methods. Linearization not only simplifies the estimation process but also allows for leveraging well-established statistical tools and methods developed for linear models.

6.5.2.1 Concept of Linearization

The core idea is to create a linearized version of the response variable, known as the *working response* or *pseudo-response*, denoted as $\tilde{y}_i$. This pseudo-response is designed to approximate the original nonlinear relationship in a linear form, facilitating easier estimation of model parameters. The conditional mean of this pseudo-response is expressed as:

$$E(\tilde{y}_i \mid \boldsymbol{\alpha}) = \mathbf{x}_i'\boldsymbol{\beta} + \mathbf{z}_i'\boldsymbol{\alpha}$$

Here:

- $\mathbf{x}_i$ is the design matrix for fixed effects,
- $\boldsymbol{\beta}$ represents the fixed effect parameters,
- $\mathbf{z}_i$ is the design matrix for random effects,
- $\boldsymbol{\alpha}$ denotes the random effects.

In addition to the conditional mean, it is essential to estimate the conditional variance of the pseudo-response to fully characterize the linearized model:

$$\text{Var}(\tilde{y}_i \mid \boldsymbol{\alpha})$$

This variance estimation ensures that the model accounts for the inherent variability in the data, maintaining the integrity of statistical inferences.

6.5.2.2 Application of Linearization

Once linearized, the model structure closely resembles that of a linear mixed model, allowing us to apply standard estimation techniques from LMMs. These techniques include methods such as MLE and REML, which are computationally efficient and statistically robust.

The primary difference between various linearization-based methods lies in how the linearization is performed. This often involves expanding the nonlinear function $f(\mathbf{x}, \boldsymbol{\theta}, \boldsymbol{\alpha})$ or the inverse link function $g^{-1}(\cdot)$. The goal is to approximate these complex functions with simpler linear expressions while retaining as much of the original model's characteristics as possible.

6.5.2.2.1 Taylor Series Expansion

A widely used method for linearization is the Taylor series expansion. This approach approximates the nonlinear mean function around initial estimates of the random effects. The first-order Taylor series expansion of the nonlinear function is given by:

$$f(\mathbf{x}_{ij}, \boldsymbol{\theta}, \boldsymbol{\alpha}_i) \approx f(\mathbf{x}_{ij}, \boldsymbol{\theta}, \hat{\boldsymbol{\alpha}}_i) + \left.\frac{\partial f}{\partial \boldsymbol{\alpha}_i}\right|_{\hat{\boldsymbol{\alpha}}_i} (\boldsymbol{\alpha}_i - \hat{\boldsymbol{\alpha}}_i)$$

In this expression:

- $f(\mathbf{x}_{ij}, \boldsymbol{\theta}, \hat{\boldsymbol{\alpha}}_i)$ is the function evaluated at the initial estimates of the random effects $\hat{\boldsymbol{\alpha}}_i$,

- $\frac{\partial f}{\partial \boldsymbol{\alpha}_i}\big|_{\hat{\boldsymbol{\alpha}}_i}$ represents the gradient (or derivative) of the function with respect to the random effects, evaluated at $\hat{\boldsymbol{\alpha}}_i$,
- $(\boldsymbol{\alpha}_i - \hat{\boldsymbol{\alpha}}_i)$ captures the deviation from the initial estimates.

The initial estimates $\hat{\boldsymbol{\alpha}}_i$ are often set to zero for simplicity, especially in the early stages of model fitting. This approximation reduces the model to a linear form, making it amenable to standard LMM estimation techniques.

6.5.2.2.2 Advantages and Considerations

Linearization offers several advantages:

1. Simplified Computation: By transforming complex nonlinear relationships into linear forms, linearization reduces computational complexity.
2. Flexibility: Despite the simplification, linearized models retain the ability to capture key features of the original data structure.
3. Statistical Robustness: The use of established LMM estimation techniques ensures robust parameter estimation.

However, linearization also comes with considerations:

- Approximation Error: The accuracy of the linearized model depends on how well the linear approximation captures the original nonlinear relationship.
- Choice of Expansion Point: The selection of initial estimates $\hat{\boldsymbol{\alpha}}_i$ can influence the quality of the approximation.
- Higher-Order Terms: In cases where the first-order approximation is insufficient, higher-order Taylor series terms may be needed, increasing model complexity.

6.5.2.3 Penalized Quasi-likelihood

Penalized quasi-likelihood (PQL) is one of the most popular linearization-based estimation methods for GLMMs.

The linearization is achieved through a first-order Taylor expansion of the inverse link function around current estimates of the parameters. The working response at the k-th iteration is given by:

$$\tilde{y}_i^{(k)} = \hat{\eta}_i^{(k-1)} + \left(y_i - \hat{\mu}_i^{(k-1)}\right) \cdot \frac{d\eta}{d\mu}\bigg|_{\hat{\eta}_i^{(k-1)}}$$

where:

- $\eta_i = g(\mu_i)$ is the linear predictor,
- $\hat{\eta}_i^{(k-1)}$ and $\hat{\mu}_i^{(k-1)}$ are the estimates from the previous iteration $(k-1)$,
- $g(\cdot)$ is the link function, and $\mu_i = g^{-1}(\eta_i)$.

PQL Estimation Algorithm

1. Initialization:
   Start with initial estimates of $\boldsymbol{\beta}$ and $\boldsymbol{\alpha}$ (commonly set to zeros).
2. Compute the Working Response:
   Use the formula above to compute $\tilde{y}_i^{(k)}$ based on current parameter estimates.
3. Fit a Linear Mixed Model:
   Apply standard LMM estimation techniques to the pseudo-response $\tilde{y}_i^{(k)}$ to update estimates of $\boldsymbol{\beta}$ and $\boldsymbol{\alpha}$.
4. Update Variance Components:
   Estimate $\mathrm{Var}(\tilde{y}_i \mid \boldsymbol{\alpha})$ based on updated parameter estimates.
5. Iteration:
   Repeat steps 2–4 until the estimates converge.

Comments on PQL:

- Advantages:
  - Easy to implement using existing LMM software.
  - Fast convergence for many practical datasets.
- Limitations:
  - Inference is only asymptotically correct due to the linearization approximation.
  - Biased estimates are common, especially:

    For binomial responses with small group sizes,
    In Bernoulli models (worst-case scenario),
    In Poisson models with small counts (Faraway, 2016).
  - Hypothesis testing and confidence intervals can be unreliable.

6.5.2.4 Generalized Estimating Equations

Generalized estimating equations (GEE) offer an alternative approach to parameter estimation in models with correlated data, particularly for marginal models where the focus is on population-averaged effects rather than subject-specific effects.

GEE estimates are obtained by solving estimating equations rather than maximizing a likelihood function.

Consider a marginal generalized linear model:

$$\mathrm{logit}(E(\mathbf{y})) = \mathbf{X}\boldsymbol{\beta}$$

Assuming a working covariance matrix $\mathbf{V}$ for the elements of $\mathbf{y}$, the estimating equation for $\boldsymbol{\beta}$ is:

$$\mathbf{X}'\mathbf{V}^{-1}(\mathbf{y} - E(\mathbf{y})) = 0$$

If $\mathbf{V}$ correctly specifies the covariance structure, the estimator is unbiased. In practice, we often assume a simple structure (e.g., independence) and obtain robust standard errors even when the covariance is misspecified.

6.5.2.4.1 GEE for Repeated Measures

Let y_{ij} denote the j-th measurement on the i-th subject, with:

$$\mathbf{y}_i = \begin{pmatrix} y_{i1} \\ \vdots \\ y_{in_i} \end{pmatrix}, \quad \boldsymbol{\mu}_i = \begin{pmatrix} \mu_{i1} \\ \vdots \\ \mu_{in_i} \end{pmatrix}, \quad \mathbf{x}_{ij} = \begin{pmatrix} X_{ij1} \\ \vdots \\ X_{ijp} \end{pmatrix}$$

Define the working covariance matrix of $\mathbf{y}_i$ as:

$$\mathbf{V}_i = \text{Cov}(\mathbf{y}_i)$$

Following Liang and Zeger (1986), the GEE for estimating $\boldsymbol{\beta}$ is:

$$S(\boldsymbol{\beta}) = \sum_{i=1}^{K} \frac{\partial \boldsymbol{\mu}_i'}{\partial \boldsymbol{\beta}} \mathbf{V}_i^{-1}(\mathbf{y}_i - \boldsymbol{\mu}_i) = 0$$

where:

- K is the number of subjects (or clusters),
- $\boldsymbol{\mu}_i = E(\mathbf{y}_i)$,
- $\mathbf{V}_i$ is the working covariance matrix.

6.5.2.4.2 Working Correlation Structures

The covariance matrix $\mathbf{V}_i$ is modeled as:

$$\mathbf{V}_i = a(\phi)\mathbf{B}_i^{1/2}\mathbf{R}(\boldsymbol{c})\mathbf{B}_i^{1/2}$$

- $a(\phi)$ is a dispersion parameter,
- $\mathbf{B}_i$ is a diagonal matrix with variance functions $V(\mu_{ij})$ on the diagonal,
- $\mathbf{R}(\boldsymbol{c})$ is the working correlation matrix, parameterized by $\boldsymbol{c}$. If $\mathbf{R}(\boldsymbol{c})$ is the true correlation matrix of $\mathbf{y}_i$, then $\mathbf{V}_i$ is the true covariance matrix.

Common Working Correlation Structures:

- Independence: $\mathbf{R} = \mathbf{I}$ (simplest, but ignores correlation).
- Exchangeable: Constant correlation between all pairs within a cluster.
- Autoregressive (AR(1)): Correlation decreases with time lag.
- Unstructured: Each pair has its own correlation parameter.

6.5.2.4.3 Iterative Algorithm for GEE Estimation

1. Initialization:
   - Compute an initial estimate of $\boldsymbol{\beta}$ using a GLM under the independence assumption ($\mathbf{R} = \mathbf{I}$).
2. Estimate the Working Correlation Matrix:
   - Based on residuals from the initial fit, estimate $\mathbf{R}(\boldsymbol{c})$.
3. Update the Covariance Matrix:
   - Calculate $\hat{\mathbf{V}}_i$ using the updated working correlation matrix.
4. Update $\boldsymbol{\beta}$:

$$\boldsymbol{\beta}^{(r+1)} = \boldsymbol{\beta}^{(r)} + \left(\sum_{i=1}^{K} \frac{\partial \boldsymbol{\mu}_i'}{\partial \boldsymbol{\beta}} \hat{\mathbf{V}}_i^{-1} \frac{\partial \boldsymbol{\mu}_i}{\partial \boldsymbol{\beta}}\right)^{-1} \left(\sum_{i=1}^{K} \frac{\partial \boldsymbol{\mu}_i'}{\partial \boldsymbol{\beta}} \hat{\mathbf{V}}_i^{-1} (\mathbf{y}_i - \boldsymbol{\mu}_i)\right)$$

5. Iteration:
   - Repeat steps 2–4 until convergence (i.e., when changes in $\boldsymbol{\beta}$ are negligible).

Comments on GEE:

- Advantages:
  - Provides consistent estimates of $\boldsymbol{\beta}$ even if the working correlation matrix is misspecified.
  - Robust standard errors (also known as "sandwich" estimators) account for potential misspecification.
- Limitations:
  - Not a likelihood-based method, so likelihood-ratio tests are not appropriate.
  - Efficiency loss if the working correlation matrix is poorly specified.
  - Estimation of random effects is not possible—GEE focuses on marginal (population-averaged) effects.

6.5.3 Estimation by Bayesian Hierarchical Models

Bayesian methods provide a flexible framework for estimating parameters in NLMMs and GLMMs. Unlike frequentist approaches that rely on MLE or linearization techniques, Bayesian estimation fully incorporates prior information and naturally accounts for uncertainty in both parameter estimation and predictions.

In the Bayesian context, we are interested in the posterior distribution of the model parameters, given the observed data $\mathbf{y}$:

$$f(\boldsymbol{\alpha}, \boldsymbol{\beta} \mid \mathbf{y}) \propto f(\mathbf{y} \mid \boldsymbol{\alpha}, \boldsymbol{\beta}) f(\boldsymbol{\alpha}) f(\boldsymbol{\beta})$$

where:

- $f(\mathbf{y} \mid \boldsymbol{\alpha}, \boldsymbol{\beta})$ is the likelihood of the data,
- $f(\boldsymbol{\alpha})$ is the prior distribution for the random effects,
- $f(\boldsymbol{\beta})$ is the prior distribution for the fixed effects,
- $f(\boldsymbol{\alpha}, \boldsymbol{\beta} \mid \mathbf{y})$ is the posterior distribution, which combines prior beliefs with observed data.

Advantages of Bayesian Estimation

- **No Need for Simplifying Approximations:** Bayesian methods do not require linearization or asymptotic approximations.
- **Full Uncertainty Quantification:** Provides credible intervals for parameters and predictive distributions for new data.
- **Flexible Modeling:** Easily accommodates complex hierarchical structures, non-standard distributions, and prior information.

Computational Challenges

Despite its advantages, Bayesian estimation can be computationally intensive and complex, especially for high-dimensional models. Key implementation issues include:

1. **Non-valid Joint Distributions:**
   In some hierarchical models, specifying valid joint distributions for the data, random effects, and parameters can be challenging.
2. **Constraints from Mean-Variance Relationships:**
   The inherent relationship between the mean and variance in GLMMs, combined with random effects, imposes constraints on the marginal covariance structure.
3. **Computational Intensity:**
   Fitting Bayesian models often requires advanced numerical techniques like Markov chain Monte Carlo, which can be slow to converge, especially for large datasets or complex models.

6.5.3.1 Bayesian Estimation Methods

Bayesian estimation can proceed through two general approaches:

1. **Approximating the Objective Function (Marginal Likelihood)**

The marginal likelihood is typically intractable because it requires integrating over random effects:

$$f(\mathbf{y} \mid \boldsymbol{\beta}) = \int f(\mathbf{y} \mid \boldsymbol{\alpha}, \boldsymbol{\beta}) f(\boldsymbol{\alpha}) d\boldsymbol{\alpha}$$

Since this integral cannot be solved analytically, we approximate it using the following methods:

- **Laplace Approximation**
  - Approximates the integral by expanding the log-likelihood around the mode of the integrand.
  - Provides an efficient, asymptotically accurate approximation when the posterior is approximately Gaussian near its mode.
- **Quadrature Methods**
  - **Gaussian quadrature** (e.g., Gauss-Hermite quadrature) is effective for low-dimensional random effects.
  - Approximates the integral by summing weighted evaluations of the function at specific points.
- **Monte Carlo Integration**
  - Uses random sampling to approximate the integral.
  - **Importance sampling** improves efficiency by drawing samples from a distribution that closely resembles the target distribution.

2. **Approximating the Model (Linearization)**

Alternatively, we can approximate the model itself using **Taylor series linearization** around current estimates of the parameters. This approach simplifies the model, making Bayesian estimation computationally more feasible, though at the cost of some approximation error.

6.5.3.2 Markov Chain Monte Carlo Methods

The most common approach for fully Bayesian estimation is **MCMC**, which generates samples from the posterior distribution through iterative simulation. Popular MCMC algorithms include:

- **Gibbs Sampling:**
  Efficient when full conditional distributions are available in closed form.
- **Metropolis-Hastings Algorithm:**
  More general and flexible, used when full conditionals are not easily sampled.
- **Hamiltonian Monte Carlo (HMC):**
  Implemented in packages like `Stan`, provides faster convergence for complex models by leveraging gradient information.

The posterior distribution is then approximated using the generated samples:

$$f(\boldsymbol{\alpha}, \boldsymbol{\beta} \mid \mathbf{y}) \approx \frac{1}{N} \sum_{i=1}^{N} \delta(\boldsymbol{\alpha} - \boldsymbol{\alpha}^{(i)}, \boldsymbol{\beta} - \boldsymbol{\beta}^{(i)})$$

where N is the number of MCMC samples and $\delta(\cdot)$ is the Dirac delta function.

6.5.4 Practical Implementation in R

Several R packages facilitate Bayesian estimation for GLMMs and NLMMs:

- **GLMM Estimation:**
  - `MASS::glmmPQL`—Penalized Quasi-Likelihood for GLMMs.
  - `lme4::glmer`—Frequentist estimation for GLMMs using Laplace approximation.
  - `glmmTMB`—Handles complex random effects structures efficiently.
- **NLMM Estimation:**
  - `nlme::nlme`—Nonlinear mixed effect modeling.
  - `lme4::nlmer`—Extends `lme4` for nonlinear mixed models.
  - `brms::brm`—Bayesian estimation via `Stan`, supporting NLMMs.
- **Bayesian Estimation:**
  - `MCMCglmm`—Implements MCMC algorithms for GLMMs.
  - `brms::brm`—High-level interface for Bayesian regression models, leveraging `Stan` for efficient MCMC sampling.

Example (Non-Gaussian Repeated Measurements)
Consider the case of **repeated measurements**:

- **If the data are Gaussian:** Use linear mixed models.
- **If the data are non-Gaussian:** Use nonlinear and generalized linear mixed models.

6.6 Application: Nonlinear and Generalized Linear Mixed Models

6.6.1 Binomial Data: CBPP Dataset

We will use the **CBPP dataset** from the `lme4` package to demonstrate different estimation approaches for binomial mixed models.

```
library(lme4)
#> Loading required package: Matrix
data(cbpp, package = "lme4")
head(cbpp)
#>   herd incidence size period
#> 1    1         2   14      1
#> 2    1         3   12      2
#> 3    1         4    9      3
#> 4    1         0    5      4
#> 5    2         3   22      1
#> 6    2         1   18      2
```

The data contain information about contagious bovine pleuropneumonia (CBPP) cases across different herds and periods.

1. **Penalized Quasi-likelihood**

   Pros:

- Linearizes the response to create a pseudo-response, similar to linear mixed models.
- Computationally efficient.

   Cons:

- Biased for binary or Poisson data with small counts.
- Random effects must be interpreted on the link scale.
- AIC/BIC values are not interpretable since PQL does not rely on full likelihood.

```
library(MASS)
pql_cbpp <- glmmPQL(
    cbind(incidence, size - incidence) ~ period,
    random  = ~ 1 | herd,
    data    = cbpp,
    family  = binomial(link = "logit"),
    verbose = FALSE
)
summary(pql_cbpp)
#> Linear mixed-effects model fit by maximum likelihood
#>   Data: cbpp
#>   AIC BIC logLik
#>    NA  NA     NA
#>
#> Random effects:
#>  Formula: ~1 | herd
#>         (Intercept) Residual
#> StdDev:   0.5563535 1.184527
```

```
#>
#> Variance function:
#>  Structure: fixed weights
#>  Formula: ~invwt
#> Fixed effects:  cbind(incidence, size - incidence) ~ period
↪
#>                  Value Std.Error DF   t-value p-value
#> (Intercept) -1.327364 0.2390194 38 -5.553372  0.0000
#> period2     -1.016126 0.3684079 38 -2.758156  0.0089
#> period3     -1.149984 0.3937029 38 -2.920944  0.0058
#> period4     -1.605217 0.5178388 38 -3.099839  0.0036
#>  Correlation:
#>         (Intr) perid2 perid3
#> period2 -0.399
#> period3 -0.373  0.260
#> period4 -0.282  0.196  0.182
#>
#> Standardized Within-Group Residuals:
#>        Min         Q1        Med         Q3        Max
#> -2.0591168 -0.6493095 -0.2747620  0.5170492  2.6187632
#>
#> Number of Observations: 56
#> Number of Groups: 15
```

Interpretation

```
exp(0.556)
#> [1] 1.743684
```

The above result shows how herd-specific odds vary, accounting for random effects.

The fixed effects are interpreted similarly to logistic regression. For example, with the logit link:

- The **log odds** of having a case in **period 2** are **-1.016** less than in **period 1** (baseline).

```
summary(pql_cbpp)$tTable
#>                  Value Std.Error DF   t-value      p-value
#> (Intercept) -1.327364 0.2390194 38 -5.553372 2.333216e-06
#> period2     -1.016126 0.3684079 38 -2.758156 8.888179e-03
#> period3     -1.149984 0.3937029 38 -2.920944 5.843007e-03
#> period4     -1.605217 0.5178388 38 -3.099839 3.637000e-03
```

2. **Numerical Integration with `glmer`**

   Pros:

- More accurate estimation since the method directly integrates over random effects.

   Cons:

- Computationally more expensive, especially with high-dimensional random effects.
- May struggle with convergence for complex models.

```
numint_cbpp <- glmer(
    cbind(incidence, size - incidence) ~ period + (1 | herd),
    data = cbpp,
    family = binomial(link = "logit")
)
summary(numint_cbpp)
#> Generalized linear mixed model fit by maximum likelihood
#>   (Laplace Approximation) [glmerMod]
#>  Family: binomial  ( logit )
#> Formula:
#> cbind(incidence, size - incidence) ~ period + (1 | herd)
#>    Data: cbpp
#>
#>      AIC       BIC    logLik -2*log(L)  df.resid
#>    194.1     204.2     -92.0     184.1        51
#>
#> Scaled residuals:
#>     Min      1Q  Median      3Q     Max
#> -2.3816 -0.7889 -0.2026  0.5142  2.8791
#>
#> Random effects:
#>  Groups Name        Variance Std.Dev.
#>  herd   (Intercept) 0.4123   0.6421
#> Number of obs: 56, groups:  herd, 15
#>
#> Fixed effects:
#>             Estimate Std. Error z value Pr(>|z|)
#> (Intercept)  -1.3983     0.2312  -6.048 1.47e-09 ***
#> period2      -0.9919     0.3032  -3.272 0.001068 **
#> period3      -1.1282     0.3228  -3.495 0.000474 ***
#> period4      -1.5797     0.4220  -3.743 0.000182 ***
#> ---
#> Signif. codes:
#> 0 '***' 0.001 '**' 0.01 '*' 0.05 '.' 0.1 ' ' 1
```

```
#>
#> Correlation of Fixed Effects:
#>         (Intr) perid2 perid3
#> period2 -0.363
#> period3 -0.340  0.280
#> period4 -0.260  0.213  0.198
```

Comparing PQL and Numerical Integration

For small datasets, the difference between PQL and numerical integration may be minimal.

```
library(rbenchmark)
benchmark(
    "PQL (MASS)" = {
        glmmPQL(
            cbind(incidence, size - incidence) ~ period,
            random = ~ 1 | herd,
            data = cbpp,
            family = binomial(link = "logit"),
            verbose = FALSE
        )
    },
    "Numerical Integration (lme4)" = {
        glmer(
            cbind(incidence, size - incidence) ~ period + (1
            ↪ | herd),
            data = cbpp,
            family = binomial(link = "logit")
        )
    },
    replications = 50,
    columns = c("test", "replications", "elapsed",
    ↪ "relative"),
    order = "relative"
)
#>                           test replications elapsed
#> 1                   PQL (MASS)           50    4.18
#> 2 Numerical Integration (lme4)           50    9.21
#>   relative
#> 1    1.000
#> 2    2.203
```

Improving Accuracy with Gauss-Hermite Quadrature
Setting `nAGQ` > 1 increases the accuracy of the likelihood approximation:

```
numint_cbpp_GH <- glmer(
    cbind(incidence, size - incidence) ~ period + (1 | herd),
    data = cbpp,
    family = binomial(link = "logit"),
    nAGQ = 20
)
summary(numint_cbpp_GH)$coefficients[, 1] -
    summary(numint_cbpp)$coefficients[, 1]
#>    (Intercept)       period2       period3       period4
#> -0.0008808634  0.0005160912  0.0004066218  0.0002644629
```

3. **Bayesian Approach with `MCMCglmm`**

Pros:

- Incorporates prior information and handles complex models with intractable likelihoods.
- Provides full posterior distributions for parameters.

Cons:

- Computationally intensive, especially with large datasets or complex hierarchical structures.

```
library(MCMCglmm)
#> Loading required package: coda
#> Loading required package: ape
Bayes_cbpp <- MCMCglmm(
    cbind(incidence, size - incidence) ~ period,
    random = ~ herd,
    data = cbpp,
    family = "multinomial2",
    verbose = FALSE
)
summary(Bayes_cbpp)
#>
#>  Iterations = 3001:12991
#>  Thinning interval  = 10
#>  Sample size  = 1000
#>
#>  DIC: 538.0278
#>
```

```
#>  G-structure:  ~herd
#>
#>      post.mean  l-95% CI u-95% CI eff.samp
#> herd   0.01635 4.927e-17  0.07067     20.7
#>
#>  R-structure:  ~units
#>
#>       post.mean l-95% CI u-95% CI eff.samp
#> units      1.06   0.2738     1.98    298.6
#>
#>  Location effects: cbind(incidence, size - incidence) ~
↪  period
#>
#>             post.mean l-95% CI u-95% CI eff.samp  pMCMC
#> (Intercept)   -1.5237  -2.1877  -0.9389    729.2 <0.001 ***
#> period2       -1.2603  -2.1912  -0.1930    897.7  0.018 *
#> period3       -1.3986  -2.4473  -0.3547    706.7  0.010 **
#> period4       -1.9238  -3.1983  -0.7111    487.9  0.004 **
#> ---
#> Signif. codes:
#> 0 '***' 0.001 '**' 0.01 '*' 0.05 '.' 0.1 ' ' 1
```

- `MCMCglmm` fits a residual variance component (useful with dispersion issues).

Variance Component Analysis

```
# explains less variability
apply(Bayes_cbpp$VCV, 2, sd)
#>       herd      units
#> 0.08558652 0.48190786
```

Posterior Summaries

```
summary(Bayes_cbpp)$solutions
#>             post.mean  l-95% CI   u-95% CI eff.samp pMCMC
#> (Intercept) -1.523664 -2.187707 -0.9389008 729.1546 0.001
#> period2     -1.260348 -2.191243 -0.1930167 897.7091 0.018
#> period3     -1.398638 -2.447253 -0.3547237 706.6777 0.010
#> period4     -1.923829 -3.198325 -0.7110988 487.9483 0.004
```

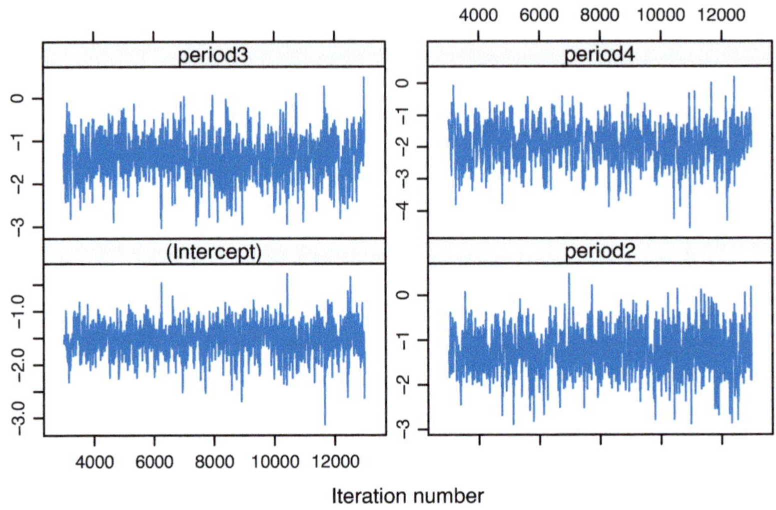

Fig. 6.1 MCMC Diagnostic

MCMC Diagnostics (Fig. 6.1)

```
library(lattice)
xyplot(as.mcmc(Bayes_cbpp$Sol), layout = c(2, 2))
```

There is no trend (i.e., well-mixed).

```
xyplot(as.mcmc(Bayes_cbpp$VCV), layout = c(2, 1))
```

For the herd variable, many of the values are 0, which suggests a problem (Fig. 6.2). To address the instability in the herd effect sampling, we can either:

- Modify prior distributions,
- Increase the number of iterations

```
Bayes_cbpp2 <- MCMCglmm(
    cbind(incidence, size - incidence) ~ period,
    random = ~ herd,
    data = cbpp,
    family = "multinomial2",
    nitt = 20000,
```

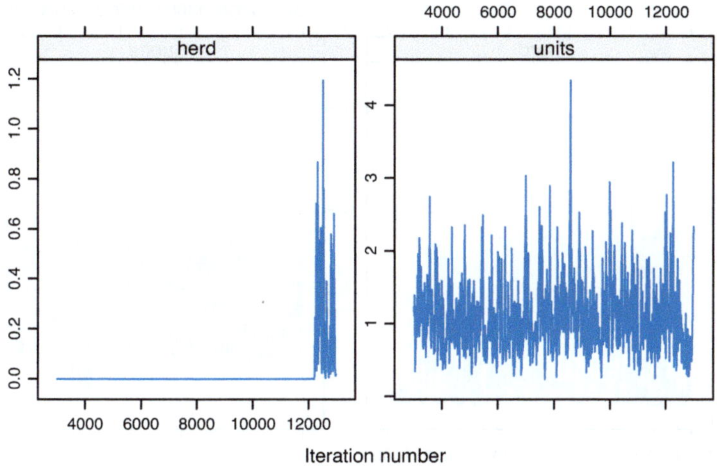

Fig. 6.2 Variable MCMC diagnostic

```
    burnin = 10000,
    prior = list(G = list(list(V = 1, nu = 0.1))),
    verbose = FALSE
)
xyplot(as.mcmc(Bayes_cbpp2$VCV), layout = c(2, 1))
```

To change the shape of priors, in `MCMCglmm` use (Fig. 6.3):

- `V` controls for the location of the distribution (default = 1)
- `nu` controls for the concentration around V (default = 0)

6.6.2 Count Data: Owl Dataset

We'll now model count data using the **Owl dataset** (Fig. 6.4)

```
library(glmmTMB)
library(dplyr)
#>
#> Attaching package: 'dplyr'
#> The following object is masked from 'package:ape':
#>
```

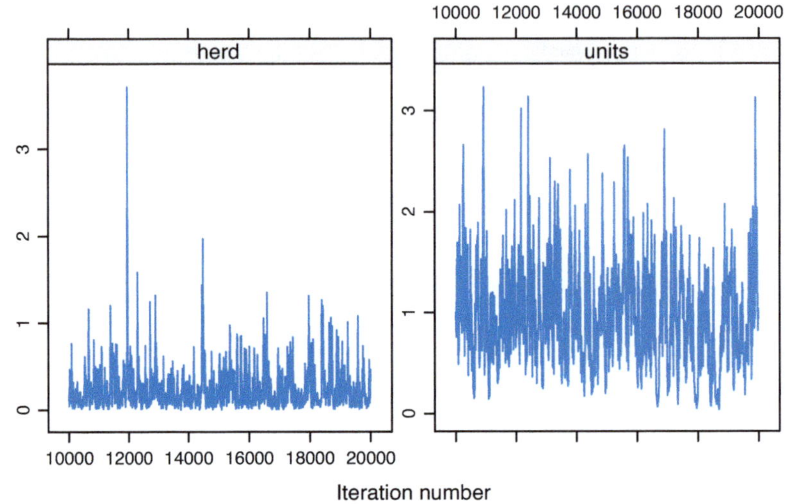

Fig. 6.3 Variable MCMC diagnostic (alternative)

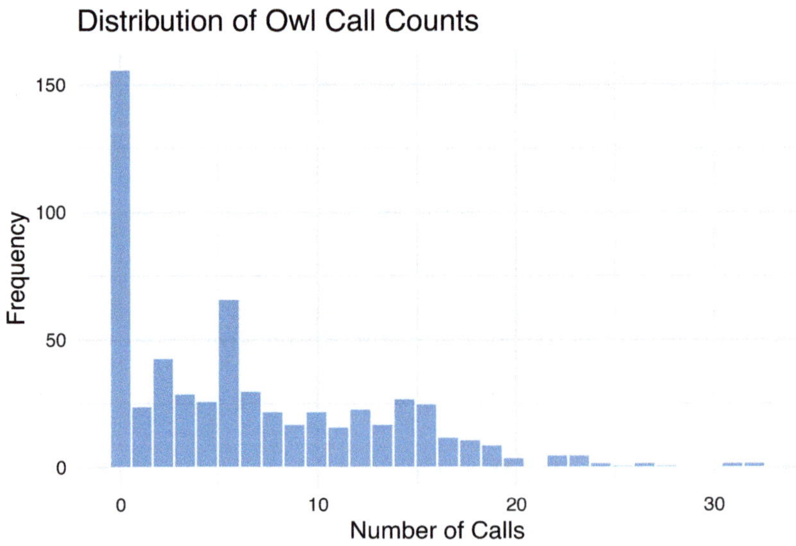

Fig. 6.4 Distribution of owl call counts

```
#>     where
#> The following object is masked from 'package:MASS':
#>
#>     select
#> The following objects are masked from 'package:stats':
```

```
#>
#>     filter, lag
#> The following objects are masked from 'package:base':
#>
#>     intersect, setdiff, setequal, union
data(Owls, package = "glmmTMB")
Owls <- Owls %>% rename(Ncalls = SiblingNegotiation)
```

1. **Poisson GLMM**

   Modeling call counts with a Poisson distribution:
   In a typical Poisson model, the Poisson mean λ is modeled as:

$$\log(\lambda) = x'\beta$$

However, if the response variable represents a rate (e.g., counts per **BroodSize**), we can model it as:

$$\log\left(\frac{\lambda}{b}\right) = x'\beta$$

This is equivalent to:

$$\log(\lambda) = \log(b) + x'\beta$$

where b represents **BroodSize**. In this formulation, we "offset" the mean by including the logarithm of b as an offset term in the model. This adjustment accounts for the varying exposure or denominator in rate-based responses.

```
owls_glmer <- glmer(
    Ncalls ~ offset(log(BroodSize)) + FoodTreatment *
 ↪  SexParent + (1 | Nest),
    family = poisson,
    data = Owls
)
summary(owls_glmer)
#> Generalized linear mixed model fit by maximum likelihood
#>   (Laplace Approximation) [glmerMod]
#>  Family: poisson  ( log )
#> Formula:
#> Ncalls ~ offset(log(BroodSize)) + FoodTreatment * SexParent
 ↪  +
#>     (1 | Nest)
#>    Data: Owls
```

```
#>
#>       AIC       BIC    logLik -2*log(L)  df.resid
#>    5212.8    5234.8   -2601.4    5202.8       594
#>
#> Scaled residuals:
#>     Min      1Q  Median      3Q     Max
#> -3.5529 -1.7971 -0.6842  1.2689 11.4312
#>
#> Random effects:
#>  Groups Name        Variance Std.Dev.
#>  Nest   (Intercept) 0.2063   0.4542
#> Number of obs: 599, groups:  Nest, 27
#>
#> Fixed effects:
#>                                      Estimate Std. Error
#> (Intercept)                           0.65585    0.09567
#> FoodTreatmentSatiated                -0.65612    0.05606
#> SexParentMale                        -0.03705    0.04501
#> FoodTreatmentSatiated:SexParentMale   0.13135    0.07036
#>                                      z value Pr(>|z|)
#> (Intercept)                            6.855 7.12e-12 ***
#> FoodTreatmentSatiated                -11.705  < 2e-16 ***
#> SexParentMale                         -0.823   0.4104
#> FoodTreatmentSatiated:SexParentMale    1.867   0.0619 .
#> ---
#> Signif. codes:
#> 0 '***' 0.001 '**' 0.01 '*' 0.05 '.' 0.1 ' ' 1
#>
#> Correlation of Fixed Effects:
#>             (Intr) FdTrtS SxPrnM
#> FdTrtmntStt -0.225
#> SexParentMl -0.292  0.491
#> FdTrtmS:SPM  0.170 -0.768 -0.605
```

- Nest explains a relatively large proportion of the variability (its standard deviation is larger than some coefficients).
- The model fit isn't great (deviance of 5202 on 594 df).

2. **Negative Binomial Model**

   Addressing overdispersion using the negative binomial distribution:

```
owls_glmerNB <- glmer.nb(
    Ncalls ~ offset(log(BroodSize)) + FoodTreatment *
↪   SexParent + (1 | Nest),
    data = Owls
)
summary(owls_glmerNB)
#> Generalized linear mixed model fit by maximum likelihood
#>   (Laplace Approximation) [glmerMod]
#>  Family: Negative Binomial(0.8423)  ( log )
#> Formula:
#> Ncalls ~ offset(log(BroodSize)) + FoodTreatment * SexParent
↪   +
#>     (1 | Nest)
#>    Data: Owls
#>
#>       AIC       BIC    logLik -2*log(L)  df.resid
#>    3495.6    3522.0   -1741.8    3483.6       593
#>
#> Scaled residuals:
#>     Min      1Q  Median      3Q     Max
#> -0.8859 -0.7737 -0.2701  0.4443  6.1432
#>
#> Random effects:
#>  Groups Name        Variance Std.Dev.
#>  Nest   (Intercept) 0.1245   0.3529
#> Number of obs: 599, groups:  Nest, 27
#>
#> Fixed effects:
#>                                     Estimate Std. Error
#> (Intercept)                          0.69005    0.13400
#> FoodTreatmentSatiated               -0.76657    0.16509
#> SexParentMale                       -0.02605    0.14575
#> FoodTreatmentSatiated:SexParentMale  0.15680    0.20512
#>                                     z value Pr(>|z|)
#> (Intercept)                           5.150 2.61e-07 ***
#> FoodTreatmentSatiated                -4.643 3.43e-06 ***
#> SexParentMale                        -0.179    0.858
#> FoodTreatmentSatiated:SexParentMale   0.764    0.445
#> ---
#> Signif. codes:
#> 0 '***' 0.001 '**' 0.01 '*' 0.05 '.' 0.1 ' ' 1
#>
#> Correlation of Fixed Effects:
#>             (Intr) FdTrtS SxPrnM
```

```
#> FdTrtmntStt -0.602
#> SexParentMl -0.683  0.553
#> FdTrtmS:SPM  0.450 -0.744 -0.671
```

There is an improvement using negative binomial considering over-dispersion

```
library(ggplot2)
ggplot(Owls, aes(x = Ncalls)) +
  geom_histogram(
    bins = 30,
    fill = "steelblue",
    color = "white",
    alpha = 0.8
  ) +
  labs(title = "Distribution of Owl Call Counts",
       x = "Number of Calls",
       y = "Frequency") +
  theme_minimal(base_size = 14)
```

3. **Zero-Inflated Model**

Handling excess zeros with a zero-inflated Poisson model:

```
library(glmmTMB)
owls_glmm <-
    glmmTMB(
        Ncalls ~ FoodTreatment * SexParent +
 ↪  offset(log(BroodSize)) +
            (1 | Nest),
        ziformula = ~ 0,
        family = nbinom2(link = "log"),
        data = Owls
    )
owls_glmm_zi <-
    glmmTMB(
        Ncalls ~ FoodTreatment * SexParent +
 ↪  offset(log(BroodSize)) +
            (1 | Nest),
        ziformula = ~ 1,
        family = nbinom2(link = "log"),
        data = Owls
    )

# Scale Arrival time to use as a covariate for zero-inflation
 ↪  parameter
```

```
Owls$ArrivalTime <- scale(Owls$ArrivalTime)
owls_glmm_zi_cov <- glmmTMB(
    Ncalls ~ FoodTreatment * SexParent +
        offset(log(BroodSize)) +
        (1 | Nest),
    ziformula = ~ ArrivalTime,
    family = nbinom2(link = "log"),
    data = Owls
)

as.matrix(anova(owls_glmm, owls_glmm_zi))
#>              Df      AIC      BIC    logLik deviance
#> owls_glmm     6 3495.610 3521.981 -1741.805 3483.610
#> owls_glmm_zi  7 3431.646 3462.413 -1708.823 3417.646
#>                 Chisq Chi Df   Pr(>Chisq)
#> owls_glmm          NA     NA           NA
#> owls_glmm_zi 65.96373      1 4.592983e-16


as.matrix(anova(owls_glmm_zi, owls_glmm_zi_cov))
#>                  Df      AIC      BIC    logLik deviance
#> owls_glmm_zi      7 3431.646 3462.413 -1708.823 3417.646
#> owls_glmm_zi_cov  8 3422.532 3457.694 -1703.266 3406.532
#>                     Chisq Chi Df   Pr(>Chisq)
#> owls_glmm_zi           NA     NA           NA
#> owls_glmm_zi_cov 11.11411      1 0.0008567362


summary(owls_glmm_zi_cov)
#>  Family: nbinom2  ( log )
#> Formula:
#> Ncalls ~ FoodTreatment * SexParent + offset(log(BroodSize))
↪  +
#>     (1 | Nest)
#> Zero inflation:            ~ArrivalTime
#> Data: Owls
#>
#>       AIC       BIC    logLik -2*log(L)  df.resid
#>    3422.5    3457.7   -1703.3    3406.5       591
#>
#> Random effects:
#>
#> Conditional model:
#>  Groups Name        Variance Std.Dev.
```

```
#>  Nest   (Intercept) 0.07487  0.2736
#> Number of obs: 599, groups:  Nest, 27
#>
#> Dispersion parameter for nbinom2 family (): 2.22
#>
#> Conditional model:
#>                                    Estimate Std. Error
#> (Intercept)                         0.84778    0.09961
#> FoodTreatmentSatiated              -0.39529    0.13742
#> SexParentMale                      -0.07025    0.10435
#> FoodTreatmentSatiated:SexParentMale 0.12388    0.16449
#>                                    z value Pr(>|z|)
#> (Intercept)                          8.511  < 2e-16 ***
#> FoodTreatmentSatiated               -2.877  0.00402 **
#> SexParentMale                       -0.673  0.50079
#> FoodTreatmentSatiated:SexParentMale  0.753  0.45138
#> ---
#> Signif. codes:
#> 0 '***' 0.001 '**' 0.01 '*' 0.05 '.' 0.1 ' ' 1
#>
#> Zero-inflation model:
#>             Estimate Std. Error z value Pr(>|z|)
#> (Intercept)  -1.3018     0.1261  -10.32  < 2e-16 ***
#> ArrivalTime   0.3545     0.1074    3.30 0.000966 ***
#> ---
#> Signif. codes:
#> 0 '***' 0.001 '**' 0.01 '*' 0.05 '.' 0.1 ' ' 1
```

`glmmTMB` can handle ZIP GLMMs since it adds automatic differentiation to existing estimation strategies.

We can see ZIP GLMM with an arrival time covariate on the zero is best.

- Arrival time has a positive effect on observing a nonzero number of calls
- Interactions are non-significant, the food treatment is significant (fewer calls after eating)
- Nest variability is large in magnitude (without this, the parameter estimates change)

6.6.3 *Binomial Example: Gotway Hessian Fly Data*

We will analyze the **Gotway Hessian Fly** dataset from the `agridat` package to model binomial outcomes using both frequentist and Bayesian approaches (Fig. 6.5).

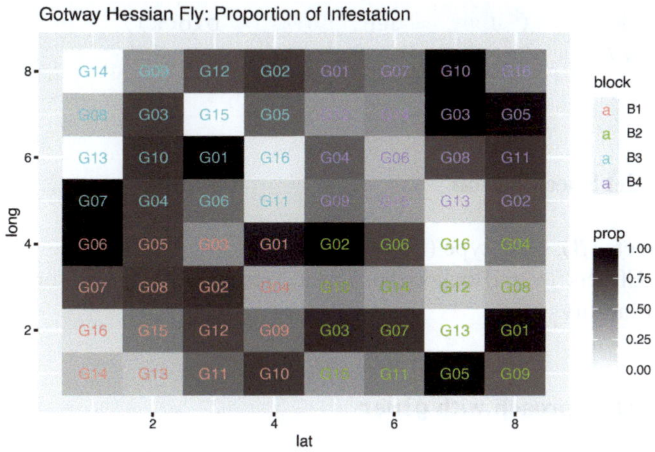

Fig. 6.5 Gotway Hessian Fly: proportion of infestation

6.6.3.1 Data Visualization

```
library(agridat)
library(ggplot2)
library(lme4)
library(spaMM)
#> Registered S3 methods overwritten by 'registry':
#>   method               from
#>   print.registry_field proxy
#>   print.registry_entry proxy
#> spaMM (Rousset & Ferdy, 2014, version 4.5.0) is loaded.
#> Type 'help(spaMM)' for a short introduction,
#> 'news(package='spaMM')' for news,
#> and 'citation('spaMM')' for proper citation.
#> Further infos, slides, etc. at
↪  https://gitlab.mbb.univ-montp2.fr/francois/spamm-ref.

data(gotway.hessianfly)
dat <- gotway.hessianfly
dat$prop <- dat$y / dat$n  # Proportion of successes
```

```
ggplot(dat, aes(x = lat, y = long, fill = prop)) +
    geom_tile() +
    scale_fill_gradient(low = 'white', high = 'black') +
```

```
    geom_text(aes(label = gen, color = block)) +
    ggtitle('Gotway Hessian Fly: Proportion of Infestation')
```

6.6.3.2 Model Specification

- Fixed Effects ($\boldsymbol{\beta}$): Genotype (`gen`)
- Random Effects ($\boldsymbol{\alpha}$): Block (`block`), accounting for spatial or experimental design variability

1. **Frequentist Approach with `glmer`**

```
flymodel <- glmer(
    cbind(y, n - y) ~ gen + (1 | block),
    data   = dat,
    family = binomial,
    nAGQ   = 5 # Using adaptive Gauss-Hermite quadrature for
     ↪  accuracy
)
summary(flymodel)
#> Generalized linear mixed model fit by maximum likelihood
#>   (Adaptive Gauss-Hermite Quadrature, nAGQ = 5) [glmerMod]
#>  Family: binomial  ( logit )
#> Formula: cbind(y, n - y) ~ gen + (1 | block)
#>    Data: dat
#> 
#>      AIC      BIC   logLik -2*log(L)  df.resid 
#>    162.2    198.9    -64.1     128.2        47 
#> 
#> Scaled residuals: 
#>      Min       1Q   Median       3Q      Max 
#> -2.38644 -1.01188  0.09631  1.03468  2.75479 
#> 
#> Random effects:
#>  Groups Name        Variance Std.Dev.
#>  block  (Intercept) 0.001022 0.03196 
#> Number of obs: 64, groups:  block, 4
#> 
#> Fixed effects:
#>             Estimate Std. Error z value Pr(>|z|)    
#> (Intercept)   1.5035     0.3914   3.841 0.000122 ***
#> genG02       -0.1939     0.5302  -0.366 0.714645    
```

```
#> genG03          -0.5408     0.5103  -1.060 0.289263
#> genG04          -1.4342     0.4714  -3.043 0.002346 **
#> genG05          -0.2037     0.5429  -0.375 0.707487
#> genG06          -0.9783     0.5046  -1.939 0.052534 .
#> genG07          -0.6041     0.5111  -1.182 0.237237
#> genG08          -1.6774     0.4907  -3.418 0.000630 ***
#> genG09          -1.3984     0.4725  -2.960 0.003079 **
#> genG10          -0.6817     0.5333  -1.278 0.201183
#> genG11          -1.4630     0.4843  -3.021 0.002522 **
#> genG12          -1.4591     0.4918  -2.967 0.003010 **
#> genG13          -3.5528     0.6600  -5.383 7.31e-08 ***
#> genG14          -2.5073     0.5264  -4.763 1.90e-06 ***
#> genG15          -2.0872     0.4851  -4.302 1.69e-05 ***
#> genG16          -2.9697     0.5383  -5.517 3.46e-08 ***
#> ---
#> Signif. codes:
#> 0 '***' 0.001 '**' 0.01 '*' 0.05 '.' 0.1 ' ' 1
#>
#> Correlation matrix not shown by default, as p = 16 > 12.
#> Use print(x, correlation=TRUE)  or
#>     vcov(x)        if you need it
```

Interpretation:

- The **fixed effects** (gen) indicate how different genotypes influence the infestation probability.
- The **random effect** for block captures variability due to experimental blocks, improving model robustness.
- **Odds Ratios:** Exponentiating coefficients helps interpret the impact on infestation odds.

2. **Bayesian Approach with MCMCglmm**

```
library(MCMCglmm)
library(coda)

Bayes_flymodel <- MCMCglmm(
    cbind(y, n - y) ~ gen,
    random  = ~ block,
    data    = dat,
    family  = "multinomial2",
    verbose = FALSE
)
summary(Bayes_flymodel)
#>
```

```
#>  Iterations = 3001:12991
#>  Thinning interval  = 10
#>  Sample size  = 1000
#>
#>  DIC: 877.3222
#>
#>  G-structure:  ~block
#>
#>       post.mean  l-95% CI u-95% CI eff.samp
#> block  0.005184 2.515e-17 0.006832    727.1
#>
#>  R-structure:  ~units
#>
#>       post.mean l-95% CI u-95% CI eff.samp
#> units     1.008   0.3516    1.853    391.3
#>
#>  Location effects: cbind(y, n - y) ~ gen
#>
#>             post.mean l-95% CI u-95% CI eff.samp  pMCMC
#> (Intercept)   1.99787  0.49381  3.36611    888.6  0.002 **
#> genG02       -0.44474 -2.17668  1.66557    863.7  0.670
#> genG03       -0.77772 -2.63723  1.05854   1000.0  0.388
#> genG04       -1.83038 -3.59102  0.20901   1000.0  0.052 .
#> genG05       -0.35082 -2.27588  1.51624   1000.0  0.716
#> genG06       -1.35434 -3.33723  0.44835   1000.0  0.168
#> genG07       -0.78886 -2.57929  1.03158    865.8  0.392
#> genG08       -2.11231 -3.89369 -0.28357    873.2  0.024 *
#> genG09       -1.93650 -3.85477 -0.09033   1000.0  0.046 *
#> genG10       -0.83054 -2.71868  1.10082    870.3  0.400
#> genG11       -1.98427 -4.03048 -0.34949    825.3  0.028 *
#> genG12       -1.96285 -3.78219 -0.11436   1000.0  0.032 *
#> genG13       -4.49358 -6.72407 -2.30588    743.8 <0.001 ***
#> genG14       -3.28225 -5.15521 -1.41681    780.2 <0.001 ***
#> genG15       -2.79892 -4.60076 -0.81913    900.8  0.002 **
#> genG16       -3.90091 -5.90755 -1.83792    872.0 <0.001 ***
#> ---
#> Signif. codes:
#> 0 '***' 0.001 '**' 0.01 '*' 0.05 '.' 0.1 ' ' 1
```

MCMC Diagnostics

- **Trace Plot:** Checks for chain mixing and convergence (Fig. 6.6).
- **Autocorrelation Plot:** Evaluates dependency between MCMC samples (Fig. 6.7).

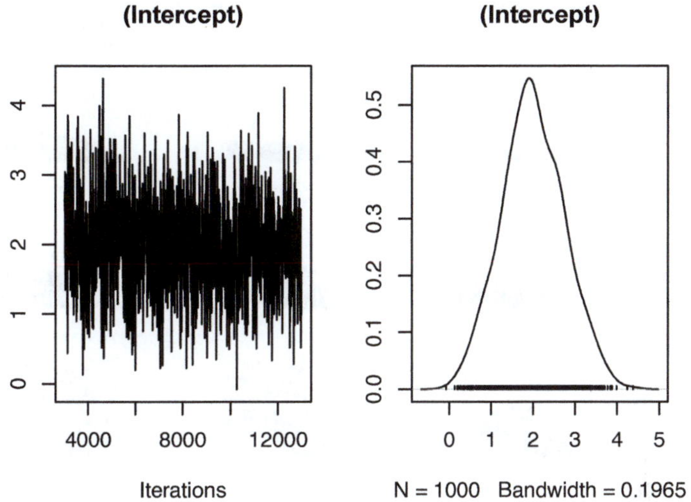

Fig. 6.6 Trace plot

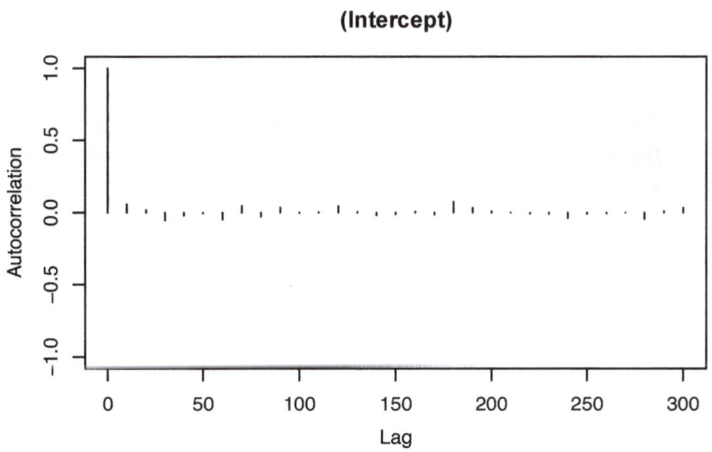

Fig. 6.7 Autocorrelation plot

```
# Trace plot for the first fixed effect
plot(Bayes_flymodel$Sol[, 1],
     main = colnames(Bayes_flymodel$Sol)[1])
```

```
# Autocorrelation plot
autocorr.plot(Bayes_flymodel$Sol[, 1],
              main = colnames(Bayes_flymodel$Sol)[1])
```

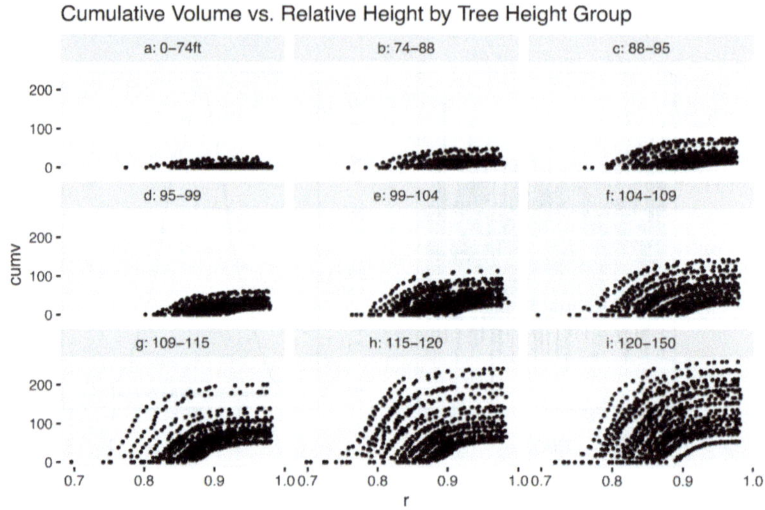

Fig. 6.8 Cumulative volume vs. relative height by tree height group

Bayesian Interpretation:

- **Posterior Means:** Represent the central tendency of the parameter estimates.
- **Credible Intervals:** Unlike frequentist confidence intervals, they can be interpreted directly as the probability that the parameter lies within the interval.

6.6.4 Nonlinear Mixed Model: Yellow Poplar Data

This dataset comes from Schabenberger and Pierce (2001) (Fig. 6.8).

6.6.4.1 Data Preparation

```
dat2 <- read.table("images/YellowPoplarData_r.txt")
names(dat2) <-
    c('tn', 'k', 'dbh', 'totht', 'dob', 'ht', 'maxd', 'cumv')
dat2$t <- dat2$dob / dat2$dbh
dat2$r <- 1 - dat2$dob / dat2$totht
```

6.6.4.2 Data Visualization

```
library(dplyr)

dat2 <- dat2 %>% group_by(tn) %>% mutate(
    z = case_when(
        totht < 74 ~ 'a: 0-74ft',
        totht < 88 ~ 'b: 74-88',
        totht < 95 ~ 'c: 88-95',
        totht < 99 ~ 'd: 95-99',
        totht < 104 ~ 'e: 99-104',
        totht < 109 ~ 'f: 104-109',
        totht < 115 ~ 'g: 109-115',
        totht < 120 ~ 'h: 115-120',
        totht < 140 ~ 'i: 120-150',
        TRUE ~ 'j: 150+'
    )
)
```

```
ggplot(dat2, aes(x = r, y = cumv)) +
  geom_point(size = 0.5) +
  facet_wrap(vars(z)) +
  labs(title = "Cumulative Volume vs. Relative Height by Tree
  ↪  Height Group")
```

6.6.4.3 Model Specification

The proposed nonlinear mixed model is:

$$V_{ij} = \left(\beta_0 + (\beta_1 + b_{1i})\frac{D_i^2 H_i}{1000}\right) \exp\left[-(\beta_2 + b_{2i})t_{ij} \exp(\beta_3 t_{ij})\right] + e_{ij}$$

where:

- b_{1i}, b_{2i} are random effects for tree i.
- e_{ij} are residual errors.

6.6.4.4 Fitting the Nonlinear Mixed Model

```
library(nlme)
#>
#> Attaching package: 'nlme'
#> The following object is masked from 'package:dplyr':
#>
#>     collapse
#> The following object is masked from 'package:lme4':
#>
#>     lmList

tmp <- nlme(
    cumv ~ (b0 + (b1 + u1) * (dbh^2 * totht / 1000)) *
        exp(-(b2 + u2) * (t / 1000) * exp(b3 * t)),
    data = dat2,
    fixed = b0 + b1 + b2 + b3 ~ 1,
    random = pdDiag(u1 + u2 ~ 1),  # Uncorrelated random
    ↪  effects
    groups = ~ tn,                 # Grouping by tree
    start = list(fixed = c(b0 = 0.25, b1 = 2.3, b2 = 2.87, b3
    ↪  = 6.7))
)
summary(tmp)
#> Nonlinear mixed-effects model fit by maximum likelihood
#>   Model: cumv ~ (b0 + (b1 + u1) * (dbh^2 * totht/1000)) *
↪   exp(-(b2 + u2) *      (t/1000) * exp(b3 * t))
#>   Data: dat2
#>        AIC      BIC    logLik
#>   31103.73 31151.33 -15544.86
#>
#> Random effects:
#>  Formula: list(u1 ~ 1, u2 ~ 1)
#>  Level: tn
#>  Structure: Diagonal
#>                u1       u2 Residual
#> StdDev: 0.1508094 0.447829 2.226361
#>
#> Fixed effects:  b0 + b1 + b2 + b3 ~ 1
#>       Value  Std.Error   DF  t-value p-value
#> b0 0.249386 0.12894686 6297   1.9340  0.0532
#> b1 2.288832 0.01266804 6297 180.6777  0.0000
#> b2 2.500497 0.05606686 6297  44.5985  0.0000
#> b3 6.848871 0.02140677 6297 319.9395  0.0000
```

```
#>  Correlation:
#>    b0     b1     b2
#> b1 -0.639
#> b2  0.054  0.056
#> b3 -0.011 -0.066 -0.850
#>
#> Standardized Within-Group Residuals:
#>           Min            Q1           Med            Q3
#> -6.694575e+00 -3.081861e-01 -8.907041e-05  3.469469e-01
#>           Max
#>  7.855665e+00
#>
#> Number of Observations: 6636
#> Number of Groups: 336
nlme::intervals(tmp)
#> Approximate 95% confidence intervals
#>
#>  Fixed effects:
#>          lower      est.     upper
#> b0 -0.003317833 0.2493858 0.5020894
#> b1  2.264006069 2.2888323 2.3136585
#> b2  2.390620116 2.5004971 2.6103742
#> b3  6.806919325 6.8488712 6.8908232
#>
#>  Random Effects:
#>   Level: tn
#>             lower      est.     upper
#> sd(u1) 0.1376068 0.1508094 0.1652787
#> sd(u2) 0.4056207 0.4478290 0.4944295
#>
#>  Within-group standard error:
#>    lower     est.    upper
#> 2.187259 2.226361 2.266161
```

6.6.4.5 Interpretation

- **Fixed Effects** (β): Describe the average growth pattern across all trees.
- **Random Effects** (b_i): Capture tree-specific deviations from the average trend.

This result is a bit different from the original study because of different implementation of nonlinear mixed models.

6.6.4.6 Visualizing Model Predictions

```
library(cowplot)

# Prediction function
nlmmfn <- function(fixed, rand, dbh, totht, t) {
    (fixed[1] + (fixed[2] + rand[1]) * (dbh ^ 2 * totht / 1000)) *
        exp(-(fixed[3] + rand[2]) * (t / 1000) * exp(fixed[4] * t))
}

# Function to generate plots for selected trees
plot_tree <- function(tree_id) {
    pred <- data.frame(dob = seq(1, max(dat2$dob), length.out = 100))
    pred$tn <- tree_id
    pred$dbh <- unique(dat2$dbh[dat2$tn == tree_id])
    pred$t <- pred$dob / pred$dbh
    pred$totht <- unique(dat2$totht[dat2$tn == tree_id])
    pred$r <- 1 - pred$dob / pred$totht

    pred$with_random <- predict(tmp, pred)
    pred$without_random <-
        nlmmfn(tmp$coefficients$fixed,
               c(0, 0),
               pred$dbh,
               pred$totht,
               pred$t)

    ggplot(pred) +
        geom_line(aes(x = r, y = with_random, color = 'With Random Effects')) +
        geom_line(aes(x = r, y = without_random, color = 'Without Random Effects')) +
        geom_point(data = dat2[dat2$tn == tree_id,], aes(x = r, y = cumv)) +
        labs(title = paste('Tree', tree_id), colour = "") +
        theme(legend.position = "bottom")
}

# Plotting for selected trees
p1 <- plot_tree(1)
```

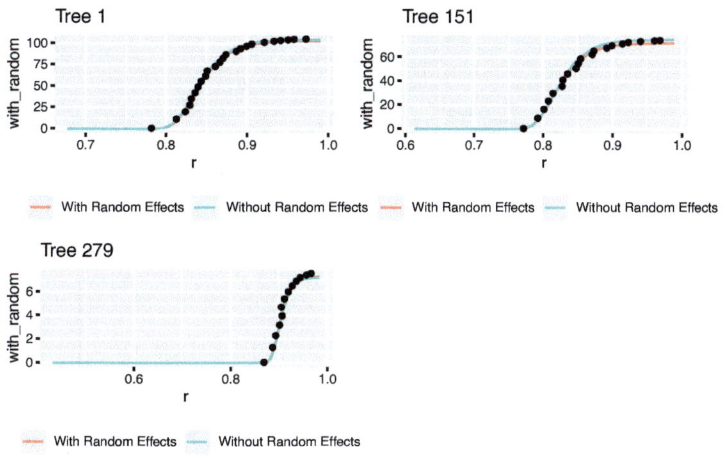

Fig. 6.9 Individual tree scatterplot

```
p2 <- plot_tree(151)
p3 <- plot_tree(279)
```

```
plot_grid(p1, p2, p3)
```

Figure 6.9 shows the scatter plot of individual trees.

- **Red Line:** Model predictions with tree-specific random effects.
- **Teal Line:** Model predictions based only on fixed effects (ignoring tree-specific variation).
- **Dots:** Observed cumulative volume for each tree.

6.7 Summary

Figure 6.10 shows relationships between different regression models.

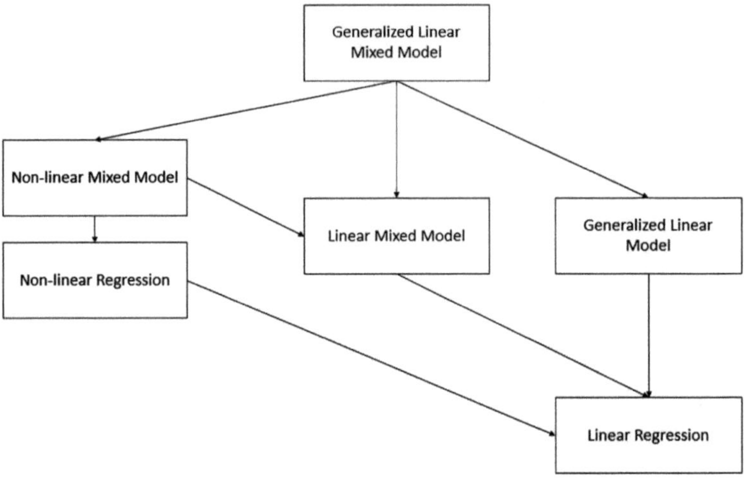

Fig. 6.10 Taxonomy of regression models

Chapter 7
Nonparametric Regression

This chapter surveys regression techniques that relax functional-form assumptions. Beginning with kernel and local polynomial estimators, we derive bias-variance trade-offs and bandwidth selection criteria. We then explore splines, generalized additive models, regression trees, random forests, and wavelet regression, emphasizing their interpretability and robustness. Multivariate nonparametrics are introduced through radial-basis functions. Confidence-interval construction via asymptotics and bootstrap methods is detailed, and a forward-looking conclusion discusses how nonparametric ideas underpin modern machine-learning algorithms, reinforcing the evolving landscape of regression analysis.

Nonparametric regression refers to a class of regression techniques that do not assume a specific functional form (e.g., linear, polynomial of fixed degree) for the relationship between a predictor $x \in \mathbb{R}$ (or $\mathbf{x} \in \mathbb{R}^p$) and a response variable $y \in \mathbb{R}$. Instead, nonparametric methods aim to estimate this relationship directly from the data, allowing the data to "speak for themselves."

In a standard regression framework, we have a response variable Y and one or more predictors $\mathbf{X} = (X_1, X_2, \ldots, X_p)$. Let us start with a univariate setting for simplicity. We assume the following model:

$$Y = m(x) + \varepsilon,$$

where:

- $m(x) = \mathbb{E}[Y \mid X = x]$ is the **regression function** we aim to estimate,
- ε is a random **error term** (noise) with $\mathbb{E}[\varepsilon \mid X = x] = 0$ and constant variance $\text{Var}(\varepsilon) = \sigma^2$.

In **parametric regression** (e.g., Linear Regression), we might assume $m(x)$ has a specific form, such as:

$$m(x) = \beta_0 + \beta_1 x + \cdots + \beta_d x^d,$$

M. Nguyen, *Regression Techniques for Data Analysis*,
https://doi.org/10.1007/978-3-032-01835-9_7

where $\beta_0, \beta_1, \ldots, \beta_d$ are parameters to be estimated. In contrast, **nonparametric regression** relaxes this assumption and employs methods that can adapt to potentially complex shapes in $m(x)$ without pre-specifying its structure.

7.1 Why Nonparametric?

7.1.1 Flexibility

Nonparametric methods can capture **nonlinear relationships** and **complex patterns** in your data more effectively than many traditional parametric methods.

- **Adaptive Fit**: They rely on the data itself to determine the shape of the relationship, rather than forcing a specific equation like $Y = \beta_0 + \beta_1 x$ (linear) or a polynomial.
- **Local Structures**: Techniques like kernel smoothing or local regression focus on small neighborhoods around each observation, allowing the model to adjust dynamically to local variations.

When This Matters

- **Highly Variable Data**: If the data shows multiple peaks, sharp transitions, or other irregular patterns.
- **Exploratory Analysis**: When you're trying to uncover hidden structures or trends in a dataset without strong prior assumptions.

7.1.2 Fewer Assumptions

Parametric methods typically assume:

- A specific functional form (e.g., linear, quadratic).
- A specific error distribution (e.g., normal, Poisson).

Nonparametric methods, on the other hand, **relax these assumptions**, making them:

- **Robust to Misspecification**: Less risk of biased estimates due to incorrect modeling choices.
- **Flexible in Error Structure**: They can handle complex error distributions without explicitly modeling them.

When This Matters

- **Heterogeneous Populations**: In fields like ecology, genomics, or finance, where data might come from unknown mixtures of distributions.

- **Lack of Theoretical Guidance**: If theory does not suggest a strong functional form or distribution family.

7.1.3 Interpretability

Nonparametric models can still offer valuable insights:

- **Visual Interpretations**: Methods like kernel smoothing provide smooth curves that you can plot to see how Y changes with x.
- **Tree-Based Methods**: Random forests and gradient boosting (also nonparametric in nature) can be interpreted via **variable importance** measures or **partial dependence** plots, although they can be more complex than simple curves.

While you don't get simple coefficient estimates as in linear regression, you can still convey how certain predictors influence the response through plots or importance metrics.

7.1.4 Practical Considerations

7.1.4.1 When to Prefer Nonparametric

1. **Larger Sample Sizes:** Nonparametric methods often need more data because they let the data "speak" rather than relying on a fixed formula.
2. **Unknown or Complex Relationships:** If you suspect strong nonlinearity or have no strong theory about the functional form, nonparametric approaches provide the flexibility to discover patterns.
3. **Exploratory or Predictive Goals:** In data-driven or machine learning contexts, minimizing predictive error often takes precedence over strict parametric assumptions.

7.1.4.2 When to Be Cautious

1. **Small Sample Sizes:** Nonparametric methods can overfit and exhibit high variance if there isn't enough data to reliably estimate the relationship.
2. **Computational Cost:** Some nonparametric methods (e.g., kernel methods, large random forests) can be computationally heavier than parametric approaches like linear regression.
3. **Strong Theoretical Models:** If domain knowledge strongly suggests a specific parametric form, ignoring that might reduce clarity or conflict with established theory.

4. **Extrapolation:** Nonparametric models typically **do not** extrapolate well beyond the observed data range, because they rely heavily on local patterns.

7.1.5 Balancing Parametric and Nonparametric Approaches

In practice, it's not always an either/or decision. Consider (Table 7.1):

- **Semiparametric Models**: Combine parametric components (for known relationships or effects) with nonparametric components (for unknown parts).
- **Model Selection & Regularization**: Use techniques like cross-validation to choose bandwidths (kernel smoothing), number of knots (splines), or hyperparameters (tree depth) to avoid overfitting.
- **Diagnostic Tools**: Start with a simple parametric model, look at residual plots to identify patterns that might warrant a nonparametric approach.

Drawbacks and Challenges

1. **Curse of Dimensionality:** As the number of predictors p increases, nonparametric methods often require exponentially larger sample sizes to maintain accuracy. This phenomenon, known as the curse of dimensionality, leads to sparse data in high-dimensional spaces, making it harder to obtain reliable estimates.
2. **Choice of Hyperparameters:** Methods such as kernel smoothing and splines depend on hyperparameters like bandwidth or smoothing parameters, which must be carefully selected to balance bias and variance.
3. **Computational Complexity:** Nonparametric methods can be computationally intensive, especially with large datasets or in high-dimensional settings.

Table 7.1 Comparison of parametric and nonparametric statistical methods

Criterion	Parametric methods	Nonparametric methods
Assumptions	Requires strict assumptions (e.g., linearity, distribution form)	Minimal assumptions, flexible functional forms
Data requirements	Often works with smaller datasets if assumptions hold	Generally more data-hungry due to flexibility
Interpretability	Straightforward coefficients, easy to explain	Visual or plot-based insights; feature importance in trees
Complexity & overfitting	Less prone to overfitting if form is correct	Can overfit if not regularized (e.g., bandwidth selection)
Extrapolation	Can extrapolate if the assumed form is correct	Poor extrapolation outside the observed data range
Computational cost	Typically low to moderate (e.g., $O(n)$ to $O(n^2)$) depending on method	Can be higher (e.g., repeated local estimates or ensemble methods)

7.2 Basic Concepts in Nonparametric Estimation

7.2.1 *Bias-Variance Trade-Off*

For a given method of estimating $m(x)$, we denote the estimator as $\hat{m}(x)$. The **mean squared error** (MSE) at a point x is defined as:

$$\text{MSE}(x) = \mathbb{E}\left[\{\hat{m}(x) - m(x)\}^2\right].$$

This MSE can be decomposed into two key components: **bias** and **variance**:

$$\text{MSE}(x) = \left[\mathbb{E}[\hat{m}(x)] - m(x)\right]^2 + \text{Var}(\hat{m}(x)).$$

where:

- **Bias**: Measures the systematic error in the estimator:

$$\text{Bias}^2 = \left[\mathbb{E}[\hat{m}(x)] - m(x)\right]^2.$$

- **Variance**: Measures the variability of the estimator around its expected value:

$$\text{Var}(\hat{m}(x)) = \mathbb{E}\left[\{\hat{m}(x) - \mathbb{E}[\hat{m}(x)]\}^2\right].$$

Nonparametric methods often have **low bias** because they can adapt to a wide range of functions. However, this flexibility can lead to **high variance**, especially when the model captures noise rather than the underlying signal.

The **bandwidth** or **smoothing parameter** in nonparametric methods typically controls this trade-off:

- **Large bandwidth** $\Rightarrow$ smoother function $\Rightarrow$ higher bias, lower variance.
- **Small bandwidth** $\Rightarrow$ more wiggly function $\Rightarrow$ lower bias, higher variance.

Selecting an optimal bandwidth is critical, as it determines the balance between underfitting (high bias) and overfitting (high variance).

7.2.2 *Kernel Smoothing and Local Averages*

Many nonparametric regression estimators can be viewed as **weighted local averages** of the observed responses $\{Y_i\}$. In the univariate case, if x_i are observations of the predictor and y_i are the corresponding responses, the nonparametric estimator at a point x often takes the form:

$$\hat{m}(x) = \sum_{i=1}^{n} w_i(x)\, y_i,$$

where the weights $w_i(x)$ depend on the distance between x_i and x, and they satisfy:

$$\sum_{i=1}^{n} w_i(x) = 1.$$

We will see how this arises more concretely in kernel regression below.

7.3 Kernel Regression

7.3.1 Basic Setup

A **kernel function** $K(\cdot)$ is a non-negative, symmetric function whose integral (or sum, in a discrete setting) equals 1. In nonparametric statistics—such as kernel density estimation or local regression—kernels serve as **weighting** mechanisms, assigning higher weights to points closer to the target location and lower weights to points farther away. Specifically, a valid kernel function must satisfy:

1. **Non-negativity**:

$$K(u) \geq 0 \quad \text{for all } u.$$

2. **Normalization**:

$$\int_{-\infty}^{\infty} K(u)\, du = 1.$$

3. **Symmetry**:

$$K(u) = K(-u) \quad \text{for all } u.$$

In practice, the **bandwidth** (sometimes called the smoothing parameter) used alongside a kernel usually has a greater impact on the quality of the estimate than the particular form of the kernel. However, choosing a suitable kernel can still influence computational efficiency and the smoothness of the resulting estimates.

7.3.1.1 Common Kernel Functions

A kernel function essentially measures *proximity*, assigning higher weights to observations x_i that are close to the target point x, and smaller weights to those farther away.

1. Gaussian Kernel

$$K(u) = \frac{1}{\sqrt{2\pi}} e^{-\frac{u^2}{2}}.$$

- **Shape**: Bell-shaped and infinite support (i.e., $K(u)$ is technically nonzero for all $u \in (-\infty, \infty)$), though values decay rapidly as $|u|$ grows.
- **Usage**: Due to its smoothness and mathematical convenience (especially in closed-form expressions and asymptotic analysis), it is the most widely used kernel in both density estimation and regression smoothing.
- **Properties**: The Gaussian kernel minimizes mean square error in many asymptotic scenarios, making it a common "default choice."

2. Epanechnikov Kernel

$$K(u) = \begin{cases} \frac{3}{4}(1 - u^2) & \text{if } |u| \leq 1, \\ 0 & \text{otherwise.} \end{cases}$$

- **Shape**: Parabolic (inverted) on $[-1, 1]$, dropping to 0 at $|u| = 1$.
- **Usage**: Known for being **optimal** in a minimax sense for certain classes of problems, and it is frequently preferred when compact support (zero weights outside $|u| \leq 1$) is desirable.
- **Efficiency**: Because it is only supported on a finite interval, computations often involve fewer points (those outside $|u| \leq 1$ have zero weight), which can be computationally more efficient in large datasets.

3. Uniform (or Rectangular) Kernel

$$K(u) = \begin{cases} \frac{1}{2} & \text{if } |u| \leq 1, \\ 0 & \text{otherwise.} \end{cases}$$

- **Shape**: A simple "flat top" distribution on $[-1, 1]$.
- **Usage**: Sometimes used for its simplicity. In certain methods (e.g., a "moving average" approach), the uniform kernel equates to giving all points within a fixed window the same weight.
- **Drawback**: Lacks smoothness at the boundaries u=1|u|=1 u=1, and it can introduce sharper transitions in estimates compared to smoother kernels.

4. Triangular Kernel

$$K(u) = \begin{cases} 1 - |u| & \text{if } |u| \le 1, \\ 0 & \text{otherwise.} \end{cases}$$

- **Shape**: Forms a triangle with a peak at $u = 0$ and linearly descends to 0 at $|u| = 1$.
- **Usage**: Provides a continuous but piecewise-linear alternative to the uniform kernel; places relatively more weight near the center compared to the uniform kernel.

5. Biweight (or Quartic) Kernel

$$K(u) = \begin{cases} \frac{15}{16}\left(1 - u^2\right)^2 & \text{if } |u| \le 1, \\ 0 & \text{otherwise.} \end{cases}$$

- **Shape**: Smooth and "bump-shaped," similar to the Epanechnikov but with a steeper drop-off near $|u| = 1$.
- **Usage**: Popular when a smoother, polynomial-based kernel with compact support is desired.

6. Cosine Kernel

$$K(u) = \begin{cases} \frac{\pi}{4}\cos\left(\frac{\pi}{2}u\right) & \text{if } |u| \le 1, \\ 0 & \text{otherwise.} \end{cases}$$

- **Shape**: A single "arch" of a cosine wave on the interval $[-1, 1]$.
- **Usage**: Used less frequently but can be appealing for certain smoothness criteria or specific signal processing contexts.

Below is a comparison of widely used kernel functions, their functional forms, support, and main characteristics (Table 7.2).

7.3.1.2 Additional Details and Usage Notes

1. **Smoothness and Differentiability**

Table 7.2 Comparison of common kernel functions for nonparametric estimation

Kernel	Formula	Support	Key characteristics
Gaussian	$K(u) = \frac{1}{\sqrt{2\pi}} e^{-\frac{u^2}{2}}$	$u \in (-\infty, \infty)$	**Smooth, bell-shaped** Nonzero for all u, but decays quickly Often the default choice due to favorable analytical properties
Epanechnikov	$K(u) = \begin{cases} \frac{3}{4}(1 - u^2) & \lvert u \rvert \leq 1 \\ 0 & \text{otherwise} \end{cases}$	$[-1, 1]$	**Parabolic shape** Compact support Minimizes mean integrated squared error in certain theoretical contexts
Uniform	$K(u) = \begin{cases} \frac{1}{2} & \lvert u \rvert \leq 1 \\ 0 & \text{otherwise} \end{cases}$	$[-1, 1]$	**Flat (rectangular) shape** Equal weight for all points within $[-1, 1]$ Sharp boundary can lead to less smooth estimates
Triangular	$K(u) = \begin{cases} 1 - \lvert u \rvert & \lvert u \rvert \leq 1 \\ 0 & \text{otherwise} \end{cases}$	$[-1, 1]$	**Linear decrease** from the center $u = 0$ to 0 at $\lvert u \rvert = 1$ Compact support A bit smoother than the uniform kernel
Biweight (quartic)	$K(u) = \begin{cases} \frac{15}{16}(1 - u^2)^2 & \lvert u \rvert \leq 1 \\ 0 & \text{otherwise} \end{cases}$	$[-1, 1]$	**Polynomial shape,** smooth Compact support Often used for its relatively smooth taper near the boundaries
Cosine	$K(u) = \begin{cases} \frac{\pi}{4} \cos\left(\frac{\pi}{2} u\right) & \lvert u \rvert \leq 1 \\ 0 & \text{otherwise} \end{cases}$	$[-1, 1]$	**Single arch** of a cosine wave Compact support Less commonly used, but still mathematically straightforward

- Kernels with **infinite support** (like the Gaussian) can yield very smooth estimates but require summing over (practically) all data points.
- Kernels with **compact support** (like Epanechnikov, biweight, triangular, etc.) go to zero outside a fixed interval. This can make computations more efficient since only data within a certain range of the target point matter.

2. **Choice of Kernel vs. Choice of Bandwidth**

- While the kernel shape does have some effect on the estimator's smoothness, **the choice of bandwidth (sometimes denoted** h) is typically more critical. If h is too large, the estimate can be excessively smooth (high bias). If h is too small, the estimate can exhibit high variance or appear "noisy."

3. **Local Weighting Principle**
   - At a target location x, a kernel function $K\left(\frac{x-x_i}{h}\right)$ down-weights data points (x_i) that are farther from x. Nearer points have larger kernel values, hence exert greater influence on the local estimate.
4. **Interpretation in Density Estimation**
   - In **kernel density estimation**, each data point contributes a small "bump" (shaped by the kernel) to the overall density. Summing or integrating these bumps yields a continuous estimate of the underlying density function, in contrast to discrete histograms.

7.3.2 Nadaraya-Watson Kernel Estimator

The most widely used kernel-based regression estimator is the Nadaraya-Watson estimator (Nadaraya, 1964; Watson, 1964), defined as:

$$\hat{m}_h(x) = \frac{\sum_{i=1}^{n} K\left(\frac{x-x_i}{h}\right) y_i}{\sum_{i=1}^{n} K\left(\frac{x-x_i}{h}\right)},$$

where $h > 0$ is the **bandwidth parameter**. Intuitively, this formula computes a weighted average of the observed y_i values, with weights determined by the kernel function applied to the scaled distance between x and each x_i.

Interpretation

- When $|x - x_i|$ is **small** (i.e., x_i is close to x), the kernel value $K\left(\frac{x-x_i}{h}\right)$ is **large**, giving more weight to y_i.
- When $|x - x_i|$ is **large**, the kernel value becomes **small** (or even zero for compactly supported kernels like the Epanechnikov), reducing the influence of y_i on $\hat{m}_h(x)$.

Thus, observations near x have a larger impact on the estimated value $\hat{m}_h(x)$ than distant ones.

7.3.2.1 Weights Representation

We can define the **normalized weights**:

$$w_i(x) = \frac{K\left(\frac{x-x_i}{h}\right)}{\sum_{j=1}^{n} K\left(\frac{x-x_j}{h}\right)},$$

so that the estimator can be rewritten as:

$$\hat{m}_h(x) = \sum_{i=1}^{n} w_i(x) y_i,$$

where $\sum_{i=1}^{n} w_i(x) = 1$ for any x. Notice that $0 \leq w_i(x) \leq 1$ for all i.

7.3.3 Priestley-Chao Kernel Estimator

The **Priestley-Chao kernel estimator** (Priestley & Chao, 1972) is an early kernel-based regression estimator designed to estimate the regression function $m(x)$ from observed data $\{(x_i, y_i)\}_{i=1}^{n}$. Unlike the Nadaraya-Watson estimator, which uses pointwise kernel weighting, the Priestley–Chao estimator incorporates **differences in the predictor variable** to approximate integrals more accurately.

The estimator is defined as:

$$\hat{m}_h(x) = \frac{1}{h} \sum_{i=1}^{n-1} K\left(\frac{x - x_i}{h}\right) \cdot (x_{i+1} - x_i) \cdot y_i,$$

where:

- $K(\cdot)$ is a kernel function,
- $h > 0$ is the **bandwidth parameter**,
- $(x_{i+1} - x_i)$ represents the **spacing between consecutive observations**.

7.3.3.1 Interpretation

- The estimator can be viewed as a **Riemann sum approximation** of an integral, where the kernel-weighted y_i values are scaled by the spacing $(x_{i+1} - x_i)$.
- Observations where x_i is **close to** x receive more weight due to the kernel function.
- The inclusion of $(x_{i+1} - x_i)$ accounts for **non-uniform spacing** in the data, making the estimator more accurate when the predictor values are irregularly spaced.

This estimator is particularly useful when the design points $\{x_i\}$ are **unevenly distributed**.

7.3.3.2 Weights Representation

We can express the estimator as a **weighted sum** of the observed responses y_i:

$$\hat{m}_h(x) = \sum_{i=1}^{n-1} w_i(x)\, y_i,$$

where the weights are defined as:

$$w_i(x) = \frac{1}{h} \cdot K\left(\frac{x - x_i}{h}\right) \cdot (x_{i+1} - x_i).$$

Properties of the weights

- **Non-negativity:** If $K(u) \geq 0$, then $w_i(x) \geq 0$.
- **Adaptation to spacing:** Larger gaps $(x_{i+1} - x_i)$ increase the corresponding weight.
- Unlike Nadaraya-Watson, the weights **do not sum to 1**, as they approximate an integral rather than a normalized average.

7.3.4 Gasser-Müller Kernel Estimator

The **Gasser-Müller kernel estimator** (Gasser & Müller, 1979) improves upon the Priestley-Chao estimator by using a **cumulative kernel function** to smooth over the predictor space. This estimator is particularly effective for **irregularly spaced data** and aims to reduce bias at the boundaries.

The estimator is defined as:

$$\hat{m}_h(x) = \frac{1}{h} \sum_{i=1}^{n-1} \left[K^*\left(\frac{x - x_i}{h}\right) - K^*\left(\frac{x - x_{i+1}}{h}\right) \right] \cdot y_i,$$

where:

- $K^*(u) = \int_{-\infty}^{u} K(v)\, dv$ is the **cumulative distribution function (CDF)** of the kernel K,
- $h > 0$ is the **bandwidth parameter**.

7.3.4.1 Interpretation

- The estimator computes the **difference of cumulative kernel functions** at two consecutive design points, effectively assigning weight to the interval between x_i and x_{i+1}.
- Observations contribute more to $\hat{m}_h(x)$ when x lies **between** x_i and x_{i+1}, with the contribution decreasing as the distance from x increases.
- This method smooths over **intervals** rather than just at points, reducing bias near the boundaries and improving performance with unevenly spaced data.

7.3.4.2 Weights Representation

The Gasser-Müller estimator can also be expressed as a **weighted sum**:

$$\hat{m}_h(x) = \sum_{i=1}^{n-1} w_i(x)\, y_i,$$

where the weights are:

$$w_i(x) = \frac{1}{h}\left[K^*\left(\frac{x - x_i}{h}\right) - K^*\left(\frac{x - x_{i+1}}{h}\right)\right].$$

Properties of the weights

- **Non-negativity:** The weights are non-negative if K^* is non-decreasing (which holds if K is non-negative).
- **Adaptation to spacing:** The weights account for the spacing between x_i and x_{i+1}.
- Similar to the Priestley-Chao estimator, the weights **do not sum to 1** because the estimator approximates an integral rather than a normalized sum.

7.3.5 Comparison of Kernel-Based Estimators

Table 7.3 shows differences between kernel-based estimators.

7.3.6 Bandwidth Selection

The choice of **bandwidth** h is crucial because it controls the trade-off between bias and variance:

Table 7.3 Comparison of kernel-based estimators

Estimator	Formula	Key feature	Weights sum to 1?
Nadaraya-Watson	$\hat{m}_h(x) = \dfrac{\sum K\left(\frac{x-x_i}{h}\right) y_i}{\sum K\left(\frac{x-x_i}{h}\right)}$	Weighted average of y_i	Yes
Priestley-Chao	$\hat{m}_h(x) = \frac{1}{h}\sum K\left(\frac{x-x_i}{h}\right)(x_{i+1} - x_i)y_i$	Incorporates data spacing	No
Gasser-Müller	$\hat{m}_h(x) = \frac{1}{h}\sum \left[K^*\left(\frac{x-x_i}{h}\right) - K^*\left(\frac{x-x_{i+1}}{h}\right)\right] y_i$	Uses cumulative kernel differences	No

- If h is **too large**, the estimator becomes **overly smooth**, incorporating too many distant data points. This leads to **high bias** but **low variance**.
- If h is **too small**, the estimator becomes **noisy** and sensitive to fluctuations in the data, resulting in **low bias** but **high variance**.

7.3.6.1 Mean Squared Error and Optimal Bandwidth

To analyze the performance of kernel estimators, we often examine the **mean integrated squared error** (MISE):

$$\text{MISE}(\hat{m}_h) = \mathbb{E}\left[\int \left\{\hat{m}_h(x) - m(x)\right\}^2 dx\right].$$

As $n \to \infty$, under smoothness assumptions on $m(x)$ and regularity conditions on the kernel K, the MISE has the following asymptotic expansion:

$$\text{MISE}(\hat{m}_h) \approx \frac{R(K)}{nh}\sigma^2 + \frac{1}{4}\mu_2^2(K)\,h^4 \int \left\{m''(x)\right\}^2 dx,$$

where:

- $R(K) = \int_{-\infty}^{\infty} K(u)^2 du$ measures the **roughness** of the kernel.
- $\mu_2(K) = \int_{-\infty}^{\infty} u^2 K(u) du$ is the **second moment** of the kernel (related to its spread).
- σ^2 is the **variance of the noise**, assuming $\text{Var}(\varepsilon \mid X = x) = \sigma^2$.
- $m''(x)$ is the **second derivative** of the true regression function $m(x)$.

To find the **asymptotically optimal bandwidth**, we differentiate the MISE with respect to h, set the derivative to zero, and solve for h:

$$h_{\text{opt}} = \left(\frac{R(K)\,\sigma^2}{\mu_2^2(K)\int \{m''(x)\}^2\,dx} \cdot \frac{1}{n}\right)^{1/5}.$$

In practice, σ^2 and $\int \{m''(x)\}^2 dx$ are unknown and must be estimated from data. A common data-driven approach is cross-validation.

7.3.6.2 Cross-Validation

The **leave-one-out cross-validation** (LOOCV) method is widely used for bandwidth selection:

1. For each $i = 1, \ldots, n$, fit the kernel estimator $\hat{m}_{h,-i}(x)$ using all data **except** the i-th observation (x_i, y_i).
2. Compute the squared prediction error for the left-out point: $(y_i - \hat{m}_{h,-i}(x_i))^2$.

3. Average these errors across all observations:

$$\mathrm{CV}(h) = \frac{1}{n}\sum_{i=1}^{n}\left\{y_i - \hat{m}_{h,-i}(x_i)\right\}^2.$$

The bandwidth h that minimizes $\mathrm{CV}(h)$ is selected as the optimal bandwidth.

7.3.7 Asymptotic Properties

For the Nadaraya-Watson estimator, under regularity conditions and assuming $h \to 0$ as $n \to \infty$ (but not too fast), we have:

- **Consistency:**

$$\hat{m}_h(x) \xrightarrow{p} m(x),$$

meaning the estimator converges in probability to the true regression function.
- **Rate of Convergence:** The mean squared error (MSE) decreases at the rate:

$$\mathrm{MSE}(\hat{m}_h(x)) = O\left(n^{-4/5}\right)$$

in the one-dimensional case. This rate results from balancing the variance term $(O(1/(nh)))$ and the squared bias term $(O(h^4))$.

7.3.8 Derivation of the Nadaraya-Watson Estimator

The Nadaraya-Watson estimator can be derived from a **density-based perspective**:

1. By the definition of conditional expectation:

$$m(x) = \mathbb{E}[Y \mid X = x] = \frac{\int y\, f_{X,Y}(x, y)\, dy}{f_X(x)},$$

where $f_{X,Y}(x, y)$ is the joint density of (X, Y) and $f_X(x)$ is the marginal density of X.
2. Estimate $f_X(x)$ using a kernel density estimator:

$$\hat{f}_X(x) = \frac{1}{n}\sum_{i=1}^{n}\frac{1}{h}K\left(\frac{x - x_i}{h}\right).$$

3. Estimate the joint density $f_{X,Y}(x, y)$:

$$\hat{f}_{X,Y}(x, y) = \frac{1}{n} \sum_{i=1}^{n} \frac{1}{h} K\left(\frac{x - x_i}{h}\right) \delta_{y_i}(y),$$

where $\delta_{y_i}(y)$ is the Dirac delta function (a point mass at y_i).

4. The kernel regression estimator becomes:

$$\hat{m}_h(x) = \frac{\int y \, \hat{f}_{X,Y}(x, y) \, dy}{\hat{f}_X(x)} = \frac{\sum_{i=1}^{n} K\left(\frac{x-x_i}{h}\right) y_i}{\sum_{i=1}^{n} K\left(\frac{x-x_i}{h}\right)},$$

which is exactly the Nadaraya-Watson estimator.

```
# Load necessary libraries
library(ggplot2)
library(gridExtra)

# 1. Simulate Data
set.seed(123)

# Generate predictor x and response y
n <- 100
x <-
    sort(runif(n, 0, 10))  # Sorted for Priestley--Chao and
    ↪  Gasser--Müller
true_function <-
    function(x)
        sin(x) + 0.5 * cos(2 * x)  # True regression function

# Add Gaussian noise
y <-
    true_function(x) + rnorm(n, sd = 0.3)
```

```
# Visualization of the data
ggplot(data.frame(x, y), aes(x, y)) +
    geom_point(color = "darkblue") +
    geom_line(aes(y = true_function(x)),
              color = "red",
```

```
            linetype = "dashed") +
    labs(title = "Simulated Data with True Regression
    ↪  Function",
         x = "x", y = "y") +
    theme_minimal()
```

```
# Gaussian Kernel Function
gaussian_kernel <- function(u) {
    (1 / sqrt(2 * pi)) * exp(-0.5 * u ^ 2)
}

# Epanechnikov Kernel Function
epanechnikov_kernel <- function(u) {
    ifelse(abs(u) <= 1, 0.75 * (1 - u ^ 2), 0)
}

# Cumulative Kernel for Gasser--Müller (CDF of Gaussian
↪  Kernel)
gaussian_cdf_kernel <- function(u) {
    pnorm(u, mean = 0, sd = 1)
}
```

```
# Nadaraya-Watson Estimator
nadaraya_watson <-
    function(x_eval, x, y, h, kernel = gaussian_kernel) {
        sapply(x_eval, function(x0) {
            weights <- kernel((x0 - x) / h)
            sum(weights * y) / sum(weights)
        })
    }

# Bandwidth Selection (fixed for simplicity)
h_nw <- 0.5  # Bandwidth for Nadaraya--Watson

# Apply Nadaraya--Watson Estimator
x_grid <- seq(0, 10, length.out = 200)
nw_estimate <- nadaraya_watson(x_grid, x, y, h_nw)
```

```
# Plot Nadaraya--Watson Estimate
ggplot() +
    geom_point(aes(x, y), color = "darkblue", alpha = 0.6) +
    geom_line(aes(x_grid, nw_estimate),
              color = "green",
              linewidth = 1.2) +
    geom_line(aes(x_grid, true_function(x_grid)),
              color = "red",
              linetype = "dashed") +
    labs(
        title = "Nadaraya--Watson Kernel Estimator",
        subtitle = paste("Bandwidth (h) =", h_nw),
        x = "x",
        y = "Estimated m(x)"
    ) +
    theme_minimal()
```

- The **green curve** is the Nadaraya-Watson estimate (Fig. 7.2).
- The **dashed red line** is the true regression function (Fig. 7.1).
- The **blue dots** are the observed noisy data.
- The estimator **smooths the data**, assigning more weight to points close to each evaluation point based on the Gaussian kernel.

Fig. 7.1 Simulated data with true regression function

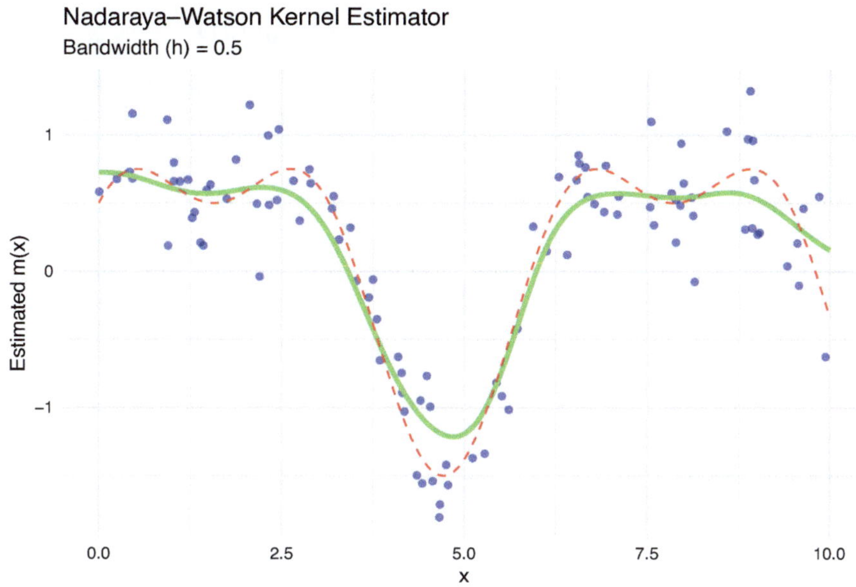

Fig. 7.2 Nadaraya–Watson kernel estimator

```
# Priestley--Chao Estimator
priestley_chao <-
    function(x_eval, x, y, h, kernel = gaussian_kernel) {
        sapply(x_eval, function(x0) {
            weights <- kernel((x0 - x[-length(x)]) / h) *
↪   diff(x)
            sum(weights * y[-length(y)]) / h
        })
    }

# Apply Priestley--Chao Estimator
h_pc <- 0.5
pc_estimate <- priestley_chao(x_grid, x, y, h_pc)
```

```
# Plot Priestley--Chao Estimate
ggplot() +
    geom_point(aes(x, y), color = "darkblue", alpha = 0.6) +
    geom_line(aes(x_grid, pc_estimate),
              color = "orange",
              size = 1.2) +
```

```
    geom_line(aes(x_grid, true_function(x_grid)),
              color = "red",
              linetype = "dashed") +
    labs(
        title = "Priestley-Chao Kernel Estimator",
        subtitle = paste("Bandwidth (h) =", h_pc),
        x = "x",
        y = "Estimated m(x)"
    ) +
    theme_minimal()
#> Warning: Using `size` aesthetic for lines was deprecated in
↪  ggplot2
#> 3.4.0.
#> i Please use `linewidth` instead.
#> This warning is displayed once every 8 hours.
#> Call `lifecycle::last_lifecycle_warnings()` to see where
#> this warning was generated.
```

- The **orange curve** is the Priestley-Chao estimate (Fig. 7.3).
- This estimator incorporates the **spacing between consecutive data points** (`diff(x)`), making it more sensitive to **non-uniform data spacing**.
- It performs similarly to Nadaraya-Watson when data are evenly spaced.

```
# Gasser--Müller Estimator
gasser_mueller <-
    function(x_eval, x, y, h, cdf_kernel =
     ↪  gaussian_cdf_kernel) {
        sapply(x_eval, function(x0) {
            weights <-
                (cdf_kernel((x0 - x[-length(x)]) / h) -
↪   cdf_kernel((x0 - x[-1]) / h))
            sum(weights * y[-length(y)]) / h
        })
    }

# Apply Gasser--Müller Estimator
h_gm <- 0.5
gm_estimate <- gasser_mueller(x_grid, x, y, h_gm)
```

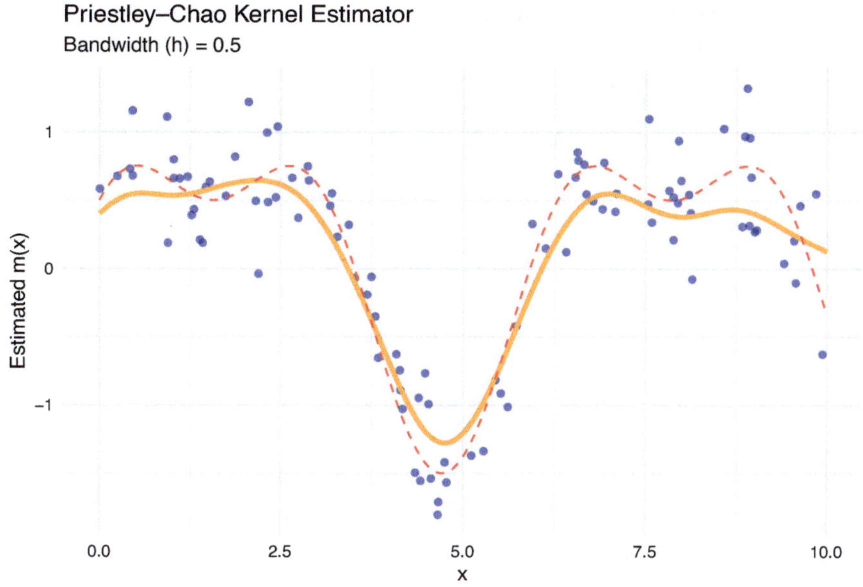

Fig. 7.3 Priestley-Chao kernel estimator

```r
# Plot Gasser--Müller Estimate
ggplot() +
    geom_point(aes(x, y), color = "darkblue", alpha = 0.6) +
    geom_line(aes(x_grid, gm_estimate),
              color = "purple",
              size = 1.2) +
    geom_line(aes(x_grid, true_function(x_grid)),
              color = "red",
              linetype = "dashed") +
    labs(
        title = "Gasser--Müller Kernel Estimator",
        subtitle = paste("Bandwidth (h) =", h_gm),
        x = "x",
        y = "Estimated m(x)"
    ) +
    theme_minimal()
```

- The **purple curve** is the Gasser–Müller estimate (Fig. 7.4).
- This estimator uses **cumulative kernel functions** to handle **irregular data spacing** and **reduce boundary bias**.
- It performs well when data are unevenly distributed.

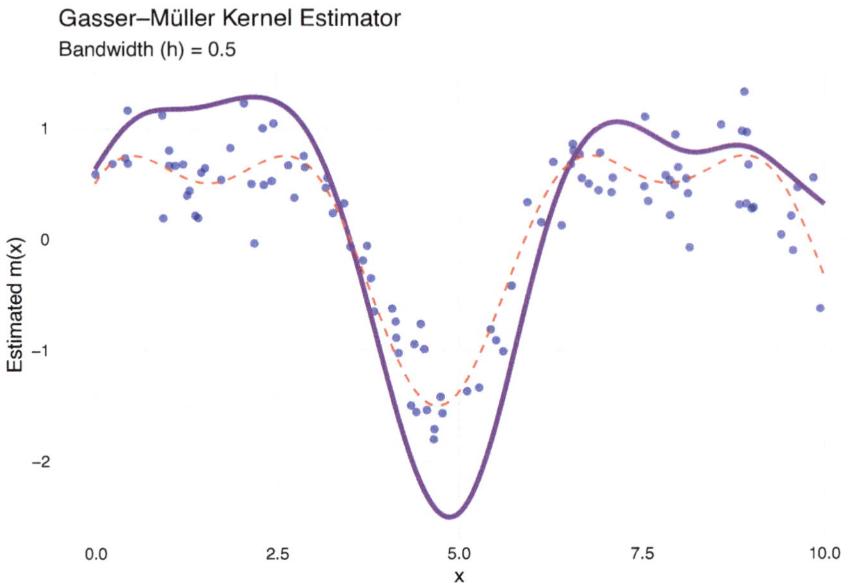

Fig. 7.4 Gasser-Müller kernel estimator

```
# Combine all estimates for comparison
estimates_df <- data.frame(
    x = x_grid,
    Nadaraya_Watson = nw_estimate,
    Priestley_Chao = pc_estimate,
    Gasser_Mueller = gm_estimate,
    True_Function = true_function(x_grid)
)
```

```
# Plot all estimators together
ggplot() +
  geom_point(aes(x, y), color = "gray60", alpha = 0.5) +
  geom_line(aes(x, Nadaraya_Watson, color =
  ↪  "Nadaraya--Watson"),
            data = estimates_df, size = 1.1) +
  geom_line(aes(x, Priestley_Chao, color = "Priestley--Chao"),
            data = estimates_df, size = 1.1) +
  geom_line(aes(x, Gasser_Mueller, color = "Gasser--Müller"),
```

```
            data = estimates_df, size = 1.1) +
  geom_line(aes(x, True_Function, color = "True Function"),
            data = estimates_df, linetype = "dashed", size =
            ↪ 1) +
  scale_color_manual(
    name = "Estimator",
    values = c("Nadaraya--Watson" = "green",
               "Priestley--Chao"  = "orange",
               "Gasser--Müller"   = "purple",
               "True Function"    = "red")
  ) +
  labs(
    title = "Comparison of Kernel-Based Regression
    ↪ Estimators",
    x = "x",
    y = "Estimated m(x)"
  ) +
  theme_minimal()
```

- **All estimators** approximate the true function well when the bandwidth is appropriately chosen (Fig. 7.5).
- The Nadaraya-Watson estimator is sensitive to bandwidth selection and assumes uniform data spacing.
- The Priestley-Chao estimator accounts for **data spacing**, making it more flexible with uneven data.
- The Gasser-Müller estimator reduces **boundary bias** and handles **irregular data** effectively.

```
# Cross-validation for bandwidth selection (for
↪ Nadaraya--Watson)
cv_bandwidth <- function(h, x, y, kernel = gaussian_kernel) {
  n <- length(y)
  cv_error <- 0
  for (i in 1:n) {
    x_train <- x[-i]
    y_train <- y[-i]
    y_pred <- nadaraya_watson(x[i], x_train, y_train, h,
↪ kernel)
    cv_error <- cv_error + (y[i] - y_pred)^2
  }
  return(cv_error / n)
}
```

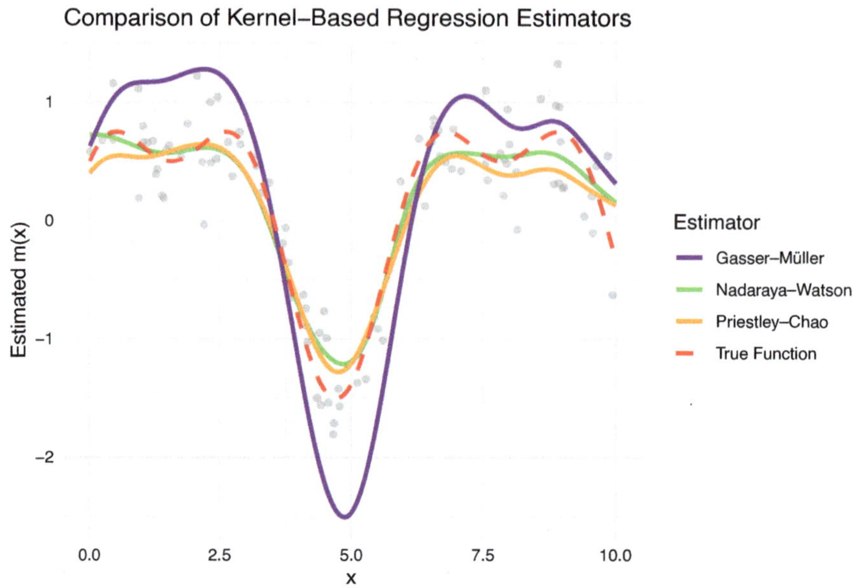

Fig. 7.5 Comparison of kernel-based regression estimators

```
# Optimize bandwidth
bandwidths <- seq(0.1, 2, by = 0.1)
cv_errors <- sapply(bandwidths, cv_bandwidth, x = x, y = y)

# Optimal bandwidth
optimal_h <- bandwidths[which.min(cv_errors)]
optimal_h
#> [1] 0.3
```

```
# Plot CV errors
ggplot(data.frame(bandwidths, cv_errors), aes(bandwidths,
↪  cv_errors)) +
    geom_line(color = "blue") +
    geom_point(aes(x = optimal_h, y = min(cv_errors)),
               color = "red",
               size = 3) +
    labs(title = "Cross-Validation for Bandwidth Selection",
         x = "Bandwidth (h)", y = "CV Error") +
    theme_minimal()
```

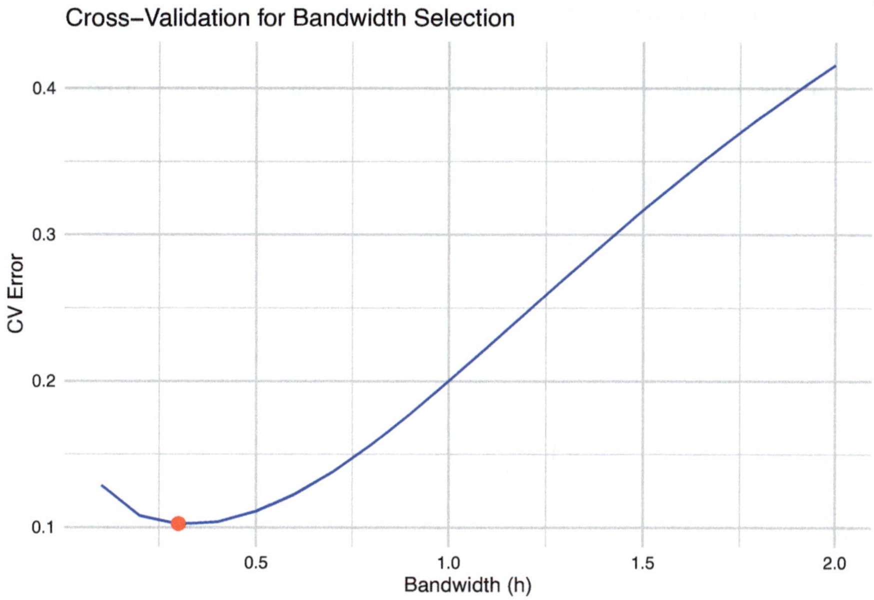

Fig. 7.6 Cross-validation for bandwidth selection

```
#> Warning in geom_point(aes(x = optimal_h, y =
↪   min(cv_errors)), color = "red", : All aesthetics have
↪   length 1, but the data has 20 rows.
#> i Please consider using `annotate()` or provide this layer
#>   with data containing a single row.
```

- The **red point** indicates the **optimal bandwidth** that minimizes the cross-validation error (Fig. 7.6).
- Selecting the **right bandwidth** is critical, as it balances **bias and variance** in the estimator.

7.4 Local Polynomial Regression

While the Nadaraya-Watson estimator is effectively a **local constant** estimator (it approximates $m(x)$ by a constant in a small neighborhood of x), **local polynomial regression** extends this idea by fitting a **local polynomial** around each point x. The advantage of local polynomials is that they can better handle boundary bias and can capture local curvature more effectively.

7.4.1 Local Polynomial Fitting

A **local polynomial regression** of degree p at point x estimates a polynomial function:

$$m_x(t) = \beta_0 + \beta_1(t - x) + \beta_2(t - x)^2 + \cdots + \beta_p(t - x)^p$$

that best fits the data $\{(x_i, y_i)\}$ within a neighborhood of x, **weighted by a kernel**. Specifically, we solve:

$$(\hat{\beta}_0, \hat{\beta}_1, \ldots, \hat{\beta}_p) = \arg\min_{\beta_0,\ldots,\beta_p} \sum_{i=1}^{n} \left[y_i - \left\{\beta_0 + \beta_1(x_i - x) + \cdots + \beta_p(x_i - x)^p\right\}\right]^2 \times K\left(\frac{x_i - x}{h}\right).$$

We then estimate:

$$\hat{m}(x) = \hat{\beta}_0,$$

because β_0 is the constant term of the local polynomial expansion around x, which represents the estimated value at that point.

Why center the polynomial at x rather than 0?
Centering at x ensures that the fitted polynomial provides the best approximation *around* x. This is conceptually similar to a Taylor expansion, where local approximations are most accurate near the point of expansion.

7.4.2 Mathematical Form of the Solution

Let $\mathbf{X}_x$ be the design matrix for the local polynomial expansion at point x. For a polynomial of degree p, each row i of $\mathbf{X}_x$ is:

$$\left(1, x_i - x, (x_i - x)^2, \ldots, (x_i - x)^p\right).$$

Define $\mathbf{W}_x$ as the diagonal matrix with entries:

$$(\mathbf{W}_x)_{ii} = K\left(\frac{x_i - x}{h}\right),$$

representing the kernel weights. The parameter vector $\boldsymbol{\beta} = (\beta_0, \beta_1, \ldots, \beta_p)^T$ is estimated via weighted least squares:

$$\hat{\boldsymbol{\beta}}(x) = \left(\mathbf{X}_x^T \mathbf{W}_x \mathbf{X}_x\right)^{-1} \mathbf{X}_x^T \mathbf{W}_x \mathbf{y},$$

where $\mathbf{y} = (y_1, y_2, \ldots, y_n)^T$. The local polynomial estimator of $m(x)$ is given by:

$$\hat{m}(x) = \hat{\beta}_0(x).$$

Alternatively, we can express this concisely using a selection vector:

$$\hat{m}(x) = \mathbf{e}_1^T \left(\mathbf{X}_x^T \mathbf{W}_x \mathbf{X}_x\right)^{-1} \mathbf{X}_x^T \mathbf{W}_x \mathbf{y},$$

where $\mathbf{e}_1 = (1, 0, \ldots, 0)^T$ picks out the intercept term.

7.4.3 Bias, Variance, and Asymptotics

Local polynomial estimators have well-characterized bias and variance properties, which depend on the polynomial degree p, the bandwidth h, and the smoothness of the true regression function $m(x)$.

7.4.3.1 Bias

- The leading bias term is proportional to h^{p+1} and involves the $(p+1)$-th derivative of $m(x)$:

$$\text{Bias}\left[\hat{m}(x)\right] \approx \frac{h^{p+1}}{(p+1)!} m^{(p+1)}(x) \cdot B(K, p),$$

  where $B(K, p)$ is a constant depending on the kernel and the polynomial degree.
- For local linear regression ($p = 1$), the bias is of order $O(h^2)$, while for **local quadratic regression** ($p = 2$), it's of order $O(h^3)$.

7.4.3.2 Variance

- The variance is approximately:

$$\text{Var}\left[\hat{m}(x)\right] \approx \frac{\sigma^2}{nh} \cdot V(K, p),$$

  where σ^2 is the error variance and $V(K, p)$ is another kernel-dependent constant.
- The variance decreases with larger n and larger h, but increasing h also increases bias, illustrating the classic **bias-variance trade-off**.

7.4.3.3 Boundary Issues

One of the key advantages of local polynomial regression—especially local linear regression—is its ability to reduce **boundary bias**, which is a major issue for the Nadaraya-Watson estimator. This is because the linear term allows the fit to adjust for slope changes near the boundaries, where the kernel becomes asymmetric due to fewer data points on one side.

7.4.4 Special Case: Local Linear Regression

Local linear regression (often called a local polynomial fit of degree 1) is particularly popular because:

- It mitigates boundary bias effectively, making it superior to Nadaraya-Watson near the edges of the data (Table 7.4).
- It remains computationally simple yet provides better performance than local-constant (degree 0) models.
- It is robust to heteroscedasticity, as it adapts to varying data densities.

The resulting estimator for $\hat{m}(x)$ simplifies to:

$$\hat{m}(x) = \frac{S_2(x)\, S_{0y}(x) - S_1(x)\, S_{1y}(x)}{S_0(x)\, S_2(x) - S_1^2(x)},$$

where

$$S_k(x) = \sum_{i=1}^{n} (x_i - x)^k\, K\left(\tfrac{x_i - x}{h}\right)$$

$$S_{ky}(x) = \sum_{i=1}^{n} (x_i - x)^k\, y_i\, K\left(\tfrac{x_i - x}{h}\right).$$

Table 7.4 Comparison: Nadaraya-Watson vs. local polynomial regression

Aspect	Nadaraya-Watson (local constant)	Local polynomial regression
Bias at boundaries	High	Reduced (especially for $p = 1$)
Flexibility	Limited (constant fit)	Captures local trends (linear/quadratic)
Complexity	Simpler	Slightly more complex (matrix operations)
Robustness to heteroscedasticity	Lower	Higher (adapts better to varying densities)

To see why the local linear fit helps reduce bias, consider approximating m around the point x via a first-order Taylor expansion:

$$m(t) \approx m(x) + m'(x)\,(t - x).$$

When we perform local linear regression, we solve the weighted least squares problem

$$\min_{\beta_0,\beta_1} \sum_{i=1}^{n} [y_i - \{\beta_0 + \beta_1(x_i - x)\}]^2 \; K\left(\tfrac{x_i - x}{h}\right).$$

If we assume $y_i = m(x_i) + \varepsilon_i$, then expanding $m(x_i)$ in a Taylor series around x gives:

$$m(x_i) = m(x) + m'(x)\,(x_i - x) + \tfrac{1}{2}\, m''(x)\,(x_i - x)^2 + \cdots.$$

For x_i close to x, the higher-order terms may be small, but they contribute to the bias if we truncate at the linear term.

Let us denote:

$$S_0(x) = \sum_{i=1}^{n} K\left(\frac{x_i - x}{h}\right),$$

$$S_1(x) = \sum_{i=1}^{n} (x_i - x)\, K\left(\frac{x_i - x}{h}\right),$$

$$S_2(x) = \sum_{i=1}^{n} (x_i - x)^2\, K\left(\frac{x_i - x}{h}\right).$$

Similarly, define

$$\sum_{i=1}^{n} y_i\, K\left(\tfrac{x_i - x}{h}\right) \quad \text{and} \quad \sum_{i=1}^{n} (x_i - x)\, y_i\, K\left(\tfrac{x_i - x}{h}\right)$$

for the right-hand sides. The estimated coefficients $\hat{\beta}_0$ and $\hat{\beta}_1$ are found by solving:

$$\begin{pmatrix} S_0(x) & S_1(x) \\ S_1(x) & S_2(x) \end{pmatrix} \begin{pmatrix} \hat{\beta}_0 \\ \hat{\beta}_1 \end{pmatrix} = \begin{pmatrix} \sum_{i=1}^{n} y_i\, K\left(\tfrac{x_i - x}{h}\right) \\ \sum_{i=1}^{n} (x_i - x)\, y_i\, K\left(\tfrac{x_i - x}{h}\right) \end{pmatrix}.$$

Once $\hat{\beta}_0$ and $\hat{\beta}_1$ are found, we identify $\hat{m}(x) = \hat{\beta}_0$.

By substituting the Taylor expansion $y_i = m(x_i) + \varepsilon_i$ and taking expectations, one can derive how the "extra" $\frac{1}{2}\, m''(x)\,(x_i - x)^2$ terms feed into the local fit's bias.

From these expansions and associated algebra, one finds:

- **Bias:** The leading bias term for local linear regression is typically on the order of h^2, often written as $\frac{1}{2}\, m''(x)\, \mu_2(K)\, h^2$ for some constant $\mu_2(K)$ depending on the kernel's second moment.
- **Variance:** The leading variance term at a single point x is on the order of $\frac{1}{n\,h}$.

Balancing these two orders of magnitude—i.e., setting $h^2 \sim \frac{1}{n\,h}$—gives $h \sim n^{-1/3}$. Consequently, the mean squared error at x then behaves like

$$\left(\hat{m}(x) - m(x)\right)^2 = O_p\left(n^{-2/3}\right).$$

While local constant (Nadaraya-Watson) and local linear estimators often have the same *interior* rate, the local linear approach can eliminate leading-order bias near the boundaries, making it preferable in many practical settings.

7.4.5 Bandwidth Selection

Just like in kernel regression, the **bandwidth** h controls the smoothness of the local polynomial estimator.

- **Small** h: Captures fine local details but increases variance (potential overfitting).
- **Large** h: Smooths out noise but may miss important local structure (potential underfitting).

7.4.5.1 Cross-Validation for Local Polynomial Regression

Bandwidth selection via **cross-validation** is also common here. The leave-one-out CV criterion is:

$$\mathrm{CV}(h) = \frac{1}{n}\sum_{i=1}^{n}\left(y_i - \hat{m}_{-i,h}(x_i)\right)^2,$$

where $\hat{m}_{-i,h}(x_i)$ is the estimate at x_i obtained by leaving out the i-th observation.

Alternatively, for local linear regression, computational shortcuts (like **generalized cross-validation**) can significantly speed up bandwidth selection.

7.4.6 Asymptotic Properties Summary

- **Consistency:** $\hat{m}(x) \xrightarrow{p} m(x)$ as $n \to \infty$, under mild conditions.

- **Rate of Convergence:** For local linear regression, the MSE converges at rate $O(n^{-4/5})$, similar to kernel regression, but with better performance at boundaries.
- **Optimal Bandwidth:** Balances bias ($O(h^{p+1})$) and variance ($O(1/(nh))$), with cross-validation as a practical selection method (Figs. 7.8 and 7.9).

```
# Load necessary libraries
library(ggplot2)
library(gridExtra)

# 1. Simulate Data
set.seed(123)

# Generate predictor x and response y
n <- 100
x <- sort(runif(n, 0, 10))  # Sorted for local regression
true_function <-
    function(x)
        sin(x) + 0.5 * cos(2 * x)  # True regression function
y <-
    true_function(x) + rnorm(n, sd = 0.3) # Add Gaussian noise
```

```
# Visualization of the data
ggplot(data.frame(x, y), aes(x, y)) +
    geom_point(color = "darkblue") +
    geom_line(aes(y = true_function(x)),
              color = "red",
              linetype = "dashed") +
    labs(title = "Simulated Data with True Regression
    ↪   Function",
         x = "x", y = "y") +
    theme_minimal().
```

See Fig. 7.7.

```
# Gaussian Kernel Function
gaussian_kernel <- function(u) {
    (1 / sqrt(2 * pi)) * exp(-0.5 * u ^ 2)
}
```

Fig. 7.7 Simulated data with true regression function

```
# Local Polynomial Regression Function
local_polynomial_regression <-
    function(x_eval,
             x,
             y,
             h,
             p = 1,
             kernel = gaussian_kernel) {
        sapply(x_eval, function(x0) {
            # Design matrix for polynomial of degree p
            X <- sapply(0:p, function(k)
                (x - x0) ^ k)

            # Kernel weights
            W <- diag(kernel((x - x0) / h))

            # Weighted least squares estimation
            beta_hat <- solve(t(X) %*% W %*% X, t(X) %*% W
↪   %*% y)

            # Estimated value at x0 (intercept term)
            beta_hat[1]
```

```
        })
    }

# Evaluation grid
x_grid <- seq(0, 10, length.out = 200)

# Apply Local Linear Regression (p = 1)
h_linear <- 0.8
llr_estimate <-
    local_polynomial_regression(x_grid, x, y, h = h_linear, p
    ↪   = 1)

# Apply Local Quadratic Regression (p = 2)
h_quadratic <- 0.8
lqr_estimate <-
    local_polynomial_regression(x_grid, x, y, h = h_quadratic,
    ↪   p = 2)
```

```
# Plot Local Linear Regression
p1 <- ggplot() +
    geom_point(aes(x, y), color = "darkblue", alpha = 0.6) +
    geom_line(aes(x_grid, llr_estimate),
              color = "green",
              size = 1.2) +
    geom_line(aes(x_grid, true_function(x_grid)),
              color = "red",
              linetype = "dashed") +
    labs(
        title = "Local Linear Regression (p = 1)",
        subtitle = paste("Bandwidth (h) =", h_linear),
        x = "x",
        y = "Estimated m(x)"
    ) +
    theme_minimal()

# Plot Local Quadratic Regression
p2 <- ggplot() +
    geom_point(aes(x, y), color = "darkblue", alpha = 0.6) +
    geom_line(aes(x_grid, lqr_estimate),
              color = "orange",
              size = 1.2) +
    geom_line(aes(x_grid, true_function(x_grid)),
```

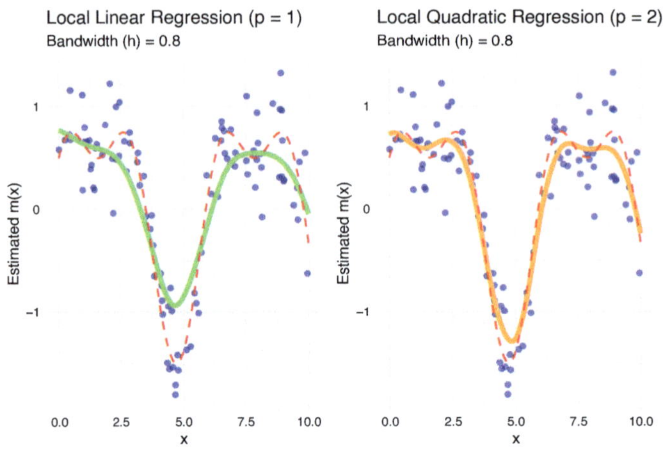

Fig. 7.8 Comparison of local regression

```
                color = "red",
                linetype = "dashed") +
    labs(
        title = "Local Quadratic Regression (p = 2)",
        subtitle = paste("Bandwidth (h) =", h_quadratic),
        x = "x",
        y = "Estimated m(x)"
    ) +
    theme_minimal()
```

```
# Display plots side by side
grid.arrange(p1, p2, ncol = 2)
```

- The green curve represents the local linear regression estimate (Fig. 7.8).
- The orange curve represents the local quadratic regression estimate.
- The dashed red line is the true regression function.
- Boundary effects are better handled by local polynomial methods, especially with quadratic fits that capture curvature more effectively.

```
# Leave-One-Out Cross-Validation for Bandwidth Selection
cv_bandwidth_lp <- function(h, x, y, p = 1, kernel =
↪  gaussian_kernel) {
```

```r
  n <- length(y)
  cv_error <- 0

  for (i in 1:n) {
    # Leave-one-out data
    x_train <- x[-i]
    y_train <- y[-i]

    # Predict the left-out point
    y_pred <- local_polynomial_regression(x[i], x_train,
↪   y_train, h = h, p = p, kernel = kernel)

    # Accumulate squared error
    cv_error <- cv_error + (y[i] - y_pred)^2
  }

  return(cv_error / n)
}

# Bandwidth grid for optimization
bandwidths <- seq(0.1, 2, by = 0.1)

# Cross-validation errors for local linear regression
cv_errors_linear <- sapply(bandwidths, cv_bandwidth_lp, x = x,
↪   y = y, p = 1)

# Cross-validation errors for local quadratic regression
cv_errors_quadratic <- sapply(bandwidths, cv_bandwidth_lp, x
↪   = x, y = y, p = 2)

# Optimal bandwidths
optimal_h_linear <- bandwidths[which.min(cv_errors_linear)]
optimal_h_quadratic <-
↪   bandwidths[which.min(cv_errors_quadratic)]

# Display optimal bandwidths
optimal_h_linear
#> [1] 0.4
optimal_h_quadratic
#> [1] 0.7
```

```
# CV Error Plot for Linear and Quadratic Fits
cv_data <- data.frame(
  Bandwidth = rep(bandwidths, 2),
  CV_Error = c(cv_errors_linear, cv_errors_quadratic),
  Degree = rep(c("Linear (p=1)", "Quadratic (p=2)"),
               each = length(bandwidths))
)
```

```
ggplot(cv_data, aes(x = Bandwidth, y = CV_Error, color =
↪  Degree)) +
    geom_line(size = 1) +
    geom_point(
        data = subset(
            cv_data,
            Bandwidth %in% c(optimal_h_linear,
↪  optimal_h_quadratic)
        ),
        aes(x = Bandwidth, y = CV_Error),
        color = "red",
        size = 3
    ) +
    labs(
        title = "Cross-Validation for Bandwidth Selection",
        subtitle = "Red points indicate optimal bandwidths",
        x = "Bandwidth (h)",
        y = "CV Error"
    ) +
    theme_minimal() +
    scale_color_manual(values = c("green", "orange"))
```

- The optimal bandwidth minimizes the cross-validation error (Fig. 7.9).
- The red points mark the bandwidths that yield the lowest errors for linear and quadratic fits.
- Smaller bandwidths can overfit, while larger bandwidths may oversmooth.

```
# Apply Local Polynomial Regression with Optimal Bandwidths
final_llr_estimate <-
    local_polynomial_regression(x_grid, x, y, h =
     ↪  optimal_h_linear, p = 1)
```

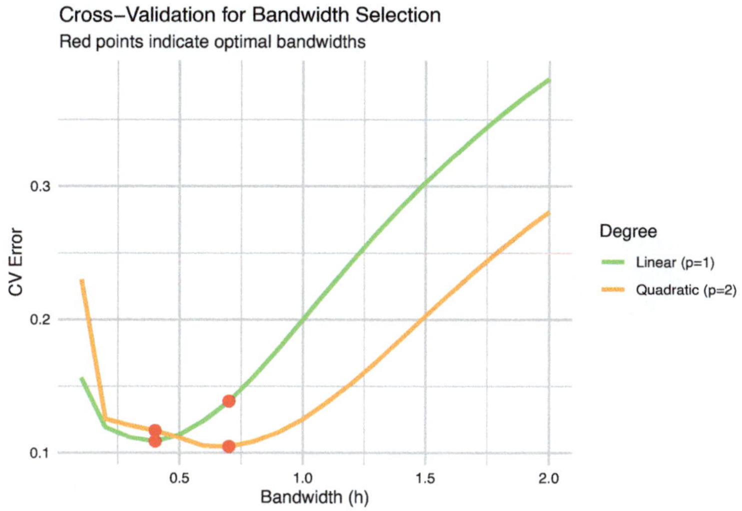

Fig. 7.9 Cross-validation for bandwidth selection

```
final_lqr_estimate <-
    local_polynomial_regression(x_grid, x, y, h =
    ↪  optimal_h_quadratic, p = 2)
```

```
# Plot final fits
ggplot() +
    geom_point(aes(x, y), color = "gray60", alpha = 0.5) +
    geom_line(
        aes(x_grid, final_llr_estimate, color = "Linear
        ↪  Estimate"),
        size = 1.2,
        linetype = "solid"
    ) +
    geom_line(
        aes(x_grid, final_lqr_estimate, color = "Quadratic
        ↪  Estimate"),
        size = 1.2,
        linetype = "solid"
    ) +
    geom_line(
        aes(x_grid, true_function(x_grid), color = "True
        ↪  Function"),
```

```
    linetype = "dashed"
  ) +
  labs(
    x = "x",
    y = "Estimated m(x)",
    color = "Legend"  # Add a legend title
  ) +
  scale_color_manual(
    values = c(
      "Linear Estimate" = "green",
      "Quadratic Estimate" = "orange",
      "True Function" = "red"
    )
  ) +
  theme_minimal()
```

See Fig. 7.10.

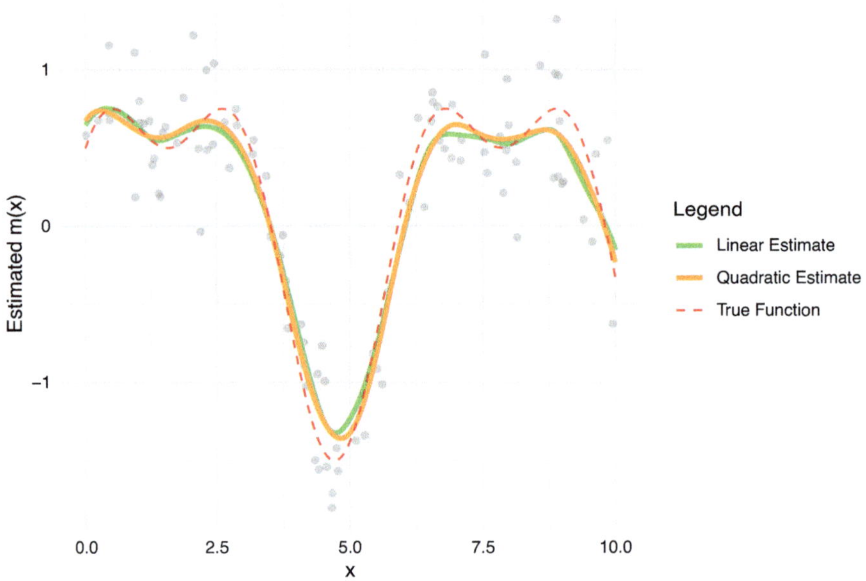

Fig. 7.10 Line comparison between linear and quadratic estimates

7.5 Smoothing Splines

A **spline** is a piecewise polynomial function that is smooth at the junction points (called **knots**). **Smoothing splines** provide a flexible nonparametric regression technique by balancing the trade-off between closely fitting the data and maintaining smoothness. This is achieved through a penalty on the function's curvature.

In the univariate case, suppose we have data $\{(x_i, y_i)\}_{i=1}^n$ with $0 \leq x_1 < x_2 < \cdots < x_n \leq 1$ (rescaling is always possible if needed). The **smoothing spline estimator** $\hat{m}(x)$ is defined as the solution to the following optimization problem:

$$\hat{m}(x) = \arg\min_{f \in \mathcal{H}} \left\{ \sum_{i=1}^{n} (y_i - f(x_i))^2 + \lambda \int_0^1 \left(f''(t)\right)^2 dt \right\},$$

where:

- The **first term** measures the **lack of fit** (residual sum of squares).
- The **second term** is a **roughness penalty** that discourages excessive curvature in f, controlled by the **smoothing parameter** $\lambda \geq 0$.
- The space $\mathcal{H}$ denotes the set of all twice-differentiable functions on $[0, 1]$.

Special Cases

- When $\lambda = 0$: No penalty is applied, and the solution interpolates the data exactly (an interpolating spline).
- As $\lambda \to \infty$: The penalty dominates, forcing the solution to be as smooth as possible—reducing to a linear regression (since the second derivative of a straight line is zero).

7.5.1 Properties and Form of the Smoothing Spline

A key result from spline theory is that the minimizer $\hat{m}(x)$ is a **natural cubic spline** with knots at the observed data points $\{x_1, \ldots, x_n\}$. This result holds despite the fact that we are minimizing over an infinite-dimensional space of functions.

The solution can be expressed as:

$$\hat{m}(x) = a_0 + a_1 x + \sum_{j=1}^{n} b_j (x - x_j)_+^3,$$

where:

- $(u)_+ = \max(u, 0)$ is the **positive part function** (the cubic spline basis function),
- The coefficients $\{a_0, a_1, b_1, \ldots, b_n\}$ are determined by solving a system of linear equations derived from the optimization problem.

This form implies that the spline is a cubic polynomial within each interval between data points, with smooth transitions at the knots. The smoothness conditions ensure continuity of the function and its first and second derivatives at each knot.

7.5.2 Choice of λ

The **smoothing parameter** λ plays a crucial role in controlling the trade-off between goodness-of-fit and smoothness:

- **Large** λ: Imposes a strong penalty on the roughness, leading to a smoother (potentially underfitted) function that captures broad trends.
- **Small** λ: Allows the function to closely follow the data, possibly resulting in overfitting if the data are noisy.

A common approach to selecting λ is **generalized cross-validation (GCV)**, which provides an efficient approximation to leave-one-out cross-validation:

$$\text{GCV}(\lambda) = \frac{\frac{1}{n}\sum_{i=1}^{n}\left(y_i - \hat{m}_\lambda(x_i)\right)^2}{\left[1 - \frac{\text{tr}(\mathbf{S}_\lambda)}{n}\right]^2},$$

where:

- $\hat{m}_\lambda(x_i)$ is the fitted value at x_i for a given λ,
- $\mathbf{S}_\lambda$ is the smoothing matrix (or influence matrix) such that $\hat{\mathbf{y}} = \mathbf{S}_\lambda \mathbf{y}$,
- $\text{tr}(\mathbf{S}_\lambda)$ is the effective degrees of freedom, reflecting the model's flexibility.

The optimal λ minimizes the GCV score, balancing fit and complexity without the need to refit the model multiple times (as in traditional cross-validation).

7.5.3 Connection to Reproducing Kernel Hilbert Spaces

Smoothing splines can be viewed through the lens of **reproducing kernel Hilbert spaces (RKHS)**. The penalty term:

$$\int_0^1 \left(f''(t)\right)^2 dt$$

defines a **semi-norm** that corresponds to the squared norm of f in a particular RKHS associated with the cubic spline kernel. This interpretation reveals that smoothing splines are equivalent to solving a **regularization problem in an RKHS**, where the penalty controls the smoothness of the solution (Fig. 7.11).

Fig. 7.11 Simulated data with true regression function

This connection extends naturally to more general kernel-based methods (e.g., Gaussian process regression, kernel ridge regression) and higher-dimensional spline models.

```
# Load necessary libraries
library(ggplot2)
library(gridExtra)

# 1. Simulate Data
set.seed(123)


# Generate predictor x and response y
n <- 100
x <- sort(runif(n, 0, 10))  # Sorted for smoother
↪  visualization
true_function <-
    function(x)
        sin(x) + 0.5 * cos(2 * x)  # True regression function
y <-
    true_function(x) + rnorm(n, sd = 0.3) # Add Gaussian noise
```

```
# Visualization of the data
ggplot(data.frame(x, y), aes(x, y)) +
    geom_point(color = "darkblue", alpha = 0.6) +
    geom_line(aes(y = true_function(x)),
              color = "red",
              linetype = "dashed") +
    labs(title = "Simulated Data with True Regression
    ↪ Function",
         x = "x", y = "y") +
    theme_minimal()
```

See Fig. 7.12.

```
# Apply Smoothing Spline with Default Lambda
# (automatically selected using GCV)
spline_fit_default <- smooth.spline(x, y)

# Apply Smoothing Spline with Manual Lambda
# (via smoothing parameter 'spar')
spline_fit_smooth <- smooth.spline(x, y, spar = 0.8)  #
↪  Smoother fit
spline_fit_flexible <-
    smooth.spline(x, y, spar = 0.4)  # More flexible fit

# Create grid for prediction
x_grid               <- seq(0, 10, length.out = 200)
spline_pred_default  <- predict(spline_fit_default, x_grid)
spline_pred_smooth   <- predict(spline_fit_smooth, x_grid)
spline_pred_flexible <- predict(spline_fit_flexible, x_grid)
```

```
# Plot Smoothing Splines with Different Smoothness Levels
ggplot() +
    geom_point(aes(x, y), color = "darkblue", alpha = 0.5) +
    geom_line(aes(x_grid, spline_pred_default$y),
              color = "green",
              size = 1.2) +
    geom_line(
        aes(x_grid, spline_pred_smooth$y),
        color = "orange",
        size = 1.2,
        linetype = "dotted"
```

```
    ) +
    geom_line(
        aes(x_grid, spline_pred_flexible$y),
        color = "purple",
        size = 1.2,
        linetype = "dashed"
    ) +
    geom_line(
        aes(x_grid, true_function(x_grid)),
        color = "red",
        linetype = "solid",
        size = 1
    ) +
    labs(
        title = "Smoothing Spline Fits",
        subtitle = "Green: GCV-selected | Orange: Smooth
        ↪  (spar=0.8) | Purple: Flexible (spar=0.4)",
        x = "x",
        y = "Estimated m(x)"
    ) +
    theme_minimal()
```

- The green curve is the fit with the optimal λ selected automatically via GCV.
- The orange curve (with `spar = 0.8`) is smoother, capturing broad trends but missing finer details.
- The purple curve (with `spar = 0.4`) is more flexible, fitting the data closely, potentially overfitting noise.
- The red solid line represents the true regression function (Fig. 7.12).

```
# Extract Generalized Cross-Validation (GCV) Scores
spline_fit_default$cv.crit  # GCV criterion for the default
↪   fit
#> [1] 0.09698728

# Compare GCV for different spar values
spar_values <- seq(0.1, 1.5, by = 0.05)
gcv_scores <- sapply(spar_values, function(spar) {
    fit <- smooth.spline(x, y, spar = spar)
    fit$cv.crit
})
```

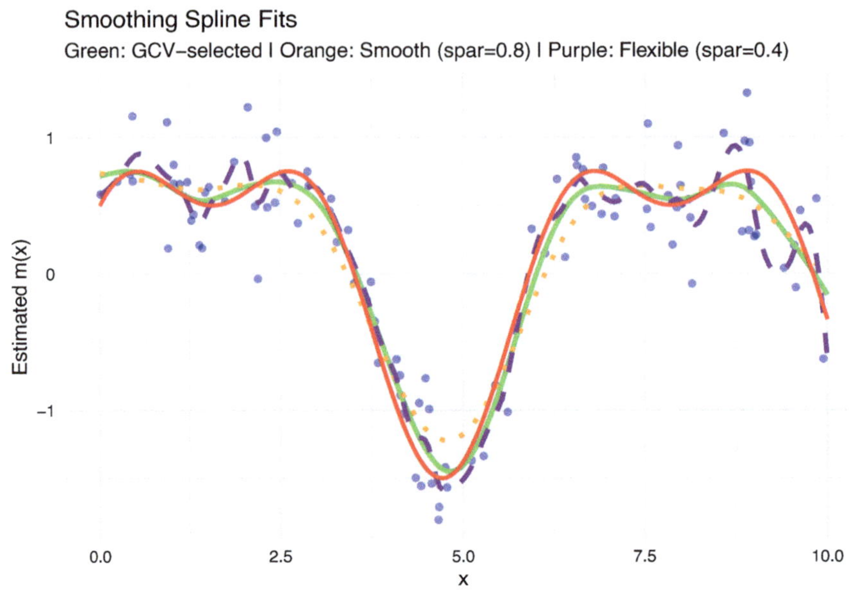

Fig. 7.12 Smoothing spline fits

```
# Optimal spar corresponding to the minimum GCV score
optimal_spar <- spar_values[which.min(gcv_scores)]
optimal_spar
#> [1] 0.7
```

```
# GCV Plot
ggplot(data.frame(spar_values, gcv_scores),
       aes(x = spar_values, y = gcv_scores)) +
    geom_line(color = "blue", size = 1) +
    geom_point(aes(x = optimal_spar, y = min(gcv_scores)),
               color = "red",
               size = 3) +
    labs(
        title = "GCV for Smoothing Parameter Selection",
        subtitle = paste("Optimal spar =", round(optimal_spar,
        ↪ 2)),
        x = "Smoothing Parameter (spar)",
        y = "GCV Score"
    ) +
    theme_minimal()
```

```
#> Warning in geom_point(aes(x = optimal_spar, y =
↪  min(gcv_scores)), color = "red", : All aesthetics have
↪  length 1, but the data has 29 rows.
#> i Please consider using `annotate()` or provide this layer
#>   with data containing a single row.
```

- The blue curve shows how the GCV score changes with different smoothing parameters (spar).
- The red point indicates the optimal smoothing parameter that minimizes the GCV score.
- Low spar values correspond to flexible fits (risking overfitting), while high spar values produce smoother fits (risking underfitting) (Fig. 7.13).

```
# Apply Smoothing Spline with Optimal spar
spline_fit_optimal <- smooth.spline(x, y, spar = optimal_spar)
spline_pred_optimal <- predict(spline_fit_optimal, x_grid)

# Plot Final Fit
ggplot() +
```

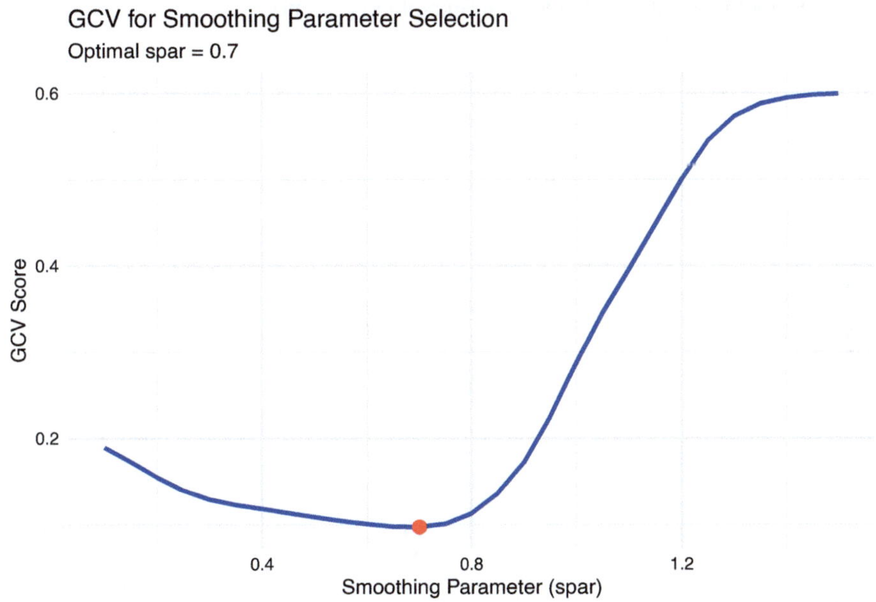

Fig. 7.13 GCV for smoothing parameter selection

```
  geom_point(aes(x, y), color = "gray60", alpha = 0.5) +
  geom_line(aes(x_grid, spline_pred_optimal$y),
            color = "green",
            size = 1.5) +
  geom_line(
    aes(x_grid, true_function(x_grid)),
    color = "red",
    linetype = "dashed",
    size = 1.2
  ) +
  labs(
    title = "Smoothing Spline with Optimal Smoothing
    ↪ Parameter",
    subtitle = paste("Optimal spar =", round(optimal_spar,
    ↪ 2)),
    x = "x",
    y = "Estimated m(x)"
  ) +
  theme_minimal()
```

See Fig. 7.14.

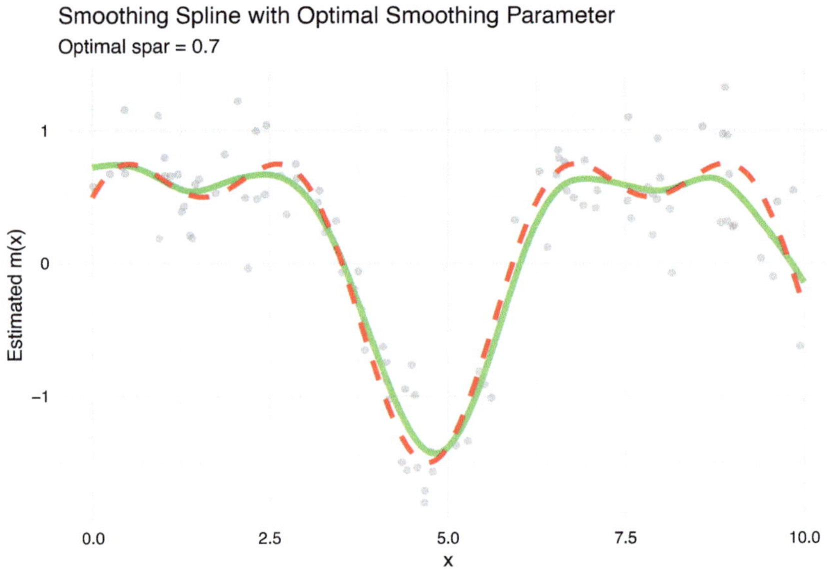

Fig. 7.14 Smoothing spline with optimal smoothing parameter

7.6 Confidence Intervals in Nonparametric Regression

Constructing **confidence intervals** (or bands) for nonparametric regression estimators like kernel smoothers, local polynomials, and smoothing splines is more complex than in parametric models. The key challenges arise from the flexible nature of the models and the dependence of bias and variance on the local data density and smoothing parameters.

7.6.1 Asymptotic Normality

Under regularity conditions, nonparametric estimators are **asymptotically normal**. For a given point x, we have:

$$\sqrt{nh}\,\left\{\hat{m}(x) - m(x)\right\} \xrightarrow{\mathcal{D}} \mathcal{N}\left(0, \sigma^2\, \nu(x)\right),$$

where:

- n is the sample size,
- h is the bandwidth (for kernel or local polynomial estimators) or a function of λ (for smoothing splines),
- σ^2 is the variance of the errors,
- $\nu(x)$ is a function that depends on the estimator, kernel, and local data density.

An approximate $(1 - \alpha)$ pointwise confidence interval for $m(x)$ is given by:

$$\hat{m}(x) \pm z_{\alpha/2} \cdot \sqrt{\widehat{\text{Var}}[\hat{m}(x)]},$$

where:

- $z_{\alpha/2}$ is the $(1 - \alpha/2)$ quantile of the standard normal distribution,
- $\widehat{\text{Var}}[\hat{m}(x)]$ is an estimate of the variance, which can be obtained using plug-in methods, sandwich estimators, or resampling techniques.

7.6.2 Bootstrap Methods

The **bootstrap** provides a powerful alternative for constructing confidence intervals and bands, particularly when asymptotic approximations are unreliable (e.g., small sample sizes or near boundaries).

7.6.2.1 Bootstrap Approaches

- **Residual Bootstrap:**
  1. Fit the nonparametric model to obtain residuals $\hat{\varepsilon}_i = y_i - \hat{m}(x_i)$.
  2. Generate bootstrap samples $y_i^* = \hat{m}(x_i) + \varepsilon_i^*$, where ε_i^* are resampled (with replacement) from $\{\hat{\varepsilon}_i\}$.
  3. Refit the model to each bootstrap sample to obtain $\hat{m}^*(x)$.
  4. Repeat many times to build an empirical distribution of $\hat{m}^*(x)$.
- **Wild Bootstrap:**
  Particularly useful for heteroscedastic data. Instead of resampling residuals directly, we multiply them by random variables with mean zero and unit variance to preserve the variance structure.

7.6.2.2 Bootstrap Confidence Bands

While **pointwise confidence intervals** cover the true function at a specific x with probability $(1 - \alpha)$, **simultaneous confidence bands** cover the entire function over an interval with the desired confidence level. Bootstrap methods can be adapted to estimate these bands by capturing the distribution of the **maximum deviation** between $\hat{m}(x)$ and $m(x)$ over the range of x.

7.6.3 Practical Considerations

- **Bias Correction:**
  Nonparametric estimators often have non-negligible bias, especially near boundaries. Bias correction techniques or undersmoothing (choosing a smaller bandwidth) are sometimes used to improve interval coverage.
- **Effective Degrees of Freedom:**
  For smoothing splines, the **effective degrees of freedom** (related to $\mathrm{tr}(\mathbf{S}_\lambda)$) provide insight into model complexity and influence confidence interval construction.

7.7 Generalized Additive Models

A generalized additive model (GAM) extends generalized linear models by allowing **additive smooth terms**:

$$g(\mathbb{E}[Y]) = \beta_0 + f_1(X_1) + f_2(X_2) + \cdots + f_p(X_p),$$

where:

- $g(\cdot)$ is a **link function** (as in GLMs),
- β_0 is the intercept,
- Each f_j is a smooth, potentially nonparametric function (e.g., a spline, kernel smoother, or local polynomial smoother),
- p is the number of predictors, with $p \geq 2$ highlighting the flexibility of GAMs in handling multivariate data.

This structure allows for **nonlinear relationships** between each predictor X_j and the response Y while maintaining **additivity**, which simplifies interpretation compared to fully nonparametric models.

Traditional linear models assume a strictly linear relationship:

$$g(\mathbb{E}[Y]) = \beta_0 + \beta_1 X_1 + \beta_2 X_2 + \cdots + \beta_p X_p.$$

However, real-world data often exhibit complex, nonlinear patterns. While generalized linear models extend linear models to non-Gaussian responses, they still rely on linear predictors. GAMs address this by replacing linear terms with smooth functions:

- **GLMs:** Linear effects (e.g., $\beta_1 X_1$)
- **GAMs:** Nonlinear smooth effects (e.g., $f_1(X_1)$)

The general form of a GAM is:

$$g(\mathbb{E}[Y \mid \mathbf{X}]) = \beta_0 + \sum_{j=1}^{p} f_j(X_j),$$

where:

- Y is the response variable,
- $\mathbf{X} = (X_1, X_2, \ldots, X_p)$ are the predictors,
- f_j are smooth functions capturing potentially nonlinear effects,
- The link function $g(\cdot)$ connects the mean of Y to the additive predictor.

Special Cases

- When g is the identity function and Y is continuous: This reduces to an **additive model** (a special case of GAM).
- When g is the logit function and Y is binary: We have a **logistic GAM** for classification tasks.
- When g is the log function and Y follows a Poisson distribution: This is a **Poisson GAM** for count data.

7.7.1 Estimation via Penalized Likelihood

GAMs are typically estimated using **penalized likelihood methods** to balance model fit and smoothness. The objective function is:

$$\mathcal{L}_{\text{pen}} = \ell(\beta_0, f_1, \ldots, f_p) - \frac{1}{2}\sum_{j=1}^{p} \lambda_j \int \left(f_j''(x)\right)^2 dx,$$

where:

- $\ell(\beta_0, f_1, \ldots, f_p)$ is the (log-)likelihood of the data,
- $\lambda_j \geq 0$ are **smoothing parameters** controlling the smoothness of each f_j,
- The penalty term $\int (f_j'')^2 dx$ discourages excessive curvature, similar to smoothing splines.

7.7.1.1 Backfitting Algorithm

For continuous responses, the classic **backfitting algorithm** is often used:

1. **Initialize**: Start with an initial guess for each f_j, typically zero.
2. **Iterate**: For each $j = 1, \ldots, p$:

   - Compute the **partial residuals**:

   $$r_j = y - \beta_0 - \sum_{k \neq j} f_k(X_k)$$

   - Update f_j by fitting a smoother to (X_j, r_j).

3. **Convergence**: Repeat until the functions f_j stabilize.

This approach works because of the additive structure, which allows each smooth term to be updated conditionally on the others.

7.7.1.2 Generalized Additive Model Estimation (for GLMs)

When Y is non-Gaussian (e.g., binary, count data), we use iteratively reweighted least squares (IRLS) in combination with backfitting. Popular implementations, such as in the `mgcv` package in R, use penalized likelihood estimation with efficient computational algorithms (e.g., penalized iteratively reweighted least squares).

7.7.2 Interpretation of GAMs

One of the key advantages of GAMs is their **interpretability**, especially compared to fully nonparametric or black-box machine learning models.

- **Additive Structure:** Each predictor's effect is modeled separately via $f_j(X_j)$, making it easy to interpret marginal effects.

- **Partial Dependence Plots:** Visualization of $f_j(X_j)$ shows how each predictor affects the response, holding other variables constant.

Example
For a marketing dataset predicting customer purchase probability:

$$\log\left(\frac{\mathbb{P}(\text{Purchase})}{1-\mathbb{P}(\text{Purchase})}\right) = \beta_0 + f_1(\text{Age}) + f_2(\text{Income}) + f_3(\text{Ad Exposure})$$

- $f_1(\text{Age})$ might show a peak in purchase likelihood for middle-aged customers.
- $f_2(\text{Income})$ could reveal a threshold effect where purchases increase beyond a certain income level.
- $f_3(\text{Ad Exposure})$ might show diminishing returns after repeated exposures.

7.7.3 Model Selection and Smoothing Parameter Estimation

The **smoothing parameters** λ_j control the complexity of each smooth term:

- **Large** λ_j: Strong smoothing, leading to nearly linear fits.
- **Small** λ_j: Flexible, wiggly fits that may overfit if λ_j is too small.

Methods for Choosing
λ_j

1. **Generalized Cross-Validation (GCV):**

$$\text{GCV} = \frac{1}{n}\frac{\sum_{i=1}^{n}(y_i - \hat{y}_i)^2}{\left(1 - \frac{\text{tr}(\mathbf{S})}{n}\right)^2}$$

   where $\mathbf{S}$ is the smoother matrix. GCV is a popular method for selecting the smoothing parameter λ_j because it approximates leave-one-out cross-validation without requiring explicit refitting of the model. The term tr($\mathbf{S}$) represents the effective degrees of freedom of the smoother, and the denominator penalizes overfitting.
2. **Unbiased Risk Estimation:** This method extends the idea of GCV to non-Gaussian families (e.g., Poisson, binomial). It aims to minimize an unbiased estimate of the risk (expected prediction error). For Gaussian models, it often reduces to a form similar to GCV, but for other distributions, it incorporates the appropriate likelihood or deviance.
3. **Akaike Information Criterion (AIC):**

$$AIC = -2\log(L) + 2tr\,(S)$$

where L is the likelihood of the model. AIC balances model fit (measured by the likelihood) and complexity (measured by the effective degrees of freedom $tr(S)$). The smoothing parameter λ_j is chosen to minimize the AIC.

4. **Bayesian Information Criterion (BIC):**

$$BIC = -2\log(L) + \log(n)tr(S)$$

where n is the sample size. BIC is similar to AIC but imposes a stronger penalty for model complexity, making it more suitable for larger datasets.

5. **Leave-One-Out Cross-Validation (LOOCV):**

$$LOOCV = \frac{1}{n}\sum_{i=1}^{n}(y_i - \hat{y}_i^{(-i)})^2,$$

where $y_i^{(-i)}$ is the predicted value for the ii-th observation when the model is fitted without it. LOOCV is computationally intensive but provides a direct estimate of prediction error.

6. **Empirical Risk Minimization:**
For some non-parametric regression methods, λ_j can be chosen by minimizing the empirical risk (e.g., mean squared error) on a validation set or via resampling techniques like k-fold cross-validation.

7.7.4 Extensions of GAMs

- **GAM with Interaction Terms:**

$$g(\mathbb{E}[Y]) = \beta_0 + f_1(X_1) + f_2(X_2) + f_{12}(X_1, X_2)$$

where f_{12} captures the interaction between X_1 and X_2 (using tensor product smooths).

- **GAMMs (Generalized Additive Mixed Models):** Incorporate random effects to handle hierarchical or grouped data (Fig. 7.15).
- **Varying Coefficient Models:** Allow regression coefficients to vary smoothly with another variable, e.g.,

$$Y = \beta_0 + f_1(Z) \cdot X + \varepsilon$$

```
# Load necessary libraries
library(mgcv)     # For fitting GAMs
#> Loading required package: nlme
```

```
#> This is mgcv 1.9-3. For overview type
↪ 'help("mgcv-package")'.
library(ggplot2)
library(gridExtra)

# Simulate Data
set.seed(123)

n <- 100
x1 <- runif(n, 0, 10)
x2 <- runif(n, 0, 5)
x3 <- rnorm(n, 5, 2)

# True nonlinear functions
f1 <- function(x)
    sin(x)                    # Nonlinear effect of x1
f2 <- function(x)
    log(x + 1)                # Nonlinear effect of x2
f3 <- function(x)
    0.5 * (x - 5) ^ 2          # Quadratic effect for x3

# Generate response variable with noise
y <- 3 + f1(x1) + f2(x2) - f3(x3) + rnorm(n, sd = 1)

# Data frame for analysis
data_gam <- data.frame(y, x1, x2, x3)

# Plotting the true functions with simulated data
p1 <-
    ggplot(data_gam, aes(x1, y)) +
    geom_point() +
    labs(title = "Effect of x1 (sin(x1))")
p2 <-
    ggplot(data_gam, aes(x2, y)) +
    geom_point() +
    labs(title = "Effect of x2 (log(x2+1))")
p3 <-
    ggplot(data_gam, aes(x3, y)) +
    geom_point() +
    labs(title = "Effect of x3 (quadratic)")
```

```
# Display plots side by side
grid.arrange(p1, p2, p3, ncol = 3)
```

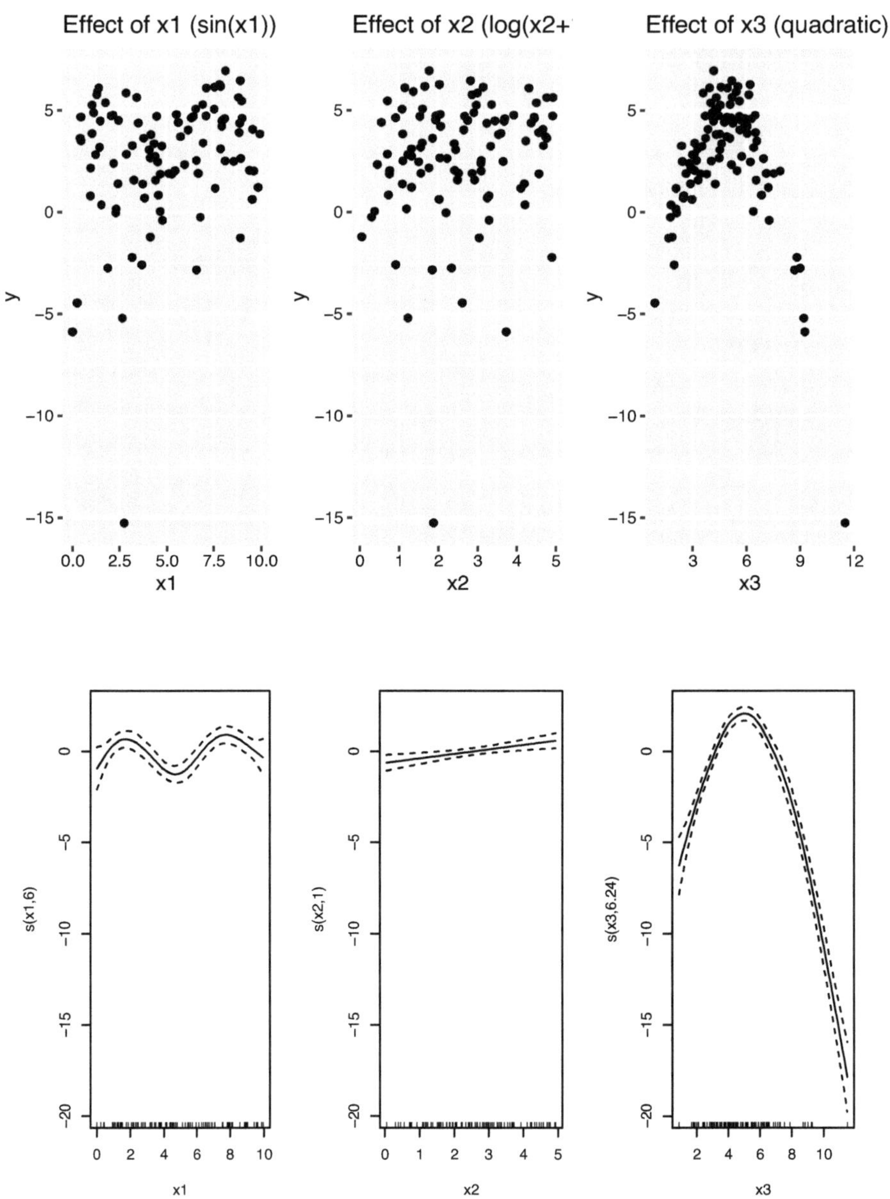

Fig. 7.15 Smooth functions

```r
# Fit a GAM using mgcv
gam_model <-
    gam(y ~ s(x1) + s(x2) + s(x3),
        data = data_gam, method = "REML")

# Summary of the model
summary(gam_model)
#>
#> Family: gaussian
#> Link function: identity
#>
#> Formula:
#> y ~ s(x1) + s(x2) + s(x3)
#>
#> Parametric coefficients:
#>             Estimate Std. Error t value Pr(>|t|)
#> (Intercept)  2.63937    0.09511   27.75   <2e-16 ***
#> ---
#> Signif. codes:
#> 0 '***' 0.001 '**' 0.01 '*' 0.05 '.' 0.1 ' ' 1
#>
#> Approximate significance of smooth terms:
#>         edf Ref.df       F p-value
#> s(x1) 5.997  7.165   7.966   5e-07 ***
#> s(x2) 1.000  1.000  10.249 0.00192 **
#> s(x3) 6.239  7.343 105.551 < 2e-16 ***
#> ---
#> Signif. codes:
#> 0 '***' 0.001 '**' 0.01 '*' 0.05 '.' 0.1 ' ' 1
#>
#> R-sq.(adj) =   0.91   Deviance explained = 92.2%
#> -REML = 155.23  Scale est. = 0.90463   n = 100
```

```r
# Plot smooth terms
par(mfrow = c(1, 3))  # Arrange plots in one row
plot(gam_model, shade = TRUE, seWithMean = TRUE)
```

```r
par(mfrow = c(1, 1))  # Reset plotting layout
```

```
# Using ggplot2 with mgcv's predict function
pred_data <- with(data_gam, expand.grid(
    x1 = seq(min(x1), max(x1), length.out = 100),
    x2 = mean(x2),
    x3 = mean(x3)
))

# Predictions for x1 effect
pred_data$pred_x1 <-
    predict(gam_model, newdata = pred_data, type = "response")
```

```
ggplot(pred_data, aes(x1, pred_x1)) +
    geom_line(color = "blue", size = 1.2) +
    labs(title = "Partial Effect of x1",
         x = "x1",
         y = "Effect on y") +
    theme_minimal()
```

See Fig. 7.16.

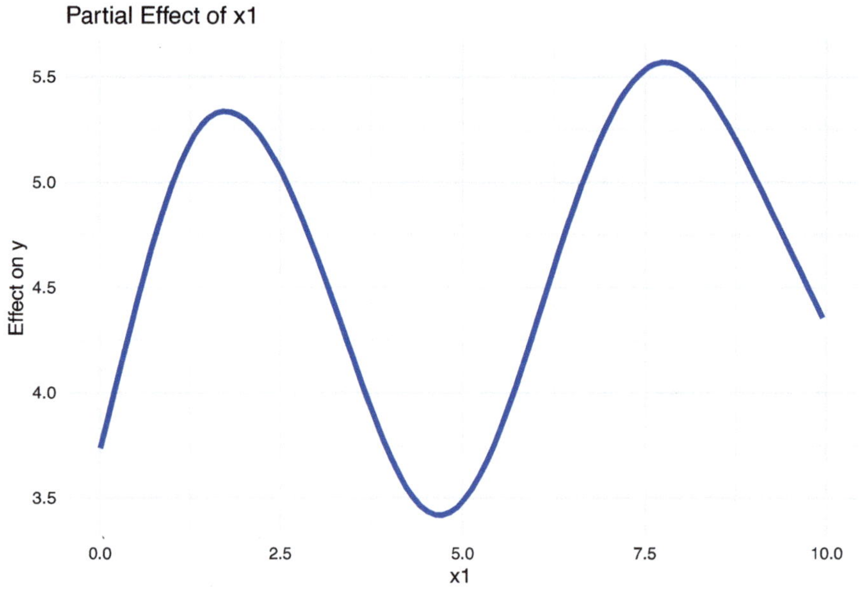

Fig. 7.16 Partial effect of x1

```
# Check AIC and GCV score
AIC(gam_model)
#> [1] 289.8201
gam_model$gcv.ubre  # GCV/UBRE score
#>     REML
#> 155.2314
#> attr(,"Dp")
#> [1] 47.99998
```

```
# Compare models with different smoothness
gam_model_simple <-
    gam(y ~ s(x1, k = 4) + s(x2, k = 4) + s(x3, k = 4),
        data = data_gam)
gam_model_complex <-
    gam(y ~ s(x1, k = 20) + s(x2, k = 20) + s(x3, k = 20),
        data = data_gam)
```

```
# Compare models using AIC
AIC(gam_model, gam_model_simple, gam_model_complex)
#>                          df      AIC
#> gam_model         15.706428 289.8201
#> gam_model_simple   8.429889 322.1502
#> gam_model_complex 13.571165 287.4171
```

- Lower AIC indicates a better model balancing fit and complexity.
- GCV score helps select the optimal level of smoothness.
- Compare models to prevent overfitting (too flexible) or underfitting (too simple).

```
# GAM with interaction using tensor product smooths
gam_interaction <- gam(y ~ te(x1, x2) + s(x3),
                       data = data_gam)

# Summary of the interaction model
summary(gam_interaction)
#>
#> Family: gaussian
#> Link function: identity
#>
#> Formula:
#> y ~ te(x1, x2) + s(x3)
#>
```

```
#> Parametric coefficients:
#>             Estimate Std. Error t value Pr(>|t|)
#> (Intercept)  2.63937    0.09364   28.19   <2e-16 ***
#> ---
#> Signif. codes:
#> 0 '***' 0.001 '**' 0.01 '*' 0.05 '.' 0.1 ' ' 1
#>
#> Approximate significance of smooth terms:
#>             edf Ref.df       F p-value
#> te(x1,x2) 8.545  8.923   9.218  <2e-16 ***
#> s(x3)     4.766  5.834 147.595  <2e-16 ***
#> ---
#> Signif. codes:
#> 0 '***' 0.001 '**' 0.01 '*' 0.05 '.' 0.1 ' ' 1
#>
#> R-sq.(adj) =  0.912   Deviance explained = 92.4%
#> GCV = 1.0233  Scale est. = 0.87688   n = 100
```

```
# Visualization of interaction effect
vis.gam(
    gam_interaction,
    view = c("x1", "x2"),
    plot.type = "contour",
    color = "terrain"
)
```

- The tensor product smooth `te(x1, x2)` captures nonlinear interactions between `x1` and `x2`.
- The contour plot visualizes how their joint effect influences the response (Fig. 7.17).

```
# Simulate binary response
set.seed(123)
prob <- plogis(1 + f1(x1) - f2(x2) + 0.3 * x3)  # Logistic
↪  function
y_bin <- rbinom(n, 1, prob)                      # Binary
↪  outcome

# Fit GAM for binary classification
```

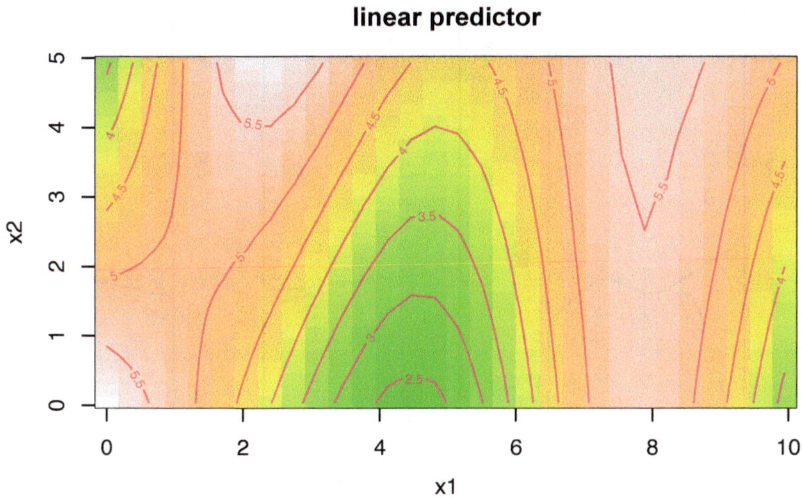

Fig. 7.17 Linear predictor

```
gam_logistic <-
    gam(y_bin ~ s(x1) + s(x2) + s(x3),
        family = binomial,
        data = data_gam)

# Summary and visualization
summary(gam_logistic)
#>
#> Family: binomial
#> Link function: logit
#>
#> Formula:
#> y_bin ~ s(x1) + s(x2) + s(x3)
#>
#> Parametric coefficients:
#>             Estimate Std. Error z value Pr(>|z|)
#> (Intercept)    22.30      32.18   0.693    0.488
#>
#> Approximate significance of smooth terms:
#>         edf Ref.df Chi.sq p-value
#> s(x1) 4.472  5.313  2.645   0.775
#> s(x2) 1.000  1.000  1.925   0.165
#> s(x3) 1.000  1.000  1.390   0.238
#>
```

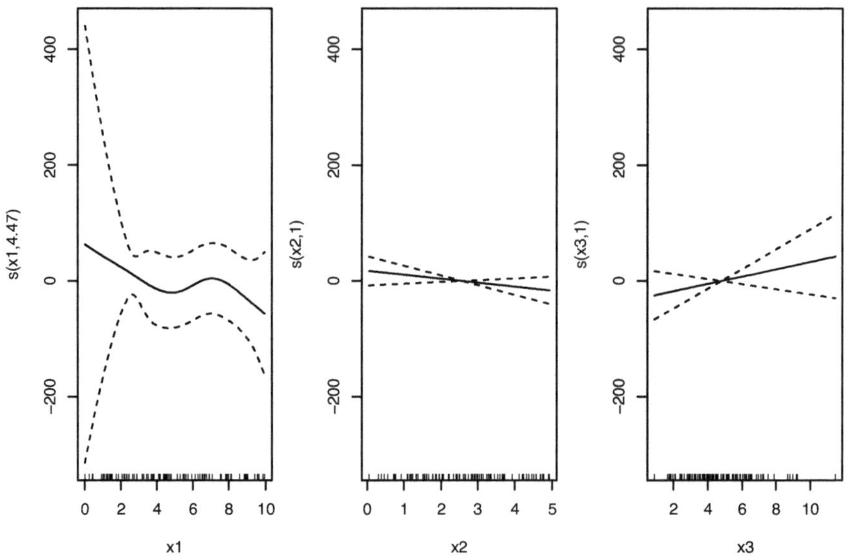

Fig. 7.18 Smooth functions of variables

```
#> R-sq.(adj) =      1   Deviance explained = 99.8%
#> UBRE = -0.84802  Scale est. = 1         n = 100
```

```
par(mfrow = c(1, 3))
plot(gam_logistic, shade = TRUE)
```

```
par(mfrow = c(1, 1))
```

- The logistic GAM models nonlinear effects on the log-odds of the binary outcome.
- Smooth plots indicate predictors' influence on probability of success (Fig. 7.18).

```
# Diagnostic plots
par(mfrow = c(2, 2))
gam.check(gam_model)
```

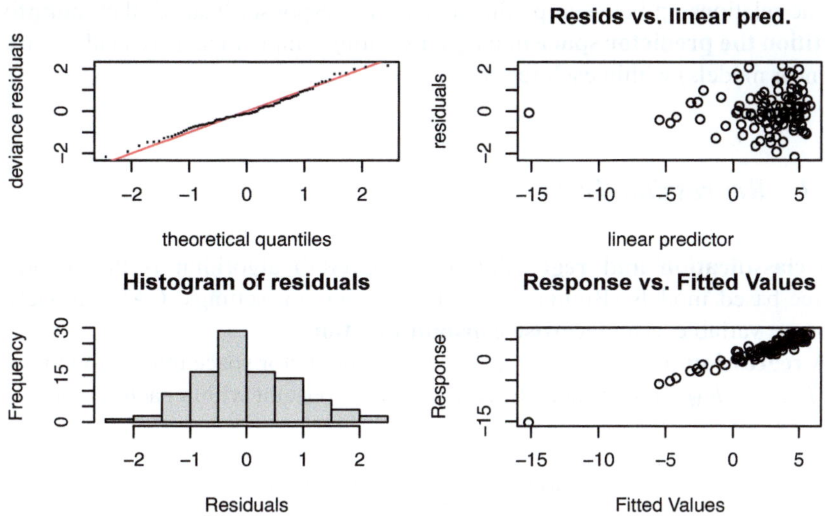

Fig. 7.19 Model diagnostics

```
#>
#> Method: REML   Optimizer: outer newton
#> full convergence after 9 iterations.
#> Gradient range [-5.387854e-05,2.006026e-05]
#> (score 155.2314 & scale 0.9046299).
#> Hessian positive definite, eigenvalue range [5.387409e-05,48.28647].
#> Model rank =  28 / 28
#>
#> Basis dimension (k) checking results. Low p-value (k-index<1) may
#> indicate that k is too low, especially if edf is close to k'.
#>
#>         k'  edf k-index p-value
#> s(x1) 9.00 6.00    1.01    0.46
#> s(x2) 9.00 1.00    1.16    0.92
#> s(x3) 9.00 6.24    1.07    0.72
par(mfrow = c(1, 1))
```

- Residual plots assess model fit (Fig. 7.19).
- QQ plot checks for normality of residuals.
- K-index evaluates the adequacy of smoothness selection.

7.8 Regression Trees and Random Forests

Though not typically framed as "kernel" or "spline," **tree-based methods**—such as **classification and regression trees (CART)** and **random forests**—are also **nonparametric** models. They do not assume a predetermined functional form for the relationship between predictors and the response. Instead, they **adaptively partition the predictor space** into regions, fitting simple models (usually constants or linear models) within each region.

7.8.1 Regression Trees

The **classification and regression trees (CART)** algorithm is the foundation of tree-based models (Breiman, 2017). In regression settings, CART models the response variable as a **piecewise constant function**.

A **regression tree** recursively partitions the predictor space into disjoint regions, $R_1, R_2, \ldots, R_M$, and predicts the response as a constant within each region:

$$\hat{m}(x) = \sum_{m=1}^{M} c_m \cdot \mathbb{I}(x \in R_m),$$

where:

- c_m is the **predicted value** (usually the mean of y_i) for all observations in region R_m,
- $\mathbb{I}(\cdot)$ is the indicator function.

Tree-Building Algorithm (Greedy Recursive Partitioning)

1. **Start with the full dataset** as a single region.
2. **Find the best split**:
   - Consider all possible splits of all predictors (e.g., $X_j < s$).
   - Choose the split that minimizes the **residual sum of squares (RSS)**:

   $$\text{RSS} = \sum_{i \in R_1} (y_i - \bar{y}_{R_1})^2 + \sum_{i \in R_2} (y_i - \bar{y}_{R_2})^2,$$

   where $\bar{y}_{R_k}$ is the mean response in region R_k.
3. **Recursively repeat** the splitting process for each new region (node) until a stopping criterion is met (e.g., minimum number of observations per leaf, maximum tree depth).
4. **Assign a constant prediction** to each terminal node (leaf) based on the average response of observations within that node.

Stopping Criteria and Pruning

- **Pre-pruning (early stopping):** Halt the tree growth when a predefined condition is met (e.g., minimal node size, maximal depth).
- **Post-pruning (cost-complexity pruning):** Grow a large tree first, then prune back to avoid overfitting. The **cost-complexity criterion** is:

$$C_\alpha(T) = \text{RSS}(T) + \alpha|T|,$$

where $|T|$ is the number of terminal nodes (leaves) and α controls the penalty for complexity.

Advantages of Regression Trees

- **Interpretability:** Easy to visualize and understand.
- **Handling of different data types:** Can naturally handle numerical and categorical variables.
- **Nonlinear relationships and interactions:** Captures complex patterns without explicit modeling.

Limitations

- **High variance:** Trees are sensitive to small changes in data (unstable).
- **Overfitting risk:** Without pruning or regularization, deep trees can overfit noise.

7.8.2 *Random Forests*

To address the high variance of single trees, **random forests** combine many regression trees to create an ensemble model with improved predictive performance and stability (Breiman, 2001).

A **random forest** builds multiple decision trees and aggregates their predictions to reduce variance. For regression, the final prediction is the **average** of the predictions from individual trees:

$$\hat{m}_{\text{RF}}(x) = \frac{1}{B}\sum_{b=1}^{B} \hat{m}^{(b)}(x),$$

where:

- B is the number of trees in the forest,
- $\hat{m}^{(b)}(x)$ is the prediction from the b-th tree.

Random Forest Algorithm

1. **Bootstrap Sampling:** For each tree, draw a bootstrap sample from the training data (sampling with replacement).
2. **Random Feature Selection:** At each split in the tree:
   - Randomly select a subset of predictors (usually $\sqrt{p}$ for classification or $p/3$ for regression).
   - Find the best split **only** among the selected features.

3. **Tree Growth:** Grow each tree to full depth without pruning.
4. **Aggregation:** For regression, average the predictions from all trees. For classification, use majority voting.

Why Does Random Forest Work?

- **Bagging (Bootstrap Aggregating):** Reduces variance by averaging over multiple models.
- **Random Feature Selection:** Decorrelates trees, further reducing variance.

7.8.3 Theoretical Insights

Bias-Variance Trade-Off

- **Regression Trees:** Low bias but high variance.
- **Random Forests:** Slightly higher bias than a single tree (due to randomization) but significantly reduced variance, leading to lower overall prediction error.

Out-of-Bag (OOB) Error

Random forests provide an internal estimate of prediction error using **out-of-bag** samples (the data not included in the bootstrap sample for a given tree). The OOB error is computed by:

1. For each observation, predict its response using only the trees where it was **not** included in the bootstrap sample.
2. Calculate the error by comparing the OOB predictions to the true responses.

OOB error serves as an efficient, unbiased estimate of test error without the need for cross-validation.

7.8.4 Feature Importance in Random Forests

Random forests naturally provide measures of **variable importance**, helping identify which predictors contribute most to the model.

1. **Mean Decrease in Impurity (MDI):** Measures the total reduction in impurity (e.g., RSS) attributed to each variable across all trees.
2. **Permutation Importance:** Measures the increase in prediction error when the values of a predictor are randomly permuted, breaking its relationship with the response.

Table 7.5 Advantages and limitations of tree-based methods

Aspect	Regression trees	Random forests
Interpretability	High (easy to visualize)	Moderate (difficult to interpret individual trees)
Variance	High (prone to overfitting)	Low (averaging reduces variance)
Bias	Low (flexible to data patterns)	Slightly higher than a single tree
Feature importance	Basic (via tree splits)	Advanced (permutation-based measures)
Handling of missing data	Handles with surrogate splits	Handles naturally in ensemble averaging
Computational cost	Low (fast for small datasets)	High (especially with many trees)

7.8.5 Advantages and Limitations of Tree-Based Methods

See Table 7.5.

```
# Load necessary libraries
library(ggplot2)
library(rpart)              # For regression trees
library(rpart.plot)         # For visualizing trees
library(randomForest)       # For random forests
#> randomForest 4.7-1.2
#> Type rfNews() to see new features/changes/bug fixes.
#>
#> Attaching package: 'randomForest'
#> The following object is masked from 'package:gridExtra':
#>
#>     combine
#> The following object is masked from 'package:ggplot2':
#>
#>     margin
library(gridExtra)

# Simulate Data
set.seed(123)

n <- 100
x1 <- runif(n, 0, 10)
x2 <- runif(n, 0, 5)
x3 <- rnorm(n, 5, 2)

# Nonlinear functions
f1 <- function(x)
```

```
  sin(x)
f2 <- function(x)
  log(x + 1)
f3 <- function(x)
  0.5 * (x - 5) ^ 2

# Generate response variable with noise
y <- 3 + f1(x1) + f2(x2) - f3(x3) + rnorm(n, sd = 1)

# Data frame
data_tree <- data.frame(y, x1, x2, x3)

# Quick visualization of data
p1 <- ggplot(data_tree, aes(x1, y)) +
  geom_point() +
  labs(title = "Effect of x1")
p2 <- ggplot(data_tree, aes(x2, y)) +
  geom_point() +
  labs(title = "Effect of x2")
p3 <- ggplot(data_tree, aes(x3, y)) +
  geom_point() +
  labs(title = "Effect of x3")
```

```
grid.arrange(p1, p2, p3, ncol = 3)
```

See Fig. 7.20.

```
# Fit a Regression Tree using rpart
tree_model <-
  rpart(
    y ~ x1 + x2 + x3,
    data = data_tree,
    method = "anova",
    control = rpart.control(cp = 0.01)
  ) # cp = complexity parameter

# Summary of the tree
summary(tree_model)
#> Call:
#> rpart(formula = y ~ x1 + x2 + x3, data = data_tree, method
↪ = "anova",
```

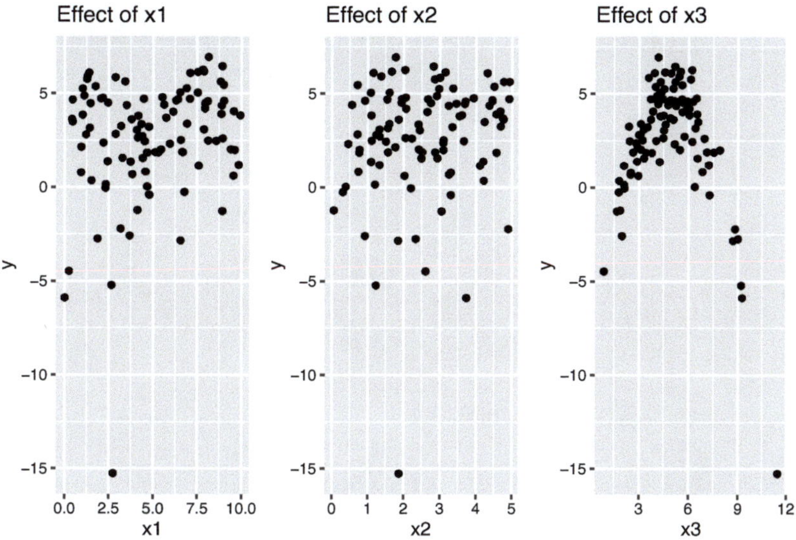

Fig. 7.20 Scatterplots of the effects of variables

```
#>     control = rpart.control(cp = 0.01))
#>   n= 100
#>
#>           CP nsplit rel error    xerror      xstd
#> 1 0.39895879      0 1.0000000 1.0134781 0.3406703
#> 2 0.17470339      1 0.6010412 0.8649973 0.3336272
#> 3 0.04607373      2 0.4263378 0.5707932 0.1880333
#> 4 0.02754858      3 0.3802641 0.5287366 0.1866728
#> 5 0.01584638      4 0.3527155 0.5061104 0.1867491
#> 6 0.01032524      5 0.3368691 0.5136765 0.1861020
#> 7 0.01000000      7 0.3162187 0.4847072 0.1861849
#>
#> Variable importance
#> x3 x2 x1
#> 91  6  3
#>
#> Node number 1: 100 observations,    complexity
↪  param=0.3989588
#>   mean=2.639375, MSE=9.897038
#>   left son=2 (7 obs) right son=3 (93 obs)
#>   Primary splits:
#>       x3 < 7.707736  to the right, improve=0.39895880, (0
↪  missing)
```

```
#>       x1 < 6.84138   to the left,  improve=0.07685517, (0
↪  missing)
#>       x2 < 2.627429  to the left,  improve=0.04029839, (0
↪  missing)
#>
#> Node number 2: 7 observations
#>   mean=-4.603469, MSE=24.47372
#>
#> Node number 3: 93 observations,    complexity
↪  param=0.1747034
#>   mean=3.184535, MSE=4.554158
#>   left son=6 (18 obs) right son=7 (75 obs)
#>   Primary splits:
#>       x3 < 2.967495  to the left,  improve=0.40823990, (0
↪  missing)
#>       x2 < 1.001856  to the left,  improve=0.07353453, (0
↪  missing)
#>       x1 < 6.84138   to the left,  improve=0.07049507, (0
↪  missing)
#>   Surrogate splits:
#>       x2 < 0.3435293 to the left,  agree=0.828, adj=0.111,
↪  (0 split)
#>
#> Node number 6: 18 observations
#>   mean=0.4012593, MSE=3.4521
#>
#> Node number 7: 75 observations,    complexity
↪  param=0.04607373
#>   mean=3.852521, MSE=2.513258
#>   left son=14 (12 obs) right son=15 (63 obs)
#>   Primary splits:
#>       x3 < 6.324486  to the right, improve=0.24191360, (0
↪  missing)
#>       x2 < 1.603258  to the left,  improve=0.10759280, (0
↪  missing)
#>       x1 < 6.793804  to the left,  improve=0.09106168, (0
↪  missing)
#>
#> Node number 14: 12 observations
#>   mean=2.065917, MSE=2.252311
#>
#> Node number 15: 63 observations,    complexity
↪  param=0.02754858
#>   mean=4.192826, MSE=1.839163
```

```
#>   left son=30 (9 obs) right son=31 (54 obs)
#>   Primary splits:
#>       x3 < 3.548257  to the left,  improve=0.2353119, (0
↪ missing)
#>       x2 < 1.349633  to the left,  improve=0.1103019, (0
↪ missing)
#>       x1 < 7.006669  to the left,  improve=0.1019295, (0
↪ missing)
#>
#> Node number 30: 9 observations
#>   mean=2.581411, MSE=0.3669647
#>
#> Node number 31: 54 observations,    complexity
↪ param=0.01584638
#>   mean=4.461396, MSE=1.579623
#>   left son=62 (10 obs) right son=63 (44 obs)
#>   Primary splits:
#>       x2 < 1.130662  to the left,  improve=0.18386040, (0
↪ missing)
#>       x1 < 6.209961  to the left,  improve=0.14561510, (0
↪ missing)
#>       x3 < 4.517029  to the left,  improve=0.01044883, (0
↪ missing)
#>
#> Node number 62: 10 observations
#>   mean=3.330957, MSE=2.001022
#>
#> Node number 63: 44 observations,    complexity
↪ param=0.01032524
#>   mean=4.718314, MSE=1.127413
#>   left son=126 (27 obs) right son=127 (17 obs)
#>   Primary splits:
#>       x1 < 6.468044  to the left,  improve=0.16079230, (0
↪ missing)
#>       x3 < 5.608708  to the right, improve=0.05277854, (0
↪ missing)
#>       x2 < 2.784688  to the left,  improve=0.03145241, (0
↪ missing)
#>   Surrogate splits:
#>       x2 < 3.074905  to the left,  agree=0.636, adj=0.059,
↪ (0 split)
#>       x3 < 5.888028  to the left,  agree=0.636, adj=0.059,
↪ (0 split)
#>
```

```
#> Node number 126: 27 observations,    complexity
↪  param=0.01032524
#>   mean=4.380469, MSE=1.04313
#>   left son=252 (12 obs) right son=253 (15 obs)
#>   Primary splits:
#>       x1 < 3.658072  to the right, improve=0.4424566, (0
↪  missing)
#>       x3 < 4.270123  to the right, improve=0.1430466, (0
↪  missing)
#>       x2 < 2.658809  to the left,  improve=0.1121999, (0
↪  missing)
#>   Surrogate splits:
#>       x2 < 2.707432  to the left,  agree=0.815, adj=0.583,
↪  (0 split)
#>       x3 < 4.010151  to the right, agree=0.593, adj=0.083,
↪  (0 split)
#>
#> Node number 127: 17 observations
#>   mean=5.25489, MSE=0.7920812
#>
#> Node number 252: 12 observations
#>   mean=3.620914, MSE=0.6204645
#>
#> Node number 253: 15 observations
#>   mean=4.988114, MSE=0.5504908
```

```
# Visualize the Regression Tree
rpart.plot(
    tree_model,
    type = 2,
    extra = 101,
    fallen.leaves = TRUE,
    main = "Regression Tree"
)
```

- Splits are made based on conditions (e.g., x1 < 4.2), partitioning the predictor space (Fig. 7.21).
- Terminal nodes (leaves) show the predicted value (mean response in that region).
- The tree depth affects interpretability and overfitting risk.

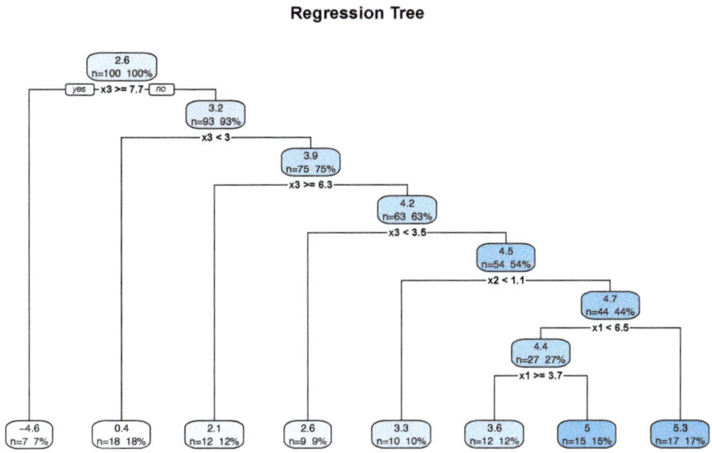

Fig. 7.21 Regression tree

```
# Optimal pruning based on cross-validation error
printcp(tree_model) # Displays CP table with cross-validation
↪   error
#>
#> Regression tree:
#> rpart(formula = y ~ x1 + x2 + x3, data = data_tree, method
↪  = "anova",
#>     control = rpart.control(cp = 0.01))
#>
#> Variables actually used in tree construction:
#> [1] x1 x2 x3
#>
#> Root node error: 989.77/100 = 9.897
#>
#> n= 100
#>
#>         CP nsplit rel error  xerror    xstd
#> 1 0.398959      0   1.00000 1.01348 0.34067
#> 2 0.174703      1   0.60104 0.86500 0.33363
#> 3 0.046074      2   0.42634 0.57079 0.18803
#> 4 0.027549      3   0.38026 0.52874 0.18667
#> 5 0.015846      4   0.35272 0.50611 0.18675
#> 6 0.010325      5   0.33687 0.51368 0.18610
#> 7 0.010000      7   0.31622 0.48471 0.18618
optimal_cp <-
    tree_model$cptable[which.min(tree_model$cptable[,
↪   "xerror"]), "CP"]
```

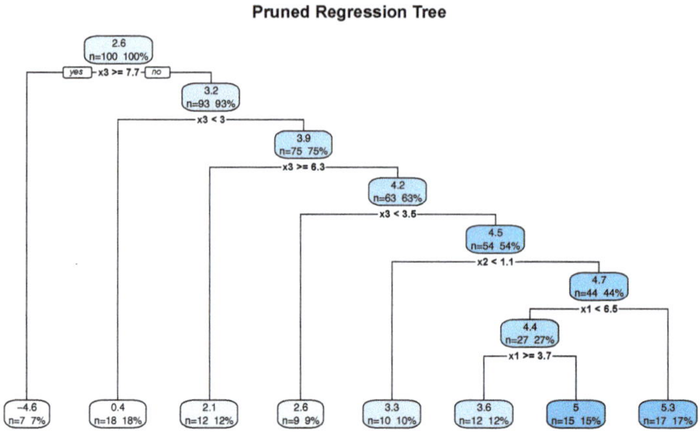

Fig. 7.22 Pruned regression tree diagram

```
# Prune the tree
pruned_tree <- prune(tree_model, cp = optimal_cp)
```

```
# Visualize the pruned tree
rpart.plot(
    pruned_tree,
    type = 2,
    extra = 101,
    fallen.leaves = TRUE,
    main = "Pruned Regression Tree"
)
```

- Pruning reduces overfitting by simplifying the tree (Fig. 7.22).
- The optimal CP minimizes cross-validation error, balancing complexity and fit.
- A shallower tree improves generalization on unseen data.

```
# Fit a Random Forest
set.seed(123)
rf_model <- randomForest(
    y ~ x1 + x2 + x3,
    data = data_tree,
    ntree = 500,
```

```
    mtry = 2,
    importance = TRUE
)

# Summary of the Random Forest
print(rf_model)
#>
#> Call:
#>  randomForest(formula = y ~ x1 + x2 + x3, data = data_tree,
↪  ntree = 500,       mtry = 2, importance = TRUE)
#>                Type of random forest: regression
#>                      Number of trees: 500
#> No. of variables tried at each split: 2
#>
#>           Mean of squared residuals: 3.031589
#>                     % Var explained: 69.37
```

- MSE decreases as more trees are added.
- % Variance Explained reflects predictive performance.
- `mtry = 2` indicates 2 random predictors are considered at each split.

```
# Plot OOB Error vs. Number of Trees
plot(rf_model, main = "Out-of-Bag Error for Random Forest")
```

- OOB error stabilizes as more trees are added, providing an unbiased estimate of test error.
- Helps determine if more trees improve performance or if the model has converged (Fig. 7.23).

```
# Variable Importance
importance(rf_model)                    # Numerical importance
↪  measures
#>      %IncMSE IncNodePurity
#> x1 10.145674     137.09918
#> x2  1.472662      77.41256
#> x3 44.232816     718.49567
```

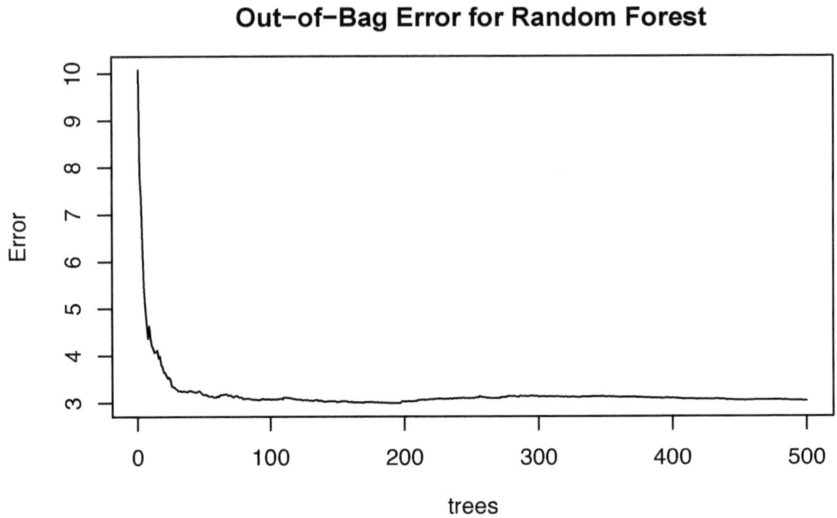

Fig. 7.23 Out-of-bag error for random forest

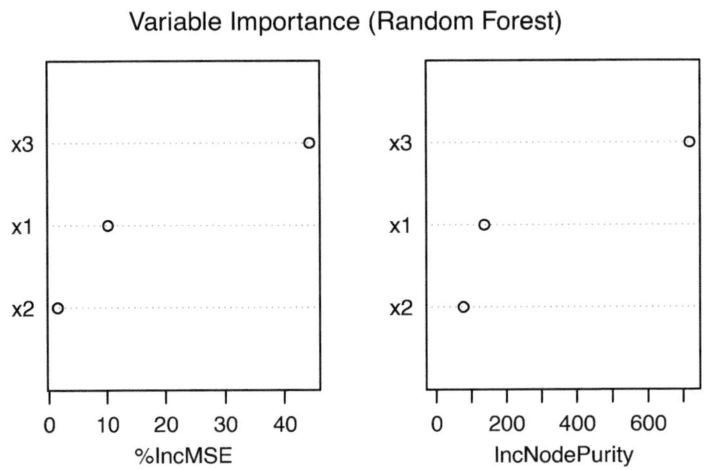

Fig. 7.24 Variable importance (random forest)

```
varImpPlot(rf_model, main = "Variable Importance (Random
↪ Forest)")
```

- Mean decrease in MSE indicates how much the model's error increases when a variable is permuted (Fig. 7.24).

- Mean decrease in node impurity reflects how much each variable reduces variance in splits.
- Variables with higher importance are more influential in the model.

```
# Predictions on new data
x_new <- seq(0, 10, length.out = 200)
test_data <- data.frame(x1 = x_new,
                        x2 = mean(x2),
                        x3 = mean(x3))

# Predictions
tree_pred <- predict(pruned_tree, newdata = test_data)
rf_pred <- predict(rf_model, newdata = test_data)
```

```
# Visualization
ggplot() +
    geom_point(aes(x1, y),
               data = data_tree,
               alpha = 0.5,
               color = "gray") +
    geom_line(
        aes(x_new, tree_pred),
        color = "blue",
        size = 1.2,
        linetype = "dashed"
    ) +
    geom_line(aes(x_new, rf_pred), color = "green", size =
    ↪   1.2) +
    labs(
        title = "Regression Tree vs. Random Forest",
        subtitle = "Blue: Pruned Tree | Green: Random Forest",
        x = "x1",
        y = "Predicted y"
    ) +
    theme_minimal()
```

- The pruned regression tree (blue) shows step-like predictions, characteristic of piecewise constant fits (Fig. 7.25).
- The random forest (green) provides a smoother fit by averaging across many trees, reducing variance.

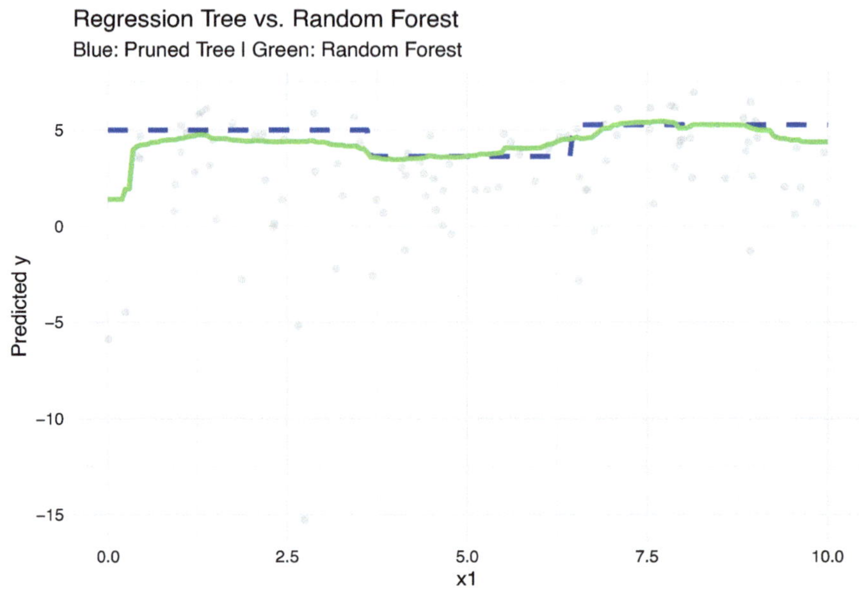

Fig. 7.25 Regression tree vs. random forest

```
# OOB Error (Random Forest)
oob_mse <- rf_model$mse[length(rf_model$mse)]  # Final OOB MSE

# Cross-Validation Error (Regression Tree)
cv_mse_tree <-
    min(tree_model$cptable[, "xerror"]) * var(data_tree$y)

# Compare OOB and CV errors
data.frame(
    Model = c("Pruned Regression Tree", "Random Forest"),
    MSE = c(cv_mse_tree, oob_mse)
)
#>                      Model      MSE
#> 1 Pruned Regression Tree 4.845622
#> 2          Random Forest 3.031589
```

- OOB error (random forest) provides an efficient, unbiased estimate without cross-validation.

- Cross-validation error (regression tree) evaluates generalization through resampling.
- Random forest often shows lower MSE due to reduced variance.

7.9 Wavelet Regression

Wavelet regression is a nonparametric regression technique that represents the target function as a combination of **wavelet basis functions**. Unlike traditional basis functions (e.g., polynomials or splines), wavelets have excellent **localization properties** in both the **time (or space)** and **frequency** domains. This makes wavelet regression particularly effective for capturing **local features**, such as sharp changes, discontinuities, and transient patterns.

A wavelet is a function $\psi(t)$ that oscillates (like a wave) but is localized in both time and frequency. The key idea is to represent a function as a linear combination of shifted and scaled versions of a mother wavelet $\psi(t)$.

Wavelet basis functions are generated by scaling and translating the mother wavelet:

$$\psi_{j,k}(t) = 2^{j/2}\,\psi(2^j t - k),$$

where:

- j (scale parameter): Controls the **frequency**—larger j captures finer details (high-frequency components).
- k (translation parameter): Controls the **location**—shifting the wavelet along the time or space axis.
- The factor $2^{j/2}$ ensures that the wavelet basis functions are **orthonormal**.

In addition to the mother wavelet $\psi(t)$, there's also a **scaling function** $\phi(t)$, which captures the low-frequency (smooth) components of the data.

7.9.1 Wavelet Series Expansion

Just as Fourier analysis represents functions as sums of sines and cosines, **wavelet analysis** represents a function as a sum of wavelet basis functions:

$$f(t) = \sum_k c_{J_0,k}\,\phi_{J_0,k}(t) + \sum_{j=J_0}^{J_{\max}} \sum_k d_{j,k}\,\psi_{j,k}(t),$$

where:

- $c_{J_0,k}$ are the **approximation coefficients** at the coarsest scale J_0, capturing smooth trends,
- $d_{j,k}$ are the **detail coefficients** at scale j, capturing finer details,
- $\phi_{J_0,k}(t)$ are the scaling functions, and $\psi_{j,k}(t)$ are the wavelet functions.

The goal of wavelet regression is to estimate these coefficients based on observed data.

7.9.2 *Wavelet Regression Model*

Given data $\{(x_i, y_i)\}_{i=1}^n$, the wavelet regression model assumes:

$$y_i = f(x_i) + \varepsilon_i,$$

where:

- $f(x)$ is the unknown regression function,
- ε_i are i.i.d. errors with mean zero and variance σ^2.

We approximate $f(x)$ using a finite number of wavelet basis functions:

$$\hat{f}(x) = \sum_k \hat{c}_{J_0,k}\,\phi_{J_0,k}(x) + \sum_{j=J_0}^{J_{\max}} \sum_k \hat{d}_{j,k}\,\psi_{j,k}(x),$$

where the coefficients $\hat{c}_{J_0,k}$ and $\hat{d}_{j,k}$ are estimated from the data.

Coefficient Estimation

- **Linear Wavelet Regression:** Estimate coefficients via least squares, projecting the data onto the wavelet basis.
- **Nonlinear Wavelet Regression (Thresholding):** Apply **shrinkage** or **thresholding** (e.g., hard or soft thresholding) to the detail coefficients $d_{j,k}$ to reduce noise and prevent overfitting. This is especially useful when the true signal is sparse in the wavelet domain.

7.9.3 *Wavelet Shrinkage and Thresholding*

Wavelet shrinkage is a powerful denoising technique introduced by Donoho and Johnstone (1995). The idea is to suppress small coefficients (likely to be noise) while retaining large coefficients (likely to contain the true signal) (Table 7.6).

Table 7.6 Comparison of nonparametric smoothing methods

Aspect	**Kernel/local polynomial**	**Splines**	**Wavelets**
Smoothness	Smooth, localized	Globally smooth with knots	Captures sharp discontinuities
Basis functions	Kernel functions	Piecewise polynomials	Compact, oscillatory wavelets
Handling of noise	Prone to overfitting without smoothing	Controlled via penalties	Excellent via thresholding
Computational cost	Moderate	High (for large knots)	Efficient (fast wavelet transform)
Applications	Curve fitting, density estimation	Smoothing trends	Signal processing, time series

Thresholding Rules

1. **Hard Thresholding:**

$$\hat{d}_{j,k}^{(\text{hard})} = \begin{cases} d_{j,k}, & \text{if } |d_{j,k}| > \tau, \\ 0, & \text{otherwise,} \end{cases}$$

2. **Soft Thresholding:**

$$\hat{d}_{j,k}^{(\text{soft})} = \text{sign}(d_{j,k}) \cdot \max\left(|d_{j,k}| - \tau,\ 0\right),$$

where τ is the **threshold parameter**, often chosen based on the noise level (e.g., via cross-validation or universal thresholding) (Fig. 7.26).

```
# Simulated data: Noisy signal with discontinuities
set.seed(123)
n <- 96
x <- seq(0, 1, length.out = n)
signal <-
    sin(4 * pi * x) + ifelse(x > 0.5, 1, 0)  # Discontinuity
    ↪  at x = 0.5
y <- signal + rnorm(n, sd = 0.2)

# Wavelet Regression using Discrete Wavelet Transform (DWT)
library(waveslim)
#> Loading required package: multitaper
#>
#> waveslim: Wavelet Method for 1/2/3D Signals (version =
↪  1.8.5)
```

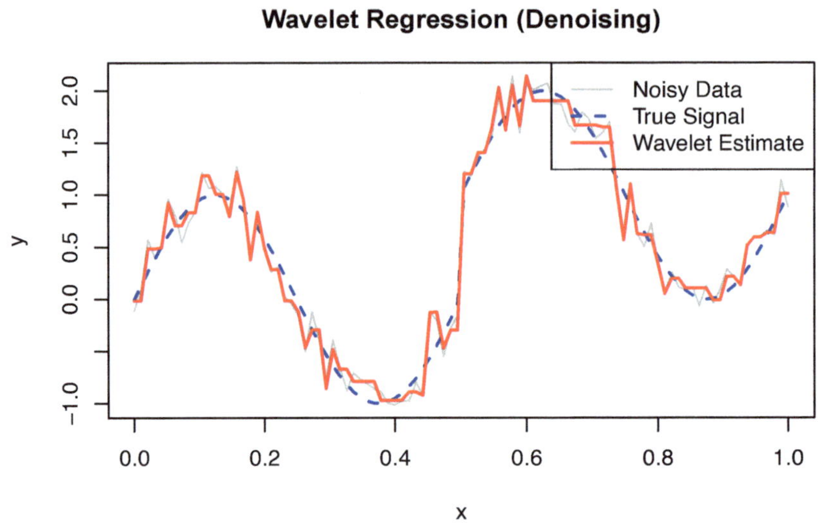

Fig. 7.26 Wavelet regression (denoising)

```
# Apply DWT with the correct parameter name
dwt_result <- dwt(y, wf = "haar", n.levels = 4)

# Thresholding (currently hard thresholding)
threshold <- 0.2
dwt_result_thresh <- dwt_result
for (i in 1:4) {
    dwt_result_thresh[[i]] <-
        ifelse(abs(dwt_result[[i]]) > threshold,
        ↪  dwt_result[[i]], 0)
}

# for soft thresholding
# for (i in 1:4) {
#     dwt_result_thresh[[i]] <-
#         sign(dwt_result[[i]]) * pmax(abs(dwt_result[[i]]) -
↪  threshold, 0)
# }

# Inverse DWT to reconstruct the signal
y_hat <- idwt(dwt_result_thresh)
```

```
# Plotting
plot(
    x,
    y,
    type = "l",
    col = "gray",
    lwd = 1,
    main = "Wavelet Regression (Denoising)"
)
lines(x,
      signal,
      col = "blue",
      lwd = 2,
      lty = 2)  # True signal
lines(x, y_hat, col = "red", lwd = 2)                # Denoised
↪  estimate
legend(
    "topright",
    legend = c("Noisy Data", "True Signal", "Wavelet
    ↪  Estimate"),
    col = c("gray", "blue", "red"),
    lty = c(1, 2, 1),
    lwd = c(1, 2, 2)
)
```

Advantages

- **Local Feature Detection:** Excellent for capturing sharp changes, discontinuities, and localized phenomena.
- **Multiresolution Analysis:** Analyzes data at multiple scales, making it effective for both global trends and fine details.
- **Denoising Capability:** Powerful noise reduction through thresholding in the wavelet domain.

Limitations

- **Complexity:** Requires careful selection of wavelet basis, decomposition levels, and thresholding methods.
- **Less Interpretability:** Coefficients are less interpretable compared to spline or tree-based methods.
- **Boundary Effects:** May suffer from artifacts near data boundaries without proper treatment.

```r
# Load necessary libraries
library(waveslim)   # For Discrete Wavelet Transform (DWT)
library(ggplot2)


# Simulate Data: Noisy signal with discontinuities
set.seed(123)
n <- 96
x <- seq(0, 1, length.out = n)

# True signal: Sinusoidal with a discontinuity at x = 0.5
signal <- sin(4 * pi * x) + ifelse(x > 0.5, 1, 0)

# Add Gaussian noise
y <- signal + rnorm(n, sd = 0.2)
```

```r
# Plot the noisy data and true signal
ggplot() +
    geom_line(aes(x, y),
              color = "gray",
              size = 0.8,
              alpha = 0.7) +
    geom_line(
        aes(x, signal),
        color = "blue",
        linetype = "dashed",
        size = 1
    ) +
    labs(title = "Noisy Data with Underlying True Signal",
         x = "x", y = "Signal") +
    theme_minimal()
```

See Fig. 7.27.

```r
# Apply Discrete Wavelet Transform (DWT) using Haar wavelet
dwt_result <- dwt(y, wf = "haar", n.levels = 4)

# View DWT structure
str(dwt_result)
#> List of 5
```

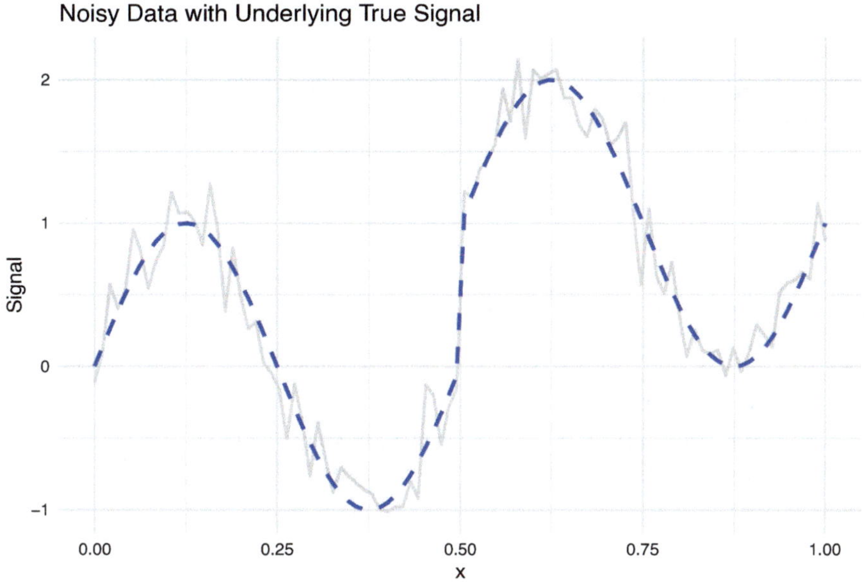

Fig. 7.27 Noisy data with underlying true signal

```
#>  $ d1: num [1:48] 0.14 -0.1221 0.3017 -0.1831 0.0745 ...
#>  $ d2: num [1:24] 0.50003 -0.06828 0.35298 0.01582 -0.00853
↪  ...
#>  $ d3: num [1:12] 0.669 0.128 -0.746 -0.532 -0.258 ...
#>  $ d4: num [1:6] 1.07 -1.766 0.752 0.663 -1.963 ...
#>  $ s4: num [1:6] 2.963 -0.159 -2.758 6.945 3.845 ...
#>  - attr(*, "class")= chr "dwt"
#>  - attr(*, "wavelet")= chr "haar"
#>  - attr(*, "boundary")= chr "periodic"
```

- `wf = "haar"`: Haar wavelet, simple and effective for detecting discontinuities.
- `n.levels = 4`: Number of decomposition levels (captures details at different scales).

The DWT output contains approximation coefficients and detail coefficients for each level.

```
# Hard Thresholding
threshold <- 0.2  # Chosen threshold for demonstration
dwt_hard_thresh <- dwt_result
```

```
# Apply hard thresholding to detail coefficients
for (i in 1:4) {
  dwt_hard_thresh[[i]] <- ifelse(abs(dwt_result[[i]]) >
↪  threshold, dwt_result[[i]], 0)
}
```

```
# Soft Thresholding
dwt_soft_thresh <- dwt_result

# Apply soft thresholding to detail coefficients
for (i in 1:4) {
  dwt_soft_thresh[[i]] <- sign(dwt_result[[i]]) *
↪  pmax(abs(dwt_result[[i]]) - threshold, 0)
}
```

- Hard Thresholding: Keeps coefficients above the threshold, sets others to zero.
- Soft Thresholding: Shrinks coefficients toward zero, reducing potential noise smoothly.

```
# Reconstruct the denoised signals using Inverse DWT
y_hat_hard <- idwt(dwt_hard_thresh)
y_hat_soft <- idwt(dwt_soft_thresh)
```

```
# Combine data for ggplot
df_plot <- data.frame(
    x = rep(x, 4),
    y = c(y, signal, y_hat_hard, y_hat_soft),
    Type = rep(
        c(
            "Noisy Data",
            "True Signal",
            "Hard Thresholding",
            "Soft Thresholding"
        ),
        each = n
    )
)
```

```
# Plotting
ggplot(df_plot, aes(x, y, color = Type, linetype = Type)) +
    geom_line(size = 1) +
    scale_color_manual(values = c("gray", "blue", "red",
    ↪ "green")) +
    scale_linetype_manual(values = c("solid", "dashed",
    ↪ "solid", "solid")) +
    labs(
        title = "Wavelet Regression (Denoising)",
        subtitle = "Comparison of Hard vs. Soft Thresholding",
        x = "x",
        y = "Signal"
    ) +
    theme_minimal() +
    theme(legend.position = "top")
```

See Fig. 7.28.

```
# Compute Mean Squared Error (MSE) for each method
mse_noisy <- mean((y - signal) ^ 2)
mse_hard <- mean((y_hat_hard - signal) ^ 2)
mse_soft <- mean((y_hat_soft - signal) ^ 2)
```

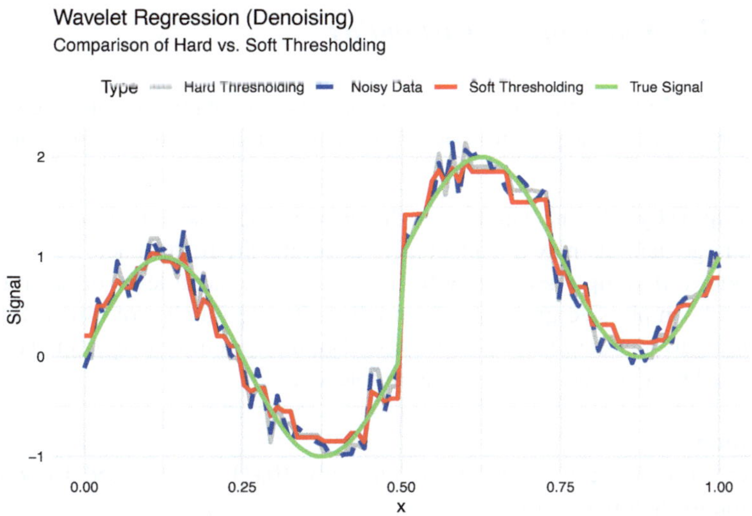

Fig. 7.28 Wavelet regression (denoising)

```
# Display MSE comparison
data.frame(
  Method = c("Noisy Data", "Hard Thresholding", "Soft
  ↪ Thresholding"),
  MSE = c(mse_noisy, mse_hard, mse_soft)
)
#>              Method        MSE
#> 1        Noisy Data 0.03127707
#> 2 Hard Thresholding 0.02814465
#> 3 Soft Thresholding 0.02267171
```

- Lower MSE indicates better denoising performance.
- Soft thresholding often achieves smoother results with lower MSE.

7.10 Multivariate Nonparametric Regression

Nonparametric regression in higher dimensions ($p > 1$) presents significant challenges compared to the univariate case. This difficulty arises primarily from the **curse of dimensionality**, which refers to the exponential growth of data requirements as the number of predictors increases.

7.10.1 The Curse of Dimensionality

The **curse of dimensionality** refers to various phenomena that occur when analyzing and organizing data in high-dimensional spaces. In the context of nonparametric regression:

- **Data sparsity:** As the number of dimensions increases, data become sparse. Even large datasets may not adequately cover the predictor space.
- **Exponential sample size growth:** To achieve the same level of accuracy, the required sample size grows exponentially with the number of dimensions. For example, to maintain the same density of points when moving from 1D to 2D, you need roughly the square of the sample size.

Illustration
Consider estimating a function on the unit cube $[0, 1]^p$. If we need 10 points per dimension to capture the structure:

- In 1D: 10 points suffice.

- In 2D: $10^2 = 100$ points are needed.
- In 10D: $10^{10} = 10,000,000,000$ points are required.

This makes traditional nonparametric methods, like kernel smoothing, impractical in high dimensions without additional structure or assumptions.

7.10.2 Multivariate Kernel Regression

A straightforward extension of kernel regression to higher dimensions is the multivariate Nadaraya-Watson estimator:

$$\hat{m}_h(\mathbf{x}) = \frac{\sum_{i=1}^{n} K_h(\mathbf{x} - \mathbf{x}_i)\, y_i}{\sum_{i=1}^{n} K_h(\mathbf{x} - \mathbf{x}_i)},$$

where:

- $\mathbf{x} = (x_1, x_2, \ldots, x_p)$ is the predictor vector,
- $K_h(\cdot)$ is a **multivariate kernel function**, often a product of univariate kernels:

$$K_h(\mathbf{x} - \mathbf{x}_i) = \prod_{j=1}^{p} K\left(\frac{x_j - x_{ij}}{h_j}\right),$$

- h_j is the bandwidth for the j-th predictor.

Challenges

- The product kernel suffers from **inefficiency** in high dimensions because most data points are far from any given target point $\mathbf{x}$, resulting in very small kernel weights.
- Selecting an optimal **multivariate bandwidth matrix** is complex and computationally intensive.

7.10.3 Multivariate Splines

Extending splines to multiple dimensions involves more sophisticated techniques, as the simplicity of piecewise polynomials in 1D does not generalize easily.

7.10.3.1 Thin-Plate Splines

Thin-plate splines generalize cubic splines to higher dimensions. They minimize a smoothness penalty that depends on the derivatives of the function:

$$\hat{m}(\mathbf{x}) = \arg\min_f \left\{ \sum_{i=1}^{n} (y_i - f(\mathbf{x}_i))^2 + \lambda \int \|\nabla^2 f(\mathbf{x})\|^2 d\mathbf{x} \right\},$$

where $\nabla^2 f(\mathbf{x})$ is the **Hessian matrix** of second derivatives and $\|\cdot\|^2$ represents the sum of squared elements.

Thin-plate splines are **rotation-invariant** and do not require explicit placement of knots, but they become computationally expensive as the number of dimensions increases.

7.10.3.2 Tensor Product Splines

For structured multivariate data, **tensor product splines** are commonly used. They construct a basis for each predictor and form the multivariate basis via tensor products:

$$\hat{m}(\mathbf{x}) = \sum_{i=1}^{K_1} \sum_{j=1}^{K_2} \beta_{ij} B_i(x_1) B_j(x_2),$$

where $B_i(x_1)$ and $B_j(x_2)$ are spline basis functions for x_1 and x_2, respectively.

Tensor products allow flexible modeling of **interactions** between predictors but can lead to large numbers of parameters as the number of dimensions increases.

7.10.4 Additive Models (GAMs)

A powerful approach to mitigating the curse of dimensionality is to assume an **additive structure** for the regression function:

$$m(\mathbf{x}) = \beta_0 + f_1(x_1) + f_2(x_2) + \cdots + f_p(x_p),$$

where each f_j is a univariate smooth function estimated nonparametrically.

7.10.4.1 Why Additive Models Help:

- **Dimensionality Reduction:** Instead of estimating a full p-dimensional surface, we estimate p separate functions in 1D.
- **Interpretability:** Each function $f_j(x_j)$ represents the effect of predictor X_j on the response, holding other variables constant.

Extensions

- **Interactions:** Additive models can be extended to include interactions:

$$m(\mathbf{x}) = \beta_0 + \sum_{j=1}^{p} f_j(x_j) + \sum_{j<k} f_{jk}(x_j, x_k),$$

  where f_{jk} are bivariate smooth functions capturing interaction effects.

7.10.5 Radial Basis Functions

Radial basis functions (RBF) are another approach to multivariate nonparametric regression, particularly effective in scattered data interpolation.

A typical RBF model is:

$$\hat{m}(\mathbf{x}) = \sum_{i=1}^{n} \alpha_i \, \phi(\|\mathbf{x} - \mathbf{x}_i\|),$$

where:

- $\phi(\cdot)$ is a radial basis function (e.g., Gaussian: $\phi(r) = e^{-\gamma r^2}$),
- $\|\mathbf{x} - \mathbf{x}_i\|$ is the Euclidean distance between $\mathbf{x}$ and the data point $\mathbf{x}_i$,
- α_i are coefficients estimated from the data.

```
# Load necessary libraries
library(ggplot2)
library(np)       # For multivariate kernel regression
#> Nonparametric Kernel Methods for Mixed Datatypes (version
↪  0.60-18)
#> [vignette("np_faq",package="np") provides answers to
↪  frequently asked questions]
#> [vignette("np",package="np") an overview]
#> [vignette("entropy_np",package="np") an overview of
↪  entropy-based methods]
library(mgcv)     # For thin-plate and tensor product splines
library(fields)   # For radial basis functions (RBF)
#> Loading required package: spam
#> Spam version 2.11-1 (2025-01-20) is loaded.
#> Type 'help( Spam)' or 'demo( spam)' for a short introduction
#> and overview of this package.
#> Help for individual functions is also obtained by adding
↪  the
```

```
#> suffix '.spam' to the function name, e.g. 'help( chol.spam)'.
#>
#> Attaching package: 'spam'
#> The following objects are masked from 'package:base':
#>
#>     backsolve, forwardsolve
#> Loading required package: viridisLite
#>
#> Try help(fields) to get started.
library(reshape2) # For data manipulation


# Simulate Multivariate Data
set.seed(123)
n <- 100
x1 <- runif(n, 0, 5)
x2 <- runif(n, 0, 5)
x3 <- runif(n, 0, 5)

# True nonlinear multivariate function with interaction
↪  effects
true_function <- function(x1, x2, x3) {
    sin(pi * x1) * cos(pi * x2) + exp(-((x3 - 2.5) ^ 2)) +
     ↪  0.5 * x1 * x3
}

# Response with noise
y <- true_function(x1, x2, x3) + rnorm(n, sd = 0.5)

# Data frame
data_multi <- data.frame(y, x1, x2, x3)

# Visualization of marginal relationships
p1 <-
    ggplot(data_multi, aes(x1, y)) +
    geom_point(alpha = 0.5) +
    labs(title = "Effect of x1")
p2 <-
    ggplot(data_multi, aes(x2, y)) +
    geom_point(alpha = 0.5) +
    labs(title = "Effect of x2")
p3 <-
    ggplot(data_multi, aes(x3, y)) +
```

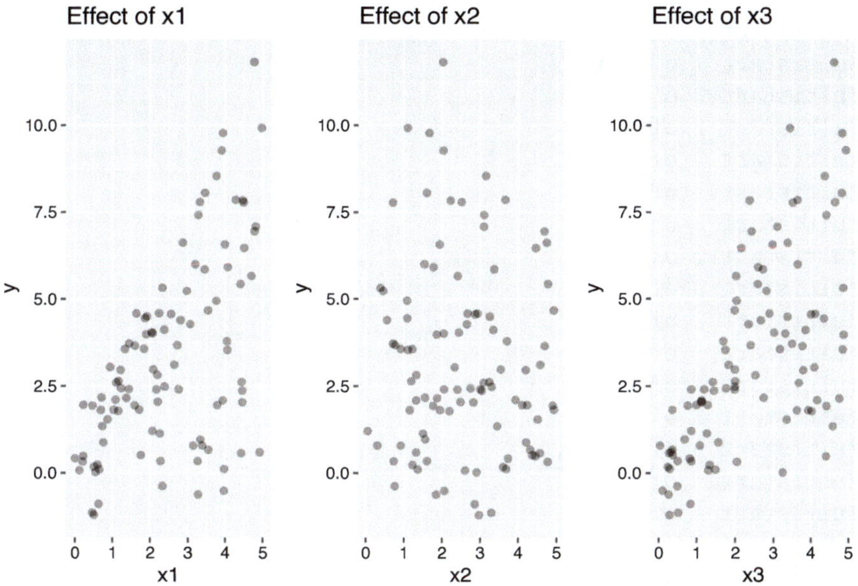

Fig. 7.29 Scatterplot

```
    geom_point(alpha = 0.5) +
    labs(title = "Effect of x3")
```

```
gridExtra::grid.arrange(p1, p2, p3, ncol = 3)
```

See Fig. 7.29.

```
# Multivariate Kernel Regression using np package
bw <-
    npregbw(y ~ x1 + x2 + x3, data = data_multi)  # Bandwidth
    ↪   selection
#>
Multistart 1 of 3 |
Multistart 1 of 3 |
Multistart 1 of 3 |
Multistart 1 of 3 /
Multistart 1 of 3 -
Multistart 1 of 3 |
Multistart 1 of 3 |
Multistart 1 of 3 /
```

```
Multistart 1 of 3 -
Multistart 1 of 3 \
Multistart 2 of 3 |
Multistart 2 of 3 |
Multistart 2 of 3 /
Multistart 2 of 3 -
Multistart 2 of 3 \
Multistart 2 of 3 |
Multistart 2 of 3 |
Multistart 2 of 3 |
Multistart 2 of 3 /
Multistart 3 of 3 |
Multistart 3 of 3 |
Multistart 3 of 3 /
Multistart 3 of 3 -
Multistart 3 of 3 \
Multistart 3 of 3 |
Multistart 3 of 3 |
Multistart 3 of 3 |
Multistart 3 of 3 /

kernel_model <- npreg(bws = bw)

# Predict on a grid for visualization
grid_data <- expand.grid(
    x1 = seq(0, 5, length.out = 50),
    x2 = seq(0, 5, length.out = 50),
    x3 = mean(data_multi$x3)
)
pred_kernel <- predict(kernel_model, newdata = grid_data)

# Visualization
grid_data$pred <- pred_kernel
```

```
ggplot(grid_data, aes(x1, x2, fill = pred)) +
    geom_raster() +
    scale_fill_viridis_c() +
    labs(title = "Multivariate Kernel Regression (x3 fixed)",
         x = "x1",
         y = "x2") +
    theme_minimal()
```

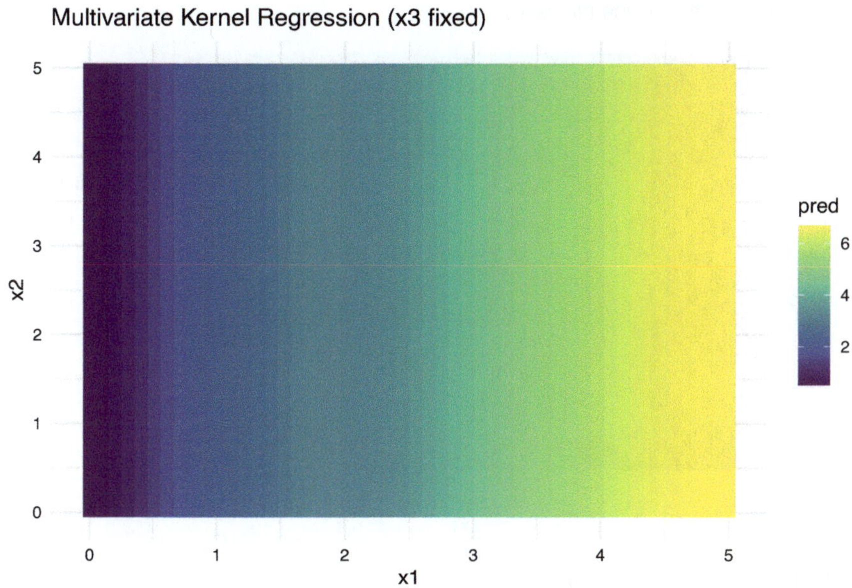

Fig. 7.30 Multivariate kernel regression (x3 fixed)

- Kernel regression captures nonlinear interactions but suffers from data sparsity in high dimensions (curse of dimensionality) (Fig. 7.30).
- Bandwidth selection is critical for performance.

```
# Fit Thin-Plate Spline
tps_model <- gam(y ~ s(x1, x2, x3, bs = "tp", k = 5), data =
↪   data_multi)
#> Warning in smooth.construct.tp.smooth.spec(object, dk$data,
↪   dk$knots): basis dimension, k, increased to minimum
↪   possible

# Predictions
grid_data$pred_tps <- predict(tps_model, newdata = grid_data)
```

```
# Visualization
ggplot(grid_data, aes(x1, x2, fill = pred_tps)) +
    geom_raster() +
    scale_fill_viridis_c() +
    labs(title = "Thin-Plate Spline (x3 fixed)",
```

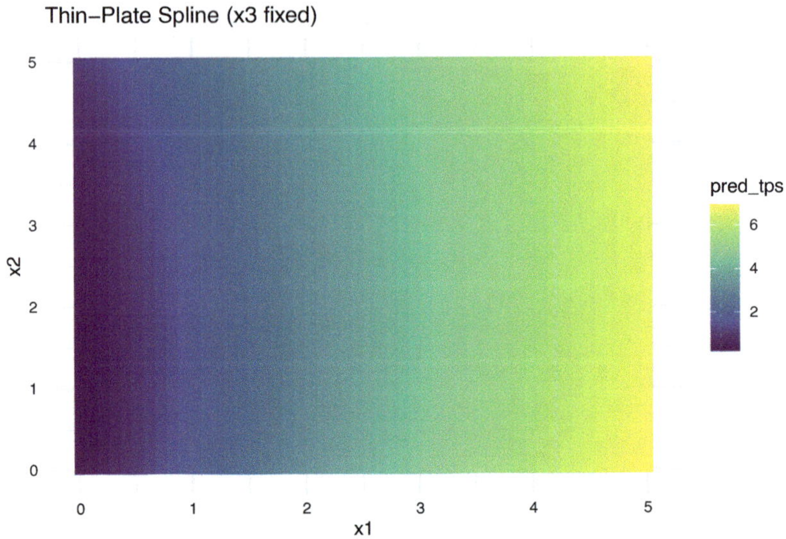

Fig. 7.31 Thin-plate spline (x3 fixed)

```
        x = "x1", y = "x2") +
    theme_minimal()
```

- Thin-plate splines handle smooth surfaces well and are rotation-invariant (Fig. 7.31).
- Computational cost increases with higher dimensions due to matrix operations.

```
# Fit Tensor Product Spline
tensor_model <- gam(y ~ te(x1, x2, x3), data = data_multi)

# Predictions
grid_data$pred_tensor <- predict(tensor_model, newdata =
↪  grid_data)
```

```
# Visualization
ggplot(grid_data, aes(x1, x2, fill = pred_tensor)) +
    geom_raster() +
    scale_fill_viridis_c() +
    labs(title = "Tensor Product Spline (x3 fixed)",
        x = "x1", y = "x2") +
    theme_minimal()
```

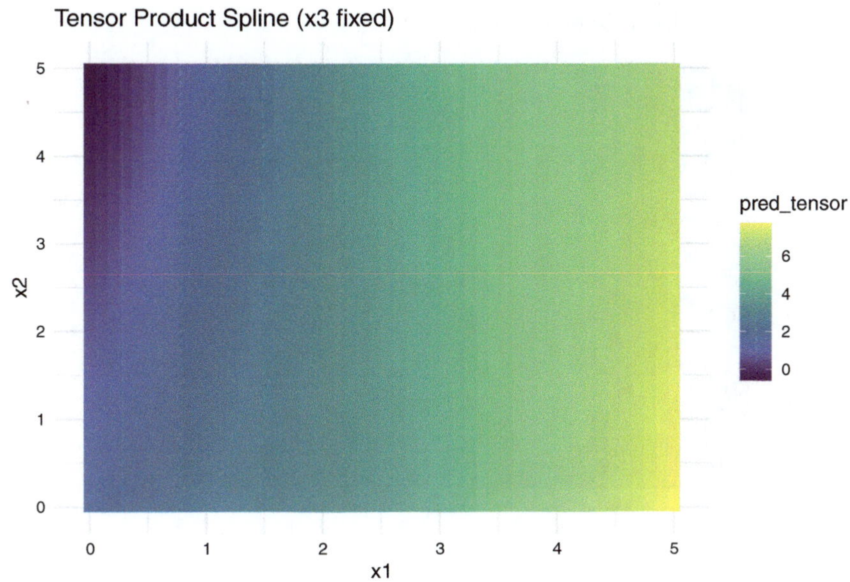

Fig. 7.32 Tensor product spline (x3 fixed)

- Tensor product splines model interactions explicitly between variables (Fig. 7.32).
- Suitable when data have structured dependencies but can lead to many parameters in higher dimensions.

```
# Additive Model (GAM)
gam_model <- gam(y ~ s(x1) + s(x2) + s(x3), data = data_multi)

# Predictions
grid_data$pred_gam <- predict(gam_model, newdata = grid_data)
```

```
# Visualization
ggplot(grid_data, aes(x1, x2, fill = pred_gam)) +
    geom_raster() +
    scale_fill_viridis_c() +
    labs(title = "Additive Model (GAM, x3 fixed)",
         x = "x1", y = "x2") +
    theme_minimal()
```

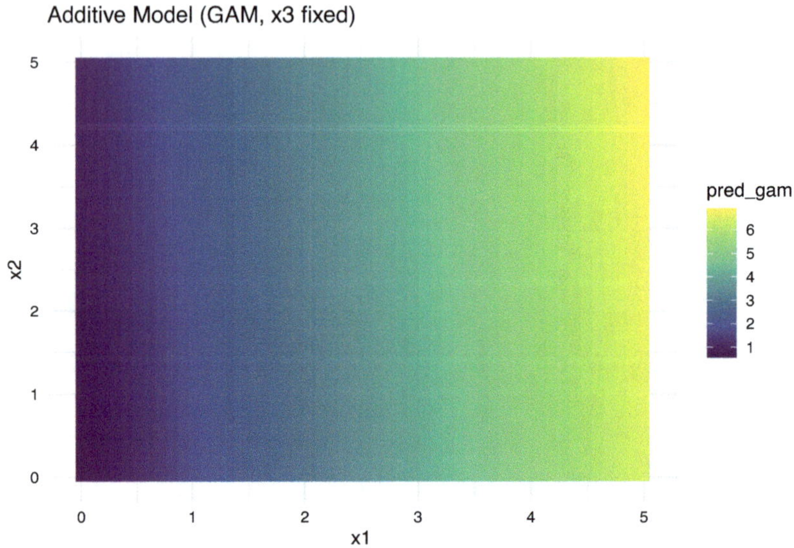

Fig. 7.33 Additive model (GAM, x3 fixed)

See Fig. 7.33.
Why GAMs help:

- Dimensionality reduction: Instead of estimating a full multivariate function, GAM estimates separate 1D functions for each predictor.
- Interpretability: Easy to understand individual effects.
- Limitations: Cannot capture complex interactions unless explicitly added.

```
# Radial Basis Function Model
rbf_model <- Tps(cbind(x1, x2, x3), y)  # Thin-plate spline
↪  RBF
#> Warning:
#> Grid searches over lambda (nugget and sill variances) with
↪  minima at the endpoints:
#>   (GCV) Generalized Cross-Validation
#>     minimum at  right endpoint  lambda  =  0.0002622876
#> (eff. df= 95.00001 )

# Predictions
grid_data$pred_rbf <-
    predict(rbf_model, cbind(grid_data$x1, grid_data$x2,
     ↪  grid_data$x3))
```

```
# Visualization
ggplot(grid_data, aes(x1, x2, fill = pred_rbf)) +
    geom_raster() +
    scale_fill_viridis_c() +
    labs(title = "Radial Basis Function Regression (x3
    ↪  fixed)",
         x = "x1", y = "x2") +
    theme_minimal()
```

- RBFs capture complex, smooth surfaces and interactions based on distance (Fig. 7.34).
- Perform well for scattered data but can be computationally expensive for large datasets.

```
# Compute Mean Squared Error for each model
mse_kernel <- mean((predict(kernel_model) - data_multi$y) ^ 2)
mse_tps <- mean((predict(tps_model) - data_multi$y) ^ 2)
mse_tensor <- mean((predict(tensor_model) - data_multi$y) ^ 2)
mse_gam <- mean((predict(gam_model) - data_multi$y) ^ 2)
mse_rbf <-
```

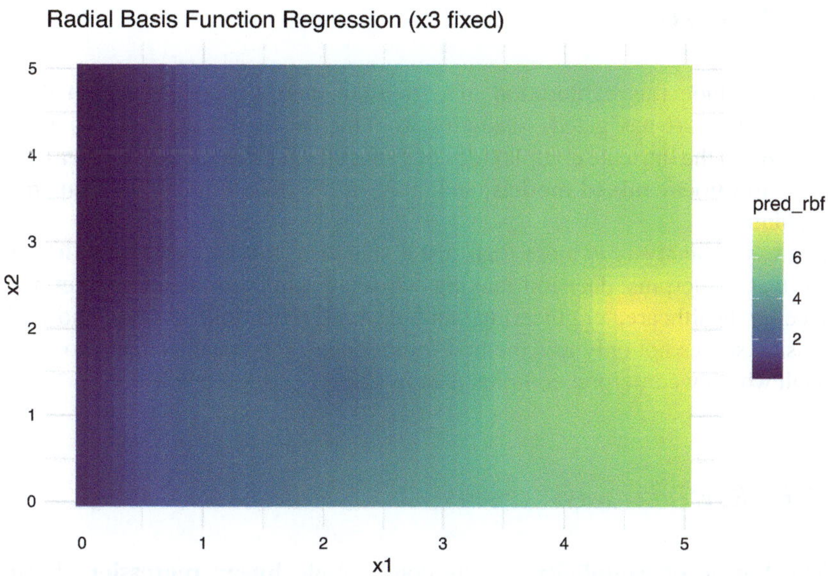

Fig. 7.34 Radial basis function regression (x3 fixed)

```
    mean((predict(rbf_model, cbind(x1, x2, x3)) -
    ↪ data_multi$y) ^ 2)

# Display MSE comparison
data.frame(
    Model = c(
        "Kernel Regression",
        "Thin-Plate Spline",
        "Tensor Product Spline",
        "Additive Model (GAM)",
        "Radial Basis Functions"
    ),
    MSE = c(mse_kernel, mse_tps, mse_tensor, mse_gam, mse_rbf)
)
#>                    Model         MSE
#> 1      Kernel Regression 0.253019836
#> 2      Thin-Plate Spline 0.449096723
#> 3  Tensor Product Spline 0.304680406
#> 4   Additive Model (GAM) 1.595616092
#> 5 Radial Basis Functions 0.001361915
```

7.11 Conclusion: The Evolving Landscape of Regression Analysis

As we conclude this exploration of regression analysis, we reflect on the vast landscape we have navigated—spanning from the foundational principles of **linear regression** to the intricate complexities of **generalized linear models, linear mixed models, nonlinear mixed models**, and, now, the flexible world of **nonparametric regression**.

Regression analysis is more than just a statistical tool; it is a **versatile framework** that underpins decision-making across disciplines—from marketing and finance to healthcare, engineering, and beyond. This journey has shown how regression serves not only as a method for modeling relationships but also as a lens through which we interpret complex data in an ever-changing world.

7.11.1 Key Takeaways

- **The Power of Simplicity:** At its core, simple **linear regression** illustrates how relationships between variables can be modeled with clarity and elegance. Mastering these fundamentals lays the groundwork for more complex techniques.

- **Beyond Linearity: Nonlinear regression** and **generalized linear models** extend our capabilities to handle data that defy linear assumptions—capturing curved relationships, non-normal error structures, and diverse outcome distributions.
- **Accounting for Hierarchies and Dependencies:** Real-world data often exhibit structures such as nested observations or repeated measures. **Linear mixed models** and **generalized linear mixed models** enable us to account for both fixed effects and random variability, ensuring robust and nuanced inferences.
- **Complex Systems, Flexible Models: Nonlinear mixed models** allow us to capture dynamic, nonlinear processes with hierarchical structures, bridging the gap between theoretical models and real-world complexity.
- **The Flexibility of Nonparametric Regression: Nonparametric methods**, such as **kernel regression, local polynomial regression, smoothing splines, wavelet regression**, and **regression trees**, provide powerful tools when parametric assumptions are too restrictive. These models excel at capturing complex, nonlinear patterns without assuming a specific functional form, offering greater adaptability in diverse applications.

7.11.2 The Art and Science of Regression

While statistical formulas and algorithms form the backbone of regression analysis, the **true art** lies in **model selection**, **diagnostic evaluation**, and **interpretation**. No model is inherently perfect; each is an approximation of reality, shaped by the assumptions we make and the data we collect. The most effective analysts are those who approach models critically—testing assumptions, validating results, and recognizing the limitations of their analyses.

Nonparametric methods remind us that **flexibility** often comes at the cost of **interpretability** and **efficiency**, just as parametric models offer simplicity but may risk oversimplification. The key is not to choose between these paradigms, but to understand when each is most appropriate.

7.11.3 Looking Forward

The field of regression continues to evolve, driven by rapid advancements in **computational power**, **data availability**, and **methodological innovation**. This evolution has given rise to a wide range of modern techniques that extend beyond traditional frameworks:

- **Machine Learning Algorithms:** While methods like random forests, support vector machines, and gradient boosting are well-established, recent developments include:

  - **Extreme Gradient Boosting (XGBoost)** and **LightGBM**, optimized for speed and performance in large-scale data environments.
  - **CatBoost**, which handles categorical features more effectively without extensive preprocessing.

- **Bayesian Regression Techniques:** Modern Bayesian approaches go beyond simple hierarchical models to include:
  - **Bayesian Additive Regression Trees (BART):** A flexible, nonparametric Bayesian method that combines the power of regression trees with probabilistic inference.
  - **Bayesian Neural Networks (BNNs):** Extending deep learning with uncertainty quantification, enabling robust decision-making in high-stakes applications.

- **High-Dimensional Data Analysis:** Regularization methods like LASSO and ridge regression have paved the way for more advanced techniques, such as:
  - **Graphical Models and Sparse Precision Matrices:** For capturing complex dependency structures in high-dimensional data.

- **Deep Learning for Regression:** Deep neural networks (DNNs) are increasingly used for regression tasks, particularly when dealing with:
  - **Structured Data (e.g., tabular datasets):** Through architectures like **TabNet**.
  - **Unstructured Data (e.g., images, text):** Using **convolutional neural networks (CNNs)** and **transformer-based models**.

- **Causal Inference in Regression:** The integration of causal modeling techniques into regression frameworks has advanced significantly:
  - **Double Machine Learning (DML):** Combining machine learning with econometric methods for robust causal effect estimation.
  - **Causal Forests:** An extension of random forests designed to estimate heterogeneous treatment effects.

- **Functional Data Analysis (FDA):** For analyzing data where predictors or responses are functions (e.g., curves, time series), using methods like:
  - **Functional Linear Models (FLM)** and **Functional Additive Models (FAM)**.
  - **Dynamic Regression Models** for real-time prediction in streaming data environments.

While these modern approaches differ in implementation, many are rooted in the fundamental concepts covered in this book. Whether through parametric precision or nonparametric flexibility, the principles of regression remain central to data-driven inquiry.

7.11.4 *Final Thoughts*

As you apply these techniques in your own work, remember that regression is not just about fitting models—it's about:

- **Asking the right questions**
- **Interpreting results thoughtfully**
- **Using data to generate meaningful insights**

Whether you're developing marketing strategies, forecasting financial trends, optimizing healthcare interventions, or conducting academic research, the tools you've gained here will serve as a strong foundation.

> As **George E.P. Box** put it: *"All models are wrong"—yet some, he noted, are still useful.* (Box, 1976)
>
> Our goal as analysts is to find models that are not only useful but also **enlightening**—models that reveal patterns, guide decisions, and deepen our understanding of the world.

Bibliography

Andrews, D. F. (1974). A robust method for multiple linear regression. *Technometrics*, *16*(4), 523–531.

Bates, D., Mächler, M., Bolker, B., & Walker, S. (2015). Fitting linear mixed-effects models using lme4. *Journal of Statistical Software*, *67*, 1–48.

Bates, D. M., & Watts, D. G. (1980). Relative curvature measures of nonlinearity. *Journal of the Royal Statistical Society: Series B (Methodological)*, *42*(1), 1–16.

Bates, D. M., & Watts, D. G. (1981). A relative off set orthogonality convergence criterion for nonlinear least squares. *Technometrics*, *23*(2), 179–183.

Beaton, A. E., & Tukey, J. W. (1974). The fitting of power series, meaning polynomials, illustrated on band-spectroscopic data. *Technometrics*, *16*(2), 147–185.

Box, G. E. (1976). Science and statistics. *Journal of the American Statistical Association*, *71*(356), 791–799.

Breiman, L. (2001). Random forests. *Machine Learning*, *45*, 5–32.

Breiman, L. (2017). *Classification and regression trees*. Routledge.

Chen, J., & Roth, J. (2024). Logs with zeros? some problems and solutions. *The Quarterly Journal of Economics*, *139*(2), 891–936.

Cramér, H. (1999). *Mathematical methods of statistics*, vol. 43. Princeton University Press.

Donoho, D. L., & Johnstone, I. M. (1995). Adapting to unknown smoothness via wavelet shrinkage. *Journal of the American Statistical Association*, *90*(432), 1200–1224.

Faraway, J. J. (2016). *Extending the linear model with R: generalized linear, mixed effects and nonparametric regression models*. CRC Press.

Gasser, T., & Müller, H.-G. (1979). Kernel estimation of regression functions. In *Smoothing Techniques for Curve Estimation: Proceedings of a Workshop Held in Heidelberg, April 2–4, 1979* (pp. 23–68). Springer.

Goldberger, A. S. (1991). *A course in econometrics*. Harvard University Press.

Greene, W. H. (1990). A gamma-distributed stochastic frontier model. *Journal of Econometrics*, *46*(1-2), 141–163.

Hampel, F. R., Ronchetti, E. M., Rousseeuw, P. J., & Stahel, W. A. (2005). *Robust statistics. Wiley Series in Probability and Statistics*. Wiley.

Henderson, C. R. (1975). Best linear unbiased estimation and prediction under a selection model. *Biometrics*, 423–447.

Hoerl, A. E., & Kennard, R. W. (1970). Ridge regression: Biased estimation for nonorthogonal problems. *Technometrics*, *12*(1), 55–67.

Huber, P. J. (1964). Robust estimation of a location parameter. *The Annals of Mathematical Statistics*, *35*(1), 73–101.

M. Nguyen, *Regression Techniques for Data Analysis*,
https://doi.org/10.1007/978-3-032-01835-9

Hurvich, C. M., & Tsai, C.-L. (1989). Regression and time series model selection in small samples. *Biometrika*, *76*(2), 297–307.

Laird, N. M., & Ware, J. H. (1982). Random-effects models for longitudinal data. *Biometrics*, 963–974.

Liang, K.-Y., & Zeger, S. L. (1986). Longitudinal data analysis using generalized linear models. *Biometrika*, *73*(1), 13–22.

Little, R. J. (1988). A test of missing completely at random for multivariate data with missing values. *Journal of the American statistical Association*, *83*(404), 1198–1202.

Lovell, M. C. (2008). A simple proof of the fwl theorem. *The Journal of Economic Education*, *39*(1), 88–91.

Magel, R. C., & Hertsgaard, D. (1987). A collinearity diagnostic for nonlinear regression: A collinearity diagnostic. *Communications in Statistics-Simulation and Computation*, *16*(1), 85–97.

McCullagh, P. (2019). *Generalized linear models*. Routledge.

Nadaraya, E. A. (1964). On estimating regression. *Theory of Probability & Its Applications*, *9*(1), 141–142.

Nelder, J. A., & Wedderburn, R. W. (1972). Generalized linear models. *Journal of the Royal Statistical Society Series A: Statistics in Society*, *135*(3), 370–384.

Priestley, M. B., & Chao, M. (1972). Non-parametric function fitting. *Journal of the Royal Statistical Society Series B: Statistical Methodology*, *34*(3), 385–392.

Rao, C. R. (1992). Information and the accuracy attainable in the estimation of statistical parameters. In *Breakthroughs in Statistics: Foundations and Basic Theory*, (pp. 235–247). Springer.

Schabenberger, O., & Pierce, F. J. (2001). *Contemporary statistical models for the plant and soil sciences*. CRC Press.

St Laurent, R. T., & Cook, R. D. (1992). Leverage and superleverage in nonlinear regression. *Journal of the American Statistical Association*, *87*(420), 985–990.

Tibshirani, R. (1996). Regression shrinkage and selection via the lasso. *Journal of the Royal Statistical Society Series B: Statistical Methodology*, *58*(1), 267–288.

Watson, G. S. (1964). Smooth regression analysis. *Sankhyā: The Indian Journal of Statistics, Series A*, 359–372.

Zou, H., & Hastie, T. (2005). Regularization and variable selection via the elastic net. *Journal of the Royal Statistical Society Series B: Statistical Methodology*, *67*(2), 301–320.

MIX
Papier aus verantwortungsvollen Quellen
Paper from responsible sources
FSC® C105338

If you have any concerns about our products,
you can contact us on
ProductSafety@springernature.com

In case Publisher is established outside the EU,
the EU authorized representative is:
Springer Nature Customer Service Center GmbH
Europaplatz 3, 69115 Heidelberg, Germany

Printed by Libri Plureos GmbH
in Hamburg, Germany